# Science With A Vengeance

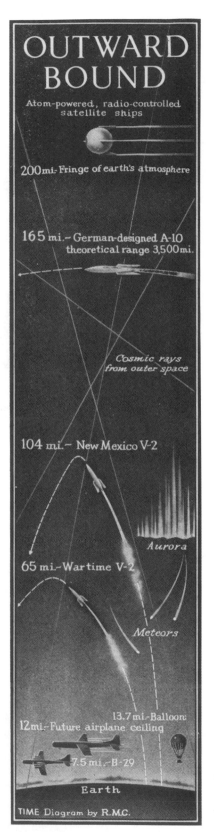

Depiction of the new high atmosphere by R. M. Chapin, from *Time* magazine, 2 September 1946. Copyright 1946 Time Inc. Reprinted with permission.

David H. DeVorkin

# Science With A Vengeance

*How the Military Created the US Space
Sciences After World War II*

With 109 Illustrations

Springer-Verlag
New York Berlin Heidelberg London Paris
Tokyo Hong Kong Barcelona Budapest

David DeVorkin
Department of Space History
National Air and Space Museum
Smithsonian Institution
Washington, DC 20560
USA

*Cover Illustration:* V-2 on launch pad, in profile. U.S. Army Air Forces. NASM SI 80-38-25.

**Library of Congress Cataloging-in-Publication Data**
DeVorkin, David H., 1944—
    Science with a vengeance : how the military created the US
space sciences after World War II / David H. DeVorkin. — 1st ed.
        p.    cm.
    Includes bibliographical references and index.
    ISBN 0-387-94137-1 (New York : alk. paper : pbk.).
    ISBN 3-540-94137-1 (Berlin : alk. paper : pbk.).
    1. Astronautics—United States—History.   2. Astronautics,
Military—United States—History.   3. V-2 rocket—Scientific
applications.   I. Title.
TL789.8.U5D48    1993
629.4'0973—dc20                              93-30683

Printed on acid-free paper.

Production managed by Christin R. Ciresi; Manufacturing supervised by Vincent Scelta.
Typeset by Impressions, A Division of Edwards Brothers, Ann Arbor, MI.
Printed and bound by Edwards Brothers, Ann Arbor, MI.
Printed in the United States of America

9 8 7 6 5 4 3 2 1

ISBN 0-387-94137-1 Springer-Verlag New York Berlin Heidelberg (pbk.)
ISBN 3-540-94137-1 Springer-Verlag Berlin Heidelberg New York

ISBN 0-387-97770-8 Springer-Verlag New York Berlin Heidelberg (hbk.)
ISBN 3-540-97770-8 Springer-Verlag Berlin Heidelberg New York

*This book is dedicated to*

*Howard DeVorkin*

*who first told me about the V-2*

# Acknowledgements

My awareness of the V-2 extends back to my early childhood when my father was employed at the Jet Propulsion Laboratory, and instilled in me a fascination for rockets, space travel, and astronomy, which, not surprisingly, all became related in my childhood dreams. The revitalization of my interest and its maturation, however, were a consequence of continued contact with historians interested in how science was transformed by the Second World War. Thus when I became interested once again in the V-2, my first thoughts centered around describing the science, namely the astronomy, that was done with it and the people who did it. But as I became familiar with the subject matter, discussing it with colleagues like Michael Dennis, Paul Forman, Karl Hufbauer, Allan Needell, and, more recently, Michael Neufeld, I realized that a more important study lurked beneath the science, one that provided me with a clearer view of how the space sciences originated in the United States.

Inspiration and insight came largely from those named above, but also through extended contact with scholars habituating the Contemporary History Seminar at the National Air and Space Museum. These seminars, faithfully organized by Needell, brought in Ron Doel, Bruce Hevly, Bill Leslie, Daniel Kevles, and dozens of other scholars whose commentary and perspective have greatly enriched my own. This cyclic stimulus, along with a graduate research seminar organized by Robert Friedel and held at the museum in the spring 1991 academic semester, greatly influenced my thinking about the relation of the space sciences to the national security state.

Beyond the published record, this book is based both on original archival documenta-tion and on oral history testimony. The citations in the Sources section do not adequately identify the many archivists whose sympathies and energies were of incalculable aid in the pursuit of this research project. Sources on Peenemünde and on the activities of Erich Regener were made available by Otto Mayr and Walter Rathjen and their staff at the Deutsches Museum, Munich; by Marion Kazemi, who kindly provided copies of material concerning Erich Regener's *Forschungstelle für Physik der Stratosphäre* held in the historical archives of the Max Planck Society; by Michael Eckert, who provided material from the Heisenberg Papers; and by Eugen Hintsches, of the Max Planck Society's Public Affairs Office, who provided copies of documents presented by Erwin Schopper at a 1988 symposium held at the Max Planck Institute for Aeronomy, Lindau.

The history of activity at the Naval Research Laboratory (NRL) came from many sources. The Naval Research Laboratory History Office, headed in sequence by David K. Allison, John Pitts, and David van Keuran, provided advice and access to NRL records held at the Washington National Records Center, and, in addition, through the indefatigable efforts of Dean Bundy in the NRL Archives Branch, access to dozens of laboratory notebooks that required declassification. NRL records in private hands also proved highly useful, especially from Ernst Krause, Charles Johnson, Francis Johnson, Milton Rosen, and Richard Tousey. The history of the Applied Physics Laboratory (APL) was pieced together from sources kindly provided by Lorence Fraser and James Van Allen and benefited greatly from the archival digging Michael Dennis performed for his doctoral dissertation. Highly useful archival records pertaining to APL and Princeton were also uncovered in the Department of Terrestrial Magnetism attic, opened for Dennis and myself by Louis Brown. Air Force Geophysics Laboratory (AFGL) historian Ruth Liebowitz kindly provided access to the papers of Marcus O'Day and introduced me to the very helpful librarians at AFGL, especially Connie Wiley, who provided access to the Laboratory's extensive contractor files. Dorothy Schaumberg of the Mary Lea Shane Archives provided copies of records from the Shane papers, and Philip Reed provided useful information from the Imperial War Museum, Department of Documents. Lee D. Saegesser and the staff of the NASA History Office kindly granted access to the Papers of Homer E. Newell. In addition to those mentioned above, many original players provided copies of documents or were willing to have their collections deposited in an appropriate archive for preservation and use. Among the latter were William Gould Dow, William Rense, S. Fred Singer, Lyman Spitzer, and William Stroud. Nelson Spencer provided both originals and copies, as did Dorrit Hoffleit and Julius Braun, who, with Tom Starkweather, provided detailed information on tracking and telemetry services at White Sands and on the history of the range.

Most of those named above provided critical advice and commentary on earlier drafts of this work. For the German period, Ernst Stuhlinger, Gerhard Reisig, Victor Regener, H.K. Paetzold, and Erwin Schopper were very helpful, as were historians Michael Neufeld and Mark Walker. For the balance of the text, earlier drafts were read and commented on by Krause, Tousey, Spitzer, Van Allen, Dow, Fraser, Singer, William Baum, Fred Whipple, Martin Harwit, John Logsdon, Gilbert Perlow, W. H. Pickering, Warren Berning, Herbert Friedman, William Rense, J. Carl Seddon, Paul R. Shifflett, John Simpson, Nelson Spencer, C. V. Strain, William G. Stroud, Clyde Tombaugh, Jesse Greenstein, Leo Goldberg, and Walter Orr Roberts. Critical commentary by Hevly, Dennis, Hufbauer, Forman, Neufeld, Doel, Needell, as well as by R. Cargill Hall, Robert Smith, John Sanderson, Joe Tatarewicz, James Capshew, C. Stewart Gillmor, Charles Ziegler, John Mauer, and many others has been of enormous value, even though at times the realization of how much corrective action was required left me overwhelmed and deeply in their debt. I can never repay the kindness and assistance provided me by five people who passed away during this project—Lorence Fraser, Leo Goldberg, Ernst Krause, DeWitt Purcell, and Walter Orr Roberts.

Many historians allowed free access to unpublished drafts and research materials. Needell provided chapters of his biographical study of Lloyd Berkner, just as Forman provided his chapters on Charles Townes' development of the maser; and Neufeld, Hufbauer, and Peggy Kidwell provided chapters or articles still in manuscript form. Dennis, Doel, and Hevly bravely let me read their dissertation drafts and research materials, all of which provided critical insight.

Oral and video histories are too numerous to mention and are listed in the sources section. What made these expensive undertakings possible, however, must be acknowledged here. The Space Astronomy Oral History Project at the National Air and Space Museum was supported by the National Aeronautics and Space Administration (NASA), NRL, and a Smithsonian Scholarly Studies grant. The Sources for the History of Modern Astrophysics Project, headed by Spencer R. Weart and a venture of the American Institute of Physics Center for History of Physics, was supported largely by the National Science Foundation. The Smithsonian Institution Videohistory Program was funded by the Alfred P. Sloan Foundation of New York. And oral history support is continuing through the archival unit headed by Martin Collins within the Department of Space History, National Air and Space Museum (NASM).

The interviewing activity has yielded a wealth of additional original documentation. For example, the complete minutes of the V-2 Panel spanning 15 years were put together from copies provided by Van Allen and

Stroud, as well as from the papers of William Dow and Homer Newell. Francis Johnson, Charles Johnson, Krause, Tousey, Van Allen, Spencer, and many of the others also provided material that is now part of their oral history files in the Department of Space History. Dow, although not interviewed, made available a highly detailed autobiographical memorandum that provided much insight into his history.

Many of those identified above furnished photographic materials, as did William E. Buchanan of APL, Fred Wilshusen of the University of Colorado, and the Max Planck Institute, Lindau. Arthur Cotts gave his valuable time and effort to organizing and processing these records and has my thanks for his devotion to this essential and demanding task. Art Cotts was also an invaluable research assistant, as were Peggy Shea, Susan Gould, Michael Dennis, Sophie Mayr, and Lisa Moscatiello. All-important translations of German documents were undertaken by Sophie Mayr, Paul Forman, Wil Baber, Marian Graham, and Michael Neufeld.

The National Air and Space Museum has not only rekindled my interest in this subject, but has provided an atmosphere conducive to pursuing it; the latter very much the result of the Museum's present director, Martin Harwit. In the past four years I have enjoyed the freedom necessary to conduct intensive research, and acknowledge with thanks Gregg Herken and the Department of Space History, who maintained a benevolent indifference to my whereabouts, allowing me to root through the rich archival and library treasures found in the Washington area and elsewhere. This activity was not without direct benefit to Museum programs, however, as it informed a recent exhibit surrounding the V-2 missile that adorns NASM's Space Hall and has helped me identify and collect a wealth of artifacts representing that era.

The technical details of making this book a reality were made tractable first by Jeff Robbins and then Tom von Foerster of Springer-Verlag and Trish Graboske of NASM. The photographic materials were processed by Mark Avino's NASM staff, Dave Gant of the NASM exhibits staff provided his excellent graphics abilities, and Vicki McIntyre once again attended to the copyediting with consummate skill and keen insight. The technical details attending to the final production of the book were completed during a sabbatical at the Institute for Advanced Study in Princeton, upon the very kind invitation of John Bahcall. Although my main activity in Princeton was to take up old interests in the history of astrophysics, the opportunity for intensive concentration without distraction helped to refine a number of key points raised in the text. I also had the enthusiastic technical assistance of Margaret Best of the IAS School of Natural Sciences, who put all the tables into final form.

As always, I owe more than can be stated to my wife Kunie and daughter Hannah, who live their own lives yet provide invaluable support and comfort by their faith in my abilities.

# Contents

## Part 2. Science with a Vengeance

**Contents**

# Abbreviations

(For abbreviations of source citations, see also section on sources.)

*AAF* Army Air Forces

*AFCRC* Air Force Cambridge Research Center (Also known as CFS, AFCRL, AFGL)

*AFCRL* Air Force Cambridge Research Laboratory (Also known as CFS, AFCRC, AFGL)

*AFGL* Air Force Geophysics Laboratory (see above)

*AGF* Army Ground Forces

*AMC* Air Materiel Command, Wright Field, Ohio

*APL* Applied Physics Laboratory, Silver Spring, Maryland

*BRL* Army Service Forces Ordnance Department, Ballistic Research Laboratory, Aberdeen Proving Ground, Maryland

*BuAer* Navy Bureau of Aeronautics

*BuOrd* Navy Bureau of Ordnance

*BuShips* Navy Bureau of Ships

*CFS* Cambridge Field Station (Also known as AFCRC, AFCRL, AFGL)

*CIOS* Combined Intelligence Objectives Subcommittee

*CIW* Carnegie Institution of Washington

*CNO* Chief of Naval Operations

*CNR* Chief of Naval Research

*CRPL* Central Radio Propagation Laboratory of the NBS

*CSAGI* Special Committee for the International Geophysical Year

*DOVAP* Doppler Velocity and Position

*DTM* Department of Terrestrial Magnetism of the Carnegie Institution of Washington

*GALCIT* Guggenheim Aeronautical Laboratory of the California Institute of Technology

*GRD* Geophysics Research Directorate (of AFGL)

*IGY* International Geophysical Year

*JCS* Joint Chiefs of Staff

*JRDB* Joint Research and Development Board (later known as RDB)

*JPL* Jet Propulsion Laboratory, California Institute of Technology

*KWG* Kaiser Wilhelm Gesellschaft

*NACA* National Advisory Committee for Aeronautics

*NASA* National Aeronautics and Space Administration

*NASM* National Air and Space Museum, Smithsonian Institution

*NBS* National Bureau of Standards, Washington, D.C.

*NDRC* National Defense Research Committee

*NOTS* Naval Ordnance Test Station, Inyokern, California

*NOTU* Naval Ordnance Test Unit, NOTS

*NRL* Naval Research Laboratory, Washington, D.C.

*OCO* Army Office of the Chief of Ordnance

*ONR* Office of Naval Research (Originally called ORI)

*ORI* Office of Research and Inventions (later known as ONR)

*OSRD*  Office of Scientific Research and Development

*RDB*  Research and Development Board (originally called the JRDB)

*SCEL*  Signal Corps Evans Laboratory, New Jersey

*SCIGY*  Special Committee for the IGY

*UARRP*  Upper Atmosphere Rocket Research Panel

*USNC*  U.S. National Committee for the IGY

*WSPG*  White Sands Proving Ground, Army Ordnance Test Station

# Chronology

## 1929

*July*  A Goddard rocket flies to 30 meters' altitude.

*December*  Members of the Carnegie Institution of Washington express interest in using Goddard's rockets for upper atmospheric research.

## 1937

*October*  Regener ousted from Stuttgart.

## 1938

*August*  Regener's institute at Friedrichshafen established.

## 1942

*July*  Von Braun meets Regener in Friedrichshafen and agrees on a research contract for upper atmospheric research.

*October*  First successful test of a V-2 at Peenemünde.

## 1943

Whipple's "Meteors and the Earth's Upper Atmosphere" published.

## 1944

*May*  Most of Regener's meteorological instruments completed.

*September*  V-2 missile bombings of London and Antwerp begin.

*November*  Von Braun orders test of Photo-Tonne; Kuiper's first intelligence report circulated.

## 1945

*January*  Regener's instruments integrated into Tonne and await a flight. Instruments packed up at Peenemünde as Regener's men are evacuated.

*February*  Arthur C. Clarke ponders the scientific use of the V-2.

*March*  Army Ordnance forms Special Mission V-2.

*April*  Mittelwerk under control of U. S. Army Ordnance.

*May*  Von Braun and key members of Peenemünde staff captured. Third Reich surrenders.

*September*  Kuiper interrogates Regener.

*October 11*  First WAC Corporal flies successfully from White Sands, reaching 75 kilometers altitude.

*December*  The Rocket Sonde Research Section formed at NRL.

## 1946

*January*  Army Ordnance arranges a series of meetings with parties interested in utilizing V-2 missiles for upper atmospheric research.

*February*  American scientific groups examine Regener instruments at Aberdeen.

*February 27*  First official meeting of the V-2 Panel

*March 15*  Static firing of V-2 no. 1 at White Sands.

*April 16*  V-2 no. 2 reaches 5.5 km. with APL cosmic-ray payload.

*Spring*  NACA Special Subcommittee on the Upper Atmosphere is formed, and by June finds knowledge of the upper atmosphere inadequate for a new extension to its Standard Atmosphere.

*May 10*   V-2 no. 3 reaches 112 km. with APL payload and disintegrates at impact.

*May 26*   V-2 no. 4 reaches 112 km. with APL payload but suffers an air burst.

*June 28*   V-2 no. 6 reaches 107 km. with NRL cosmic-ray and ultraviolet spectrograph payloads which were lost at impact. Retrieval of physical data becomes an acute problem.

*July 9*   V-2 no. 7 reaches 134 km. with NRL payload but disintegrates in flight.

*July 19*   V-2 no. 8 reaches 5 km.

*July 30*   V-2 no. 9 reaches 161 km. with APL payload. First successful warhead separation takes place, resulting in retrieval of large portions of midsection and tail.

*Summer*   NRL decides to move spectrograph to tail section.

*August 15*   V-2 no. 10 reaches 6 km. with first Princeton cosmic-ray payload.

*August 22*   V-2 no. 11 spins wildly after launch and crashes, destroying first Michigan package.

*September*   JRDB Panel on the Upper Atmosphere established.

*October 10*   V-2 no. 12 reaches 173 km. and is the first to return ultraviolet spectra of the sun.

*October 24*   V-2 no. 13 reaches 104 km. with the first APL spectrograph.

*October 25*   Meeting of Army Ordnance Advisory Committee for V-2 Firings ends with a call for the assembly of additional V-2s, subject to endorsement by the V-2 Panel.

*November 7*   V-2 no. 14 spins wildly after launch and is destroyed, along with second Princeton payload.

*December 17*   V-2 no. 17 reaches 182 km. during first night launch to perform artificial meteor experiment.

## 1947

*January*   "Tentative Standard Properties of the Upper Atmosphere" published by NACA.

*April 1*   V-2 no. 22 returns successful APL solar spectra, but Greenstein's spectrograph fails to operate.

*April*   Krause testifies before the JRDB Panel on the Upper Atmosphere that V-2 Panel needs as many V-2s as are "practical or economically feasible."

*May*   First Corporal E flies successfully.

*July*   The JRDB Panel asks Whipple to chair a working group to identify "reasonable objectives in upper atmosphere research."

*July*   Passage of National Defense Reorganization Act.

*July 29*   V-2 no. 30 reaches 160 kilometers with APL cosmic-ray payload that provides profile of cosmic-ray plateau.

*Summer/Fall*   Budgets tighten within the military research and development sector; the JRDB endorses V-2 Panel activities.

*November 24*   First operational Aerobee flown from White Sands, reaches 56 km. with APL cosmic-ray payload.

*November*   Ernst Krause retires as chairman and member of the V-2 Panel.

*December*   RDB Guided Missiles Committee adopts compromise position that "there is no unwarranted duplication at present in the development of upper air research vehicles."

## 1948

*January 22*   NRL cloud chamber flies on V-2 no. 34 to 158 km. and returns tracks of high-energy cosmic rays.

*February*   V-2 Panel becomes the Upper Atmosphere Rocket Research Panel (UARRP), at the suggestion of Holger Toftoy.

*October*   Goldberg and Spitzer confront the poor quality of spectroscopic data returned by rocket, and discontinue their ONR astrophysical consulting group.

## 1949

*January*   Air Force offers MX-774 as an upper atmospheric research vehicle. By March the RDB rules against the MX-774 in favor of Viking.

*February 24*   Bumper-WAC no. 5 flies to 390 km. from White Sands carrying BRL DOVAP system for ionospheric research.

*March 17*   First firing of an Aerobee (APL) from shipboard.

*May*   Van Allen realizes that APL does not wish to continue upper atmospheric research.

*May 3*   Viking 1 reaches 80 kilometers.

*June 23*   First successful flight of a photoelectric spectrometer aboard APL Aerobee no. 14 reveals continuous ozone profile.

*September*   Viking 2 launch.

*September 29*   V-2 no. 49 carries first successful NRL ionospheric package to 150 km., as well as Friedman's first X-ray photon counter.

## 1950

*February*   Viking 3 launch.

*Spring*   National Security Council calls for rearmament in anticipation of war in Korea.

*April*   An international year for cooperative geophysical research discussed by Berkner, Chapman, and others during a dinner party given at Van Allen's home.

*April/May*   NRL fails to win UARRP endorsement for Viking.

*May*   Viking 4 launch from U.S.S. *Norton Sound* close to the equator, reaches 170 kilometers.

*Sept. 7/8*   At 26th meeting of the UARRP, members express doubts about the future of scientific rocketry, and decide that the panel can no longer justify new vehicles for basic research.

*December*   Rocket Research Branch formed at NRL to explore feasibility of converting Viking into a missile.

*December*   Van Allen leaves APL.

*Dec. 11/12*   T-Day at White Sands includes Viking 6, USAF Aerobee no. 9, and Signal Corps Aerobees 14, 15, and 16, as well as numerous balloon launches and concentrated meteor observations by Harvard.

## 1951

*January 25*   APL's Aerobee no. 20 produces refined ozone profile with a continuously recording photoelectric spectrometer.

*February 6*   APL launches its last Aerobee.

*April*   Van Allen, now at Iowa, urges the panel to produce a new "Standard Scientific Atmosphere."

*April 12*   The first Colorado biaxial pointing control flies on USAF Aerobee no. 11, but flight is aborted due to a fuel leak.

*August 14/15*   At 29th meeting of the UARRP, members compare upper atmospheric data and prepare their extension of the NACA Standard Atmosphere.

*Fall/Winter*   Whipple reviews state of knowledge of the upper atmosphere and revises his model of the types of meteorites causing meteor trails.

## 1952

*Summer*   Van Allen tests Rockoon at White Sands.

*Sept. 19*   Last V-2 flies from White Sands.

*September*   First season of Rockoon firings from shipboard near Thule, Greenland.

*Late Fall*   National Academy appoints a U.S. National Committee for the IGY, headed by UCLA's Joseph Kaplan and managed by Hugh Odishaw.

*December 12*   First fully successful flight of a Colorado biaxial pointing control aboard USAF Aerobee no. 33 returns an image of solar Lyman Alpha.

*December*   "Rocket Panel Atmosphere," published in the *Physical Review*.

## 1953

*July*   Second season of Rockoon firings, confirms the detection of auroral radiation.

*August*   Oxford Conference.

## 1954

Detailed planning for the IGY commences. The UARRP provides plans and budgets for rocket research.

## 1956

*Jan. 26/27* Tenth anniversary meeting of the UARRP is held to discuss the scientific uses of earth satellites.

## 1957

*October* 165 Aerobees had flown by the time Sputnik 1 orbits.

# 1

# Introduction

On 22 October 1946, there was tension in the darkness of a photo lab at the U.S. Naval Research Laboratory, on the outskirts of Washington, D.C. Physicists gingerly removed a strip of 35-millimeter photographic film from a heavy steel cassette which 12 days earlier had flown 173 kilometers into space in a spectrograph aboard a V-2 missile. Richard Tousey, head of the group that had built the spectrograph, knew from the telemetry record of the flight that the instrument had worked correctly, but the payoff would be on the film once it was developed in the Navy darkroom. He and his group hoped to see the fingerprints of the ultraviolet spectrum of the sun, a region forever blocked from view by the earth's atmosphere.

Search teams took six days to find the spectrograph at the crash site in the desolate Tularosa Basin of southern New Mexico, where the Army had established its White Sands Proving Ground. Tousey knew that moisture, light leaks, and wind-blown sand could have spoiled the film and its precious emulsion. This was Tousey's second try; the first spectrograph was never found after its missile crashed the previous June. Now at least they had recovered the film, but would it show anything?

As soon as the film was in the fixer, someone turned on the lights and Tousey peered at the wet photographic strip. It looked completely blank, but spaced along its length was a series of tiny gray smudges, each hardly a few centimeters long. Under a magnifying glass, they became tiny solar spectra and revealed portions of the heretofore blocked ultraviolet spectrum of the sun. After nine months of intense effort, Tousey and his team had tasted their first success in capturing a new view of the sun from rockets.

Richard Tousey was not a solar physicist in 1946; he was a branch head within the Physical Optics Division of the Naval Research Laboratory (NRL). Trained as a vacuum ultraviolet spectroscopist, Tousey was a specialist in laboratory optical techniques, spectroscopy, and the limits of vision. He knew his way around a laboratory and how to make spectrographs work.

Tousey's solar spectra excited astronomers interested in the sun and caught the attention of physicists and meteorologists studying absorption processes in the earth's high atmosphere. When the venerable Princeton astronomer Henry Norris Russell first saw a Tousey solar spectrum, he wrote to a younger colleague: "These rocket spectra are certainly fascinating. My first look at one gives me a sense that I was seeing something that no astronomer could expect to see unless he was good and went to heaven!"[1] Astronomers had long wondered what the ultraviolet spectrum of the sun would look like beyond the atmospheric barrier. In the late 1930s, according to a longtime associate, Russell would sometimes muse about it with his students: "A bunch of us got to talking about what the ultraviolet solar spectrum would be like. It was always our dream, but at that time we thought we'd never live to see such a thing."[2]

But neither Russell nor his astronomical progeny were involved in securing the first solar spectrum from space. Nor did any of those prominent in ionospheric research, cosmic-ray physics, or atmospheric physics before World War II participate directly in the exploration of those realms with rockets. None of the pioneer space scientists like Tousey, his NRL colleague Herbert Friedman, or Iowa's James A. Van Allen, harbored research interests related to the upper atmosphere before entering the field. When they did enter, it was to perform research with rockets.

It is this technological context that motivates this book. The scientific problems addressed with the V-2 existed before the war. However, a whole new set of workers, with talents and alignments quite distinct from those of the people traditionally active in upper atmospheric research, were the first to use the rockets for scientific investigations. As a result, an entirely new way to conduct science, constructed around the rocket, emerged after World War II within a world being redefined in terms of national security needs. From this new way of doing science evolved a new definition of the scientist.

As astrophysicist Martin Harwit has aptly remarked, "scientists are quick to pick up and put to good use any successful new research tool." Harwit felt that "were this not so, whole classes of potentially effective approaches would be permitted to go unused—a waste quite uncharacteristic of most scientific efforts."[3] Along with historians, Harwit recognizes that although this model reasonably describes incremental change, it does not indicate how revolutionary technologies are typically introduced into a field of science: when specialists migrate, bringing their new tools and techniques with them.[4] Nor does it describe adequately the complex means through which the modern space sciences emerged out of the highly technical tool-building cultures that existed within military laboratories and were stimulated in colleges of engineering and experimental physics after the war by military funding.[5] A principal goal of this book, then, is to supply this larger context for the origins of the space sciences in the United States during the V-2 era.

What I call the V-2 era, the period in which scientists first used V-2 rockets for upper atmospheric research, began in Germany during World War II, continued in the United States and Soviet Union at the close of World War II in 1946, and lasted roughly through the planning for the International Geophysical Year in 1954. This period has already been described from several distinct perspectives. First, the scientists involved have provided richly detailed technical reviews of their efforts and of the scientific results obtained and include accounts of personal experiences as well as commentary on events and processes.[6] Some popular writers primarily interested in rocket development have acknowledged the use of the V-2 for scientific research.[7] Historians, practitioners, and science writers have also examined the scientific use of rockets in the development of artificial satellite research.[8] Some have contributed useful perspective on the political and military aspects of the post-Sputnik space program, placing it, along with the space sciences, within the framework of national security.[9] Institutional histories of the Naval Research Laboratory's early promotion of space science and the Jet Propulsion Laboratory's transformation from a military contract laboratory to a NASA entity have also increased our understanding of the military origins of the space sciences.[10] Several disciplinary histories, for X-ray astronomy, planetary astronomy, and solar physics, have identified the communities that eventually formed to conduct space research.[11]

The present study draws on these sources but is distinct from them in its exploration of the social forces that promoted scientific research with rockets, in its identification of the groups that responded successfully to these forces, and in its attention to those groups whose involvement was only temporary. Although this book examines the research that was done, it leaves much of the technical detail to the review articles and memoirs cited above. Instead, this book places scientific research with rockets within the world of guided missile research and development. Even though this is not entirely a new idea, the use of this frame of reference requires some elaboration. As one entrepreneur for the Navy in space observed in 1959:

*In fact, rocket exploration of the upper atmosphere began in the U.S. Department of Defense largely because of the increasing importance of guided missiles and the expectation of very high altitude flight. It was important to know the medium through which such missiles and craft would fly and the environmental conditions they would encounter.*[12]

The impression this statement gives is that the scientists already studying the upper atmosphere were the ones to use the rockets. Strengthening this impression is the 1980 reflection by a former NASA associate administrator, Homer Newell, who, in defining the various parts of the space sciences, argued that "the parts of space science devoted to astronomy," for instance, "remain a part of the discipline of astronomy, and space scientists using rockets for astronomical research continue to view themselves as astronomers."[13] Although closer to reality in 1980 than 1946, Newell failed to make the distinction. We show here that in the beginning, even though a few astronomers, meteorologists, and cosmic-ray physicists tried to become involved, none persevered.

The point is that scientists do not respond in simple, algorithmic ways to new opportunities. In the fiercely entrepreneurial world of postwar American science, there was no direct relationship between perceived gaps in knowledge and the means to close those gaps. The groups that formed to conduct research with rockets had to invest a great deal of manpower, time, and talent to meet enormous technical challenges that frustrated any rapid harvesting of scientific data. These had to be special people, working for a special patron: one interested in the ability to perform technical experiments with rockets as much as the results of those experiments. The patron also had to be keenly interested in extending the relationship with science that had blossomed during World War II.[14]

Recent scholarship shows that government and military patronage has altered the direction and character of scientific research, and, if it ever had any, reduced its autonomy.[15] One recent observer notes that although American scientists may feel "free to pursue the details of their work as they wish, yet the broad outlines of their work are defined by the Government."[16] Because the networks of power are so complex, and the response patterns of scientist entrepreneurs so much a part of that complexity, the challenge here will be to see how the forces at play shaped the individuals and working groups that created the space sciences in the V-2 era.[17]

The circumstances surrounding scientific research with missiles did not develop overnight in the wake of World War II. In an earlier study of manned scientific ballooning in America, I explained how the American military came to control stratospheric balloon ascents in the 1930s and 1940s and the types of compromises and commitments scientists made as they became part of those ascents. In conducting scientific research with manned balloons, the scientists had to learn not only how to prepare their instruments for operation under unusual conditions, but how to work as part of a larger effort whose missions and goals were distinctly different from their own.[18]

The present study examines this process of alignment in an era when the vehicle for research and the motivations for it were initiated and funded by the military and were available to scientists through no other means. Those who entered such activities either had to adapt to a new research environment utterly unlike anything that had previously existed, or they were raised in that world during World War II and inevitably had to find the means to establish themselves in the world of science.

Our exploration will be limited to the American experience in the development of upper atmosphere and space research. It was, however, a German legacy that made such research possible in the United States following World War II. The first vehicle to provide access to the uppermost regions of the atmosphere and to near space was the German V-2 missile, created at incalculable human cost by a massive wartime engineering design and production effort. The V-2, Vergeltungswaffe zwei, or Vengeance Weapon-2, was the name Nazi propagandists gave to Germany's 14 meter tall, 12,800 kilogram liquid-fueled missile. Called by its creators the A4, the V-2 was first flown successfully at Peenemünde in October 1942 and then launched by the thousands on civilian targets in Belgium and England in 1944 and 1945.

The Germans had appreciated the importance of upper atmospheric research to the operation of a ballistic missile system well before the missile was operational. Within the maelstrom of war, a small group of physicists led by Erich Regener was invited to prepare instruments to fly in the warheads of A4 missiles, to be test fired vertically from Peenemünde. Regener provided instruments that would measure the character of the medium through which the missile traveled, adapting devices that his group had built for upper atmospheric research with balloons. The scientific agenda Regener created for the A4, defined by the interests of his Peenemünde and Luftwaffe patrons, reappeared in the American era after the war. But it was also anticipated to a large degree in 1930 in the United States when a group of scientists was asked to advise Robert H. Goddard on what scientific research should be done with the liquid-fueled rockets he wanted to build. Thus both Goddard's efforts and experiences, and those of Regener, set the stage for what happened after the war in the United States when the vehicle for research was also the object of that research.

The first part of the book deals mainly with the military context of upper atmospheric research: the military interest in seeing that such work was being done, the formation of appropriate groups in military laboratories capable of doing the work, and the development of the technical and managerial infrastructure required to get the work done. We also introduce the people who became involved in the effort, and establish a general profile for typical participants at various levels within each group hierarchy: the entrepreneurs, the managers of science and engineering, and the workers on the shop floors. We will pay closest attention to how each group dealt with a very new environment for research—White Sands Proving Ground—and then follow the fortunes of the group leaders

as they were organized by Army Ordnance into a "V-2 Panel" to become an advocacy group for the conduct of upper atmospheric research with rockets, especially for the means to continue doing so once the V-2s ran out.

In reconstructing the world of the late 1940s, we must appreciate that it was a time when postwar patriotism was strong and an "ethic of responsibility" to "help maintain the national defense" existed to a degree greater than today.[19] Even so, it was a time when most established scientists, even those who claimed the high moral ground of patriotic responsibility, did not wish to "fulfill that responsibility as subordinates to the military."[20] As we identify the individuals and groups that formed to conduct scientific research with rockets, then, we must remain sensitive to those characteristics that differentiated these scientists from their more traditional brethren in academia, a goal which will carry through the remainder of the book.

The latter half of the book examines the specific problems each group addressed, particularly the technical, professional, and managerial obstacles they faced as they explored the use of rockets for studying the sun, cosmic rays, the upper atmosphere, and the ionosphere. One of the challenges of studying the sun from rockets, for example, was technical design, including detector choice, data retrieval, and mode of instrument stabilization. Technical choices were severely inhibited, however, by the nature of the rocket as vehicle; this factor, and the required high level of technical commitment, made it very difficult for traditional astronomers to get involved. Those who tried met with resistance and failure, yet their experiences offer critical insight into what made the rocket groups unique as they persevered with solar studies. In cosmic-ray research, the general lack of scientific success of all groups drove some out of rocketry and caused others to migrate into more productive areas like geophysics. In atmospheric physics, the rocket scientists felt compelled to demonstrate the authority of their findings in the face of an established model for the atmosphere created by traditional means. In ionospheric physics, the rocket groups had to overcome a greater degree of technical difficulty and therefore attracted less attention, even when their data helped to change the dominant model of the ionosphere that had existed for 25 years. Although we will explore each area of research by highlighting those specific issues identified here, we will do so knowing that these issues were faced, more or less, by workers in all areas of upper atmospheric research.

The book ends as planning for the International Geophysical Year begins. The leading entrepreneurs for scientific rocketry played an important role in the development of the IGY, and in the creation of a new and vast geopolitical scientific enterprise which in the intervening years has given the space sciences much of their contemporary civilian patina. Probing through that coating to the original structure is the purpose of this book.

## Notes to Chapter 1

1. Russell to Leo Goldberg 10 May 1947. HNR/P.
2. Sitterly OHI, p. 51. SHMA/AIP. Chapter 11 reviews astronomers' periodic frustrations with atmospheric absorption.
3. Harwit provided this definition to identify the range of tools he wished to bring to his study of how "a science like astronomy should progress." But it applies equally well to his descriptions of scientists working within their chosen problem areas. Harwit (1981), p. 9.
4. Typical case studies of migration into astronomy include Edge and Mulkay (1976); Gilbert (1976); Hirsh (1983), chap. 3; DeVorkin (1985, 1987b); and Hufbauer (1991). General studies include Gieryn and Hirsh (1983, 1984). Wise (1985), p. 240, notes that astronomy may be a special case, as it is "especially dependent on ways of seeing."
5. On the concept of NASA as a tool-building culture, see McCurdy (1989). Naugle (1987) describes NASA as an uneasy alliance of at least three distinct elements: the rocketry groups we examine here, university groups devoted to upper atmosphere and cosmic-ray physics using balloons and aircraft, and engineering and technical groups that constituted the National Advisory Committee on Aeronautics. To these were soon added aerospace engineering groups from Army Ordnance. See Newell (1980); Koppes (1982); and McDougall (1985).
6. Newell (1980) provides a broad and insightful introduction, from the 1940s through 1970. His earlier works (1953, 1959), as well as his many technical NRL reviews, contribute to the technical history behind this present study. Van Allen (1983, 1990) provides a personal narrative of his contributions to magnetospheric physics, as does Rosen (1955) for the history of Viking; and Tousey (1953b, 1963, 1964, 1967, and 1986) and Friedman (1960) for ultraviolet and X-ray solar research. Additional contributions can be found in the bibliography. Outside the boundaries of this work are the postwar contributions of European and Soviet workers. See the articles by Jean Corbeau, S. N. Vernov, and L. A. Vedeshin, contained in Lattu (1989); articles by E. B. Dorling, G. S. Ivanov-Kholodnyi and L. A. Vedeshin, B. A. Mirtov and L. A. Vedeshin, and E. Vassy contained in Ordway (1989); as well as Ivanov-Kholodnyi and Nikol'skii (1969); A. E. Chudakov Vernov (1960); Shternfeld (1959); [USSR Academy of Sciences] (1958); and Massey and Robins (1986).
7. Most relevant here are Kennedy (1983); Ordway and Sharpe (1979); Hallion (1977); and von Braun and Ordway (1975). More than the others, Vaeth (1951) concentrates on the science.
8. Among the best, see Hall (1963). A more focused study of early interest in developing reconnaisance satellites by the Air Force can be found in Davies and Harris (1988). For Viking and Vanguard, see Rosen (1955) and Hagen (1963).
9. My sense of the national security environment derives from Forman (1987), and from Koppes (1982) in the context of ballistic missile development. The sweeping political history by McDougall (1985) has become a standard of interpretive reference against which the social forces promoting space activities are now being examined. An insightful British perspective can be found in Lovell (1973). See also Medaris and Gordon (1960) for a personal view of the role of scientific research in the ballistic missile program of the United States, and Hirsh (1983), chap. 6, for the debt modern X-ray astronomy owes to national weapons laboratories.

10. See Hevly (1987) for NRL and Koppes (1982, 1989) for JPL.
11. Hirsh (1983), Tatarewicz (1990), and Hufbauer (1991), respectively.
12. Truax (1959), p. 45.
13. He said the same applied to cosmic-ray physicists and geophysicists in remarks designed to demonstrate that space science did not constitute a discipline. Newell (1980), p. 13.
14. This relationship has been characterized as the permanent mobilization of science. See Komons (1966); Bush (1970), pp. 51, 52, 67 and 303; Kevles (1975), pp. 20–47; Sapolsky (1979); England (1982); Seidel (1983), pp. 375–400, 376, n. 2; Rearden (1984), pts. I and II; Stares (1985), chap. 2; and Forman (1987). It runs counter to the erroneous impression favored by Green and Lomask (1970), p. 14 n. 25, that such research areas as the composition of the ionosphere "were less matters for the Pentagon than for the National Academy . . ." which they justified by a garbled citation linking the V-2 Panel to the National Academy of Sciences in the 1940s and early 1950s. They correctly state that the panel fostered strong ties with the Department of Defense.
15. Roland (1985b), pp. 269–70 identifies several studies by historians, policy analysts, and practitioners on shifts in direction and methodology. Forman (1987), looking at the physics community, and Mukerji (1989), examining oceanographers, have both equated the loss of autonomy with dependency on funding that is controlled by policies external to the needs of the discipline. This was an early fear of astronomers just after World War II, as pointed out in DeVorkin (1991).
16. Nelkin (1990), p. 18, in a review of Mukerji (1989). See also Nelkin (1972), p. 11.
17. Newell (1980), p. 11, defines space science as "those scientific investigations made possible or significantly aided by rockets, satellite, and space probes." The term is modified here to imply that in its early years the conduct of space science was *defined* by the use of vehicles like rockets, as well as balloons.
18. DeVorkin (1989a).
19. The first quotation is from Nelkin (1972), p. 7; the second from Kevles (1978), p. 340. Nelkin added, p. 34, that "in the immediate postwar years, when national foreign policy was far more widely accepted than it is today, the costs of involvement were also more acceptable." Roland (1985b), pp. 267–268, especially n. 82, traces the growing tensions since World War II within the scientific community, providing strong bibliographic evidence that the tension was not confined to science.
20. Kevles, ibid.

# 2
# Establishing the High Atmosphere as a Site for Research

In 1892, astronomer Solon Bailey sought a deeper meaning to what he and his mules had recently accomplished, establishing the world's highest meteorological station on the Peruvian stratovolcano El Misti:

> I think I may claim to have taken mules higher than they ever went before. . . . Perhaps this may prove to be a step in an evolutionary process, from the primary dependence on personal endurance to the time when scientists will regularly ascend to great heights for meteorological study, by means of captive balloons or flying machines. I hope our mules on our next expedition will appreciate their honorable position as connecting links, and exert themselves accordingly.[1]

In Bailey's time, sporadic manned balloon flights had returned data on meteorological and electrical conditions in the atmosphere for over a century, and new unmanned automatic balloonsondes promised to achieve greater heights with less effort and with no danger to human life. But meteorologists were just then learning how to adapt their instruments to operate automatically under the harsh conditions found at great heights.

Meteorologists, physicists, and astronomers like Bailey wanted to know the nature of the high atmosphere, and they took advantage of whatever vehicles were available, making compromises along the way. Understanding the constitution and structure of the atmosphere, the origin and nature of the cosmic ray, how and why radio waves propagate such long distances, and the amount and character of solar radiation that hits the earth were well-established goals of science long before rockets became available. Not surprisingly, rocket pioneers like Robert H. Goddard and Hermann Oberth realized that there was a potential market for their dreams.[2]

## The Promise of Goddard's Rockets

The legend of Robert Goddard is well-known. Less known is that in his efforts to gain support he had to deal with a set of enthusiastic elite scientists who had been asked by his supporters to provide advice on the value of rocketry to science. Goddard, of course, was familiar with the demands of patronage; when he approached the Naval Consulting Board and the Smithsonian Institution in 1915 and 1916 for support to build his rockets, he told the former that the rockets could be used as vehicles for carrying bombs, and the latter that they could carry scientific instruments into space.[3]

The Smithsonian Astrophysical Observatory's Charles Greeley Abbot was intrigued by the vision shown by this unknown physics professor from Worcester, Massachusetts. Goddard's descriptions of his experiments with solid-fueled rockets were as much value to ordnance, Abbot reasoned, as they were to the types of upper atmosphere research he claimed could be done. With Abbot's backing, the Smithsonian supported Goddard's experimentation in the hope that meteorological instruments might be carried to altitudes of hundreds of kilometers.[4]

Goddard made much of the potential application of rocketry to the study of the up-

limit of balloon; 20 miles
Moon
limit of atmosphere;
200 miles
EARTH
700 miles
6 miles/s
1 lb
50 lls (H)

Robert H. Goddard at
Clark University, circa
1924. NASA.

per atmosphere. In his classic paper, "A Method of Reaching Extreme Altitudes," written and modified several times between 1914 and December 1919, Goddard pointed out an obvious use would be in determining "the density, chemical constitution, and temperature of the atmosphere, as well as the height to which it extends."[5] He also speculated that auroral phenomena and the ultraviolet spectrum of the sun could be studied with this new tool. Goddard's promises kept Abbot's enthusiasm high, and through Abbot, other astronomers supported Goddard's work. The preeminent entrepreneur George Ellery Hale provided a safe place for Goddard's tests in the canyons above Pasadena, California, and additional funding came from the Signal Corps and Army Ordnance.[6]

Throughout the 1920s, Goddard continued his experiments, turning to liquid-fueled motors. In 1925, his first model managed to lift itself for 10 seconds. By March 1926, a 5 kilogram rocket flew about 12 meters, and one did a bit better the following month. Even though Abbot and the Smithsonian demanded, as they provided each incremental payment, that they soon see a successful me-

teorological rocket, Goddard continued doggedly on, and his promises were undiminished. By July 1929, a 3.5-meter-long rocket flew to 30 meters' altitude.

This latest flight attracted much publicity, and the attention of an American hero. After his May 1927 solo flight across the Atlantic, Charles A. Lindbergh became a passionate spokesman for aviation research and, after meeting Goddard, became his agent into the world of private philanthropy. Lindbergh first approached a number of industrialists, but he was directed to John C. Merriam, director of the Carnegie Institution of Washington (CIW), to lobby on Goddard's behalf.[7] Merriam thought enough of the idea to ask John A. Fleming, acting director of the Carnegie's Department of Terrestrial Magnetism (DTM), to see what interest DTM scientists might have in using rockets as vehicles for research. Merriam also contacted Abbot to see how the Carnegie might cooperate with the Smithsonian.[8]

Goddard visited Washington for meetings with scientists from the CIW, the Weather Bureau, the Naval Research Laboratory, and the Smithsonian interested in the upper atmosphere and found them all enthusiastic. Writing to physicist Gregory Breit at New York University in December 1929, Fleming stated that he was excited about the "plans to forward the exploration of the upper air by means of rockets along experimental lines of Goddard."[9] E. O. Hulburt of the NRL argued that the ideas of what to do were "quite obvious and are no doubt exactly those which would occur to anyone who considers the matter." Hulburt listed the obvious targets: temperature, pressure, composition, and degree of ionization were first priorities, and then wind patterns, diurnal and seasonal variations, ozone and hydrogen content (atomic and molecular).[10] Hulburt was careful to separate scientific research from missile development: "These are all suggestions towards gathering information about the atmosphere, the projectile being regarded merely as a tool." But, ever sensitive to the military advantages of rocketry, Hulburt added that "the behavior of the projectile in flight, its speed, orbit, etc., might yield information of interest and value to certain departments of the Army and the Navy."[11]

Hulburt's listing of scientific priorities reflected the common opinion of those Fleming canvassed for Merriam's response to Lindbergh (Table 2.1). They agreed that "the rocket-method is at present the only one giving promise of successful application for investigations of the high atmosphere"[12] and suggested instruments to build that would not only meet these needs but would also serve in applied fields like radio reception and acoustic sound ranging. In particular, "the behavior of the rocket and projectiles from it during the flight and character of the orbits would give information not only useful in the study of physical problems but also useful to the Army and Navy."[13]

Fleming's circle of scientists had heard from Goddard that his new liquid-fueled rockets would reach 120 kilometers altitude but would cost at least $100,000 to develop. Lindbergh suggested that if the Smithsonian and the CIW could find the first $15,000, he could find the remaining $85,000 from his circle of philanthropists. This became a blueprint for Mer-

**TABLE 2.1** Scientific Program for Goddard's Rockets

— Terrestrial magnetism: magnetic fields at great heights.
— Terrestrial electricity: determination of electron density and ion density and recombination rates; study of his nature of penetrating radiation (cosmic rays), intensity of ionizing agents (UV, X-ray, corpuscular agents) from celestial objects.
— Radiotelegraphy: which frequencies get through, attenuation.
— Astrophysics and atomic physics: photos of UV and X-ray spectra of the sun, the origin of the green coronal line, auroral lines.
— Physical processes in the upper atmosphere at low pressures, the chemical composition of the upper atmosphere, ozone.

*Source:* Scientific Agenda Suggested by Fleming for Goddard's Rockets. Adapted from "Memorandum by Mr. J. A. Fleming, acting director, Department of Terrestrial Magnetism, relative to Dr. Goddard's rocket program," 10 December 1929. RG/CIWA.

riam's proposal to the Trustees of the Carnegie Corporation, but in addition, both the CIW and Smithsonian decided that they needed a scientific advisory committee to assist Goddard in the preparation of the scientific agenda for his rockets.[14]

Goddard knew that he had to keep the pressure on to obtain the full amount of funding quickly. In March 1930 he reminded Merriam that even though he was farther along than any known Europeans in building liquid-fueled rockets, he needed full funding to stay ahead: "The lead at present is in America, and I, for one, would dislike to see leadership pass to some other country, owing to our inaction here."[15] Seeing applications in science, as well as "unquestionable applications in aeronautics and national defense," Goddard wanted to know how and when "the Colonel" was going to be able to arrange for the larger grant. Colonel Lindbergh came through in June 1930, securing support from Daniel Guggenheim. Lindbergh had been an agent of the Daniel Guggenheim Fund for the Promotion of Aeronautics, Inc., and was therefore able to directly broker the scientists' endorsements to obtain $25,000 per year for two years for Goddard, with the possibility of another two-year grant at the same level.[16]

Continued Guggenheim support, however, carried the same requirement as the Carnegie money: it would depend on the findings of a scientific advisory committee.[17] Lindbergh turned once again to Merriam to form the committee, and Merriam agreed to head it. The Committee on Study of Investigations of the Upper Atmosphere consisted of Lindbergh, Abbot, Fleming, Walter Sydney Adams (director of the CIW's Mount Wilson Observatory), Robert A. Millikan (who had been a Guggenheim Fund trustee), Arthur Kennelly (codiscoverer of the ionosphere), C. F. Marvin (Chief of the Weather Bureau), and two high-ranking military associates of Lindbergh's.[18] Merriam drew advice and assistance from Fleming and a young CIW physicist, Merle Tuve, who, with Abbot and Adams, constituted a core committee.

If Goddard had succeeded at one thing, it was in exciting the hopes of his committee. While Goddard worked through the rest of 1930 setting up his western station near Roswell, New Mexico, and refining his propulsion and stabilization systems, the committee started to design instruments for his rockets. Goddard's encouraging progress reports to Merriam convinced the committee that the high atmosphere would be accessible by the summer of 1931.[19]

But Goddard's letters were not frequent, nor were they really informative. They also grew more distant when Merriam decided in October 1930 that Goddard needed collaborators: "however successful the rocket flight, its ultimate value scientifically would depend upon correlated experiments and observations to be made by associated investigators." "One man," Merriam argued, could not have mastery over the whole scientific field as well as over rocketry.[20]

A still-hopeful committee met in February 1931 to review Goddard's progress. Daniel Guggenheim had passed away in the interim, but the Guggenheim family agreed to follow through if the committee's evaluation report was positive. Goddard's upbeat reports convinced the committee that all was well; Abbot felt that an ultraviolet solar spectrograph and air samplers should be built immediately to accompany Goddard's first rockets.[21] Adams, who was going to build the spectrograph, reflected the mood of the members, writing to Fleming, "I feel more and more certain that something good is going to come out of this, especially in a scientific way, and all of us who can contribute in any way . . . should devote as much thought to the subject as possible."[22]

But throughout 1931, little was heard from Goddard, and when news did come, it was not encouraging. Goddard knew that milestone dates had been missed, letters unanswered, and promises broken. He thus went on the defensive in a December 1931 report arguing that before he was willing to think about the design of instruments, he had to solve many remaining problems before his rockets would work properly. More frank now, Goddard reminded his scientific advisers that: "the liquid-propellant rocket is a new problem in research, and is entirely unlike designing and constructing a special engine of a type in common use."[23] Reality had set in: Goddard was a long way from providing an operational rocket.

Despite the delays, the committee remained positive, hoping to use Goddard's rockets sometime soon. As head of Carnegie's DTM, Fleming was then a member of the Commission for the Polar Year, 1932–1933, and testified before the House Foreign Affairs Committee that small rockets made by Goddard might be used to loft scientific instruments to 80 kilometers.[24] The committee was heartened when, in May 1932, Goddard successfully flew a gyroscopically stabilized rocket and was planning for another flight. But Goddard was running out of money, spending at the rate of $2,000 per month.[25] In late May, a more sober committee met to assess the situation. They agreed that Goddard had made "real progress" in the first two years on propulsion and guidance, in the face of what C. F. Marvin sympathetically called "innumerable failures and large expenditures of time and money."[26]

The committee issued a favorable report, but it now called for closer contact between Goddard and CIW scientists, to overcome present obstacles and aid further development. Specifically, they hoped that instrumentalist John Anderson might have a chance to visit Roswell and critically evaluate the mechanical stabilization systems Goddard was developing.[27] This suggestion, however, did not please Goddard one bit. Worse yet, the committee's report failed to convince the Guggenheims to renew their support. The initial grant was canceled, forcing Goddard to return to Clark University for a year as professor of physics.[28]

Only through Lindbergh's continued efforts was support restored in August 1934. The new Daniel and Florence Guggenheim Foundation provided $18,000 with a promise of another equal sum the following year. Goddard was elated; leaving Worcester in haste, he wrote to Merriam that he and Lindbergh no longer needed the committee's services "until I am able to reach a considerable altitude and need cooperation in the construction and use of instruments."[29]

With funding assured, Goddard now felt independent of the oversight of scientists. He continued on with his experimentation throughout the 1930s, but his rockets never reached altitudes, or an operational status, that made them competitive vehicles.[30] Thus it is not surprising that neither Merriam nor any member of his committee pursued their relationship with Goddard further. Although among them were several who truly needed data available by no other means, none campaigned for upper air research with rockets, and none attempted to seek other means of building such devices in the 1930s.[31]

## Upper Atmosphere Research in the 1930s

Most of the scientific problems Merriam's committee planned to address with Goddard's rockets were also being studied using balloons and indirect means in the 1930s. Balloons were just more limited in how high they could fly. Manned scientific flights had reached as high as 22 kilometers, and unmanned sondes 30 kilometers.

In the 1930s, the upper atmosphere was studied both by theorists and by observers in physics and meteorology. Groups in England, Germany, Norway, the Soviet Union, and the United States developed theoretical models to describe the heat balance of the atmosphere in its various layers, its wind patterns, and especially its electrical and magnetic characteristics. A major goal was to understand why the ionosphere existed, and why it behaved as it appeared to do from extensive radio and geomagnetic studies initiated in the 1920s. Refined and sophisticated ground-based techniques for probing the upper atmosphere employing radio reflection, acoustic probing and searchlight scattering schemes, sensitive photoelectrically equipped spectroscopic observations, and mountain-top cosmic-ray and weather stations provided a wealth of direct and indirect data on a wide range of phenomena under the heading of terrestrial and atmospheric magnetism and electricity, known also as geomagnetism and solar-terrestrial physics. Edward Appleton, Sydney Chapman, Julius Bartels, F. W. Paul Götz, Erich Regener, Lars Vegard, John Fleming, and others formed the nuclei of research schools in mathematical and experimental physics and geophysics devoted to understanding the constitution and composition of the atmosphere and the solar influences on it. Götz and Regener in Germany, Vegard in Norway, Chapman and Bartels in England, and Fleming in the United States—all were experienced in geomagnetic research. Chapman's influence came as well from his theoretical studies of nonuniform gases and Bartels was able to apply powerful statistical procedures to the study of magnetic variations. Appleton was trained in physics and possessed practical experience in vacuum-tube technology, which he first applied to problems in radio reception and then to atmospheric research. The majority of these leaders were therefore trained in either mathematical or experimental physics and pursued means of examining the atmosphere indirectly from the ground, although many of them did not fail to exploit ways of examining the atmosphere directly using probes of various kinds when they were available.[32]

Knowledge of the source and maintenance of the ionosphere, beyond having direct application in society, also was critical to a better understanding of the character of terrestrial magnetism. After the first detailed probing of the ionosphere in the 1920s by Breit and Tuve at the DTM, followed closely by Appleton and Barnett in England, the structure of the ionosphere itself was revealed through continued radio probing by Appleton and J. A. Ratcliffe, who found that more than one layer of ionization was present.[33] In 1931, Sydney Chapman suggested that a number of discrete layers of ionization did exist and that each was produced by a different portion of the ultraviolet spectrum of the Sun.[34] Chapman's model provided a means to analyze the character and stability of the ionosphere; dependent somehow on solar radiation, it varied with season and latitude, as well as with time of day and night, and with the observed 11-year variation of sunspot activity—correlations that were accepted but not satisfactorily understood.[35] Many fundamental assumptions, however, remained unsubstantiated. There was as yet no precise knowledge of how efficiently hydrogen, oxygen, and nitrogen in the earth's upper atmosphere absorbed solar radiation, nor was there knowledge of the concentrations of these elements and their compounds, or of the nature of the solar spectrum itself. Through the 1930s, these unknowns made it difficult to interpret ionospheric radio probes.[36]

In the absence of direct data on composition, physicists were left to speculate on the mechanisms causing dissociation in the high atmosphere and the manner in which these mechanisms promoted ionospheric activity. For example, it was important to know which frequencies would get through and which would be blocked, but especially how the so-called layers or regions of the ionosphere themselves were formed[37] (figure, p. 13).

But what created these layers? The popular view in the 1930s was that ultraviolet solar radiation at wavelengths less than 175 nanometers caused molecular oxygen to become dissociated at altitudes greater than 90 kilometers, and that the dissociation became complete at about 130 kilometers.[38] Thus, in this region, a mixture of O, $O_2$, and $N_2$ would be present, but, above 130 kilometers, the mixture reduced to O and $N_2$ alone. While radiophysics specialists and geophysicists observed these layers and speculated on their structure, they also wanted to know, as did astronomers, which portion of the solar energy spectrum was responsible for the ionosphere's changing behavior. Various solar theories were suggested by Meghnad Saha, T. L. Eckersely, O. R. Wulf, and L. S. Deming, among many others, most of whom believed that ultraviolet radiation was the responsible agent.[39] Among this group, a subset argued that solar activity in the Lyman Alpha region was the chief source for both the creation and maintenance of the earth's ionosphere. In 1922, J. J. Hopfield, then at Berkeley, concluded that the earth's atmosphere is moderately transparent in the region near 121.6 nanometers, the position of Lyman Alpha, the spectral line arising from the ground-state transition of hydrogen to its first excited state. Lyman Alpha radiation should thus be able to penetrate, he speculated, to the lowest atmospheric layers.[40] Hopfield was soon shown to be too optimistic, and there were also a few dissenting voices that argued that radiation more intense than ultraviolet sunlight was required to create and sustain the observed degree of ionization in the E and F regions.[41] The stronger voice was that of the Norwegian Lars Vegard, and the weaker was E. O. Hulburt of the Naval Research Laboratory, who in 1938 examined the ionization in the E and F regions due to photoionization of both molecular oxygen and nitrogen but posed both ultraviolet and X-ray radiation as possible sources.

In addition to Vegard and Hulburt, a number of physicists interested in ionospheric and cosmic-ray research conjectured in the late 1930s that X-ray radiation was emitted from the sun, at least when solar flares disrupted the ionization of the upper atmosphere. J. H.

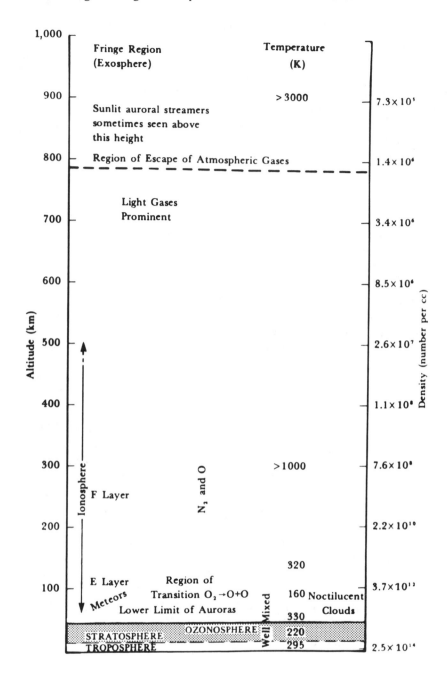

The structure of the upper atmosphere, circa 1946, as reconstructed in Newell (1980), fig. 1. NASA.

Dellinger of the National Bureau of Standards linked solar flare activity to radio fadeouts, whereas T. H. Johnson of the Bartol Research Foundation of the Franklin Institution and Serge Korff of New York University observed that ionization in the earth's atmosphere was quickly disrupted during solar flare events. From the rapidity and magni-tude of the terrestrial disturbances, they calculated that X rays of wavelengths between 0.01 and 0.15 nanometers were involved.[42]

In 1939, Hulburt reviewed the decade's work on E region ionization and established theoretically the form and value for the photoelectric recombination coefficients that governed half of the ionization balance in the ionosphere, the other half being solar radiation itself.[43] He found good general agreement for the mixture of gases he assumed to exist; indeed, his value for the recombination coefficient worked well after the war when X radiation was found in the solar flux from rocket observations.[44] Hulburt, however, was sensitive to the limits of his theoretical predictions before the war, both because of uncertainties in the temperature and composition of the high atmosphere (recent observations had shown that the density of oxygen molecules in the upper atmosphere was higher than suspected) and because of inadequacies and assumptions in the theory of opacity as it was then understood. Concerning the dominant theory of opacity, expressed by the formulas of H. Kramers, Hulburt remarked "One sometimes has the feeling that the applicability of this formula has been strained beyond the breaking point."[45]

## Astronomers' Interest in the Ultraviolet

In the 1930s, astronomers like Edison Pettit, Donald Menzel, and Meghnad Saha wanted to know the intensity of Lyman Alpha, partly to determine the energy source driving the sun's chromosphere and prominences, and partly to better understand how solar radiation controlled the ionosphere.[46] Saha, renowned for his contributions to both the theory of the solar atmosphere and of the organization of science in India, was keenly aware that the blocked ultraviolet thwarted the detailed verification of many of his theories; as early as 1920, he dearly wished to be able to see the resonance lines in the solar spectrum.[47] Saha and his colleague, the atmospheric physicist S. K. Mitra, both wanted to find which part of the solar spectrum had sufficient energy to cause the earth's ionosphere to change.

After lecturing extensively on the subject in India, Saha brought his ideas and predictions to the United States in 1937 for the Harvard Summer School in Astronomy.[48] Thinking along Hopfield's lines, that the absorbing ozone layer could be breached by high-flying manned balloons, Saha called for the establishment of a stratosphere solar observatory for synoptic observations of the resonance lines in the solar ultraviolet: "If these lines could have been observed," he told his Harvard audience, "the problem of hydrogen excitation in the Sun and stars would probably have received complete elucidation, and the problems of stellar atmospheres would have been nearer solution."[49] Saha's impassioned address stimulated many young students and faculty at Harvard to think about how one might gain access to the far ultraviolet. Dorrit Hoffleit, Leo Goldberg, and others recall in particular Saha's influence on their later thinking during the V-2 era.[50]

One of the most surprising discoveries in the late 1930s was that the solar corona, the sun's outermost atmospheric layer, is extremely hot, in the millions of degrees. In 1931, as a result of his extensive study of the solar chromosphere at Lick Observatory, Donald Menzel concluded that the transition region between the solar photosphere and the corona existed at a temperature about 1,000K below that of the lower photospheric regions.[51] By 1935 however, Menzel, now at Harvard, had concluded that coronal temperatures could be as high as 20,000K on the basis of observations of ionized helium.[52] And by the end of the 1930s, Menzel and others had gathered clues to the existence of a high-temperature coronal region and, by inference, the presence of an abnormal amount of extreme ultraviolet and soft X-ray flux from the region below the corona, "based on the fact that in the spectrum of the solar prominences and solar eruptions appear the lines of helium and ionized helium."[53]

But Menzel had another clue, just emerging as the growing war in Europe was shutting down all open lines of communication. Bengt Edlén, a young Swedish physicist, was on the verge of solving the decades-old mystery of the chemical identification of about 20 lines in the solar corona, showing them to be due to highly ionized atoms behaving unusually under extremely high temperature and low pressure.[54] Edlén's discovery, along with

Bernard Lyot's earlier studies of thermal line broadening, by 1941 strengthened Menzel's speculation that the corona was an extremely hot place.[55] To Saha and others, this meant that the corona could well be the source of line emission capable of controlling the ionosphere. Verification, as it turned out, required access to the resonance transitions for these atoms, which lay in the far ultraviolet.

The problem of the nature of the solar ultraviolet continuum and the type of radiation present beyond the atmospheric cutoff therefore had many sidelights interesting to a wide range of scientists. Some, like S. K. Mitra, wanted to know the character of the solar spectrum to learn more about the earth's atmosphere, whereas others, like Menzel, were interested in the sun itself, even though as a result of his later war work, Menzel became keenly aware of the military value of solar-terrestrial relationships.

One way or another, gaining access to the solar ultraviolet regions was only a dream before the means were available to rise through and above enough of the atmosphere to have a look at it. Saha proposed a stratospheric balloon observatory, the popular technology of the day, but no one in astronomy ever thought of proposing the use of rockets, unless it was suggested for them by the people building them, like Robert Goddard.

## Scientific Ballooning in the 1930s

A few people did attempt in situ aerial studies using balloons, primarily for atmospheric sampling, aerial mapping, photographing the solar ultraviolet, and detecting cosmic rays. As with Goddard's program, manned balloon flights required scientific underwriters to help justify their great comparative expense. As I have shown in a companion study to this volume, many prominent American scientists lent their names and instruments to these spectacles in the stratosphere.[56] All experimental physicists or trained in physics, these men hoped to obtain from these flights a wide array of data useful to science. The most compelling problem that they addressed was the elusive nature of the cosmic ray.

In the 1930s, many physicists studied the penetrating power of cosmic rays from mountaintop observing posts, with lake-bed detector arrays, or with balloon-born detectors, and large groups formed around Robert A. Millikan at Caltech, Arthur Holly Compton at Chicago, W. F. G. Swann at the Bartol Research Foundation of the Franklin Institute in Philadelphia, and Erich Regener in Germany.[57] Ever since their detection during manned balloon flights in 1912 by the Austrian physicist Victor F. Hess,[58] the celestial nature and character of cosmic rays were highly controversial. By the mid-1920s, Millikan, originally a skeptic, had finally become convinced that the radiation was celestial and therefore named them cosmic rays. But the question remained: What were they? Were they charged particles or energetic photons? The answer came from extensive latitude surveys first conducted by J. Clay and confirmed on a worldwide scale by Compton and his colleagues. Cosmic rays were affected by the earth's magnetic field; they therefore had to be charged particles moving at near-relativistic velocities in space.

Although there was continuing interest in the origin of cosmic rays, physicists found that their energetic penetrating properties made them excellent probes of the nature of matter. Everyone knew, however, that what they were catching in their detectors were not the original cosmic rays, called "primaries," but the secondary by-products of the collisions of primaries with particles in the earth's atmosphere. Thus, for many reasons, observing and analyzing primaries remained a goal for cosmic-ray physicists at the end of the decade, but then the war put an end to all research.

By the late 1930s, unmanned balloons were the preferred way to send instruments into the high atmosphere in the United States. Significant progress had been made in making rubber balloons available and reliable largely through the efforts of the Weather Bureau and Navy, which needed them for daily weather observations. Also, radio telemetry could track sondes in flight and retrieve data from them during flight. Radio telemetry was just emerging in France and the USSR at the beginning of the decade, but by 1936 it was well established as an experimental technique at the University of Chicago and the National Bureau of Standards and was gaining popularity at many other sites.

More than any of the other physicists discussed thus far who were engaged in upper atmosphere research, Erich Regener tied his instruments to balloons. Ballooning for Regener formed the technical nexus between his contributions to cosmic-ray and solar physics and to understanding the composition and structure of the upper atmosphere. Regener applied balloon technology and refined it in the process. Thus, even though Regener's use of balloons was part of a long-standing program of research, he was also interested in the vehicle for research. This dual alignment served him well as he became the first person to prepare instruments for a rocket that was capable of leaving the atmosphere. The problems that he chose to address stemmed directly from his own interests and were quite similar to those proposed by Goddard's erstwhile scientific advisory committee. But Regener's agenda was also influenced by the existence of the missile and what was needed to make it operational as a weapon of war.

## Notes to Chapter 2

1. S. I. Bailey to E. C. Pickering, 2 October 1893, quoted in Jones and Boyd (1971), p. 316. E. C. Pickering Director's Correspondence, HUA.

2. On scientific motives for ballooning, see Friedman (1989), p. 48ff. In the 1920s, Oberth (1972), pp. 453–62, speculated on the scientific value of extended access to space by rocket. He envisioned space telescopes, instruments to detect the preferential direction of the ether, and physical and physiological experiments. On Oberth, see Winter (1983) and Neufeld (1990).

3. R. H. Goddard, quoted in Jones (1965), p. 241; see also Goddard and Pendray (1970), p. 170 n. [Goddard, to "President," Smithsonian Institution, 27 September 1916]. On Goddard's proposals to the Navy, see Christ-

man (1971), p. 45, who has noted that after World War I, "Goddard was the perennial exception to the general trend of scientific dissociation from weaponry and national defense."

4. Jones (1965), p. 245; Goddard and Pendray (1970), p. 174. On Abbot's interests see DeVorkin (1990).

5. Robert H. Goddard (1919). Reprinted in [Goddard] (1970), pp. 340–41.

6. During and after World War I, Abbot's main enthusiasm for Goddard's rockets was for their ordnance potential. See, for instance, Durant (1974), p. 59; and Abbot to W. S. Adams, 30 November; 24 December 1918. WSA/H; and Wright (1966), p. 299. An excellent review of Goddard's military work is provided in Christman (1971). Hale had thought of photographing the solar corona with an automatic camera on a balloonsonde and saw Goddard's efforts leading to the same ends. See George Ellery Hale to Evalina C. Hale 28 June [1914]. GEH/CIT, Roll 68, frames 538–40. I am indebted to Don Osterbrock for pointing out this reference.

7. Hallion (1977), pp. 174–75.

8. John A. Fleming to John C. Merriam, 27 November 1929; Merriam, "Memorandum of telephone conversation with Lindbergh," 30 November 1929; Merriam, "Memorandum of conversation with Dr. Abbot," 2 December 1929. RG/CIWA.

9. John Fleming to Gregory Breit, 3 December 1929. DTM files: D1 File "Breit, Gregory." Attic file cabinet 13, 8688.2.

10. He suggested that photographic film, covered with black paper, could be used to "record the presence of x-rays" and that photographic paper covered with a silver screen could record the "presence of certain regions of ultra-violet light." E. O. Hulburt to John Fleming, "Suggestions Submitted December 5, 1929, by Dr. E. O. Hulburt. . .," 6 December 1929. RG/CIWA.

11. Ibid.

12. "Memorandum by Mr. J. A. Fleming, Acting Director, Department of Terrestrial Magnetism, relative to Dr. Goddard's rocket program," 10 December 1929. RG/CIWA.

13. Ibid. At the time, there was no independent Air Force.

14. [Merriam], "Memorandum Regarding Goddard Rocket Project," 11 December 1929. RG/CIWA.

15. Godard to Merriam, 29 March 1930. RG/CIWA.

16. On the Gugenheim fund, see Hallion (1977), especially pp. 174–77 regarding support for Goddard.

17. "The Daniel Guggenheim Fund of Clark University for Exploration of the Upper Atmosphere," clipped to Lindbergh to Merriam, 2 June 1930. RG/CIWA. See also Dewey (1962), pp. 96–97.

18. J. C. Merriam, "Memorandum of Conversation with Dr. Goddard," 7 June 1930; "The Daniel Guggenheim Fund of Clark University for Exploration of the Upper Atmosphere," clipped to Lindbergh to Merriam, 2 June 1930. RG/CIWA. The Mount Wilson Observatory was funded by the CIW.

19. Goddard to Merriam, 14 July 1930; Merriam to Goddard, 29 July 1930; Fleming to Merriam, 25 October 1930. See also Goddard to Merriam, 5 November 1930; Fleming to Merriam, 9 January 1931. RG/CIWA.

20. [Merriam] "Extract from memorandum of conversation between Colonel Lindbergh and Mr. Merriam, October 21, 1930." RG/CIWA.

21. Minutes, "Advisory Committee on the Goddard Rocket Project," 20 February 1931. RG/CIWA. Several members of Adams's staff were capable of constructing a spectrograph. These included John Anderson, who had been involved in 1927, and Harold Babcock and Edison Pettit. Merriam later acknowledged that these three should get involved. Merriam to Adams, 18 March 1931. WSA/H.

22. Adams to Fleming, 24 March 1931. RG/CIWA.

23. Goddard to Merriam, 16 December 1931, with appended "Report on Rocket Work at Roswell, New Mexico." 15 December 1931. RG/CIWA. Copy also in WSA/H.

24. This never happened. Fleming, testimony at hearings, House Committee on Foreign Affairs, 72d Congr., 1st ses. 26/27 January, 2 February 1931.

25. Goddard to Merriam, 10 May 1932. RG/CIWA. Merriam called a meeting of the committee within one week "to decide whether the rocket work is to continue." Merriam to Members of the Advisory Committee, 17 May 1932. WSA/H.

26. "Memorandum Regarding Meeting of Goddard Rocket Committee," 25 May 1932. RG/CIWA. Quotation from Marvin to Merriam, 3 June 1932. RG/CIWA.

27. John A. Anderson was a particularly adept instrument designer at Mount Wilson with specialties in mechanisms and optics. In 1927 he started to think seriously about scientific rocketry after reading about Goddard's work. After a chance meeting with Gregory Breit and Merle Tuve of the Carnegie's Department of Terrestrial Magnetism, and E. O. Hulburt of the Naval Research Laboratory in April 1927, Anderson discussed how rockets might be used "for the purpose of exploring the upper atmosphere." G. Breit, M. A. Tuve, and E. O. Hulburt, to J. A. Anderson, 17 June 1927. RG19, S70–1(5), NRL/NARA. I am indebted to Bruce Hevly for bringing this letter to my attention. Anderson's interest in examining Goddard's work is inferred from this, as well as from comments in W. S. Adams to Merriam, 26 May 1932 (WSA/H) and Wallace Atwood, "To Members of the Advisory Committee," 14 June 1932. RG/CIWA and WSA/H.

28. Wallace Atwood, president of Clark University, indicated that, after consultation with Lindbergh and others, Mrs. Daniel Guggenheim referred the matter to Harry Guggenheim, but that "a situation has arisen which made it necessary for Harry Guggenheim . . . to admit that it is impossible for them to provide the funds for this experimental work during the next year. It is their wish, however, that the matter be reviewed with them a year later." Atwood to Members of the Advisory Committee, 14 June 1932. RG/CIWA. Dewey (1962), pp. 109–13, and Hallion (1977), p. 176, note that the reasons for the funding lapse were a combination of the financial vagaries of the Depression, and the fact that the Daniel Guggenheim estate was still not settled. See also Merriam to Atwood, 5 July 1932. RG/CIWA.

29. Goddard to Merriam, 25 August 1934. RG/CIWA.

30. Some rockets carried small thermographs, barographs, and cameras, but they were more for assessing rocket performance. On 31 May 1935 a Goddard rocket ascended to 2,200 meters in a semicontrolled flight above Roswell, and in a flight on 9 August 1938, both the rocket and a barograph were successfully recovered by Goddard in collaboration with the National Aeronautic Association. See Siry (1950), p. 410; Dewey (1962), pp. 127–29; and Christman (1971) for an evaluation of Goddard's later efforts.

31. Mount Wilson's Edison Pettit, for instance, wished to know the intensity and character of the solar Lyman Alpha line lying at 121.6 nanometers to find out what drove solar prominences. He noted in 1932 to Harvard's Donald Menzel: "As to Lyman [Alpha], we all want to know how much energy, if any, the sun has in the Lyman region." Edison Pettit to Donald Menzel, 1 November 1932. HUG 4567.5.2, DHM/HUA.

32. See Akasofu, Benson, and Haurwitz (1968); entries in *American Men and Women of Science*. For general reviews of research in the 1930s, see Haurwitz (1936); Fleming (1939); Chapman and Bartels (1940).

33. Breit and Tuve, with the aid of E. O. Hulburt of the Naval Research Laboratory, set up radio pulse transmitters and receiving stations a few kilometers apart and found two separate signals in the received radiation field. Breit and Tuve (1926); Mitra (1948), pp. 142–43. See also Tuve (1974), p. 2079; and Villard (1976), p. 847. Dennis (1990) has explored the origins of Tuve's and Breit's interest in ionospheric studies. On Hulburt and NRL's traditional interest in radio propagation, see Allison (1981), p. 57; Hevly (1987); and Hulburt OHI, pp. v–vii. HONRL.

34. Mitra (1948), p. 257ff.; Chapman (1931). See also Gillmor (1981), p. 106, who provides a brief but cogent introduction to early ionospheric research, and to the perceptive argument that these layers were artifacts of the process of observation. See also Chap. 16.

35. See Stetson (1934) and DeVorkin (1990). Chapman's theory had been anticipated by the Dutch astronomer Antonie Pannekoek, who in 1925 used Saha's theory of thermal ionization as modified by E. A. Milne to describe the ionic density distribution within the earth's atmosphere on the basis of the assumption that the ionizing source was solar continuum radiation. On Pannekoek's contributions, see Mitra (1948), pp. 263–64.

36. C. Stewart Gillmor has been responsible for a good deal of the historical scholarship on the early history of ionospheric studies by radio. See, for instance Gillmor (1981, 1986, 1989).

37. Gillmor (1981), p. 104ff., explains how these layers were recognized and how their reality was taken too literally at times. An excellent contemporary review of layer concepts, and the phenomenon of refraction, by someone less convinced of their reality, is presented by Berkner (1939).

38. Mitra (1948), p. 267; Haurwitz (1936), p. 353. In addition to these reviews, Kuiper (1947), pp. 8–10, provides a valuable survey of the portions of the solar spectrum thought to influence the ionosphere, at the outset of the V-2 era.

39. Saha (1937b), p. 155; Haurwitz (1936), p. 353.

40. See, for instance, Martyn et al. (1937), p. 604; and Hopfield (1922), p. 523.

41. Hirsh (1983), p. 14; Hulburt (1938), p. 350.

42. Johnson and Korff (1939). See also Vegard (1938, 1939, 1957); and Hirsh (1983), pp. 14; 152. See also Hunter (1942–43), p. 8. Vegard linked this abnormal ionization to solar eruptions in 1937, at about the same time as J. H. Dellinger similarly showed that these solar bursts, later called flares by astronomers, were responsible for radio fadeouts. See Dellinger (1937), p. 1253.

43. Hulburt (1939).

44. The credit given to Hulburt by his colleagues for the suggestion that X radiation stimulated the ionosphere might stem from this work, and not from any direct statement on the role of X radiation. See Hevly (1987); Byram, Chubb, and Friedman (1954a), p. 275.

45. Remark based upon Norwegian Svein Rosseland's 1936 opinion of Kramer's 1923 law of opacity. Rosseland (1936), p. 194, quoted by Hulburt (1939) p. 347.

46. See Edison Pettit to Donald Menzel, 1 November 1932. HUG 4567.5.2, DHM/HUA.

47. Saha (1921), p. 149.

48. On the Harvard Summer Schools, see DeVorkin (1984).

49. Saha (1937a), p. 240.

50. Leo Goldberg to Lyman Spitzer, 29 January 1947. LG/HUA; Hoffleit (1949b), p. 7; Goldberg (1981), pp. 15–16. See also Ghosh and Gupta (1978).

51. Menzel (1931), p. 303.

52. *Harvard College Observatory Circular* 410, 1935.

53. Menzel (1939b), p. 6.

54. Eldén's work and influence upon solar physics has been examined in detail by Hufbauer (1991), pp. 113–15; and Hufbauer, "Breakthrough on the Periphery: Bengt Edlén and the Identification of the Coronal Lines," forthcoming.

55. Temperatures in excess of 2,000,000K were discussed. Menzel (1939b). See also Saha

*Collected Works*, pp. 160–01; and on the significance of these findings, see Menzel to Struve, 29 April 1941. SP/AIP; Bowen to Edlén, 26 April 1941. ISB/CIT.

56. The bulk of the discussion in this section is taken from DeVorkin (1989a). Names included Robert A. Millikan, Arthur Holly Compton of Chicago; Lyman Briggs, director of the National Bureau of Standards; William Swann, director of the Bartol Research Foundation of the Franklin Institute; and Brian O'Brien, director of the Rochester Institute of Optics.

57. There is a large literature on cosmic-ray physics in the 1930s. See, for instance, Sekido and Elliot (1985); Winckler and Hoffman (1966); Galison (1987); Auger (1945); and LePrince-Ringuet (1950). On Millikan, see Kargon (1981, 1982). On the latitude effect, see De Maria and Russo (1987), and Compton (1933, 1936). On cosmic-ray ballooning, see Ziegler (1986) and DeVorkin (1989a).

58. Hess carried three small electroscopes to some 5 kilometers and found that the residual ionization inreased with height, which ruled out a terrestrial source. LePrince-Ringuet (1950), pp. 90–1; Ziegler (1986), pp. 69–92.

# Part 1. Military Origins

# 3

# Erich Regener and the V-2

History remembers Peenemünde, an isolated northern fishing village on the Baltic shore, as the birthplace of the world's first long-range ballistic missile system. Walter Dornberger, chief of the German Army Ordnance Department rocket development group in Berlin since 1935, built Peenemünde with the help of the Luftwaffe and moved the larger part of his 90-man staff from the Kummersdorf Proving Ground south of Berlin to the remote site in 1937. Within five years, the Army Experimental Station Peenemünde grew to a work force of 5,000 engineers, technicians, and scientists, including a captive force of Russian and Polish prisoners.[1]

The main goal of this enormous project was to design and test long-range missiles, as well as to explore broadly the use of rocket power in manned aircraft. Its greatest achievement was the creation of a 14-meter-high missile weighing 12,800 kilograms when fueled, which was capable of delivering a metric ton warhead to sites 200 to 300 kilometers away. Known to its creators as the A4 (Aggregat 4, or the fourth design in the Assembly or Booster series) it was called Big Ben by British Intelligence. By September 1944, it had become feared by the world as the V-2 missile, or Vengeance Weapon 2, as it was called by Nazi propagandists (see figure, p. 25).[2]

## Peenemünde Creates the V-2

The V-2 (the operational name for the experimental A4) was unlike its nominal predecessor, the winged drone V-1 pulse jet created by the German Air Force. Both delivered one metric ton of explosives, but the V-1 was a slow noisy air breather that could be shot out of the skies above London, Antwerp, or Brussels, even though it did divert Allied air defense resources. The V-2 was a supersonic rocket that fell on its target without warning. The first successful long-range test occurred on 3 October 1942, when an A4 reached an altitude of 85 kilometers and a range of 192 kilometers in some 296 seconds of flight.[3] The weapon system was put into use in September 1944, and over 1,000 missiles fell on England alone in the next six months. The V-2 was limited to delivering conventional explosives along largely unguided ballistic trajectories and therefore was incapable of pinpointing targets or incapacitating an enemy fortification. Even so, the V-2 achieved Hitler's goal of producing a terror weapon. No warning prepared neighborhoods in Antwerp, London, and their environs; no shelter was possible before the explosion. Everyone within many kilometers of the intended target was equally likely to be hit, and they would never know what hit them.

Before and after the first successful test launch of an A4, its creators were faced with daunting technical, managerial, and political problems. Although support from the German Army's high command was strong, shortages of raw materials and manpower retarded progress as Peenemünde periodically was reduced in priority. It was only one among many projects competing for resources.[4] Dornberger, along with his young associate Wernher von Braun (figure, p. 26), knew that acquiring adequate manpower was critical to bring the V weapon to operational status for use in the war, to say nothing of what would be needed for mass production after 1942. Their solution was to seek out alternative sources of labor and expertise, turning to ci-

The launch of a V-2 from Peenemünde, circa 1943. Deutsches Museum, Munich.

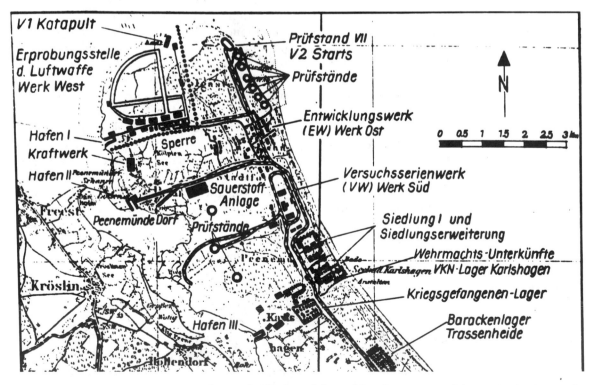

Army Ordnance Station Peenemünde, on the shores of the Baltic, showing test firing areas for V-1 and V-2 missiles. NASM SI 79–12336.

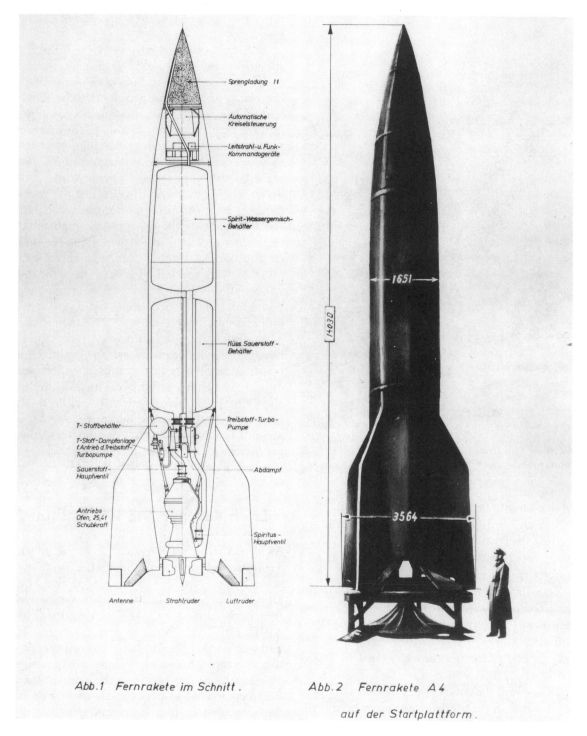

German Army Ordnance diagram of an advanced V-2 on its launch stand. Dimensions are in millimeters. Deutsches Museum, Munich, and NASM SI 88–8215.

**Wernher von Braun briefing Wehrmacht officers at Peenemünde circa 1943. NASM SI 78–5935.**

tant, they offered money and draft exemptions.

At first, recruitment forays were limited to specific engineering fields of direct assistance to missile development. But by the early 1940s, the range of technical problems to be solved had widened, as had the net cast by von Braun and his colleagues. For example, they needed to determine the electrical charging effect that the rocket exhaust and passage of the rocket through the upper atmosphere would have on the rocket fuselage, the effect of the rocket exhaust itself on radio communications, and the character of the medium through which the rocket traveled, which would enable them to improve ballistic trajectory calculations.

As early as the fall of 1941, when the Advanced Projects Office at Peenemünde was contemplating a two-stage intercontinental ballistic missile system, it was clear that such a capability required improved tables of atmospheric temperature, pressure, and composition to predict both ballistic trajectories and gliding characteristics.[8] Thus, in the late spring of 1942, before any A4 was fired successfully, von Braun and his staff made a tour around Germany to give briefings to leading professors of physics, chemistry, and engineering, among them Erich Regener.

## Erich Regener and His Institute

As director of the Physical Institute of the Technische Hochschule Stuttgart, Erich Regener (figure, p. 27) was one of the best-known experimental physicists in Germany.[9] Regener's laboratory was a magnet for those wishing to learn how to construct and fly balloons that could carry spectrographs, thermographs, barographs, and a wide range of cosmic-ray devices into the stratosphere. Regener and his students made their little instruments work like loyal and trustworthy robots, returning reliable data from heights as great as 30 kilometers.[10]

Regener had a deep and long-standing interest in the character and extent of the ozone distribution in the earth's atmosphere and had, since 1928, been studying cosmic radiation.[11]

vilian scientists and engineers in both academia and industry, and to military units for technically trained personnel. Cheap forced labor was plentiful in the prisoner of war and concentration camps.[5]

Von Braun, who had started as Dornberger's technical assistant in 1932, by 1937 was technical director of development of Peenemünde's long-range missile program (figure, p. 26)[6] After secrecy loosened in 1939, Dornberger and von Braun started contracting German scientists and engineers to perform work in their home laboratories and institutions.[7] Von Braun and his staff designed technical briefings for groups at universities and institutes of technology throughout Germany, showcasing the research the Army Ordnance project was already supporting in areas such as aerodynamics. More impor-

**Erich Regener in the late 1940s. Max Planck Society. NASM SI 89–19726.**

His students, such as Georg Pfotzer, used Geiger counters in their cosmic-ray studies and in 1935 found the region in the atmosphere where cosmic-ray interactions were at a maximum.[12] With his son Victor, Regener built and flew balloon-borne spectrographs to determine the distribution of ozone in the stratosphere (figure, p. 28). Regener also contributed to the technique of high-altitude air sampling, showing that the oxygen/nitrogen ratio did not change at altitudes to 28 kilometers.

During the late 1930s Regener experienced political and personal problems, in part because he had cosigned the 1936 Heisenberg-Wien-Geiger memorandum denouncing the Aryan physics movement in Germany. More serious, he refused to divorce his wife Victoria, a Russian Jew.[13] Regener became a target of scheming Stuttgart colleagues and soon lost his post. His son Victor escaped Germany only two days before the border was closed, going first to Italy, and then to the United States and the University of Chicago. Regener, however, chose to stay and to do all

he could to continue his research.[14] Like many of his colleagues, Regener fell afoul of the National Socialists in the late 1930s and lost not only his job, but also his sources of funding.[15] Regener's political fate thus set the stage for his eventual contract with Peenemünde to build a scientific package for a V-2 flight. What Regener set out to do was very ambitious and complex, however, in sharp contrast to the efficiencies he employed in his balloon research. In the end, this contributed to the stillbirth of what came to be known as the Regener-Tonne for the V-2.

## Contracting for Survival

On 5 October 1937, the day he was ousted, Regener issued a plea to his colleagues for advice and aid. Outwardly confident and claiming that his difficulties were due to jealous colleagues at Stuttgart, Regener said his only concern was to maintain his group and its research momentum. The first advice he received was to attach himself to the Kaiser

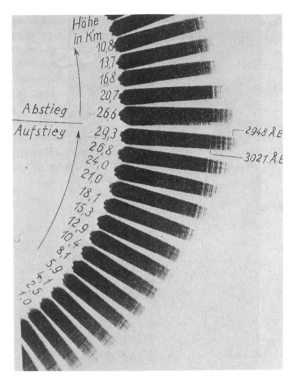

**Solar spectra obtained by Erich and Victor Regener from a balloon flight in 1934. Maximum ultraviolet penetration to 287 nanometers occurred at a maximum altitude of 29.3 kilometers. The spectrograph used circular photographic plates that rotated within a cassette. Exposures of thin wedges of the emulsion ranged up to 10 minutes, and a timing mechanism allowed some 40 exposures of the solar spectrum to be obtained during a balloon flight. From Regener and Regener (1934a), p. 791.**

Wilhelm Institute for Physics in Berlin, under Peter Debye, and that he place his staff and facilities under their protection.[16] The institute was part of the venerable Kaiser Wilhelm Gesellschaft (KWG), a far-flung scientific society founded in 1911 that retained vestiges of autonomy since it gained only a portion of its support from the state.[17] Within two weeks, Regener decided to move operations to his little field station on the shores of the Bodensee (Lake Constance near Friedrichshafen), all the while hoping that the Institute for Physics would soon have him as a member.[18]

The path Regener had plotted for himself turned out to have many ruts, and it was quite

some time before he was again secure. When the KWG offered to purchase the Bodensee field station, the rector of the Technische Hochshule Stuttgart refused to sell it, claiming that it was a necessary part of his institution's research facility, needed by remaining staff.[19] Since the rector added that this decision had the endorsement of the Reich Minister for Education, KWG General Director Ernst Telschow counseled Regener to do what he could to get Hermann Göring and the Luftwaffe behind him; if he could do this, he would have no trouble retaining his Bodensee laboratory.[20] Telschow suggested that Regener ask friends at the Rechlin Luftwaffe facility to intervene with the Reich Minister for Aviation (Göring), getting him to state that any interruption in Regener's work through the loss of his equipment would "cause significant damage to the progress of upper atmosphere research." Regener already had several small contract projects with the Air Ministry and the Luftwaffe and found that his work for them insured that he would retain the use of the Bodensee laboratory at least until 31 March 1938.[21] But he needed more than an endorsement. Power-bloc politics required commitments.

The duel between the KWG, teamed with the Air Ministry, and the Technische Hochschule Stuttgart, teamed with the Education Ministry, continued through the spring of 1938, as Regener left Stuttgart in disgust and rented a house in Friedrichshafen. With washrooms for workshops, Regener and his dwindled but loyal staff continued to fly balloonsondes and immerse cosmic-ray devices in the Bodensee to study the atmosphere and search for a better understanding of the energy spectrum of the high-energy particles that were bombarding the earth.[22]

In March 1938, Peter Debye in Berlin and other colleagues suggested that Regener deserved his own facility and his own institute.[23] Friends guided Regener's case through the Scientific Senate of the Air Ministry's Weather Service. These contacts also managed to stall Regener's permanent removal from the Bodensee until, in late May, Göring's office took action.[24]

Göring, as Reich Minister for Aviation and Supreme Commander of the Luftwaffe, responded directly and called the disputants to a meeting in Berlin. There it was decided that Regener would have his own institute within the KWG, centered at his new Bodensee facility, which would be funded almost entirely by the Deutsche Versuchsanstalt für Luftfahrt (German Research Establishment for Aviation). The KWG senate readily agreed and added a modest stipend to the primary support of the Air Ministry research arm. Over the next two months, Regener's new Forschungsstelle für Physik der Stratosphäre was created specifically to study the physics of the stratosphere, broadly construed, as well as perform "research into the means for reaching greater heights in the stratosphere" beyond those normally accessible by balloon.[25] Regener was saved.

## Research at Friedrichshafen

Salvation brought with it closer alignment with the interests of the Air Ministry and the Luftwaffe. Regener had received contracts from the Deutsche Versuchsanstalt für Luftfahrt before, but now he was fully funded by Air Ministry interests for basic research useful to wartime needs.[26] His research reports reflected this new patronage and his awareness of its needs, especially his mandate to seek means for flying as high as possible.

In 1940, Regener suggested that daily weather forecasts might be improved by studying ozone concentrations. He argued that ground-level tropospheric ozone originated at great heights and was brought down by turbulence in the atmosphere. Studying ozone profiles would yield information on turbulence. Therefore, Regener reasoned, it was important to continue his balloonsonde flights and to fly higher than conventional sondes could go. A complete picture of how ozone is produced and maintained required measurements at altitudes of 45 to 90 kilometers.[27]

In the early 1940s, Regener's Forschungsstelle grew to a staff of about 40 people as he secured direct contracts from the Luftwaffe for designing diagnostic flight instruments.[28] As the war came closer to Friedrichshafen, bringing Allied air raids, Regener dispersed his staff to outlying stations, mainly to a site near Weissenau, north of Lake Constance. There they pursued a wide range of pure and applied research, studying cosmic radiation, ultraviolet absorption, and ozone concentrations using balloonsondes, on the one hand, and developing atmospheric sensors of all types that the Luftwaffe could produce and use, studying the conditions under which water could remain fluid well below the freezing point, and examining procedures for measuring and monitoring solar flare activity, on the other.[29]

Regener's second charge from the Air Ministry, to develop methods to send devices to altitudes beyond those normally accessible by balloon, was also an extension of his earlier work. In the mid-1930s, Regener had demonstrated that there was a practical limit of some 25 to 30 kilometers altitude for standard sealed rubber balloons. He had experimented with cellophane cells for greater lift, but found them equally susceptible to the harsh conditions of the stratosphere. Low temperature, increased ultraviolet radiation, and to some extent the increased ozone all conspired, he argued, to destroy balloon fabrics, whether they were of rubber or cellophane.[30] In the early 1940s, Regener and his group, exploiting good contacts at I. G. Farben/Agfa in Munich, found that new plastic substances made especially good balloon skins. A later Allied intelligence report based on interrogations of Regener in the summer of 1945 stated: "These materials are unaffected by moisture, ultraviolet light or cold and are somewhat similar in appearance and structure to cellophane." Very strong and initially flexible when made into thin films, these plastics were fabricated into balloons that would reach 30 kilometers easily. The largest Regener built were 400 cubic meters in volume, but these were difficult to launch.[31]

Regener and his group also planned to launch small rockets or rocket-type projectiles from balloons in their continuing effort to break the 30 kilometer barrier. His new plastic balloons had sufficient lift to carry "hydrogen cannon" (Wasserstoffkanone) to some 25 kilometers, where the cannon would then shoot its payload to heights of 50 kilometers or more.[32] This ingenious idea, reborn in the United States after the war as James Van Allen's Rockoon, was dropped when V-2 missile warhead berths were offered by Peenemünde.[33]

## The V-2 Comes to Friedrichshafen

On 8 July 1942, during one of his periodic visits to Friedrichshafen, von Braun met Regener and his associates Alfred Ehmert and Erwin Schopper. Those from Peenemünde included von Braun, Ernst Steinhoff, head of the Instruments, Guidance, and Measurements Department; Reinhold Strobel, a ballistician concerned with the characteristics of the upper atmosphere; Gerhard Reisig, who had worked on instrumenting small A3 and A5 rockets to measure temperature and pressure, and several others.[34] Knowing that his A4s were soon going to be flying and would require predictive atmospheric models for accurate ballistic trajectories, von Braun asked Regener to build instruments to measure the altitude profiles of temperature, pressure, and density in the atmosphere at rocket heights. This was a wartime contract, for wartime purposes. Regener respected the limitations set by von Braun, and so he emphasized gathering meteorological and aeronomical data. It gave him the opportunity, however, to pursue his keen interest in ozone as a weather predictor using ultraviolet spectroscopy.

Von Braun's minutes of the meeting rationalized the contract relationship. Although the work would be interesting to the physicists at the institute, it was needed by the Army Research Center at Peenemünde for ballistic trajectory calculations, for predicting frictional heating of hypersonic projectiles in flight and for accurate reconnaissance of vehicle performance, especially altitude measurements.[35] He identified four specific types of instruments Regener would build: quartz barographs, recording thermometers, an ultraviolet spectrograph, and air samplers. Von Braun's report mentioned that Regener was free to develop additional instruments for the A4, but Regener apparently did not do so, leaving cosmic-ray instruments, for example, to balloonsondes.[36]

## Designing and Outfitting Regener's Tonne

Regener knew how to make things work under balloons floating in the stratosphere. Naturally, he now designed a system to mimic this habitat as closely as possible. His initial idea was to design a protective container (or Tonne) that would allow the instruments to function in an environment far more hostile than any balloon or balloonist ever encountered (figure, p. 31). After the rocket reached its peak altitude, the container would be ejected by a small explosive charge, and, buoyed by a parachute, it would drift down while the self-recording instruments did their work. If he could lick the problems of instrument integration, explosive separation, deployment of the parachute under near-space conditions, stabilized parachute flotation, and safe retrieval, the rest—that is, making the instruments themselves work—was simple. He had done that part of the work before. Thus the instruments themselves were less a problem than the design of the container and instrument systems.

Peenemünde initially provided some 25,000 reichmarks and matériel such as a complete parachute section from an A5 and radio beacons.[37] Von Braun also supplied Helmut Weiss, who became responsible for sending all the equipment to Regener's group and was their contact for future requests after March 1943. Gerhard Reisig, another von Braun staff member detailed to Regener's project for a year, recalled that from the very beginning and until he was temporarily transferred away from Peenemünde in the spring of 1943, Regener's progress was confined to preliminary concept studies and to developing the basic

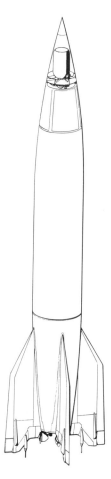

**Regener's Tonne pictured in the warhead section of a V-2. The instruments were all in the container, and its flotation and parachute sections sat at its base. Deutsches Museum, Munich.**

technical layout.[38] Mechanical prototypes were made in Regener's shops at the institute, and there were visits to Friedrichshafen by von Braun's staff, occasionally by Army officers, and several visits to Peenemünde by members of Regener's staff.[39]

Through the summer and fall of 1943, Regener's staff and their new colleagues wrestled with the Tonne's technical design problems. Regener was not satisfied with the A5 parachute system and searched for a better design, as well as a softer ejection mechanism that would release his canister without jarring its contents.[40] After months of toiling over various alternatives, the group decided the canister should be ejected at the peak of the

missile trajectory, after cover panels protecting the Tonne in the warhead area were pushed away by spring-loaded bolts. Then, a parachute would open by compressed air and tenderly drag the Tonne away from the rocket, falling tail first in space.[41] Even though they settled on this system, Regener and his staff continued searching for better alternatives; indeed they never seemed satisfied with any aspects of Tonne design.

Since members of the group were absorbed with the aerodynamics of the Tonne and with keeping it stable in free-fall, they continued to look at parachute designs. Meanwhile, Regener was working on a design for his instrument complement.[42] Given the expected great range of descent velocity, even with the single parachute system, Regener decided by May 1943 that three different types of temperature-measuring devices had to be used, each capable of measuring temperature accurately over narrow overlapping ranges. He also decided that the Tonne had to have at least two barographs, but he was still trying to decide what instruments were needed for the ultraviolet ozone spectrograph observations. H. K. Paetzold, who had been recalled from active duty for Regener's project because of his training in optical design, knew that the observing time for ultraviolet solar radiation would be far less than what had been available in Regener's balloon flights some 10 years earlier. He and Regener decided that instead of trying to increase the photographic speed of the spectrograph optical system, they had to build a special light collector to pump more light into the system; a simple diffusing screen would not collect enough light.[43]

The scientific instruments remained the least of their problems. Nobody knew how to make these instruments comfortable on a rocket, or in a canister falling from space. As a first step, they decided to produce several designs for the Tonne geometry, itself largely determined by how their instruments would nest together, and on the limitations of the V-2 warhead. After a full year of effort, progress on the Tonne and its instruments at Peenemünde and Friedrichshafen remained slow. Both sites came under Allied attack in late

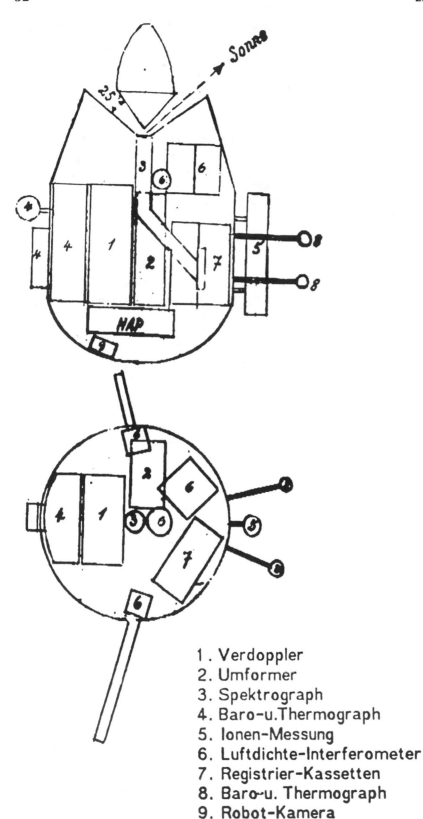

Schematic of one configuration for the Regener-Tonne, showing the placement of instruments. Note in particular the light entrance aperture for the spectrograph. No date, circa 1945. Erwin Schopper, Max Planck Institute for Aeronomy, Lindau.

1. Verdoppler
2. Umformer
3. Spektrograph
4. Baro-u.Thermograph
5. Ionen-Messung
6. Luftdichte-Interferometer
7. Registrier-Kassetten
8. Baro-u. Thermograph
9. Robot-Kamera

1943, which led to drastic changes. Peenemünde activity decentralized, sending production underground at Nordhausen and Regener's staff to Weissenau and other local sites.[44] Continual Allied bombardments as well as the increasing drain of staff caused by mounting war demands could not but hamper Regener's progress.[45] Reisig left in the spring of 1943 to return to his primary duties at Peenemünde. Paetzold continued with the Regener project, but he and others were often called away to work on more urgent tasks, such as the development of Wasserfall, an antiaircraft missile.[46]

## Regener's Instruments

Simple barographs and thermographs had flown on a small A3 rocket in December 1937 from the Griefswalder Oie.[47] Reisig had helped develop these instruments, but they required careful modifications for use in the larger, higher-flying and higher-velocity A4 missile because each one had to be integrated into the canister without disturbing the other instruments. Regener, in collaboration with Reisig and a meteorologist at the University of Berlin who was also a technical officer in the Army Ordnance Research Division, decided that two barographs and five thermographs were required to handle the expected extremes in pressure and temperature, as well as the nonlinear pressure distribution that would be encountered.

They developed high-precision barographs for the pressure range between 0 and 6 millimeters of mercury, and lower precision devices sensing between 0 and 760 millimeters of mercury. The fine pressure range was handled by a small evacuated chamber containing a thin copper-beryllium membrane resting on a support when the outside pressure was nearly one atmosphere. The coarser pressure range was measured by a C-shaped quartz-bourdon tube barograph designed by Regener for his balloon flights (figure, p. 34).[48] Both barographs were placed in sealed chambers within the Regener Tonne, with openings to the outside to allow them to come into

equilibrium with the air external to the rocket. The placement of these instrument holes in the canister remained one of the most difficult integration problems, requiring a series of wind-tunnel tests in Göttingen, followed by several redesigns of the system.

One of the thermographs was mechanical and the rest were electrical, each designed to operate during different but overlapping parts of the trajectory of the Tonne. Two measured ambient temperature, and two the skin temperature of the missile and Tonne. The fifth thermograph was a reference temperature thermostat.

Regener planned to build two very different spectrographs. One would measure ozone absorption in the earth's atmosphere, using the solar spectrum as a background source, and the other, made for the astronomer Karl O. Kiepenheuer, would examine the solar Lyman Alpha spectral region near 120 nanometers; Lyman Alpha, the resonance line in hydrogen's ground-state spectrum, was then thought to be a solar feature responsible for the maintenance or periodic destruction of the radio-reflecting properties of the earth's ionosphere. During the war Kiepenheuer established a network of coronographic observatories to look for correlations between solar activity and changes in the ionosphere's radio-reflecting properties. He found some, and even achieved a measure of success in predicting radio fadeouts, but the exact mechanism of solar influence remained unknown. If variations in the far ultraviolet Lyman Alpha line were sufficiently strong to modify the earth's ionosphere, Kiepenheuer had to know their character to improve his predictive models. Hearing of Regener's plans to observe the sun from space, he turned to Peenemünde.[49]

Both spectrographs were prism and lens systems based upon a standard Cornu design. The ozone spectrograph needed only quartz optics and photographic recording, just like Regener's balloon spectrographs—in other words, appropriate technology for the spectral region between 200 and 300 nanometers where ozone absorption dominated. The Kiepenheuer spectrograph, however, required

Quartz barograph for the Regener Tonne. The flexure of the evacuated quartz tube caused by pressure changes was amplified by a spring hinge joint and a mirror system and recorded photographically. Koppl (1947), 850; Paetzold, Pfotzer, Schopper (1974), 181. Deutsches Museum, Munich.

lithium fluoride optics and special Pohl-type halogenid crystals that turned color or changed in internal electrical resistance when exposed to far ultraviolet solar radiation.[50]

The two spectrographs also shared similar problems. Regener applied proven and familiar balloon devices, such as clock-timers and photographic platens, but now his instruments had to be enclosed in protective cases (figure, p. 36), and he had to find a way to pump more sunlight into the system in less flight time.[51] Regener's staff developed an all-reflecting light-funnel system for Kiepenheuer's spectrograph, and a complex combination of reflection and diffusion surfaces for the quartz spectrograph (figures, p. 35)[52]

Save for the spectrographs and the air samplers, all the other instruments used photographic drums driven by clockwork to record their data.[53] Each instrument had slightly different requirements, but the general characteristics of the drums were similar. Tiny mirrors set on movable membranes in the barographs and thermographs directed beams of light to rotating photographic drums, which were calibrated by timing marks. True to his traditional mode of scientific ballooning, Regener apparently never considered using telemetry to send data to earth during a flight.

His Peenemünde colleagues, however, developed several radio Doppler experiments and were interested in studying the effects of missile exhaust on radio propagation.

## Preparations During the Final Months of the War

Amidst air raids, food shortages, and staff departures for direct war work, Regener doggedly continued flying plastic balloon-sondes, measuring cosmic-ray intensities under lakes, and preparing for a flight on a missile.[54] By May 1944, most of the meteorological instruments had been completed, although the devices had yet to be integrated into the Tonne.[55] In early October 1944, Regener appeared confident that all parts of the Tonne would soon be ready for shipment to Peenemünde for final wind-tunnel tests and integration into a V-2.[56] But technical problems still plagued the Tonne system; in early November his group was still redesigning the Tonne configuration, arranging for additional wind-tunnel tests, and going through what seemed to be an endless series of design iterations.[57]

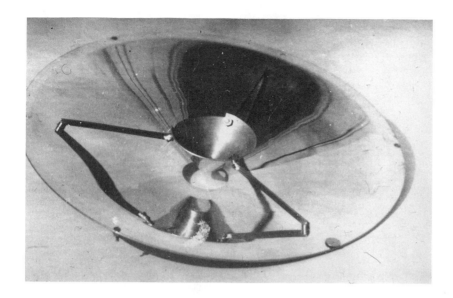

Optical system of Regener light funnel: Sunlight collected by a large concave mirror was concentrated onto a central cone, which fed the light through the hole in the first mirror. This was presumably the system used for Kiepenheuer's Lyman Alpha spectrograph, but it could also have been used with the diffusing optics illustrated below for either system. Erwin Schopper, Max Planck Institute for Aeronomy, Lindau.

Diffusing hemisphere that fed the slit of Regener's spectrograph. Deutsches Museum, Munich.

Regener's rocket spectrograph, save for the light collector, was a rugged version of Regener's earlier balloonborne spectrograph using the same circular cassette and optical design. Deutsches Museum, Munich.

Aware of these delays, Wernher von Braun decided in November 1944 that the Tonne system itself had to be tested without the instruments on board. He ordered that two V-2s be fired vertically carrying a Tonne with a simple aerial camera as payload instead of Regener's full complement of instruments.[58] The Photo-Tonne would go through all the steps planned for the real Tonne, including ejection and parachute descent, and the camera would record how the whole system behaved.[59] Von Braun's plan was prudent, for Regener's instruments were still not ready in December. Wind-tunnel tests continued through mid-December in Göttingen, while the instruments themselves underwent final calibration and adjustment in Friedrichshafen just before the new year. Finally, with everything patched together, on 3 January 1945, Schopper, Paetzold, and Rossner brought the instruments to Peenemünde, and on 6 January, the Tonne was picked up in Göttingen and sent to Peenemünde for final integration.[60] Schopper recalls that although they still would have liked to make improvements in the aerodynamics of the Tonne, the instruments were all ready—"they all worked excellently, with good calibration"—and so installation began.[61] But during final integra-

tion, the instruments still did not fit perfectly, so Schopper, Paetzold, and Peenemünde technicians hastily fabricated new housings, which delayed the final installation, adjusting, and testing until 18 January 1945 in Peenemünde.[62] But by that time, Schopper recalls, panic had set in; fearing a Russian attack, Peenemünde was being evacuated.

On von Braun's orders, Regener's men left in haste without the instruments or Tonne and in three days were back in Friedrichshafen. "We were on one of the last trains that left Berlin," Schopper added. "We got on only because we had papers showing that we were important persons."[63]

## War's End and an Assessment

When they left amidst growing chaos, Regener's men knew only that von Braun's remaining staff were still planning to launch their Tonne, but they also must have known that the chances for a flight were dim, even though Regener's group was continuing to refine the instruments in Friedrichshafen.[64] By the end of January 1945, Peenemünde was hardly the place to contemplate the structure of the upper atmosphere. The SS had ordered evacuation deep into Germany, to the area around Nordhausen and the dreadful Mittelwerk, where the V-2s were mass-produced by forced laborers who were being systematically worked to death.[65]

The last V-2 test-fired from Peenemünde was on 19 February 1945, and the last fired in battle was on 28 March 1945.[66] By 2 May critical Peenemünde staff members, including von Braun and Dornberger, had been located in Bavaria and were taken by the American Seventh Army's 44th Infantry Division.[67] The Third Reich surrendered to the West only five days later, and on 9 May to the Soviets. With the war's end vanished all chances for Regener's group to see its efforts bear fruit in Germany.

Regener failed to fly his instrument package on a V-2, as his Peenemünde and Friedrichshafen colleagues have argued, because of wartime delays and postponements.[68] The continued need to flight-test the V-2 missile itself, the great pressure to get it into production, and later, shortages of everything from fuel to manpower, all conspired against Regener.[69] The Tonne was far from ready, according to von Braun's November 1944 memorandum, by the time that war production pressures dominated. Any research program at the time would have had to justify itself on the basis of immediate practicality, and Regener's certainly could not qualify in January 1945.[70]

Beyond wartime pressures, the leading factor contributing to Regener's failure was the sheer complexity of the Tonne itself. Just as the V-2 cannot be thought of as a single device, but as a complex weapons system that created an industry dispersed across Germany requiring long supply lines and constant communication between hundreds of groups responsible for production, testing, integration, and firing, Regener's Tonne was a system requiring a network that extended far beyond his Friedrichshafen laboratory.[71] Bringing the instruments and the Tonne to perfection took time, especially when it was not the sole task at hand. And Regener's group obviously took every opportunity to be sure that everything was perfect. Indeed, they were still fiddling with integration problems and instrument details in January and February 1945.

The technical problems that Regener was trying to solve in short order took years to overcome in the United States after the war. In November 1946, Applied Physics Laboratory physicist James Van Allen considered recovery over water "impossible" for aerial photography projects; the first successful land recovery of a warhead from a V-2 by parachute came only in February 1947.[72] Even though V-2 testing was taking place both at Peenemünde (with water landings) and near Blizna, Poland, where an SS training camp was converted to testing the missiles by bombing the Polish landscape, Regener was not free to choose where to fly his instruments.[73]

Americans would have the comparative luxury of doing a little at a time, in stages, in peacetime. Regener worked only under wartime conditions. In a sense, it is remarkable that Regener got as far as he did, or that he was able to continue working on the Tonne right up to the end.

## Regener's Fate

With his Friedrichshafen buildings destroyed in February 1945, Regener and some of his staff remained at Weissenau, under French control. He was interrogated there several times by Alsos and British intelligence gatherers interested in German ionospheric research.[74] The French allowed Regener and some 40 of his staff to retain surviving instruments and buildings at various outstations, and to continue some lines of war-related work.[75]

Regener was then called back to Stuttgart to rebuild his heavily damaged institute as its professor and director.[76] He left Ehmert in charge at Friedrichshafen, while Schopper took over a high-voltage laboratory at Hecken, also under institute auspices. Since Regener and his colleagues never abandoned ballooning and ground-based cosmic-ray studies, the loss of the war and their chances of gaining access to large rockets did not end their programs.

With Paetzold, Ehmert, and others, Regener continued to conduct balloon-borne spectroscopic investigations of the ozone layer with improved lightweight instrumentation reminiscent of their prewar and wartime flights.[77]

At the end of the war, the Kaiser-Wilhelm Gesellschaft was in a shambles, its surviving elements under the control of relocation or evacuation boards controlled by Allied forces.[78] After it was restored as the Max-Planck-Gesellschaft, Regener, at age 67 in 1948, was elected its vice president; indeed, Telschow proudly defended the KWG for its protection of Regener.[79] In November 1952, the old Forschungsstelle was converted officially into the Max-Planck-Institut für Physik der Stratosphäre, which continues today as the MPI für Aeronomie, in Lindau.

Regener's positive response to the offer from Peenemünde to conduct research on rockets was a coming together of the engineering needs associated with building a ballistic missile system and the interests of a scientist who had studied the upper atmosphere for almost 15 years. When von Braun promised berths on V-2s, Regener was already charged by his military patrons with searching for means of gaining access to higher altitudes. That the two interests came together in the depths of the war is more than circumstantial; indeed, the V-2 was created for war, and its developers realized that its full elaboration required the ingenuity and talent of people like Regener. Everything Regener proposed to fly on the V-2 was of direct military interest.

The framework von Braun created in Germany to use a military missile to conduct scientific observations was repeated in America immediately after the war. Not only were the same observations proposed and carried out with captured V-2 missiles at White Sands Proving Ground in New Mexico, but they were supported by the same presumptions: that such studies would inform ballistic missile development. Funded entirely by the military, and largely confined to institutions engaged in military research, either on contract to the military as in Regener's case or within the military itself, the institutional context for upper atmospheric research with the V-2 in the United States was defined by Regener's experience in Germany.

Although knowledge of Regener's efforts may have influenced U.S. Army Ordnance policy regarding the instrumenting of captured missiles, it had less effect on American scientific groups, beyond helping to identify the types of problems that could be examined. American programs were well under way by the time participating scientists and their military patrons learned the details of Regener's instruments from interrogation reports and captured documentation. When pieces of the Tonne surfaced at the Aberdeen Proving Grounds, the American scientists had already formed much of their scientific agenda for the V-2, and it paralleled Regener's closely.

# Notes to Chapter 3

1. On Kummersdorf and the origins of German interests in rocketry and missile development, see Winter (1983), chap. 4, especially pp. 51–54; Ordway and Sharpe (1979); and Ehricke (1950). Michael J. Neufeld (1991) is currently examining Peenemünde as a government-organized laboratory within the context of Third Reich politics. I am indebted to Neufeld for many helpful discussions and for his advice on numerous points in this chapter.

2. On the origins of Peenemünde and German missiles, see Dornberger (1963); Ordway and Sharpe (1979), chaps. 3 and 4; and Kennedy (1983). On the name of the weapon, see McGovern (1964), p. 39; Ordway and Sharpe (1969), pp. 161, 192.

3. Ordway and Sharpe (1979), p. 41; Kennedy (1983), p. 18. See also G. Reisig to the author, 28 March 1985.

4. See, for instance, Ordway and Sharpe (1979), pp. 61–63; Dornberger (1963), pp. 396–97. Michael Neufeld (1991) has made the work of Hölsken (1984) on vengeance weapon development accessible and has clarified many issues surrounding problems with the mass production of the V-2.

5. Dornberger (1963), pp. 394–95; Gerhard Reisig OHI, p. 37rd. On Dornberger's methods of acquiring manpower, see Neufeld (1991).

6. On von Braun's responsibilities under Dornberger and his value as a system builder, see Neufeld (1991). Although von Braun represented himself as a civilian in his capacity at Peenemünde, he was also known to be an honorary member of the SS since 1940. See Bower (1987), p. 254.

7. Reisig OHI, pp. 55–56rd. On manpower problems and Peenemünde's various solutions for them, see Ordway and Sharpe (1979), pp. 35–37; and Neufeld (1991).

8. Noted in Ludwig Roth, "Zweistufenaggregat (180/30) A10/A9," n.d., circa Fall 1941. Peenemünde Correspondence, Vol. 14 (1940). K/DMA. The A9/A10, planned as a two-stage ballistic missile, would skim the upper atmosphere like a glider, thus extending its range to reach the east coast of the United States. See Ordway and Sharpe (1979), pp. 56–57. Roth called for observations of temperature, pressure, and composition at altitudes in excess of 100 kilometers, and Regener was first on his list, followed by the German Weather Service and the Army Meteorological Service, to build the necessary instruments. Other capabilities desired included aerodynamics studies at Mach 15, studies of heat transfer at Mach 15 for reentry at heights greater than 100 kilometers, and research on ballistic trajectories and glide theory. I am indebted to Michael Neufeld for providing the Roth memorandum.

9. The 55 year old Regener was singled out by Walther Gerlach of the University of Munich, who argued with Dutch-American physicist Samuel Goudsmit in 1936 that with active research such as Regener's on the upper atmosphere, even under National Socialism German physics had not been neglected. Beyerchen (1977), p. 169.

10. Carmichael (1985), pp. 106–7; DeVorkin (1989a).

11. Pfotzer (1985), pp. 78–79.

12. Members of his staff also identified a suspected neutron component in cosmic-ray spectra and studied the hard component of the cosmic-ray spectrum. In addition to Geiger counters, they also were pioneers in the use of stacks of photographic emulsions to record the tracks and nuclear collisions of these high-energy particles. Pfotzer (1972), p. 220; DeVorkin (1989a).

13. On the Aryan physics movement see Beyerchen (1977); and Walker (1989), pp. 61–

73. Gerhard Reisig recalls that Regener was removed from his university post because of his wife. Reisig OHI, p. 61rd. See also Ernst Stuhlinger to the author, 10 January 1985; Erwin Schopper OHI, pp. 6–7rd; and Macrakis (1989), p. 312. Fraunberger (1975) states that Regener's wife had to leave Germany along with his son and daughter. Schopper denies this.

14. Regener's daughter and physicist son-in-law escaped to Australia. Victor Regener to the author, 24 August 1984. Regener's removal from Stuttgart was in the form of a forced retirement based on the Civil Service Law of 7 April 1933. This Law for the Reinstitution of a Professional Civil Service was announced on 7 April 1933 and excluded those of "foreign race" from civil service. Later the law was extended to politically unreliable civil servants and was amended many times. See *Reichsgesetzblatt* 34 (part 1), Berlin, 7 April 1933, pp. 175–77; ibid., 74, 1 July 1933, pp. 433–34. I am indebted to Marion Kazemi for providing a copy of the announcement. See also Hintsches (1988), pp. 3–4.

15. These included support from the Deutsche Notgemeinschaft der Wissenschaft as well as the Friends of the Technische Hochschule Stuttgart. Pfotzer (1985), pp. 82. The Deutsche Notgemeinschaft der Wissenschaft was created in part because of changes in the relation between science and the state in Weimar Germany, and because post-WWI inflation in Germany and the resulting poverty of German science had to be countered. See Macrakis (1989), pp. 42–43.

16. Regener to "Dear Colleague," 5 October 1937. ER/MPG. Regener reacted like many others who were being harassed. Macrakis (1989), p. 317, concludes that many scientists who stayed in the Third Reich during the war were "serious scientists attempting to continue their work undisturbed." She found that "outward accommodation and withdrawal into one's scientific work was most common." From interrogations after the war there appeared to be a variety of motives for staying. See Walker (1989); Byerchen (1977); Weiner (1969); and, for instance, Gerard Kuiper to Fisher, "Policy for Post-War German Scientists," 9 May 1945, KP/UA.

17. The Kaiser Wilhelm Institute for Physics tried to remain self-governing in the late 1930s, but was subject to government demands. Dutch

physicist Peter Debye, its director, was told he had to become a German citizen when the institute started to pursue military research in applied nuclear fission. Debye declined to do so and left Germany. See Walker (1989), pp. 19–20. On the KWG, see Gerwin (1986); Staab (1986); and especially Macrakis (1989), pp. 96ff., who argues that the KWG managed to maintain a degree of insularity.

18. Regener to Laue, 18 October 1937. ER/MPG. The field station had been supported by the Robert Bosch Foundation and by a Württemburg industrialist, but this support was long gone as well.

19. Rector of the TH Stuttgart (Stortz) to the Director of the KWG (Telschow) 25 November 1937. ER/MPG.

20. Along with his post at the KWG, Telschow had a second desk in Göring's office and worked to direct KWG research to Göring's wishes. See Macrakis (1989), pp. 153–54.

21. Telschow to Regener, 13 December 1937; fragment, P. A. Planck to Carl Bosch, 17 December 1938. ER/MPG.

22. See Regener (1949), p. 54. This report was written in November 1940.

23. Debye to Telschow, 11 March 1938. ER/MPG.

24. Regener to Telschow, 30 March 1938; Reich Education Ministry to KWG, 22 April 1938. ER/MPG.

25. [Office of the] Reichsminister für Luftfahrt and the Supreme Commander of the Luftwaffe (Göring) to Regener in Friedrichshafen (signed by Muller), 10 August 1938. Records of the Senate meeting and the earlier meeting in Berlin are from a fragment, "Minutes of the meeting of the Senate of the Kaiser Wilhelm Gesellschaft on the 30th of May 1938," and Reichsminister für Air/Supreme Commander of the Luftwaffe (Lorentz, as Göring's representative) to KWG, 25 May 1938. ER/MPG.

26. Earlier contracts supported Alfred Ehmert's measurements of the oxygen content of the stratosphere. See Bartels and Ehmert (1961), p. 18.

27. Regener (1949), p. 56. Regener recalled that while his ultraviolet spectroscopic studies of solar radiation from balloons had thus far failed to reveal the window at 210 nanometers, success was possible if he could fly instruments higher. Regener's view of meteorology was standard. See Friedman (1989), p. 244.

28. In a report on the construction of a sensitive wire thermograph, Regener showed how it could be put into large-scale production, but did not hesitate to discuss how it could be modified for use at greater heights. Erich Regener, "Bericht über einen Drahtthermographen mit mechanischer Registrierung," *Deutsche Luftfahrtforschung* 1254 (Zentrale für wissenschaftliches Berichtwesen über Luftfahrtforschung (ZWB), 25 June 1940), DMA. Erwin Schopper's first assignment was to develop sensitive hygrometers and moisture warning devices for aircraft. Schopper, an earlier member of Regener's Stuttgart group who had been in the Luftwaffe, was detailed to Friedrichshafen in 1940 upon Regener's request, and continued on contract to 1945. Schopper OHI, pp. 8–9, 12–13.

29. Notes from E. Schopper lecture, 24 November 1988, at the MPI for Aeronomy, Lindau; Bartels and Ehmert (1961), pp. 18–19.

30. Pfotzer (1972), p. 226; DeVorkin (1989a).

31. The plastics had brand names of Viniphol and Peraphol. See Alsos report by W. R. Piggott, a member of Sir Edward Appleton's ionospheric intelligence-gathering unit in Germany (assistant director of intelligence (ADI), Science Air Ministry). Piggott visited Regener's Weissenau laboratory in the summer of 1945 and examined the new plastics he was using for his balloon flights. "Memorandum," 1 October 1945. Gerard Kuiper also visited Regener briefly, and reported on much the same observations, although he had also been in close contact with Piggott. See G. Kuiper, "Report to Scientific Chief, Alsos Mission, HQ, G2, USFET on Solar and Ionospheric Research in Germany," 25 September 1945. KP/UA.

32. Bartels and Ehmert (1961), p. 19. Piggott also reported that "The latest development, which was incomplete, was a system of carrying a small hydrogen driven rocket to great heights by the balloon and then firing it in a vertical direction." "Memorandum," 1 October 1945. KP/UA.

33. Ibid.; and Kuiper (1946), p. 276.

34. Wernher von Braun, *Niederschrift über die Besprechung*: "Entwicklung einer Apparatur zur atmosphärischen Höhenvermessung für A4," 8 July 1942. K/DMA. Pages 1 and 3 appear in Klee and Merk (1965), pp. 111–13. See also Wernher von Braun, "Survey of the development of liquid rockets in Ger-

many and their future prospects (by Professor von Braun)," Section A3, pp. 66–72, in Fritz Zwicky, "Report on Certain Phases of War Research in Germany" (Pasadena, Calif.: Aerojet Engineering Corporation, 1 October 1945). Air Materiel Command, Summary Report F-SU-3-RE, released January 1947, and Zwicky interrogation of Helmut Weiss, p. 107.

35. Ibid.

36. According to Stuhlinger, von Braun lacked familiarity with cosmic-ray physics. More likely, there was either less direct connection between this type of research and making rockets, or Regener realized that a rocket was not an appropriate platform from which to study cosmic rays. See Stuhlinger to the author, 10 January 1985; Paetzold to the author, 1 November 1984; see also Reisig OHI, pp. 56–57rd.

37. The parachute system included ejection devices, a compressed air tank for inflating the parachute in near space to prevent rupture in the lower atmosphere, small parachutes, a float capable of carrying 50 kilograms of equipment, scale drawings, and data on the accelerations produced by the mechanism on the ejected cargo.

38. Reisig to the author, 30 November 1984.

39. Reisig was one of several Peenemünde staff members who acted as technical liaison with Regener's group. He was familiar with the problems of electrical and electronic integration for the A4, and helped Regener with initial requirements studies. Reisig's primary responsibilities were at Peenemünde, which limited his trips south, so a Regener staff member, Walter Rau, later came north to discuss integration requirements in Peenemünde. Reisig OHI, p. 65rd.

40. The parachute was built by Erwin Madelung at the Graf Zeppelin testing station near Stuttgart, on contract to Peenemünde. It was Regener's original single 20 meter parachute with a ray-shaped set of tubes ending in a circular ring for expansion by compressed air. This system was tested in January 1945 and was found to open evenly during a fall of less than 20 meters. Hintsches (1988), pp. 19–20.

41. See E. Schopper, "Neiderschrift über eine Besprechung," 5 April 1943. Schopper notes, MPI für Aeronomie. Weiss, Krauss, and Hälsen attended from Peenemünde, with Regener, Ehmert, Schopper, Paetzold, and Poetzelberger from Friedrichshafen. Engineers

from Darmstadt and Ainring were also present. See also E. Schopper, "Niederschrift über eine Besprechung," 6 May 1943, p. 2, kindly provided by H. K. Paetzold. The parachute itself had to be at least 20 meters in diameter to handle the Tonne and had to be opened by inflation both rapidly and symmetrically. Various inflation designs were debated, and finally Regener came up with a six-spoked wheel design that won out at the conclusion of the meetings in March and April 1943. "Besprechungsnotiz betreffend Einbauten in das A4-Aggregat zur Höhenvermessung am 3. und 4.6.1943 in Friedrichshafen." Fragment in Schopper notes, MPI for Aeronomy. These meetings included a specialist in parachutes from the Zeppelin works in Stuttgart. See also Hintsches (1988), p. 20.

42. The group wanted models tested in wind tunnels and wanted to have construction sketches of how all of the instruments would fit in the Tonne. E. Schopper, "Niederschrift über eine Besprechung," 6 May 1943, p. 7.

43. E. Schopper, "Niederschrift über eine Besprechung," 6 May 1943, p. 5. Paetzold to the author, 1 November 1984.

44. Friedrichshafen was believed to be a site for bomber production as early as May 1935. See Bower (1987), p. 22. During the war, it became one of the decentralized sites for liquid oxygen (LOX) factories for V-2 propellants, and was also well-known as the site of major Zeppelin and Maybach plants, where V-2 LOX and fuel tanks were manufactured. These were bombed in March and on 27–28 April 1944 by U.S. 8th Air Force and RAF raids. See Ordway and Sharpe (1979), pp. 83, 94–95, 108, for a discussion of the 600 bomber RAF raid on Peenemünde during the night of 17–18 August 1943.

45. Schopper recalled in particular that bombing raids severely restricted the amount of time anyone could work on any project. There were also shortages of supplies. Schopper OHI, p. 4rd. Schopper felt that their project had highest priority, but at least one classification above SS priority existed in the latter part of the war, and industry was deluged with them. See Walker (1989), pp. 101–02.

46. Paetzold to author, 1 November 1984; see also Reisig OHI, p. 69rd.

47. Reisig to the author, 30 November 1984, p. 11; see also: "Entwicklung eines Barothermographen," 24 April 1936; and Hans

Mueller, "Ergibnis der Besprechung am 9.7.36," G. Reisig OHI working files SAOHP; Reisig OHI, p. 42rd.

48. The excursion of the quartz tube caused by pressure changes was transmitted through a spring hinge joint to a mirror system that amplified the quartz tube deflection by a factor of 110 times. This deflection was photographically registered. E. Schopper, "Verzeichnis der am 4. Jan. 45 überbrachten Geräte," 25 January 1945 ("Manifest of the devices transported on 4 January 1945," written in Friedrichshafen); E. Schopper, H. K. Paetzold, Mechanic Rossner, Herr Hälsen, "Bericht über den Einbau der instrumente der Forschungsstelle . . ." EW 12/45g/I. Conf. p. 2. K/DMA. See also Paetzold, Pfotzer, Schopper (1974), p. 181.

49. Hufbauer (1991), pp. 123–124, describes the context within which Kiepenheuer became involved in Regener's project. See Kiepenheuer (1948), pp. 261ff.; Kuiper (1946), pp. 274–75. During the early part of the war, Kiepenheuer was stationed at Rechlin, the main proving ground for the German Air Force, about 115 kilometers southwest of Peenemünde, and was engaged in organizing the ionospheric network. From that post he was aware of progress at Peenemünde, as well as Regener's activities. See S. A. Goudsmit, "Joint Communications Board Report," 1 January 1945; Gerard P. Kuiper, "Rockets, Ionosphere and Stratosphere Research," Combined Intelligence Objectives Sub-Committee Item 4, File IX-3, November 1944. Air Document Index (TECH), NASM Technical files. See also: S. A. Goudsmit, Scientific Chief, Alsos Mission, "Excerpt from Intelligence Report 20 November 1944," JX/WP 31 of "Combined Wave Propagation Committee of the Combined Communications Board." John H. Dellinger File, "Rockets" folder, NBS/NARA.

50. Pohl (1938); Kuiper (1946), pp. 275–76.

51. Hans Karl Paetzold recalls that Regener needed "another specialist, and a watchmaker or expert in fine mechanisms" to build these devices. Paetzold, a private in a replacement unit of the German Army, was transferred to Peenemünde and assigned to Regener to develop the spectrograph, assisted by a mechanic named Rossner. Paetzold to the author, 1 November 1984. On draft exemptions in the Third Reich, see Walker (1989), pp. 76–77,

124–25. See also Gerard Kuiper to Fisher, "Policy for Post-War German Scientists," 9 May 1945; and Kuiper, "Comments on German War Research in Physics," and "German Papers on Uranium," 6 May 1945. KP/UA.

52. For the Kiepenheuer device, several sets of mirrors with diffcrent vertical acceptance angles had to be made to accommodate different ranges of solar altitude that would be encountered in different seasons of the year. Gerard Kuiper noted that the acceptance angle of Kiepenheuer's spectrograph was a full quadrant. Kuiper (1946), p. 276.

53. See Regener (1939).

54. W. R. Piggott "Memorandum," 1 October 1945. KP/UA. See also Schopper OHI, pp. 2, 4–5rd.

55. See E. Schopper, (fragment), "Kurzer Bericht . . ." ("Short report concerning the status of apparatus for meteorological experiments for high altitude launch") HAP 91449, 20 May 1944. Schopper files, MPI Lindau.

56. E. Schopper, H. K. Paetzold, and Rossner, "Bericht über den Einbau der Instrumente . . ." EW 12/45g. PP/DMA.

57. The Tonne turned out to be too heavy for the parachute system. New rocket cones and frames made of magnesium or aluminum alloy were hastily fabricated by Zeppelin, a new contractor to the project, and were not ready until 12 December. Two days later the completed Tonne, with the instrument dummies, was sent to Göttingen for new wind-tunnel tests. The confusion and frustration of these many design changes, which also reflect a general lack of communication between the industrial contractors, Peenemünde, and Regener's group, are revealed in [Pabst], "Trip Report of Designer Pabst, 2/10–16/10 1944," 18 October 1944; and [WvB] "Record of Meeting 5 October 1944," 8 November 1944. File RH8/v.1271, Federal Archives, Military Archives, Freiburg.

58. Wernher von Braun, "Technische Vorplanung für Schiessen mit Photo- und Registriertonne," 22 November 1944. PP/DMA, copies in NASM technical files.

59. The firings were planned for a small island slightly north of Peenemünde and were to be under the direction of General Rossmann, who had succeeded General Dornberger at Peenemünde. Von Braun later implied, during Allied interrogations, that two experimental vertical test flights took place and served to

demonstrate that the system worked only to altitudes of 60 kilometers, far short of its capability. Wernher von Braun, "Survey of the development of liquid rockets in Germany and their future prospects (by Professor von Braun)," Section A3, pp. 66–72, in Fritz Zwicky, "Report on Certain Phases of War Research in Germany," 1 October 1945, pp. 68–69. According to an "Experimental Test" report by Schüssele, a Peenemünde engineer, the Photo-Tonne was test dropped from an HE 111 bomber flying at 8,000 meters on 23 December 1944. The parachute apparently worked and the Tonne was recovered. It is not likely that any vertical test flights were conducted after that date, unless they were for air burst studies, which had been a continuing problem. See [Schüssele], "Experimental Test Report," 23 December 1944. Peenemünde Correspondence, vol. 11, K/DMA.

60. See E. Schopper, "Verzeichnis der am 4. Jan. 45 überbrachten Geräte," 25 January 1945 ("Manifest of the devices transported on 4 January 1945," written in Friedrichshafen); E. Schopper, H. K. Paetzold, Mechanic Rossner, Herr Hälsen, "Bericht über den Einbau der instrumente der Forschungsstelle. . ." EW 12/45g/I. Conf. p. 2. Copies from PP/DMA and Schopper. See also Reisig to author, 28 March 1985; see also Schopper OHI, p. 14rd.

61. Schopper OHI, pp. 14–15rd.

62. E. Schopper, H. K. Paetzold, Mechanic Rossner, Herr Hälsen, "Bericht über den Einbau der instrumente der Forschungsstelle. . ." EW 12/45g/I. Conf. p. 2.

63. Schopper OHI, p. 15rd. The Russians did not overrun Peenemünde until May, although rumors of their imminent arrival were rampant from January on. Schopper's recollection of when they left is confirmed in Paetzold to Weiss, 19 January 1945. NASM microfilm collection, file FE 679, roll 48.

64. Ibid. Schopper notes that he was aware that the Tonne was taken to Mittelwerk in the evacuation of Peenemünde. When he knew this is not clear. Schopper OHI, p. 15rd. Back in Weissenau, Schopper reported on the January installations and final tests. He felt that a number of revisions and refinements were still required. All of the precision instruments, such as the quartz barograph, had to be fine-tuned and developed further he felt, in collaboration with the firm of Hartmann

und Braun of Frankfurt, which was also responsible for producing up to five barographs for later planned flights. Through February, other contracts were still being prepared with firms in Berlin for small revisions to various parts of the Tonne. Manpower for the continual adjustment, readjustment, and reintegration of these new parts was to be provided by two mechanics promised by von Braun, sometime in February 1945, but this never happened.

65. Ordway and Sharpe (1979), pp. 254–70.

66. Ibid.; Kennedy (1983), p. 52.

67. Ordway and Sharpe (1979), chap. 14; Kennedy (1983), p. 52.

68. One curious exception to this picture is from Gerard Kuiper, who reported from his Alsos survey that the Tonne was ready for flight in July 1944. This, however, has not been confirmed by German documentation of the period. Kuiper (1946), p. 276. It does raise the curious question if an earlier Tonne had been destroyed in the Allied air raids on Peenemünde in August 1943, as von Braun associate Ernst Stuhlinger has also implied. Stuhlinger to the author, 10 January 1985; Stuhlinger to Victor Regener, n.d., in Stuhlinger file, NASM. But again, there is no confirming evidence to support the possibility that a Regener spectrograph had been completed by that time.

69. Irving (1965), p. 288; Klee and Merk (1965), p. 112. Certainly increasing wartime pressures in late 1943 and throughout 1944, when the A4 went from prototype to production as the V-2, made the transition far from smooth. It diverted many of their best design engineers into production areas and certainly did not result in more A4 rockets available for vertical test flights. And once deployed, the missile continued to self-destruct in flight; it has been estimated that of the 3,600 launches against Britain and continental targets, some 25 percent exploded in the air. Ordway and Sharpe (1979), pp. 78–79. The Germans later argued that the premature explosions were caused by a weak midsection in the V-2 structure. See James Van Allen to E. J. Workman, 27 May 1946, APL Archives, Cabinet 8. I am indebted to Michael Dennis for providing this archival reference. See also Reisig OHI, pp. 69–71rd.

70. Regener also experienced staff attrition. In December 1944, Helmut Weiss argued that

Paetzold's attention was needed by groups engaged in developing Wasserfall. The spectrograph Paetzold was still nurturing did not require his attention, Weiss argued, and was compromising these other activities, which were far closer to immediate wartime needs and interests. See H. Wiess to Ernst Steinhoff, 14 December 1944. Peenemünde records FE 679, NASM microfilm.

71. On V-2 production requirements, see Ordway and Sharpe (1979), chap. 5, p. 89.

72. James A. Van Allen, "High Altitude Research at the Applied Physics Laboratory of the Johns Hopkins University, A Brief Summary," 10 April 1947, p. 3. JVA/UI; quoted in Appendix F: "Sounding Rockets as Vehicles for Long Range Aerial Reconnaissance," 15 November 1946. His Appendix J noted that developments at Wright Field with a Franklin Institute parachute technique (Project Blossom) looked "very desirable." On American plans for parachute systems, see various reports of the V-2 Panel, especially V-2 Panel Report no. 12 (1 October 1947), p. 9, wherein first discussions appeared for planning a Regener-type stabilized container for spectrographs and cloud chambers with a parachute and beacon system. WAC Corporals were recovered by parachute, but they could not carry the payloads the Americans wanted to develop. See Leo Goldberg to Lyman Spitzer, 25 February 1947; Spitzer to Goldberg, 20 February 1947; 1 March 1947. LG/HUA.

73. Ordway and Sharpe (1979), pp. 80, 130–31; Kennedy (1983), p. 32.

74. R. A. Beth obtained information from Regener on 13 September 1945, as noted in marginalia on Erwin Schopper's report "Übersichtsbericht über die zur Vermessung der hohen Atmosphäre entwickelten Instrumente." Copy in PP/DMA, and in Schopper's personal files. On 1 October 1945, W. R. Piggott, part of Sir Edward Appleton's Air Ministry ionosphere intelligence team, visited Weissenau and obtained information on Re-

gener's activities. See Kuiper (1946); and Chapter 4 herein.

75. This included Walter Rau's studies of ways to keep water liquid at low temperature. On the general fate of the institute after the war, see Bartels and Ehmert (1961), pp. 17–19. Regener's son Victor remained in the United States, and it was some time before he and his father were back in communication. Gerard Kuiper helped Victor obtain information about his father, although in late 1945 it was limited to several recent publications. See Victor Regener to Gerard Kuiper, 28 November 1945. World War II folder, KP/UA. Werner Heisenberg helped Erich Regener learn of his son's activities during the war. See Heisenberg to Regener, 18 February 1943. General Correspondence, January 1941–42, K-Z, Heisenberg Archive. As late as January 1947, Regener was asking Heisenberg for information about his son in America. See Regener to Heisenberg, 23 January 1947. General Correspondence, Heisenberg Archive.

76. See "Erich Regener," *Poggendorff Handwörterbuch*, pp. 696–97; Paetzold, Pfotzer, and Schopper (1974), pp. 185–86; letters, Heisenberg-Regener circa October 1946. Heisenberg Archives.

77. E. Regener, H. K. Paetzold, and A. Ehmert (1954), pp. 202–07. Bartels and Ehmert (1961), pp. 19–21, note that Regener's staff restored research teams in cosmic-ray and atmospheric physics, adding studies of atmospheric ice, atmospheric electricity, and ionospheric radio propagation, concentrating on understanding the transition region from the stratosphere into the ionospheric D layer.

78. Staab (1986), pp. 23ff.; Macrakis (1989), pp. 286–300.

79. Macrakis (1989), pp. 307–08, 312, notes that in his effort to paint the KWG as "politically clean," Macrakis feels that only Regener's case had veracity. Telshow argued in January 1946 that the KWG had become a haven for many prominent scientists like Werner Heisenberg and Erich Regener who were either harassed or ousted from their university posts.

# 4

# The V-2 Travels West

*I hope you have a little time to give this some thought. Its a screwy program, but here are the V-2's and we should do something about them.*

—J. Allen Hynek to Jesse
Greenstein, 24 April 1946.[1]

How the Allies discovered that the Germans were developing a long-range rocket and how they tried to counter the development as well as exploit it is a story told and retold.[2] A combination of Allied missions and operations called variously Wildhorn, Crossbow, Overcast, Backfire, and finally Paperclip, searched for intelligence on the German missile among other targets, and then, as the war drew to a close, captured the ordnance itself and the expertise that built it.[3] By April 1945 a mountain of captured V-2 parts were under the control of United States Army Ordnance technical teams headed by Col. Holger N. Toftoy. His teams scoured the Mittelwerk and encountered the great human toll taken by the Germans in the effort to produce the dreaded vengeance weapon.[4]

There was a great deal of speculation in the West about the new missile before much was known about its capabilities. But as speculation turned to hard facts in 1945, a few people in Great Britain and the United States began to think about the implications of the missile for science, as well as for the conduct of future wars.

## The Significance of the V-2

When Willy Ley's upbeat book, *Rockets: The Future of Travel beyond the Stratosphere*, appeared in 1944, one reviewer commented, "Thinking people the world over are rocket-conscious to a degree never before reached. This consciousness of rockets is induced not by the idea suggested by the subtitle, but by the grim fact of wanton destruction."[5] Ley, a rocket enthusiast who fled National Socialism in the early 1930s, looked only to the bright future rockets held for mankind. But as summer turned into fall in 1944 and V-2 missiles began to fall in silence on Paris, Antwerp, and London, his optimism may have seemed hard to sustain. The first hits in the outlying London boroughs of Brentford-Chiswick and in Epping on the wet and overcast evening of 8 September raised the specter of instantaneous death without warning or defense. The V-2 at best was a countercity missile; everyone in the vicinity of London, Antwerp, or any other major city had an equal chance of being hit.[6]

As terrifying as the missile was to contemplate as a weapon of war, there were Londoners who, with Willy Ley, were fascinated by the technology. *Time* magazine for 27 November 1944 reported that for "Britain's famed jack-of-all-sciences," J. B. S. Haldane, the postwar future of the V-2 was in space. If fired vertically, *Time* reported, the V-2 could reach more then 300 kilometers, and "It could take photographs . . . [of] the sun and perhaps other heavenly bodies. . . . For the cost of a day of war, it should be practicable to send a series of rockets round the moon and photograph its far side."[7] R. V. Jones, one of Britain's most notable sleuths, when asked by the *U. S. Eighth Air Force Magazine* to tell

**45**

V-2 on display at war's end in Washington D.C. at 12th and Pennsylvania, N.W. NASM Science Service collection, SI 90–378.

what he knew about the V-2, pointed out in September 1944 that "in no other way can we get free of the earth's atmosphere [for] astrophysical studies." Jones added, "Sooner or later someone will seriously try to reach the moon—-and succeed," but he also knew full well that "military applications are bound to be made, whatever the limits imposed by treaties, and we should do well to keep an eye on the possibilities."[8]

Arthur C. Clarke, then deeply engaged in RAF radar flight control research, looked to a time when missiles like the V-2 might be turned to science. In the February 1945 edition of *Wireless World*, Clarke pointed out that, "It will not have escaped your readers' notice that the German long-range rocket projectile known as V-2 passes through the E layer on its way from the Continent. If it were fired vertically without westward deviation it could reach the F1 and probably the F2 layer."[9] Clarke added that instruments of all kinds could be flown, data could be telemetered back to ground stations, and if the entire rocket were to be parachuted to earth ("besides being appreciated by the public!"), the whole apparatus could be used again.

In late 1944 and early 1945, as Clarke and his fellow visionaries daydreamed about the peaceful uses of a large rocket, Regener's efforts remained unknown in the West. But hints about what was being attempted in Friedrichshafen started to reach Allied intelligence units.

Scene in Antwerp after a V-2 strike. Far less destructive than conventional area bombings, V-2 strikes still created stultifying psychological impact. Because it could not be precisely guided, anyone within kilometers of the missile's target had an equal chance of being hit without warning. U.S. Army Signal Corps files, NARA, courtesy Frederick I. Ordway.

## Learning More About the V-2 and Regener

As the Allies moved into Italy and France in the summer of 1944, scientific intelligence units swarmed just behind, searching for information on the organization of German science and its wartime products.[10] Military and civilian personnel within military agencies were followed by civilian specialists to identify, cache, and retrieve scientific and technical plunder.

U.S. Army Ordnance competed with its allies to capture what it could of the V-2. Since the summer of 1943, stimulated by reports that the Germans were developing a long-range missile, Army Ordnance formed a rocket development division within its research and development section and stepped up its experimentation in rocketry, which until then had been confined to contracts at the Guggenheim Aeronautical Laboratory of the California Institute of Technology (GALCIT).[11] GALCIT could not satisfy the Army's accelerated needs, so the Army contracted with General Electric to form Project Hermes. Hermes, led by Richard W. Porter, was just getting under way when Maj. Gen. Gladeon M. Barnes, then chief of research and development for Army Ordnance, decided that Hermes would also participate in the evaluation of the V-2.[12]

Holger Toftoy, at the time chief of the Army Ordnance Department's Enemy Equipment Intelligence Section, was asked to form Special Mission V-2 in March 1945 to supply Project Hermes with parts sufficient to reconstruct 100 V-2 missiles for testing at the newly created White Sands Proving Ground in New Mexico. In April, news reached Toftoy that the Mittelwerk had been overtaken and was ripe for the plunder, but he also learned that within a few weeks the area would be given over to the Soviet Army.[13] He had to act fast. Starting in August 1945, with the aid of the 144th Army Ordnance Company, Toftoy removed as much as his staff could find from the Soviet area and eventually transported more than 360 metric tons of V-2 parts to the United States. This was the largest shipment of weapons Toftoy's mission brought back, not including the mountains of documentation and the German personnel that eventually came along with it.

## Navy Intelligence

Joining Toftoy in Europe in the summer of 1945 was R. W. Porter and his Hermes group. They also crossed paths with a contingent from the Naval Research Laboratory that included Ernst H. Krause, head of the NRL's Communications Security Section, who was an expert on guided missile countermeasures. Much of NRL's subsequent enthusiasm for guided missile development and upper atmosphere research was the result of Krause's efforts.

In late 1944, after examining captured German air-to-ground missiles during a Mediterranean tour, Krause had become convinced that German rocketry was far ahead of anything the Allies had.[14] Now he was back in the spring of 1945 to Germany itself to see firsthand the extent of German missile development. Krause and Porter met in Garmisch-Partenkirchen for a round of intelligence gathering on missile guidance and control. They knew the general size and range of the V-2 and something of the family of

German missiles that had been used during the war. But they had little information about the guidance systems themselves, which were Krause's interest.[15] Krause found the Germans most anxious to sell their product. They told him that they had a 5,000 kilometer rocket on the drawing board that could be used in the Pacific, and that they were ready to help out: "if only the Americans would now get behind [the Germans] . . . they'd bomb Tokyo from Washington state."[16] But Krause wanted to know more about Germany's guidance techniques, and so he asked about inertial systems, gyro drift rates, and overall accuracy.

Krause's experiences in Europe, especially his contacts with Toftoy and Porter and the world of German missile research, gave him a sense of what future wars would be like and what he had to do to bring the Navy up to speed in guided missile research. He planned to turn much of NRL's attention to the development of guided missiles and as a first step managed to convert his old Communications Security Section into a guided missiles research and development unit.[17]

In December 1945, Krause's chief ally was his new NRL director, Cmdr. Henry A. Schade, who like Krause was fresh from scientific and technical intelligence adventures in Europe. As historians David Allison and Bruce Hevly have shown, NRL at the time was determined to break out of its tradition as a military testing laboratory to become a major player in military research and development.[18] Krause's agenda not only fit perfectly into this new philosophy, but played a part in bringing it about.[19]

Schade in early 1945 was the U. S. Navy representative on the Combined Intelligence Priorities Committee and also was the Navy's representative to the Alsos mission in Italy and Germany. With Samuel A. Goudsmit, chief scientist for Alsos, Schade wanted the Navy to have its own technical initiative in Europe and not be dependent on the British or on interservice agencies.[20] Schade led the Naval Technical Mission and had some 200 scientific and technical experts scour Europe.[21] He arrived in Oberammergau on 1 May 1945

Scene at the entrance to the underground caverns at the Mittelwerk. A liberated French concentration camp prisoner identifies V-2 fuel tanks (covered with camouflage) for American soldiers. NASM SI 79–12338.

with, as Tom Bower relates, "impatient zeal and apparently limitless resources." Schade knew just whom to interrogate and what to capture and, like Krause, focused on guided missile technology.

Bolstered by his experience in Europe the previous summer, Schade advanced Krause's goal to make missile research and development part of the institutional mission of postwar basic research at NRL. Both Schade and Krause realized that for NRL to be credible as a basic research laboratory—a place where pure scientific research was to further institutional goals—developing missiles would not be enough. Indeed, the first formal proposals for research into the upper atmosphere with missiles would come from NRL. They were

based, as we shall see, on what Krause's staff learned by reading Gerard Kuiper's intelligence reports about Regener.

## Gerard Kuiper and the Alsos Mission

While Krause and Schade worked fully within the sphere of the Navy, civilian interrogators working for Alsos were also rooting through Europe. Alsos was organized in late 1943 within the Office of the Assistant Chief of Staff of the Joint Chiefs of Staff (JCS), out of an early fear of a German atomic bomb, possibly carried by rocket.[22] As Alsos chief scientist Goudsmit later argued: "We had obtained intelligence data on the V-1 and V-2. What final use could they be to the Germans unless they were meant to carry atomic explosives?"[23]

Beyond assessing Germany's progress in making an atomic bomb, Alsos also was empowered to find out as much as possible about German science and technology: how it was organized and what it produced.[24] In his effort to recruit scientists as intelligence gatherers, Goudsmit, an emigré physicist from Leiden, naturally turned to European Americans.[25] One such person was Gerard P. Kuiper, a former Leiden student, astronomer at the Yerkes Observatory, and a member of Harvard's Radio Research Laboratory working on radar countermeasures.[26] Kuiper became part of Alsos planning in June 1944 and received his travel orders in August. By October Kuiper was in London, interviewing French astronomers and by November he was in Paris piecing together what he could about German solar and ionospheric research, radar and radio propagation, and the fate of astronomical colleagues.[27]

Kuiper confirmed a suspicion that "the Germans are much interested in ionosphere research, probably in connection with rocket control."[28] Through French astronomers who had been in contact with Karl O. Kiepenheuer, Kuiper found out that the German's had a network of solar observatories across Europe capable of predicting radio fadeouts owing to the solar disruption of the ionosphere. Kiepenheuer had also told French astronomers that "his German colleagues were quite excited about some new possibilities; they planned to send up instruments to an altitude of 30 miles to record the ultraviolet radiation of the sun." And there were even plans afoot to send instruments to 100 kilometers by rockets. Kuiper reported: "Some of the instruments installed in the rockets were designed by Regener (according to other intelligence reports, Regener worked on similar subjects at a laboratory near Friedrichshafen, on Lake Constance)."[29] Kuiper's French contacts believed that Regener's research was connected with radio propagation, although Kuiper and Goudsmit felt it was more likely

connected with the radio control of rockets. As a result, Regener's name was linked to guided missile research by Allied intelligence in late 1944.[30]

Kuiper's report was sent to the Joint Communications Board Countermeasures and Wave Propagation Committees in January 1945, and from there it was circulated to the National Bureau of Standards (NBS) and NRL.[31] It was, however, based only on indirect evidence gained from French astronomers who knew of Kiepenheuer's interests. Kuiper therefore did not know the fate of the Regener-Tonne but assumed that something had been flown.[32]

Kuiper returned to Europe in the spring and through the late summer of 1945 to interrogate German scientists directly, collecting data on high-frequency radar, acoustic proximity fuses, and the German nuclear fission program.[33] Joining up with a combined intelligence unit from the British Air Ministry and members of Sir Edward Appleton's ionospheric research intelligence-gathering team, Kuiper interrogated Kiepenheuer directly in June, visited his solar observatory network, and came away with a far clearer picture.[34] Through Kiepenheuer, Kuiper found out more about Regener and the Lyman Alpha spectrograph, and Regener became his next target.[35]

With W. R. Piggott from Appleton's group, Kuiper visited Regener at Weissenau on several occasions in September and October 1945. There they learned about Regener's continued high-altitude ballooning, his application of AGFA plastic skins for balloons, his attempts to produce instruments for V-2 rockets, and the scope of upper atmospheric research he and his team had been conducting throughout the war. Piggott was impressed with Regener's instrumentation, noting that "the standard barometer and thermometer used by Regener [were] particularly neat and did not suffer from hysteresis or zero drift difficulties."[36] Kuiper was more interested in the Lyman Alpha spectrograph.

After VJ Day Kuiper continued preparing detailed technical reports and, as restrictions were lifted, a review of German astronomy during the war.[37] Harvard astronomer Donald Menzel was one of the first to read Kuiper's draft review. Menzel had founded Harvard's High Altitude Observatory at Climax, Colorado, which monitored solar activity and the solar corona and had provided valuable data for predicting ionospheric disturbances, along the very same lines Kiepenheuer had in Germany. This placed Menzel squarely in the center of wartime ionospheric research in Washington working at the Interservice Radio Propagation Laboratory of the National Bureau of Standards. At the end of the war he was also chairman of the Wave Propagation Committee under the Joint Chiefs of Staff. Menzel therefore had access to security information, had seen Kuiper's more detailed classified Alsos reports, and now that the war had ended was actively pursuing means to reestablish his research momentum.[38]

After reading Kuiper's draft in November 1945, Menzel immediately relayed it to Merle Tuve and Edward O. Salant at the Applied Physics Laboratory (APL), recalling,

> *In our discussions of the problem of the V-2 rocket, I mentioned that I thought the Germans had been doing something. Dr. G. P. Kuiper . . . on his mission to Germany discovered alot about their activities. I enclose herewith the complete statement concerning the attempt to record UV radiation. The idea seems to be very good.*[39]

Menzel distributed copies of his Salant letter and Kuiper's transcript to E. O. Hulburt of NRL, to Leo Goldberg at Michigan, and to H. Gordon Dyke of the Navy's Office of Research and Invention (ORI, which later became ONR). To Goldberg, Menzel added that Regener's "ideas are excellent. Certainly the report should be called to the attention of those who are working in the field."[40]

Thus, well before Kuiper's article appeared in print, key American astronomers and physicists at both APL and NRL had read Kuiper's digest of Regener's work. Astronomers meeting in New York in February 1946 were fascinated by what Kiepenheuer and Regener had tried to do after Yerkes Ob-

servatory chairman Otto Struve read a few lines from Kuiper's draft at their society dinner.[41] Subsequently, several astronomers planned to get involved directly. In addition, both Hulburt and Tuve, two of Robert Goddard's erstwhile scientific advisers, took steps to get their institutions involved.

## Fritz Zwicky and CIOS

Caltech astrophysicist Fritz Zwicky, who was also director of research at the Aerojet Engineering Corporation near Los Angeles, was a team member within the vast Combined Intelligence Objectives Subcommittee (CIOS), formed in August 1944 by the Combined Chiefs of Staff. CIOS provided technical assistance for special intelligence-collecting units called T-forces that had been organized under the Allied expeditionary forces. The T-forces were charged with capturing and defending potentially useful facilities, and CIOS was asked to assess the value of the booty.[42]

Swiss by birth and training and harboring an interest in jet propulsion as intense as his hatred for Germans, Zwicky was a natural for the job and became a formidable and indefatigable interrogator. He was sent to Britain in April 1945, and then to many European cities to interrogate Germans who had worked on various phases of jet and rocket propulsion. But his primary target was the Peenemünde wind-tunnel chief Rudolph Herrmann and his large staff, who had been responsible for wind-tunnel tests of the V-2 and Wasserfall.[43]

Zwicky was also one among many interrogators of Wernher von Braun and Walter Dornberger at Garmisch-Partenkirchen. Both provided Zwicky with prepared statements as well as spontaneous interviews. Von Braun was particularly anxious to convince Zwicky of the great potential of long-range rockets. He talked at length about the planned successors to the A4 and what their capabilities might have been. And knowing that he was talking to an astrophysicist, he assured Zwicky that what Regener started in Germany "could not be carried out on account of military events. It could be done in a short time however, with some of the A4 rockets still at hand."[44]

Von Braun proselytized for the military and scientific uses of long-range rockets. Improved models might carry people as well as bombs intercontinental distances; they could lift space stations into orbit, and send a man to the moon. Zwicky, however, was neither interested in dreams nor in their German promoters. For Zwicky, the V-2 was "a technical achievement of high order due less to the activity of any individual genius than to the determined and enthusiastic cooperation of a large number of only moderately competent technical individuals."[45]

Soon after Zwicky returned to the United States and to Caltech, he planned to conduct hyperballistic aerodynamics experiments using artificial meteors launched from V-2s in flight.[46] Zwicky's proposal to Army Ordnance was to examine the effect of hypersonic air friction on shaped charges in the upper atmosphere, to improve knowledge of reentry ballistics. But he also wanted to be the first person to send an artificial body into space, into orbit, and possibly even to the moon.[47] Never one to mince words, Zwicky campaigned in a professional astronomical journal to exploit the rocket as a vehicle for scientific research:

> The prediction may thus be ventured, that as long as they are tied to the surface of the earth, the 200-inch telescope, the new powerful Schmidt telescopes, and devices as yet unborn will be heavily overshadowed by the potentialities of rocket research.[48]

## What the Interrogations Stimulated

Many who became interested in using V-2s for the scientific exploration of the upper atmosphere recall that they were intrigued by intelligence reports like Kuiper's, which went significantly beyond speculation, popular news accounts, or the articles and books by Willy Ley and G. Edward Pendray. The existence of the V-2 itself was enough for most of them to speculate on how rockets might be used for science, but knowledge of what Regener tried to do revealed as well that upper atmospheric research had already been closely linked with the V-2.[49]

Richard Silberstein, a young radio engineer and member of the Interservice Radio Propagation Laboratory of the National Bureau of Standards, was also one of the early readers of Kuiper's November 1944 report and concluded from it in February 1945 that Germans had used V-2 rockets for sounding the ionosphere. This led him to suggest, as did Arthur Clarke, that "rockets could be used to study the physical and radio propagation properties of that region, and even of higher regions, if we assume that the apparatus load in the rocket used for ionospheric studies would be much lighter than the present load of explosives carried by the V-2."[50]

Silberstein proposed to study the propagation characteristics of the ionosphere by fitting the rocket with radio receivers and transmitters that could convert high-frequency ground signals and rebroadcast them back to the ground.[51] Silberstein submitted his detailed suggestions to the U.S. Joint Communications Board through his NBS superior, but nothing came of it.[52]

While Zwicky, Kuiper, and Silberstein remained on the periphery of V-2 activities in the United States, Ernst Krause headed for its center. Before Schade came to NRL, Krause had taken the initiative to establish missile research and allied scientific programs at NRL. Along with other scientific and technical managers at NRL, Krause and his group had reviewed Kuiper's November 1944 Alsos report and were generally aware of Regener's plans to instrument vertical firings of V-2 rockets. Krause recalls, however, that this information was no revelation; it was an obvious thing to do with such vehicles. Still, he recalls that knowledge of what the Germans had tried to do had some influence on NRL's postwar planning.[53]

In February 1946, NRL and APL staff were part of a scientific group that met at the Aberdeen Proving Grounds, about 100 kilometers north of Washington, D.C., at the mouth of the Susquehanna River, to have a look at some peculiar and somewhat mysterious devices. At least two of Regener's instruments reached Aberdeen sometime in December 1945 and were brought to the attention of prospective American rocket scientists during a meeting at the Naval Research Laboratory on 16 January 1946.[54] They found that the instruments were not completely intact and by April learned that the people at Aberdeen were trying to put Regener's spectrograph into working order.

By then Krause had already discussed deciphering "the Regener head for the V-2" with G.E.'s R. W. Porter. Krause and his APL counterparts had been trying to gain as much information as possible about the geometry of the German V-2 warheads and had seen two of them already. But here was an instrumented warhead that could tell them much more about how the Germans might have been conducting upper air research with V-2 rockets. Krause's staff only found that "the method of measurement that was indicated by [Regener's] instruments was somewhat confusing."[55] At least one thing was clear to all: film recovery was only possible if the delicate instrument was dropped from the rocket by parachute.

J. Allen Hynek at APL, however, remained excited by Regener's spectrograph, especially by its sophisticated light collector. Hynek, an astronomer from Ohio State University and best remembered today as an exponent for the serious examination of UFO phenomena, was then a consultant at APL, and knew that one of the main problems with the design of any rocket spectrograph was how to get as much sunlight into the system as possible. He liked Regener's ideas but thought that an astronomical colleague, Jesse Greenstein of the Yerkes Observatory, could design an even better system for astronomical research: "I hope you have a little time to give this some thought. Its a screwy program, but here are the V-2's and we should do something about them."[56]

In sum then, the existence of a rocket with the capabilities of the German V-2 not only changed how people thought about the conduct of future wars, but stimulated planning for the study of the upper atmosphere. American scientists learned about Regener's efforts through a number of paths, including interrogation reports, correspondence networks, and, by 1946, published accounts. The very fact of Regener's attempt stimulated some early excitement and proposals, even though specific plans were already afoot in the United States well before the details of Regener's instruments became available.

Knowing simply that the Germans had tried to do something with their missiles seemed to be enough; Regener's scientific agenda would be duplicated in the United States by the Navy at NRL and APL, by the Army at Aberdeen, by the Signal Corps, by the Army Air Forces, and by a number of military contractors on university campuses. But this activity meant only that the agenda, as we have shown, was obvious. Less obvious was who would carry out that agenda, and how it would be coordinated, for it constituted a completely new way to conduct scientific research, not only in its technical complexity and scale, but also in its context within postwar national security.

## Notes to Chapter 4

1. J. Allen Hynek to Jesse Greenstein, 24 April 1946. JG/CIT.
2. See, for instance: McGovern (1964); Lasby (1975); Jones (1978), chap. 45; Ordway and Sharpe (1979); Kennedy (1983); and Hinsley (1988), chaps. 55–56.
3. Project Overcast was created by the Joint Chiefs of Staff to acquire the critical talent and technology useful to winning the war in the Pacific. After VJ Day, this project became Paperclip, which combined military exploitation with—as Gimbel (1990), p. 448 notes— a civilian exploitation program by bringing key German rocket scientists to the United States. Approved only in March 1946, when it received its title, Paperclip was active by the latter part of 1945. The best of the earlier work

on this controversial subject is Lasby (1975), see esp. pp. 79–81. See also Ordway and Sharpe (1979); and McGovern (1964). The best recent effort is Gimbel (1986, 1990), who demonstrates that although the effect was the same, looser procedures for the importation of Nazi scientists into the United States were based on shifting policies rather than a conspiracy, as others have contended.

4. On early reports, see, for instance, Holger Toftoy, "Preliminary Report on German Rocket—V2," 25 September 1944, *EOT Technical Intelligence Report* 23; L. F. Woodruff and M. S. Hochmuth, "Preliminary Report on V-2 Long Range Rockets," 10 April 1945. "ETO Technical Intelligence Reports." Box 2631Q, RG 165.79, OCO/NARA. These are representative of Ordnance Intelligence documents that brought back technical information on the V-2 rockets. On Toftoy's central role in bringing V-2s and their creators back to the United States, see Lasby (1975); Ordway and Sharpe (1979); and McGovern (1964). On conditions at Mittelwerk that led to the deaths of more than 20,000 forced laborers taken from concentration camps, see Michel (1973).

5. Gingrich (1944), p. 360.

6. The concept of countercity and counterforce targeting has recently been examined as both a policy and a technical issue by MacKenzie (1990), chap. 3. Observers in central London recall single explosions that, since they were not associated with any other sounds, often were mistaken for demolitions to clear bombed streets and fell damaged buildings. Howard DeVorkin to the author, 15 October 1989. Others recall feeling relieved on hearing the sound, because that meant they survived. Tom Gehrels to the author, 31 July 1990. For detailed impressions and recollections of the psychological impact of the V-2 upon London residents, see Longmate (1985). On impressions of the V-2's military ineffectiveness, see Dyson (1979), pp. 108–09. Ordway and Sharpe (1979), pp. 203–05, provide a tabulation of V-2 firings on Antwerp, London, Liege, Paris, Norwich, Ipswich, and on numerous French sites. The nominal maximum velocity for a V-2 was 1,600 meters per second, barely below the minimum hypersonic velocity of five times the speed of sound.

7. "What Is V-2?" [Science] *Time* 44 (27 November 1944), p. 88. Weart (1988), p. 83, highlights another section of this same *Time* piece that speculated about the purpose of the V-2, to deliver Hitler's next surprise, atomic bombs.

8. R. V. Jones, quoted in Jones (1978), pp. 459–60.

9. Clarke (1945), p. 58.

10. See Baxter (1946), chap. 26; and Goudsmit (1947).

11. Koppes (1982), p. 18.

12. Porter recalls that General Electric was sought out because of its success with turbine technology, but since the groups that had developed the turbines were wholly engaged with jet propulsion problems, others inexperienced in such technologies were brought in to learn the craft, and they all "plunged immediately into intensive studies of rocket science and engineering." Richard W. Porter, quoted in "Prologue—1945–1954," *Challenge* 4 no.1 (General Electric Missile and Space Division, Spring 1965), p. 2. JPL/L historical files 5–614. See also R. Porter OHI.

13. On the general operation and hurdles Toftoy faced, see Ordway and Sharpe (1979), p. 278ff.

14. Ernst Krause OHI, pp. 29–34rd. On Krause's wartime missile countermeasures work, see Hevly (1987), p. 64ff.

15. Krause OHI, p. 37rd.

16. Krause OHI, p. 39rd. The Porter mission, although not directly under Overcast, still represented the same interests, as far as the Germans were concerned, which included capturing technology that could be used to shorten the war in the Pacific. See Lasby (1975), pp. 79–81; Bower (1987), pp. 120–25, p. 214. Still, some interrogators, such as the astronomer Gerard Kuiper, became indignant over the Germans' quick willingness to promote their services. Kuiper saw this attitude as a "lack of social responsibility," which "makes an advanced German science particularly dangerous." Kuiper to Major Fisher: "The Future of German Science," 30 June 1945, emphasis in original. Alsos folder, KP/UA.

17. Krause was recalled from Germany abruptly in June 1945, when NRL realized that it was in a domestic race to attract and capture the best American scientists and engineers who were then preparing to leave wartime laboratories for the academic world and industry.

18. It had just been transferred from the Bureau of Ships, its home since 1941, to the new Of-

fice of Research and Invention (ORI), created by Adm. Harold G. Bowen, long a promoter of basic research within the Navy, and at NRL. See Allison (1981), pp. 161–76; and Hevly (1987), pp. 66–72, 347.

19. Hevly (1987), pp. iii, 6, 54–55, 79.

20. Schade was a naval architect who held engineering degrees from MIT and the Technische Hochschule, Berlin. Hevly notes that Schade "agitated for an independent Navy technical mission to scour liberated Europe for devices that might aid in the Navy's Pacific war." Hevly (1987), p. 102. See also Schade (1946), p. 83; Bush (1970), p. 317; Lasby (1975), p. 22; and Bower (1987), p. 70.

21. Bower (1987), pp. 76–77.

22. Goudsmit (1947), p. 26, states that the code name Alsos was Greek for "Groves." Leslie Groves was the military head of the American Manhattan Project.

23. Goudsmit (1947), p. 13. Walker (1989), pp. 153–65, 204–21, has shown that Alsos teams found by the spring of 1945 that the Germans were incapable of producing the bomb, even though Goudsmit continued to exploit the rationalization. Both Walker (1989) and Goldberg (1989) provide insight into how the ideologies of Allied scientific intelligence units shaped their impressions of Germany's capability to produce a bomb. Useful to the present discussion, however, is the fact that until the spring of 1945, members of these units did fear a German atomic surprise. Astronomer Gerard Kuiper, a member of Alsos's scientific team, wrote from Europe to a colleague at the Radio Research Laboratory: "One is again surprised to see quotations from U.S. senators who think that the war will be over 'within a few days'." It would be wiser to worry about the chance we still have of losing it if certain high explosives are developed in time. This possibility may, incidentally, be one reason the Germans are not giving in. Few people here expect an immediate collapse." Kuiper to F. E. Terman, 13 March 1945. Alsos folder, KP/UA. On the fear among American physicists that they were in a race with the Germans for the bomb, see Weart (1988), pp. 83, 92ff. See also Ordway and Sharpe (1979), p. 107.

24. Consisting of elements of Army and Navy intelligence, Alsos was also a part of the Office of Scientific Research and Development's (OSRD) Office of Field Service, established in October 1943, which maintained responsibility for the scientific side of the mission. Baxter (1946), p. 413; Lasby (1975), pp. 16–17; and Walker (1989), p. 153.

25. Goudsmit was a professor at the University of Michigan until the United States entered the war, whereupon he moved to MIT's Radiation Laboratory. On Goudsmit becoming Alsos scientific head, see Goldberg (1989).

26. Kuiper since July 1943 had worked on flack systems to confuse enemy radar. See F. E. Terman to Otto Struve, 4 December 1945. SP/AIP. See also Kuiper to Dirk Brouwer, 30 January 1944. KP/UA.

27. See Gerard Kuiper to Otto Struve, 22 June 1944; 7 August 1944; 19 October 1944; 10 December 1944. SP/AIP.

28. S. A. Goudsmit, "Excerpt from Intelligence Report 20 November 1944," JX/WP 31 of "Combined Wave Propagation Committee of the Combined Communications Board." J. H. Dellinger file, "Rockets" folder, NBS/NARA. This contained Gerard P. Kuiper's report, "Rockets, Ionosphere and Stratosphere Research," Combined Intelligence Objectives Subcommittee, Item 4, File IX-3, 20 November 1944. Copy in NASM file "Captured German and Japanese Records," Air Document Index (TECH), and from J. H. Dellinger file, "Rockets" folder, NBS/NARA.

29. Goudsmit, "Excerpt from Intelligence Report 20 November 1944," p. 2. Parenthetic statement appears in Kuiper's original 20 November 1944 report.

30. The connection between the V-2 and radio control, indeed the idea that the V-2 was radio-guided, stemmed from intelligence gathered from the crash of an A4 in Sweden in September 1944. This particular example had been fitted with the Wasserfall guidance system for tests. See Jones (1978), p. 431; Dornberger (1958); Kennedy (1983), p. 37.

31. H. F. Schenk to the JCB, "Rockets, Ionosphere, and Stratospheric Research," January 1945 Report JX/CM 54; JX/WP 31, which transmitted S. A. Goudsmit, "Excerpt from Intelligence Report 20 November 1944"; Gerard P. Kuiper report "Rockets, Ionosphere and Stratosphere Research," Combined Intelligence Objectives Subcommittee, November 1944. Copy in T. H. Dellinger file, "Rockets" folder, NBS/NARA.

32. Much of Kuiper's early conclusions were based on third-party interrogations, and turned out

to be not reliable. See correspondence between Kuiper and Heinz Fischer, circa May 1952. KP/UA.

33. See George R. Eckman (Lt. Col. MI) to Maj. R. A. Fisher "Interrogation of Dr. Ernst Lübcke" 27 April 1945; Report 1045-MR-15 June 16, 1945, OSRD; Kuiper to Major Fisher, "The Future of German Science," 30 June 1945; "Comments on German War Research in Physics," and "German Papers on Uranium," 6 May 1945; Kuiper to Doris Gilbert (Div 15 OSRD) Paris, 25 May 1945; and letters, F. G. Houtermans (in Göttingen) to Kuiper, 9 July 1945; 12 June 1946. Chron. files and Alsos folders, KP/UA.

34. Kuiper added that he was bringing back a Nazi flag and a whip from a concentration camp near Göttingen to remind him of the conditions he saw there. Kuiper to Struve, 6 June 1945. SP/AIP.

35. See undated rough draft, Kuiper "German Solar Research in connection with the Ionosphere," n.d., circa August 1945. See also: Gerard P. Kuiper, "Report to Scientific Chief, Alsos Mission, HQ, G2, USFET on Solar and Ionospheric Research in Germany," 25 September 1945. Alsos folder, KP/UA.

36. W. R. Piggott, "Memorandum," 1 October 1945. Alsos folder, KP/UA. Regener's quartz barograph was later described as "An ideal instrument, particularly for the measurement of low pressures at high altitudes." Koppl (1947), p. 850.

37. See Kuiper (1946). Kuiper's summer letters reported only on European colleagues and their politics. See, for instance, Kuiper to Otto Struve, 6, 7, and 18 June 1945, and similar letters through the summer to Henry Norris Russell (8 June 1945) and to Struve. See also Struve to Bart Bok, 8 August 1945. SP/AIP. On what Alsos members could reveal and what was restricted, see Alan T. Waterman (OSRD, Chief, Office of Field Service) to Alsos members, 29 December 1945, p. 2. Alsos folder, KP/UA; and Goudsmit to Scientific Members of Alsos, re "Alsos Reports," 3 July 1945. Alsos folder, KP/UA.

38. In August 1945, Menzel had argued that "if the U.S. [was] to maintain pace with other nations in the field of the relationship between the Sun and communications," the military services had to continue to support coronal studies. Menzel to Op-20-G, Office of Naval Research, 7 August 1945. HUG

4567.5.2, DHM/HUA; Kuiper to Struve, 12 November 1945. SP/AIP. See also D. H. Menzel to Capt. H. T. Engstrom, n.d., circa 1945; Menzel to L. F. Safford, CNO, 26 June 1941. HUG 4567.11, DHM/HUA.

39. Menzel to E. O. Salant, 20 February 1946. DHM/HUA.

40. See Struve to Menzel, 15 February 1946. SP/AIP; the quotation is from Menzel to Leo Goldberg, 20 February 1946. LG/HUA.

41. Most attending the American Astronomical Society meetings wanted to know more about the fate of colleagues and institutions in Germany, but at the same time, as Struve reported back to Kuiper, "some of your Dutch compatriots raised a lot of fuss because they thought the wording of some of the sentences was Pro-German." Struve to Kuiper, 30 January, 7 February 1946. SP/AIP. Willem Luyten was annoyed, along with Bart Bok, Jan Schilt, and the Belgian George Van Biesbroeck.

42. CIOS was created to deal with problems of interservice and international competition in the scramble to acquire intelligence; it combined elements of all the military arms of the United States, the U.S. State Department and the OSRD, and parallel agencies in Britain, including the Admiralty, Air Ministry, and Aircraft Production. Lasby (1975), pp. 18–19. On interservice competition, see p. 17. A lucid overview of Allied intelligence teams can be found in Gimbel (1986), pp. 435–37.

43. See Fritz Zwicky, "Report on Certain Phases of War Research in Germany" (Pasadena, California: Aeroject Engineering Corporation, 1 October 1945). Air Materiel Command, Summary Report F-SU-3-RE, released January 1947 as a restricted document. See also Lasby (1975), pp. 35–36.

44. Wernher von Braun, "Survey of the Development of Liquid Rockets in Germany and Their Future Prospects," prepared text reprinted in Fritz Zwicky, "Report on Certain Phases of War Research in Germany" (Pasadena, California: Aerojet Engineering Corporation, 1 October 1945), pp. 66–72, 68.

45. Zwicky, ibid., p. 8. On von Braun as a system builder, see Neufeld (1991).

46. See Zwicky (1946, 1947a, 1947b).

47. Zwicky (1946), p. 261.

48. Zwicky (1947b), p. 65. This pronouncement must have been maddening to Zwicky's colleagues at Caltech, who were just then readying the 200 inch telescope for operation.

49. See, for instance: oral histories, Dorrit Hoffleit (SHMA/AIP); Ernst Krause (SAOHP/NASM); Milton Rosen (SAOHP/NASM). See also R. O. Redman to Otto Struve, n.d., early 1945. SP/AIP.

50. Richard Silberstein, "A Proposal for the Use of Rockets for the Study of the Ionosphere." 25 February 1945. Willy Ley Collection, "Rockets-Second World War A-4 Operational Use," NASM technical files. Also provided by R. Silberstein, now of Boulder, Colorado, with supporting documents.

51. Ibid., p. 2.

52. Silberstein recalls, however, that his boss, Newbern Smith, told him weeks later that the board read the proposal, but did not act on it, because it thought the ideas were not new. This was indeed the impression created by Kuiper's report, which was received by the board through Goudsmit just after 1 January 1945. See Silberstein OHI. Smith later represented the National Bureau of Standards on advisory panels that coordinated the American use of German V-2 rockets for science, but never took an active or leading role.

53. Krause OHI, added commentary, pp. 46–47rd. Hevly (1987), p. 102, notes that these intelligence reports played a role in NRL's planning for postwar years, as Milton Rosen, a member of Krause's staff, argues in his *Viking Rocket Story*. See Rosen (1955), pp. 17–18, and Rosen OHI, p. 47rd; Newell (1980), pp. 33–34.

54. W. B. Klemperer (Douglas Aircraft Research Laboratory, Project Rand) to T. H. Johnson, 11 December 1945, quoted in R. Ladenburg memorandum, 8 January 1946. Minutes of 16 January 1946 meeting of the V-2 Rocket Panel, Burnight notes, MAT/LC. C. V. Strain, "Record of Consultative Services," 16 April 1946. NRL/NARA.

55. Phone conversation, 12 February 1946 "Record of Consultative Services," E. Krause with R. W. Porter and Mr. McAllister, G. E. NRL/NARA.

56. J. Allen Hynek to Jesse Greenstein, 24 April 1946. JG/CIT.

# 5

# Organizing for Space Research

In the fall of 1945, enough parts to construct some 100 V-2 missiles were shipped to the White Sands Proving Ground, then under construction. At the same time, the mountains of documentation on guided missiles that had been discovered were being organized and were awaiting the assistance of von Braun's rocketry experts, now heading for Fort Bliss in Texas (figure, p. 60). Army Ordnance's overall plan for building a tactical ballistic missile system under Project Hermes was now gathering steam, but it still had to conform to the budgets and policies set down by the Joint Chiefs of Staff.[1] That policy emphasized interservice cooperation, which to some may have seemed an intrusion on the autonomy of the services, but to Holger Toftoy it meant an opportunity to seek creative means to fill in gaps that Project Hermes could not cover.

## The Military Context: Toftoy's Vision

On 30 October 1945, in accordance with a Joint Chiefs of Staff (JCS) directive, Army Ordnance invited the Navy to inspect its new launching facilities at White Sands and offered the facility for test firings of Navy missiles on a cost-sharing basis.[2] At about the same time, Toftoy asked one of his assistants, Col. James B. Bain, to coordinate interservice activities within Ordnance's Rocket Development Division, which included the use of the V-2s for scientific research. By the end of 1945, Bain found a number of groups in the Army, Navy, and in their various contract laboratories that wanted to use rockets to explore the upper atmosphere. He met with members of the Jet Propulsion Laboratory,

with Navy groups in the Bureaus of Aeronautics and Ordnance (BuAer and BuOrd), and with groups in the Signal Corps. Bain had to work fast because the Army planned to begin firing the first V-2 missiles in the spring of 1946 and to finish by early 1947.[3] He told them all that the first static test of a captured V-2 rocket was scheduled for February 1946 and that Army Ordnance was planning to arrange for "a number of instrumentation tests . . . during the V-2 test program." If they wanted to get involved, they were invited to a joint conference sometime in January.[4]

Toftoy wanted all parties interested in using rockets for upper air research organized into one coherent and manageable body, separate from Ordnance but subject to its review. Army Ordnance, as with the Army Air Forces' Air Staff at the time, was far from comfortable with the prospect of entering "the rarefied atmosphere of fundamental science" and thought a surrogate body, represented by the single voice of a civilian panel, should conduct basic science with rockets.[5] The Navy, in contrast, was farther along in the management of basic research in late 1945, but lacked the rockets. Through its newly established Office of Research and Inventions (ORI), the Navy used surplus funds to support basic research in the absence of a national policy.[6] All that was needed was a mechanism to bring the Army, Navy, and their scientific contractors together in the warhead of a rocket. Bain was charged with creating that mechanism.

What Toftoy set in motion was a part of the Army Ordnance vision of how to develop its ballistic missile capability. In October 1945, Army Ordnance chief of research and development, Maj. Gen. Gladeon M. Barnes, elu-

Just as he had for the Wehrmacht and the SS, von Braun advised Army Ordnance in America. Here von Braun is with Col. Holger Toftoy at Fort Bliss. NASM SI A4075A.

cidated Army Ordnance policy at a JCS interservice conference on ballistic missile development: "The policy of the Ordnance Department in guided missiles development is to place contracts with suitable commercial concerns and scientific institutions . . . rather than to attempt to build and staff new laboratories of its own."[7]

At a JCS conference in June 1946, Toftoy elaborated on this policy, arguing that Ordnance should not rush into the production of a V-2-type missile with a too naive or limited design: "More progress will result from stressing basic research and component development for the next 5 years or so, than by hurrying into large-scale production improved existing missiles or early models of promising new weapons."[8] Sensitive to considerable resistance within Army circles, both Toftoy and Barnes stressed a developmental program that included basic research; in October 1945, Toftoy identified both the short- and long-term goals of Project Hermes:

> 1. *To gain experience in the handling and firing of high-velocity missiles. This will include learning what to do as well as what not to do in designing rockets of this nature.*
> 2. *To check certain research data now available and to conduct new high-altitude research such as radio and radar transmission.*
> 3. *To provide a vehicle for testing our own control apparatus.*
> 4. *To assist in the development of countermeasures, and*
> 5. *To expedite the development testing of instrumentation required in our development* [*of a Hermes missile*].[9]

It was not hard to see how upper atmospheric research along the lines Regener had pursued could be incorporated into this agenda. In June 1946, with the scientific groups already organized and operating, Toftoy was more explicit:

> *The primary aim of the Ordnance rocket program is to produce free rockets and controlled missiles that meet military requirements, but the field is so*

*new we are placing emphasis on fundamental and basic research. Our program has been planned not only to provide the necessary basic knowledge from all fields of science and apply this knowledge to missile design, but also to conserve personnel and funds and to prevent unwarranted duplication.*[10]

In almost every public statement he made in 1946 and 1947, Toftoy rationalized his overture to scientific groups to exploit the V-2 as a means not only to build better missile systems, but to keep scientific groups together and actively involved in some form of rocketry. As did others, like Army Air Force General Hap Arnold or Navy Admiral Harold Bowen, Toftoy, supported by Barnes, wanted to establish a permanent relationship between civilian scientists and defense research.[11] While Arnold, Bowen, and their successors and civilian counterparts argued interminably over who would control defense research, Toftoy, like the Manhattan Engineer District's legendary commander, Gen. Leslie R. Groves, or the Naval Research Laboratory's new superintendent, Comdr. H. A. Schade, acted to ensure that well-defined and immediate research needs were met.[12] This was the context within which James Bain formed the V-2 Rocket Panel in early 1946.

## Interservice Conferences: Looking for a Ride

On 4 January 1946, Bain brought together representatives from the Navy's Bureau of Aeronautics, the Army's Ballistic Research Laboratory, the Signal Corps, and scientists from Princeton and Harvard. On 7 January, he held another meeting at NRL with Ernst Krause's new Rocket Sonde Research Section. At both meetings, Bain outlined the Army's firing program for the V-2 rockets, which by now included static tests in February and 25 launches starting in April, with approximately five rockets to be launched per month. Although parts for up to 100 V-2 missiles had been shipped, the Army found that it could not construct more than 25 missiles because many critical components were missing or were damaged.[13]

At the 4 January meeting, after Bain described the Army's program, scientists and their representatives identified various institutional needs and personal interests. Thomas H. Johnson, a cosmic-ray physicist who was chief of the Ballistic Measurements Laboratory of the Ballistic Research Laboratory (BRL) at Aberdeen during the war, and who had brought the existence of Erich Regener's instruments to the attention of his colleagues, outlined how BRL would determine rocket trajectories to assist those who were interested in flying scientific instruments. Johnson voiced no personal scientific interest in the V-2, whereas others, like senior Princeton physicist Rudolf Ladenburg, did. The various agency and university representatives told Army Ordnance that they were interested in a wide array of problems, including the nature of cosmic rays, the ultraviolet spectrum of the sun, and the density, pressure, temperature, and composition of the high atmosphere, along with its dust content. The techniques that would be used to attack these problems included the use of artificial meteors and radio pulse techniques for probing the high atmosphere and infrared recording techniques. Military representatives from the Army Air Forces and the Signal Corps added that they wished to gain experience in the ground tracking of missiles, in missile operations and performance, in the development of rocket and antirocket devices, and in the guidance and control of missiles.

The scientists at the 4 January 1946 meeting were excited by the possibility of using the V-2 rockets. Military officers, in contrast, saw problems in the merger. The Signal Corps representative suggested that 20 V-2 rockets should be fired quickly to satisfy the immediate needs of Ordnance, and then the remaining five could be dedicated to scientific purposes. He did not want to see the White Sands program delayed or the scientists rushed in their studies of the rockets on the one hand, or the atmosphere on the other: "there is great difficulty to keep personnel together at White Sands for a long period of time," he argued, but he also knew that it would not be possible "to develop and manufacture any [scientific] instruments in a few months."[14] Also, some of the V-2 parts were quickly deteriorating in the desert sands and had to be used soon.

Bain pointed out that the Army's time and funds were limited, so the firing schedule had to be "completed as soon as possible . . . by November 1946." The plan therefore was to test-fire the first 20 rockets on the Army's schedule. Scientific instruments would be flown on an "opportunity" basis—that is, once the instruments were ready, they would be installed on the next scheduled round.[15] At the close of the meeting, Bain asked that anyone interested in using the V-2 rockets write to him by January 15, "outlining their plans, space and weight required, etc."

Although some of the military representatives were skeptical of the plan, the scientists, led by Ladenburg and Harvard College Observatory Director Harlow Shapley, remained enthusiastic. While Shapley provided only rhetoric, Ladenburg offered concrete proposals in his capacity as representative for both Princeton and the Applied Physics Laboratory (APL).[16]

Both APL director Merle A. Tuve and Ladenburg had been aware since early December of what Barnes and Army Ordnance were planning and were anxious to see APL or Princeton, its major telemetry contractor, take part in the rocket firings.[17] Since early 1945, through its anti-aircraft missile contract (Project Bumblebee), APL had been involved in missile development. Both Tuve and Ralph E. Gibson of APL chaired technical panels of the Joint Chiefs of Staff Guided Missiles Committee, and so they were aware of the Army's deliberations. Ladenburg hoped that younger Princeton staff, like Myron Nichols or John A. Wheeler, might become involved. Nichols in particular, an assistant professor of physics and specialist in thermionic techniques, had been the liaison between Princeton's Palmer Physical Laboratory and APL and was anxious to maintain its vigorous telemetry development program.

On 7 January, Bain repeated his review of the Army's firing schedule for the benefit of Krause and other interested NRL staff, and Krause then outlined NRL's hopes and plans.[18] Krause also offered NRL as the site for a general meeting where Bain would finalize plans. It would take place one day after Bain's deadline for proposals, on 16 January.

## Establishing the V-2 Panel

Forty-one people from 12 institutions attended the 16 January organizing meeting, but only a few voices dominated the discussion.[19] By the end of the meeting, the restrictive plan of reserving five rockets for scientific instrumentation out of the first 20 or 25 had been scrapped, and a "V-2 Rocket Panel" was established to coordinate the scientific use of all launches. Although Bain was the center of attention and made the ultimate decisions, Comdr. J. S. Warfel of the Navy's Office of Research and Invention and Ernst Krause of NRL eventually led the discussion, which soon became bogged down in a morass of conflicting interests that took more than a month to unravel.

At the outset, at Bain's suggestion, all agreed on the need for some coordinating body. A small panel of participating scientists would oversee all activities related to the scientific use of the V-2s and would report directly to Bain. The next task was to decide who would be on the V-2 Panel and who would direct it. It was here that Ernst Krause made his bid, and that the proceedings became murky.

Before the 16 January meeting, Krause and Schade convinced ORI that NRL should be "coordinator of all of the Navy's program . . . in the physics of the upper atmosphere."[20] APL, however, had already received the endorsement of its parent body, the Bureau of Ordnance, to "look into the situation" of upper atmospheric research, and at the meeting Warfel added that the "Navy is interested in setting up a scientific board for the overall coordination of upper atmosphere studies." Possible candidates for the board included NRL and APL, by far the two largest laboratories represented at the meeting. James Van Allen of APL was making plans to be involved; he had contacted Toftoy and was supported by Merle Tuve.[21] But Van Allen was still serving out his final months of military service and would not be able to build up a team for some months.

Krause, however, had already secured NRL's commitment to support the overall program. At the meeting, Krause established NRL as a provider of facilities in concert with Army Ordnance and demonstrated that his institution was in a better position than APL, Princeton, or any of the others to take the lead. They had defined their mission, had obtained substantial institutional support, and were poised to coordinate the research effort.[22] Nonetheless, Bain did not designate anyone to lead at the 16 January meeting. He had originally wanted General Electric to handle the coordinating functions, but at the time General Electric was trying to cope with a major strike.[23] Eventually, the question of where to base operations blended into the question of who would chair the panel and which institutions would be members. Membership and the chairmanship remained open at the conclusion of the meeting.[24]

In spite of Bain's hesitation, Krause quickly assumed a leadership role. He prepared a report on the 16 January meeting and by the next meeting, nine days later at Columbia University, was able to announce that NRL would provide telemetry services and American-built warheads, if needed. Bain called the second meeting for 25 January and invited only the principals (Krause, Myron Nichols of Princeton, Shapley, and K. H. Kingdon and Volney Wilson from General Electric), charging them to resolve several issues: how panel members might exploit the V-2 warhead space, what the ultimate official panel would look like, and what powers it would have.[25]

Building on sentiment at the 16 January meeting, Krause argued that the panel should have responsibility for the "complete equipping and assembly of warheads" and would decide when each group's experiment would fly, within the Army's schedule. Nichols insisted that other scientists not affiliated with the panel be given a chance to participate, although outsider requests would remain under panel authority. He added that the "war heads are not considered to be exclusively assigned to the work of the responsible laboratories."[26] Everyone at the meeting eventually agreed that each group of experimenters would be responsible for the complete warhead for specific flights and that others who wished to participate would work through the designated "official" group. In like manner, any group not represented on the panel would have to work through one of the panel members as agent.[27]

Although firings had not yet started in New Mexico, Kingdon reported at this second meeting that the V-2 Panel could expect at least 10 missiles, which could be divided equally between NRL and Princeton. But GE could not yet provide the 30-channel telemetry systems it had promised and turned to Krause, who repeated his offer of 25 to 30 10-channel pulse-telemetry devices, developed during the war for JB-2 pilotless aircraft, an American version of the German V-1.[28] The remainder of the meeting was devoted to each institution's plans for equipping its V-2 warheads and what services each required or could provide. On this more collegial note, the still unofficial V-2 Panel decided to meet monthly during the first several months of V-2 launches and again asked General Electric to supply a technical aide as soon as possible; meanwhile, coordination remained within Krause's group.[29]

In the weeks following the 16 and 25 January meetings, Krause made significant progress in centralizing both facilities and institutional support at NRL. He had part of his group begin preparing the telemetry units, negotiated with the Naval Gun Factory to build new warheads, and worked with his ORI patrons to plan on-site facilities at White Sands.[30] Thus Krause became the person that Toftoy and Bain happily referred to for all V-2 Panel matters, especially as new members were proposed by service and government agencies. Bids from the National Advisory Committee for Aeronautics (NACA), from Purdue University, and from Caltech were all

passed to Krause. Between 25 January and the first official V-2 Panel meeting on 27 February, the panel grew as others made their case with the Army. Krause soon learned that the NACA did not wish to be a participating member but wanted to have an observer attend V-2 Panel meetings.[31]

Even so Krause felt the membership was getting out of control, and so he acted to confirm NRL's leadership of the V-2 Panel before its first official meeting.

## Competition and the Growth of the V-2 Panel

Krause's fears over the ultimate complexion and governance of the V-2 Panel deepened in mid-February. On 11 February, Krause learned that the Army Air Forces had contracted with the University of Michigan to engage in upper air studies, and that Dr. William G. Dow would be Michigan's and the AAF's representative on the Panel. The following day, Krause learned from Bain that Marcel Golay of the Signal Corps Engineering Laboratory in Bradley Beach, New Jersey, would represent Signal Corps interests.[32] And by February, Krause knew that APL would soon enter as a major player.

When Van Allen returned to civilian life in mid-February, APL made its bid for Panel membership. Van Allen's nascent High Altitude Research Group would work on problems in cosmic-ray physics, solar spectroscopy, and properties of the upper atmosphere. With Navy Bureau of Ordnance backing, APL also decided to become, along with NRL, a leading provider of services, including telemetry, instrument integration, and an all-new American-made sounding rocket capable of economical access to the upper atmosphere, later named the Aerobee. APL thus competed with NRL for leadership of the V-2 Panel. Predictably, there was some initial resistance from NRL at first.[33]

Krause felt that the Army was compromising its own goal of creating an effective and efficiently organized pool of talent to exploit the V-2 missile payload space. He thus petitioned ORI to keep APL off the panel by establishing clear priorities for rocket procurement and research that favored NRL.[34] APL, Krause contended, wanted to build its own launch vehicle and therefore was "not particularly [interested] in getting into the V-2 program, so it is possible that they would not be interested in the present program."[35] This was, of course, not true. Krause and NRL knew that Tuve had managed to snag a complete V-2 rocket, warhead and all, and were studying it at APL's Silver Spring, Maryland, site in anticipation of using ones like it at White Sands.

Krause's posturing before ORI was a response to pressures he feared would threaten his entire program. He had committed NRL to providing both telemetry and warheads and was evermore bogged down in the details of organizing the panel because GE could still not supply services or a full-time secretary for its use.[36] At ORI, Krause was ready to barter; if APL were to become a member of the panel, it would have to supply something as well. What Krause wanted was a part of APL's domestic rocket development program.

NRL had its own plans to build a large sounding rocket, later called the Viking. But APL was already far ahead with its more modest plan to contract with the Aerojet Engineering Corporation, in Pasadena, California, for a small liquid-fueled sounding rocket capable of sending 100 kilograms to 100 kilometers. Whatever the motive for developing the Aerobee (a combination of the names of its provider, APL's Project Bumblebee, and its maker, Aerojet), it was to be a simple vehicle for sending payloads to heights comparable to those V-2 rockets could reach. As such, it was a highly desirable vehicle.

Krause learned that the APL would receive its first Aerobee in September 1946, and 20 by February 1947, at a cost of $20,000 each. Knowing that the Aerobee would therefore be flying years before the larger and more complex Viking, Krause asked ORI to allow NRL to contract for Aerobees for its own use. This was a reasonable goal for Krause, because at the time only 25 V-2 missiles were being constructed out of the parts shipped to

White Sands. And given the present firing schedule, they would be expended by the end of 1946 or at the latest by early 1947. To establish a viable research program, both the NRL and APL groups had to have something to fly their instruments on, or at least they had to be sure that when the V-2s did run out, something acceptable would be ready to take their place. Obviously interested in having a reason to procure a piece of APL's Bureau of Ordnance-funded Aerobee, ORI was happy to grant Krause's request, so that "everybody gets a square deal." ORI would fund NRL's new rocket and buy Krause's group some of the Aerobees now under development.[37]

Between 13 and 27 February, Krause's group met a number of times with Van Allen's APL group and established a working relationship acceptable to ORI, BuAer, BuOrd, and the Army. The basic agreement was that APL could join the panel, and NRL would get at least 10 Aerobees. APL even agreed to allow NRL to share in some of the first Aerobees off the assembly line, if the V-2s ran out as early as January 1947.[38] When ORI learned about APL's offer, it told Krause to work closely with Van Allen to procure the Aerobees as soon as possible. ORI's and APL's willingness to get NRL involved was understandable; they both would benefit from an increased order for Aerobees, as this would drive down the unit price, which in fact did end up at some $18,500 per vehicle. Although NRL-APL relations improved after these negotiating sessions, the process was a long one and created competitive tensions that would last for some time.

By late February, not everything had been ironed out, but a great deal of communication had taken place between the principal players. Myron Nichols of Princeton openly accepted APL's dominant role, and several outside agencies and interests, such as the NACA, the Signal Corps, and the Army Air Forces were also accommodated. APL and NRL were cooperating—to the extent that their patrons, BuOrd, BuAer, and ORI allowed them to—and so all was ready for the first official meeting of the panel.

While the Army and its General Electric Hermes contractors feverishly worked at White Sands to ready the first static tests, 15 people met at Princeton University on 27 February to formalize the V-2 Panel. They had to elect a chairman and ratify decisions made during the previous month about how they would organize themselves.[39] They also learned more about the construction of the warhead, argued over assigned space on the first round of rocket firings scheduled for April through June, and critiqued each other's research programs.[40] NRL, APL, GE, NACA, the Signal Corps, the University of Michigan, Wright Field, the AAF, and Harvard were now represented. And General Electric could finally provide a part-time technical aide and secretary, George K. Megerian.

At the outset of the meeting, Krause was unanimously elected chairman and was assigned both technical and administrative responsibilities. NRL, APL, GE, Princeton, Harvard, Michigan, and the Signal Corps were designated active voting members of the V-2 Panel, whose functions were then twofold: "To advise [a] super-advisory panel on matters relating to technical phases of the tests," and "To supervise the design and construction of the necessary scientific equipment."[41] The "super-advisory panel," yet to be formed by Army Ordnance, became the agent that decided how many V-2 missiles would ultimately be constructed and made available to the V-2 Panel.

Krause announced formally that NRL would supply both warheads and telemetry for at least the first 25 flights. The first six missiles were to meet General Electric's needs. The remaining four panel members received five berths each: one in every fifth launch. Even before any one of its members had instrumented a single rocket, the panel asked Army Ordnance for a second round of 25 V-2 launches over a longer interval of time, if the Army found a way to reconstruct more missiles. It also hoped that these test flights could be designed to ensure, to the extent possible, that their instruments would survive, that telemetry channels for scientific information would take priority over house-keeping telemetry, and that adequate ground-based communications and tracking facilities would be in place.[42] These were the types of concerns the panel repeatedly addressed throughout its life: making sure that steps were taken to provide the technical and administrative infrastructure required to conduct scientific research with rockets properly.

The 27 February 1946 meeting constituted the V-2 Panel's first official meeting. Panel members were in place, and, save for astronomer Fred Whipple, who was designated Harvard's representative, they were engaged in designing and building instruments. Although none of the military laboratories had as yet identified specific funding, NRL, APL, and the Signal Corps were able to carry out their initial plans with general funding sources left over from wartime budgets and knew that their patrons—such as ORI, BuAer, or BuOrd—were making plans to establish funding lines. Michigan was already contracting with both the Signal Corps and with the Army Air Forces for this activity, and Princeton was able to fund its work as part of APL's Bumblebee program. At the outset, then, funding was not a major concern for most panel members.

## Planning for the Science

While the representatives on the V-2 Panel were establishing the basic channels through which the panel would operate, members within each group were planning and preparing devices to fit into the V-2 warheads. The Army had slipped to early March the first static tests of a V-2 at White Sands, with actual flights starting either later that month or in April, so the pressure was still on to have instrument packages ready.

The same problems that Regener had faced now had to be tackled by the American scientists. Warheads had to be constructed at the Naval Gun Factory, and instruments had to conform to them, as well as to NRL's new telemetry systems, all yet unbuilt. And eventually, the warheads—with their instrument packages, battery power systems, and telemetry systems—all had to be attached to the V-2 missiles at White Sands. Conducting scientific observations with these vehicles required far more than the infrastructure of a typical college laboratory. It would take a carefully coordinated and vertically integrated system of services to produce a fully ready warhead, consisting of calibrated and tested instruments that could be turned over to a group of Army contractors who would then stick it on top of a missile bound for the uppermost regions of the atmosphere. Army Ordnance had to inform the scientists of its constraints and deadlines for work at White Sands, and each scientific group had to communicate its needs to the Army. Coordination fell to the V-2 Panel.

The panel supervised and promoted a research agenda that itself presented little controversy. Most of the scientific goals set out at the first planning meetings in January 1946 were being pursued by panel member groups with interests in radio propagation in the ionosphere; atmospheric composition, pressure, temperature and density; cosmic-ray studies; atmospheric absorption; sound propagation; solar ultraviolet studies; and various types of biological studies.[43] Panel members discussed their own instrumentation plans and critiqued those of their colleagues. Through late February and into the spring, Nichols' physics colleagues at Princeton, directed by John A. Wheeler, designed sophisticated shielded arrays of Geiger counters to identify the character of cosmic-ray particles before they hit the earth's atmosphere. NRL's Rocket Sonde Research Section developed similar detector arrays and started to plan for cloud chambers and photographic emulsion stack flights. They also built gas-sampling devices for the warhead, temperature sensors for the warhead skin, stagnation pressure gauges at the nose, and a wide array of radio wave propagation experiments, including X-band transmission through the rocket exhaust. On top of this, NRL was also preparing the telemetry systems, and a group within NRL's Optics Branch, headed by Richard Tousey, was hard

at work on its first spectrographs, as were J. Allen Hynek, J. J. Hopfield, and Harold Clearman at APL. Similar activities were taking place at Michigan and in the Signal Corps labs, as all rushed to ready their instruments for the first instrumented flights.

Panel members knew that there was much duplication of goals. Clearly, they were entering new and uncharted territory, and the scientific goals were new to all of them. What each panel member brought to the effort was a specific technical expertise required to meet those goals. Each would adapt familiar laboratory styles and experiences to develop designs for rocket instruments for measuring cosmic-ray flux, solar ultraviolet intensities, or ion densities. Iterative improvements in instrument design could be based on knowledge gained collectively, using the V-2 Panel as a forum for comparison. Sophisticated technical goals, such as possible Regener-type ejection mechanisms for both the retrieval and exposure of various experiments, were as yet premature for most panel members, mainly because there was too little time before the first firings to design and construct such devices.[44]

At the formative meetings in January, military observers felt compelled to identify the types of data they would like to see obtained; at one point, a Dahlgren representative seemed to feel that he had to convince the civilian scientists that long-range ordnance would benefit greatly from a better knowledge of the densities and pressures in the upper atmosphere. But as each panel member articulated his group's goals, the military observers soon realized that the scientists needed little prodding to make these connections. There were few instruments or experiments valued by the military representatives that the scientists had not already considered. A particularly astute representative from the Signal Corps, Lt. Col. Harold Zahl, was the first to express complete confidence that his contractors, as well as those for the Navy, more than satisfied then known Signal Corps requirements. Military goals had indeed become scientific goals in the warhead of a V-2 missile.

# The Jet Propulsion Laboratory Symposium

As the V-2 Panel was preparing itself to be Army Ordnance's coordinating agent, Krause and other panel members received an invitation from Caltech to attend a four-day Guided Missiles and Upper Atmosphere Symposium, to be held from 13 to 16 March. This was the first meeting to explore upper atmospheric research within the context of missile development. Technical sessions on aerodynamic theory, fundamental research on combustion and gas dynamics, and the launching of guided missiles complemented sessions on the physics of the upper atmosphere. After three days of formal sessions, a fourth day of roundtable discussion groups addressed technical problems of building instruments for upper atmosphere research vehicles, means of returning data by physical recovery or radio, techniques for slow descent by parachute, and the relative effectiveness of balloons versus rockets as vehicles.

The symposium had actually been an afterthought; the Launching, Aerodynamics, and Ballistics Panels of the Guided Missiles Committee of the Joint Chiefs of Staff were meeting at Caltech, a leading Army Ordnance contractor, and members of Caltech's JPL staff, such as William H. Pickering, thought that this was a good chance to explore issues surrounding a fully integrated national scientific program combining military, government, industry, and the academic community.[45] Through Clark B. Millikan, acting chairman of the JPL Executive Board, Pickering had little trouble securing the participation of prominent Caltech and University of California at Los Angeles professors in the geosciences and atmospheric physics, including Joseph Kaplan, J. A. Bjerknes, H. D. Babcock, and Beno Gutenberg.[46]

William Pickering, as coordinator for the scientific sessions for the March symposium, visited NRL, APL, and BuAer in February 1946 to invite them to Pasadena. Krause quickly agreed to attend because he was then planning to have one or two members of his staff spend several months at Caltech to study techniques in rocket development, and this would be a good chance to make contacts. It would also provide Krause a chance to describe NRL's programs to an important audience.[47] Van Allen harbored parallel interests, of course. APL was planning for the Aerobee through the services of Aerojet and Douglas Aircraft, both close to JPL. And Van Allen also saw the symposium as an important opportunity to air APL's ambitious plans for the upcoming V-2 flights.

As a result of his training in balloon-borne cosmic-ray research under Robert Millikan and Victor Neher at Caltech in the 1930s, Pickering had a desire to exploit JPL's 4.8-meter-long WAC Corporal missile for upper atmospheric research. The first WACs fired in 1945 and 1946 could send 11 kilograms of payload to 50 kilometers.[48] The March symposium thus was intended as a sounding board for the many who were interested in such research, as a means of advertising his own plans for the WAC Corporal, and as a professional and institutional mixer for those engaged in rocket development. It brought JPL, Pickering reminisced in 1972, "into contact with an equally broad spectrum of potential scientific users," as well as with the "R&D groups of Southern California's aircraft industry" and "various armed-service research groups."[49] Indeed, the JPL symposium allowed Krause and Milton Rosen to become familiar with Caltech and JPL, where Rosen would spend six months planning for Viking. And it brought together people who were soon to deliberate over the conduct of upper atmospheric research with rockets on the Joint Research and Development Board of the Joint Chiefs of Staff, or within various NACA committees. Unfortunately, the symposium did not increase interest in using the WAC, nor, as we shall see, did it provide Pickering with the institutional backing he needed to engage directly in upper atmospheric research.

Pickering's failure to make JPL a charter player on the V-2 Panel highlights the importance of institutional commitment to what became a new way to conduct scientific research. Panel members representing active groups, each backed by a specific institution or agency, still needed a coordinating lobby to secure space in missile warheads, telemetry services at White Sands, and ballistic tracking data from the Ballistic Research Laboratory. The V-2 Panel became that lobby, but it was also given the authority to decide who would fly what and when, subject only to the Army making the V-2 berths available. It was thus both a coordinating and communicating body and a locus of political power. It also became, by design, the organizational site for an emerging technical culture that embodied, in the early NASA years, a "set of beliefs, norms and practices reinforcing the . . . ability to perform at very high levels of reliability."[50] A significant number of the scientists who later became a part of NASA learned their trade within the member groups of the V-2 Panel. From the beginning, each panel member had to have a plan that centered on building specific instruments and appliances that would perform reliably and provide scientific observations from rockets.

Although the individuals who were representatives on the V-2 Panel did not spend much time on the shop floor, they supervised others who did, in what was intended to be a hands-on tool-building venture wherein engineering virtuosity determined specific scientific goals. It was a world dominated by design, testing, verification, and flight evaluation. This world was centered in the technical shops and laboratories of each member group, it was displayed at V-2 Panel meetings, where each member described his institutional research goals, and it was put to the test in the desert at White Sands.

## Notes to Chapter 5

1. The overall postwar Office of the Chief of Ordnance (OCO) budget in June 1945 called for $22 million for Army facilities and $28 million for industry contracting per year, which were drastic reductions from wartime spending. In the former, Aberdeen got $7.5 million, the Frankford Arsenal $1.75 million, the Ordnance Tank Arsenal $5 million, Picatinny Arsenal $2.5 million, while White Sands got a total of $3 million for construction and $1 million per year for operations. See [L. H. Campbell, Jr., Lt. Gen. Chief of Ordnance, Research and Development Service, OCO] "Plan for Ordnance Department Post War Research and Development," 15 June 1945. Box A763, entry 646A, RG 156, OCO/NARA. The Army Ordnance contracting budget included $3 million for Hermes, from its inception through 1946 and $300,000 to Bell Telephone Laboratories for what became Nike. R. P. Haviland, "Informal Conference between Representatives of Army Ordnance and BuAer on Rocket Development Program," 7 January 1946. Meeting held 21 December 1945. BuAer engineering division intelligence memorandum F41(1) A19(8) Aer-E-313-RPH. Jet Propulsion Laboratory History files, JPL-5–494. Koppes (1982), p. 34, notes that JPL's GALCIT program received level funding just above $2 million for fiscal 47/48 and for fiscal 49/50, and during fiscal 1946–48, OCO spent a total of $130 million on ordnance research and development, of which $29 million went to long-range guided missile development. On early conservative attitudes regarding rocket technology, see Hall (1963), pp. 426–30; and on the lack of interest in strategic ballistic missile development, see MacKenzie (1990), pp. 95–97.

2. See Brown et al. (1959), pp. 22–23. Ordnance also solicited participation by the Army Air Force, Signal Corps, and the Navy's Office of Research and Inventions, Bureau of Aeronautics, Bureau of Ordnance, and Bureau of Ships, as well as university contract laboratories. See "Facilities for Full Navy Participation in Guided Missile Tests at White Sands, New Mexico, to be Completed by January 1947." Joint War Department—Navy Department Press Release, 24 October 1946. V-2 Press Release folder, Krause file, SAOHP/NASM; "Informal Conference between Representatives of Army Ordnance and BuAer on Rocket Development Program," 7 January 1946 memorandum. BuAer File 741(1), JPL/L 5–494.

3. In December 1945, Bain met with Harvey Hall and R. P. Haviland of the Bureau of Aeronautics Engineering Division, who were heading up a committee looking into the feasibility of earth satellites and manned space stations. Bain also was in contact with Frank Malina and C. B. Millikan of JPL, with R. W.

Porter of GE's Hermes project, and with representatives of the Bell Telephone Laboratories' Army Ordnance Nike project. See "First Meeting, Space Rocket Committee." JPL/L 5–370; R. P. Haviland, "Rockets—Comments Concerning Use of Development of," 10 August 1945, p. 6. JPL/L 5–381a-1. See also Hall (1963).

4. R. P. Haviland, "Informal Conference between Representatives of Army Ordnance and BuAer on Rocket Development Program," 7 January 1946, p. 2. Meeting held 21 December 1945. BuAer engineering division intelligence memorandum F41(1) A19(8) Aer-E-313-RPH. JPL/L 5–494.

5. See Komons (1966), p. 9, for a discussion of the Air Staff's initial opposition to engaging in basic research.

6. ORI became the patron of the Naval Research Laboratory as well as the Special Devices Division from the Bureau of Aeronautics. See The Bird Dogs (1981), p. 99. The Bird Dogs, the group of Navy officers that participated in the formation of ORI, recalled that "the Navy found itself the sole government agency with the power to move into the void created by the phasing out of the OSRD." Although this was certainly the case in light of the eventual failure of the Research Board for National Security, continued support of nuclear research by the Army did much to fill the void as well. See Komons (1966), pp. 6–7; Kevles (1975); and Sapolsky (1990).

7. [Gladeon M. Barnes], "Section I: Long-Range Rocket-Powered Guided Missiles," n.d., circa October 1945, pp. 10–16; 10. Box A763, "History, Planning for Demobilization Period." See also [Research and Development Office, OCO], "Outline of Ordnance Department Plans for Postwar Research and Development," 1 November 1945 p. 10. Box 763, RG 156, OCO/NARA.

8. Transcript of meeting, "Joint Army-Navy Meeting on Army Ordnance Research and Development, the Pentagon, 26 June 1946," p. 43. Box 767A, RG 156, OCO/NARA.

9. Transcript of meeting, "Joint Army-Navy Meeting on Army Ordnance Research and Development, the Pentagon, 1 October 1945," p. 40. Box 767A, RG156, OCO/NARA. MacKenzie (1990), p. 104, identifies the Army's limited interests in guided missiles.

10. Transcript of meeting, "Joint Army-Navy Meeting on Army Ordnance Research and Development, the Pentagon, 26 June 1946." p. 37. Box 767A, RG 156, OCO/NARA.

11. Komons (1966), pp. 1–2; The Bird Dogs (1981), p. 99.

12. Grove's patronage of E. O. Lawrence's Radiation Laboratory continued undeterred while AEC policy developed. See Seidel, (1983), p. 389.

13. R. Ladenburg, "Notes on Meeting . . ." 8 January 1946. MAT/LC. By mid-1946, the situation would improve as General Electric and the Army learned how to reconstruct the missing parts.

14. Ibid., pp. 2–3. MAT/LC.

15. Ibid., p. 3. MAT/LC.

16. Ladenburg is remembered today for his research on anomalous dispersion in gases, nuclear physics, and atmospheric chemistry. He had examined sources of atmospheric opacity due to oxygen and ozone in the far ultraviolet, and his wartime work at BRL brought Ladenburg into close contact with T. H. Johnson and with efforts there to unravel the growing mountain of intelligence reports on German technology, within which lay Regener's plans and paraphernalia. On Ladenburg's activities at BRL during the war, see "Minutes, Scientific Advisory Committee," 5 February 1945. BRL folder, Historical files, RG 156, OCO/NARA. Meetings were held on 2–3 February 1945, at BRL. In 1931, Ladenberg left his position as head of the physics division at the Kaiser Wilhelm Institute for Physics in Berlin to become professor of physics at Princeton. After the 1933 academic purges, Ladenburg worked to save his hapless physics colleagues in Germany. See Shenstone (1973). On Ladenburg's activism in support of German colleagues, see Weiner (1969), pp. 215–16.

17. Tuve had advised Walter Orr Roberts in early December of Army Ordnance's intentions of testing the V-2s at White Sands, as well as the possibility that Barnes would be sympathetic to the use of the warhead space for scientific observations. See Roberts to Shapley, 11 December 1945, p. 2. "Roberts File," HUG 4567.5.2, DHM/HUA. In late 1945, Ladenburg prepared a number of internal reviews of the state of knowledge of the atmosphere. These reports were used first at Princeton to determine what could be gained by access to the upper atmosphere and near space. They were then circulated to Tuve and

to the Army, as Ladenburg sought further information on Regener's work from Kuiper. See Kuiper to Ladenburg, 26 March 1946; Ladenburg to Kuiper, 22 April, 1 May 1946. KP/UA.

18. See Ernst Krause, "General Introduction," in M. A. Garstens, et al., eds., *Upper Atmosphere Research Report* 1, NRL Report R-2955, 1 October 1946, pp. 1–2.

19. Those institutions sending more than one representative included NRL, APL, BRL, Naval Proving Ground, BuAer, BuOrd, BuShips, CNO, and ORI. Sending one representative were Army Ordnance (James G. Bain), the Army Air Corps, Princeton University (M. H. Nichols), and General Electric. See "List of Attendance V-2 Conference," in H. A. Schade to J. G. Bain, "Summary of Minutes of V-2 Meeting Held 16 January 1946—Forwarding of," 6 February 1946. Box 32, NARA; see also M. H. Nichols, "Notes on the January 16 meeting at the Naval Research Laboratory for further discussion on the use of the Army's V-2 missiles," 23 January 1946. Box 118, MAT/LC. Both documents contain a summary of the 16 January meeting, prepared by T. R. Burnight of the NRL Rocket Sonde Research Section and by Myron Nichols of Princeton. See also James A. Van Allen OHI, pp. 160–61rd.

20. M. H. Nichols, "Notes on the January 16 meeting at the Naval Research Laboratory for further discussion on the use of the Army's V-2 missiles," 23 January 1946. Box 118, MAT/LC.

21. Van Allen OHI, p. 155rd.

22. "Changes in Organization of Guided Missiles Research Program," Laboratory Order No. 47–45 (17 December 1945), Ernst Krause files, SAOHP/NASM. Copies at HONRL. Michael Aaron Dennis has developed a full discussion of Krause's planning for missile research at NRL in an unpublished draft "Making Space: Sounding the Territory of the Upper Atmosphere Research Archipelago, 1944–1946."

23. On the strike, see Director, NRL to Bain, 11 February 1946, p. 1. Box 32, NRL/NARA.

24. Burnight minutes p. 5; Nichols minutes p. 4.

25. Shapley was absent. W. L. Pryor to J. G. Bain, "Report of V-2 Panel Meeting," 11 February 1946. Box 32, NRL/NARA. No information has been found as to why the meeting was held at Columbia, save for the fact that it was a convenient in-between site for the chosen members from Schenectady, Cambridge, and the Washington area.

26. Ibid.

27. M. H. Nichols, "Notes on V-2 Panel Meeting Held January 25, 1946," 4 February 1946. MAT/LC.

28. See J. N. Davis, "The NRL Pulse Telemetering System," in *Bumblebee Report* No. 42 "Princeton Telemetry Symposium," December 1946, pp. 36–37.

29. M. H. Nichols, "Notes on V-2 Panel Meeting Held January 25, 1946," 4 February 1946. MAT/LC.

30. See Ernst Krause, "Record of Consultative Services," 12 February 1946, phone conversation with R. W. Porter, 11 February. NRL/NARA.

31. NACA had been thinking about extending its atmospheric tables first to 32 kilometers and then to 160 kilometers but was not able to act quickly enough to assume a central role in the rocketry. R. E. Littell of NACA headquarters phoned Krause on 20 February to discuss how NACA might benefit from the early flights; specifically, he wished to interact with some of the experimenters. As a result, Krause and Littell agreed to have Calvin Warfield and others visit NRL and possibly attend future panel meetings as observers. See Krause, "Record of Consultative Services," 20 February 1946. Box 32, NRL/NARA. Warfield and Buckley did visit on 26 February and discussed the atmospheric data they desired with Ralph Havens, M. Garstens and Nolan R. Best of the Rocket Sonde Research Section. See N. R. Best, "Record of Consultative Services," 27 February 1946. Box 32, NRL/NARA.

32. Army Ordnance added the University of Michigan and the Signal Corps only after vigorous petitions from Lt. Col. Harold A. Zahl, chief scientist of the Signal Corps' Engineering Laboratory at Ft. Monmouth, New Jersey, on behalf of Dow and Golay.

33. See C. H. Smith, "Record of Consultative Services," 26 February 1946, conference held 13 February 1946. Box 32, NRL/NARA. On APL's activities, see "Record of Consultative Services," 21 February 1946. Box 32, NRL/NARA.

34. Smith, 13 February notes, ibid.

35. Ibid., pp. 1–2.

36. Army Ordnance's promised German warheads were still at sea. Krause, "Record of

Consultative Services," 12 February 1946, phone conversation with R. W. Porter, 11 February. NRL/NARA. GE employees were still on strike during a period of deep labor unrest. See Noble (1984), p. 25. On APL's V-2, see Krause, "Record . . ." 12 February 1946, p. 2; and "Record of Consultative Services," 20 February 1946. Meeting date: 18 February. NRL/NARA.

37. C. H. Smith, "Record of Consultative Services," 26 February 1946, p. 3. Meeting date: 13 February. Box 32, NRL/NARA. There would be more negotiations with ORI and with the Navy's Bureaus of Ordnance and Aeronautics. Legal aspects of procurement were not fully known; ORI representatives wondered if "one government agency [could tie down another government agency in the same manner as it can tie down an outside contractor[?]" And a BuOrd official wearily felt that "such a complicated contract . . . would cause excessive negotiating whenever a different design of rockets or even a minor change was made." Smith, pp. 3–4.

38. C. H. Smith, "Record of Consultative Services," 20 February 1946, meeting held 18 February 1946. C. H. Smith, "Record of Consultative Services," 21 February 1946. Meeting date: 19 February. Box 32, NRL/NARA.

39. In its initial report, the panel referred to itself simply as "the Panel." By the third meeting in May, it called itself the "V-2 Upper Atmosphere Panel" and then "Upper Atmosphere Research Panel" as the panel broadened its scope. With the depletion of the V-2 inventory, the panel renamed itself the "Upper Atmosphere Rocket Research Panel" (UARRP, as it came to be known) and went through other name changes until its demise in 1961. These name changes reflect changes in emphasis and domain, but we will call it the V-2 Panel in the first part of this book, changing to UARRP in sections dealing with later years.

40. G. K. Megerian to James G. Bain, "V-2 Report No. 1," 7 March 1946. V2/NASM. With the exception of Harvard, those attending the first meeting were, like Krause, entrepreneurial manager-scientists each actively developing upper atmosphere research programs. Attendance at later meetings broadened to include members of each group actively involved in the experimentation itself.

41. G. K. Megerian to James G. Bain, "V-2 Report No. 1," 7 March 1946, p. 3. V2/NASM. These notes, prepared by Megerian, were reviewed by Krause and then sent out as minutes of the meeting and as a report to Bain at the Pentagon. The "super-advisory" panel was soon formed to report to General Barnes of Army Ordnance, Toftoy's superior, on the status of all V-2 activities. This "Advisory Committee for V-2 Firings" consisted of representatives from Army Ordnance (such as Toftoy and Barnes), BuOrd, Signal Corps, NACA, BRL, White Sands Proving Ground, and General Electric. It held meetings at irregular intervals. See Charles F. Green, "Meeting of Advisory Committee for V-2 Firings, 25 October 1946," GE Aeronautics and Marine Engineering Division, 30 October 1946, Report No. 145. Army/GE Project Hermes (V-2) USMR(1946–1951). Project Hermes File, NHO.

42. G. K. Megerian to James G. Bain, "V-2 Report No. 1," 7 March 1946, p. 11. V2/NASM.

43. Burnight notes, p. 4.

44. Based upon deliberations at the first three V-2 Panel meetings, Megerian minutes. On duplication of effort by Office of Scientific Research teams, see DeVorkin (1980).

45. On the JPL program, see Koppes (1982).

46. Clark B. Millikan, Acting Chairman, JPL Executive Board, to "Dear Sir," 11 February 1946. JPL/L. Copy in Box 32, NRL/NARA.

47. Krause also relished the chance to inspect California missile launch facilities at Pt. Mugu and Inyokern, where NRL had specific interests. J. J. Bartke to Frank J. Malina, 21 February 1946. NRL/NARA.

48. Pickering told Harvey Hall of BuAer that he hoped to perform velocity-of-sound experiments at altitudes up to 60 kilometers. See Harvey Hall to J. W. Crowley, 5 February 1946. File 5–377, JPL/L. See also W. H. Pickering OHI no. 1, pp. 4, 7–8; no. 2, pp. 10–11, 18–20.

49. He concluded that "the lines of communication represented and reinforced at this meeting, and maintained by the [V-2 Panel], continued to be effective in the next decade of aerospace growth in the United States." Pickering and Wilson (1972), p. 410.

50. Howard McCurdy (1989), p. 301, describes early NASA teams in this manner, but it applies equally well here to the groups that later formed parts of NASA.

# 6

# Members of the V-2 Panel

Through the spring of 1946, as the V-2 Panel was being established and as new institutions applied, pressured, and bartered for membership, each of the member institutions of the panel hastened to build their instruments for the impending V-2 flights. The Naval Research Laboratory was far ahead of the rest; but the rest were catching up fast, including the Applied Physics Laboratory, the University of Michigan, the Ballistic Research Laboratory, the Signal Corps, and Princeton. The Army Air Forces entered a few months later when the Cambridge Field Station of the Air Materiel Command formalized contracts with several universities.

In this chapter, we identify the major groups formed in the first years to conduct upper atmospheric research with rockets. We discuss the backgrounds and interests of the group leaders and how they justified scientific rocketry to themselves and their institutions. We begin with an overview of the new technical culture that was evolving.

## Group Structures

Four types of institutional arrangements emerged in the first years: 1. Those within the sphere of military research and development: NRL, the Ballistic Research Laboratory (BRL), the Evans Signal Laboratory of the Signal Corps (SCEL), and the Cambridge Field Station (CFS), later known as the Air Force Cambridge Research Laboratory (and later Center) of the Air Materiel Command (AFCRL or AFCRC). 2. Those within civilian contract laboratories for one of the services (APL). 3. Those within university departments contracting with one or more of the military services (Harvard, Michigan, Colorado, Utah, Boston, and Rhode Island). 4. Those in university departments at Princeton and Chicago subcontracting with military contract laboratories.

Group size ranged from one or two persons to more than 90 organized in a distinct hierarchy at NRL. Although 300 to 400 people became involved within the various American military agencies and groups examined here, in the first four years only about 65 of them published technical laboratory reports or scientific papers. They headed teams of engineers, technicians, and military service staff whose names do not appear in publication lists as coauthors but are mentioned in laboratory reports and in oral histories.[1]

Groups within large agencies such as NRL included many technical and nonprofessional staff who were needed to maintain base or campus laboratories or to assist with field activities at White Sands. Of the some 90 people in Krause's Rocket-Sonde Research Section in 1946, 30 to 40 had identifiable professional roles developing scientific instruments or housekeeping systems such as telemetry, pointing controls, and associated electronics.[2] Of these, fewer than 20 were involved with scientific aspects of the program and only about six played a controlling or defining role. APL had only about 20 people, but more played a defining role although James Van Allen conceived most of APL's upper atmospheric projects.

The NRL and APL groups were the largest and most complex of the panel members. They also provided services, technology, and funding for other groups, as did the AFCRL and the Signal Corps a little later. Both the AFCRL and the APL also supported university projects by scientists not on the panel and so brought in a wider array of talent that will bear notice here.

Hierarchies had also emerged in the larger contract groups on university campuses at Princeton, Colorado, and Michigan. Senior professors defined the research, junior faculty directed it, and technicians and graduate students carried it out in the usual fashion. In the academic laboratories—departments of physics, electrical engineering, or aeronautical engineering—the department chairman was typically the group leader. In a few cases, academic contract laboratories managed their programs through a research department or an engineering research station. The smaller groups, mainly those at the AFCRL contract laboratories or in the Signal Corps, were less stratified, as their functions and duties centered more on the instruments and less on services.

## Group Membership Profiles

The group leaders who proposed, designed, and built instruments were mainly experimental physicists with experience in optics, electronics, and radio and mechanical engineering. Of the 20 NRL professionals, most had a bachelor's or master's degree in physics and experience in electronics, or were trained or experienced in electrical or mechanical engineering. A few—such as Krause, Gilbert Perlow, and Serge Golian—held doctorates in nuclear physics, although only the last two had actual experience in a fully functioning research atmosphere before coming to NRL. Homer Newell, with training in mathematics, acted as a physicist at NRL, and Eric Durand held a Johns Hopkins doctorate in physics with a specialty in spectroscopy but during the war had worked in radio and electronics at MIT's Radiation Laboratory. After a short term at Los Alamos in 1946, Durand came to NRL under Krause and helped expand nuclear research and solar spectroscopy.

Richard Tousey's NRL optics group included specialists in electronics, but his staff consisted mainly of optical technicians and engineers trained in mechanical engineering. Tousey, with a doctorate in physics, determined the overall agenda in his group and adhered closely to techniques he was familiar with from his training and experience in vacuum ultraviolet spectroscopy as a student of Harvard's Theodore Lyman.

APL's James Van Allen was trained in nuclear physics at Iowa, coming into contact with geophysical problems there and during his postgraduate experience at the Carnegie Institution of Washington's Department of Terrestrial Magnetism (DTM). Van Allen's High-Altitude Research Group included people who, like himself, had developed a successful radio proximity fuze, APL's major wartime achievement.[3] Van Allen added electronics and mechanical engineers such as Lorence W. Fraser, Russell S. Ostrander, and Clyde T. Holliday, as well as DTM physicist Howard A. Tatel who served as Van Allen's chief collaborator in cosmic-ray studies. The senior physicist John J. Hopfield, who came to APL from the National Bureau of Standards in the last year of the war, oversaw the development of spectroscopic instruments for solar research. Hopfield was a senior vacuum ultraviolet spectroscopist with a background similar to Tousey's, but Hopfield had also examined the extreme ultraviolet transmission of the earth's atmosphere during his long career. Van Allen also had the assistance of astronomer J. Allen Hynek, who had been Merle Tuve's assistant during the war. Through Hynek, Van Allen contracted with Yerkes astronomer Jesse Greenstein to build a solar spectrograph to complement those being developed at APL by Hopfield, Harold Clearman, and Hynek.

Marcus O'Day's own core professional staff at the Cambridge Field Station (CFS, later known as AFCRC and AFCRL) was composed almost exclusively of Ph.D. physicists recruited from the wartime radar development groups based at Harvard and MIT. Although his in-house staff had acted as radio engineers and electronics specialists, they rapidly developed speciality groups that paralleled those at NRL and formed a technical infrastructure able to handle the university groups O'Day contracted with in departments of physics and electrical engineering across the country. One of his largest contracts was with the University of Michigan, where William G. Dow and his staff were predominantly electrical engineers and specialists in vacuum-tube electronics.[4] William Dow was a senior vacuum-tube development engineer who had worked mainly at the MIT Radiation Laboratory during the war. Dow was drawn into upper atmospheric research as part of Michigan's effort to develop an antiballistic missile system for the Air Force, called Project Wizard, and was soon able to secure Signal Corps support for related campus research activities.

Among the dozen other CFS contractors, the University of Colorado physics department entered in 1948 primarily to build devices called pointing controls to stabilize solar instruments on rockets. This demanding technical goal precipitated a long-term development program that produced no scientific publications in the first four years. Quite typical of the other Air Forces contractors, Colorado's effort was led by a senior department chairman, W. B. Pietenpol, who had no strong research and publication history and oversaw a group of junior physicists knowledgeable in electronics and mechanical engineering.

Within the first decade of the rocket flights that constitute the V-2 era, the professional profiles of these groups changed very little. In 1954, the NRL Technical Information Division identified workers in upper atmospheric research with rockets; among some 60 scientists, engineers, and technicians engaged in such work within its Optics, Electronics, and Radio Divisions, two-thirds were called physicists, and the remaining third electronics specialists, electrical and electronics engineers, radio specialists, and mathematicians. None were identified as astronomers, meteorologists, or ionospheric physicists.[5] Many of those classified as physicists in 1954 were, as we have noted, trained in some electronics-related activity, many had backgrounds in optics and the vacuum ultraviolet, and a few in nuclear physics.

None of these people had established career objectives in the fields of science they planned to attack with rocket-borne devices. Unlike Erich Regener, who had adapted his own research group and its instruments to the V-2, none of the American groups planning to conduct research with rockets had existed earlier as upper atmospheric research teams. Homer Newell of NRL recalled that in late 1945 and early 1946, Krause's team spent weeks lecturing to each other about the upper atmosphere, those who were lecturers staying only a page ahead of the others in their reading in what he described as a period of "intense self-education."[6]

First they had to scramble to conceptualize what they were going to do, and then they had to determine how they were to do it. That these groups accomplished both goals in a matter of months in the spring of 1946 suggests that finding useful things to do with rockets was not too difficult, and that those who did the work or directed it knew how to build devices quickly and efficiently. But integrating the two—the intellectual background and the technical talent—proved to be quite a challenge.

Only at Princeton was there a strong preexisting interest in the field they wished to pursue with rockets. Others who did have appropriate backgrounds, such as Harvard's Fred Whipple, Harry Wexler from the Weather Bureau, or Leo Goldberg of Michigan, did wish to get involved directly in 1946 but ended up acting as advisers rather than as practitioners. Senior scientists with established careers and traditional responsibilities, such as Rudolf Ladenburg at Princeton, Gerard Kuiper at Chicago, or Joseph Kaplan at UCLA, were interested in the progress of the groups, but like Harvard's Donald Menzel, they decided not to become directly involved themselves.

With few exceptions, the people who entered upper atmospheric research with rockets were young scientists experiencing organized research for the first time during the war. Among the leaders in 1946, Krause, Van Allen, and Myron Nichols averaged 32 years of age. Krause's group averaged 31, and Van

Allen's 35, since Hopfield at 55 raised the average slightly. Marcus O'Day was 49, and those in his six-member AFCRL group averaged only 28, with four holding doctorates in radio electronics, engineering physics or electronics, and spectroscopy. William Dow was 51, but aside from one senior technician, his group average was also 28. Pietenpol was 55, whereas most of his staff were in their 20s or early 30s. How these people were led into rocketry—their arguments for doing so, and the people to whom they turned to get the job done—-demands closer investigation if we are to understand the transformation of the technical culture that paved the way for research with rockets.

## The Naval Research Laboratory

Ernst Krause (figure p. 77) came to NRL after obtaining his doctorate in 1938 at the University of Wisconsin with a specialty in experimental nuclear physics. Krause worked in the Communication Security Section of NRL's Radio Division, applying radar pulse transmission and reception techniques to telemetry.[7] During the war, activities at NRL naturally intensified, and after 1941 Krause's responsibilities grew quickly. He managed groups working on electronic jamming devices, and after 1943 he began moving into electronic guidance, control, and radio countermeasures, heading up the new Guided Missiles Subdivision.[8] By war's end, Krause's group had developed remote control guidance systems for missiles such as the American version of the German V-1, called the JB-2 Loon.[9] Activities such as these gave Krause's staff access to intelligence reports, including those covering the V-2 and Regener, and, as noted in Chapter 4, his tours in 1944 and in the summer of 1945 convinced him that NRL should shift its research toward guided missile development after the war.[10]

The NRL that Krause returned to in the Autumn of 1945 was rapidly changing.[11] In June 1945, its Radio Division split into four new scientific divisions, one containing Krause's subdivision.[12] The *Washington Post* saw this reorganization as a critical step toward Cold War preparedness in a world living in the shadow of atomic bombs, radar, "and other scientific gadgets [that have] made obsolete much of our fighting equipment." The next war, the *Post* predicted "if there is one—will be fought mainly by the scientists who are now training in their laboratories for it."[13]

Krause was designated leader of an advanced projects group for missile development and became NRL's coordinator of guided missiles.[14] In October 1945, as coordinator, Krause headed a Committee on Guided Missiles to advise an ad hoc steering committee composed of research division heads that had been charged with planning for long-term research at NRL. Krause's committee became a clearinghouse for information, but it also provided Krause with a means to assess the local talent available for guided missile development.[15] During the war, Krause commanded as many as 150 people; now he had to take steps to keep the best ones and to attract others. He would soon find his entré.

By late November 1945, while undergoing more reorganizations and shifts, NRL came under the new Office of Research and Inventions (ORI) and looked forward to different relations with the Navy bureaus (Aeronautics, Ordnance, and Ships) it had served.[16] In early December 1945, incoming Director H. A. Schade developed a plan to demonstrate to ORI and to the bureaus that NRL was a "research laboratory covering all fields of science and technology" and had to be thought of as a civilian institution.[17] He also felt that NRL must establish better public relations; its scientific staff should publish their research openly and have more contacts with the outside world. Schade pushed for a basic research agenda informed by military needs in nuclear power, radar-related electronics, countermeasures, and guided missile development.

Caught up in the general spirit, Krause's countermeasures group at NRL in the fall of 1945 puzzled over what to do in the postwar

Ernst Krause in temporary quarters at White Sands, circa 1946. U.S. Navy photograph, Ernst Krause collection. NASM SI 83–13868.

era.[18] As Milton Rosen recalled, they pondered all the ways they could apply their experience to guided missile research. Among other ideas, Rosen suggested they perform research on the upper atmosphere with missiles. Rosen remembered reading allied intelligence reports on Regener that "showed what could be done."[19] He also knew that NRL's tradition in ionospheric research, fostered by A. Hoyt Taylor and E. O. Hulburt, was "an asset."[20] Krause endorsed the idea, and from that point on, Rosen remembers thinking, "Krause would carry the ball. He would establish the project, he would get support for it. We knew he would succeed; he always had."[21] Clearly, both Krause and his group knew that, one way or another, missiles were in their future. Upper atmospheric research fit neatly as a response to NRL's new direction.[22] It was not, however, Krause's highest priority.

## Krause's Plan

Even though the proposal Krause prepared as a result of his staff deliberations emphasized basic research with rockets, its chief thrust was guided missile research, with the V-2 a symbol of what had to be accomplished. Indeed, the V-2 symbolized for Krause both activities. His ambitious proposal, presented to NRL Director Schade on 3 December 1945, was a blueprint for restructuring the laboratory's priorities to include basic research within the context of guided missile development and called for the redirection of a substantial fraction of NRL's manpower. Such a commitment required forceful arguments.

Krause challenged NRL to commit half of its applied research manpower to the development of guided missiles, at a level equivalent to one-fourth its total research budget, and to reserve for its basic research effort the study of the upper atmosphere.[23] Applied research into missile guidance, propulsion, operational analysis, aerodynamics, and ordnance required several hundred people at least; whole NRL divisions would have to move directly into the new enterprise.[24]

The basic research Krause envisioned as part of the guided missile program was split into five broad categories: the ionosphere; atmospheric pressure, density, and temperature; cosmic rays; and solar physics. Each was justified by what it could contribute to the development of guided missiles. Cosmic-ray research, Krause argued, might reveal "staggering knowledge of how mass is converted into energy," whereas the atmospheric physics studies would gather knowledge about the "medium through which missiles of the future are sure to go."[25] Performing basic research with rockets would provide practical experience in launching, guiding, and tracking rockets through the ionosphere. "Such knowledge," Krause added, "could be as dangerous as it is useful. It could very conceivably lead to the destruction of the earth." He concluded, "Such awesome military applications were imposing enough to warrant a vigorous program."[26]

Capitalizing on the fact that NRL was determined to establish itself as a research institution of the highest caliber, Krause argued that quality scientists needed a reason to stay with NRL in peacetime. His personal NRL experience thus far had been 98 percent administration and 2 percent research, Krause claimed, and this was stultifying for a practicing physicist:

> I am perfectly willing to sacrifice myself on such a bureaucratic altar during wartime, but I feel that my value to the Navy and to the nation will deteriorate very rapidly unless I gain more experience in the laboratory. This is a feeling which I share with many people . . . at NRL. The proposed project can provide us with the necessary experience.[27]

Invoking all the issues familiar to a scientific recruiter, Krause made his case for basic research. Upper atmospheric research with rockets would be interesting and visible and would garner public respect and scientific le-

gitimacy. It would also demonstrate NRL's viability as a guided-missile center.

Suitably impressed by Krause's presentation on 3 December, Schade brought the proposal to NRL's division superintendents, directing them to estimate the manpower they could donate to NRL's guided missile program.[28] Their deliberations took less than a week, during which time Krause negotiated for considerably less than his original proposal had specified. On 17 December 1945, Schade issued two laboratory orders defining and establishing both the Rocket-Sonde Research Subdivision (soon renamed a section) and the reorganization and expansion of the Guided Missiles Research Program, both within one of the reorganized radio divisions called the Electronic Special Research Division.

Krause did not receive anything near the manpower he requested to start up a sizable guided missile effort. Although the restructuring did increase NRL's guided missile activity, it did not all come under Krause. Nor did Krause obtain the manpower to conduct all aspects of upper atmospheric research.[29] The negotiations resulted in the creation of a Rocket-Sonde Research Section, whose functions were to "investigate the physical phenomena in and the properties of the upper atmosphere with a view to supplying knowledge which will influence the course of future military operations." It would be "basic research and [would] include development of the necessary techniques, instrumentation and devices required to carry out the function."[30]

## The Rocket-Sonde Research Section

Krause's new section embodied NRL's new self-image. In his public support for the permanent mobilization of science, Commodore Schade matched Holger Toftoy. Schade ardently linked national security to basic research and highlighted his laboratory's efforts to study the atmosphere as an important ingredient in postwar naval research. In a news release promoting NRL's first V-2 launch in June 1946, Schade argued: "The security of our nation demands that we maintain leadership in basic scientific research. The upper atmosphere is a new frontier, the exploration of which can now be undertaken with profit."[31]

To Krause, upper atmospheric research meant developing not only guided missiles but also specific instrumentation. Electronics expertise pervaded his group, and it was the dominant technology brought to bear on the group's study of ionospheric radio propagation, the characteristics of the atmosphere, and cosmic rays. He also created two support groups to develop telemetry systems and a new sounding rocket for the Navy.

Although the availability of the V-2 as a vehicle supporting the Rocket-Sonde Research Section's agenda was formally stated only in mid-January 1946 by the Army, Krause felt that by the time of his proposal in early December, he knew this was going to happen and that "the formation of Rocket-Sonde certainly envisioned the use of V-2s."[32] In his 3 December 1945 proposal, Krause pointed out, "There is now available to us a new tool, the rocket, which already has reached altitudes four times as high as a balloon and ten times as high as an airplane."[33] These were capabilities beyond JPL's small WAC Corporal and were closer to the capabilities of the V-2. Homer Newell, who was part of the early discussions, later rationalized that they knew the WAC was "fully developed" and might well have been their workhorse if the V-2 had not been available.[34]

This point is important for two reasons. First, even though he did not receive the support he requested, Krause remained deeply committed to guided missile research at NRL, and the scale of rocket he hoped to develop matched the capabilities of the V-2. Second, his staff advocated upper atmospheric research only within the context of missile development. The V-2 symbolized, at the very least, what was possible not only to develop, but to use as a vehicle.

The missile Krause envisioned for NRL was, indeed, equal in capability and scale to the V-2. He directed Rosen and C. H. Smith to begin planning for a large sounding rocket in early 1946. Rosen explored sources for suitable vehicles and found that Aerojet, an industrial offspring of the Jet Propulsion Laboratory, was now building the V-2-sized Corporal E for the Army.[35] Accordingly, Krause negotiated with JPL staff to have Rosen study rocket propulsion and aerodynamics there and at Caltech.[36]

While Rosen and Smith attended to the sounding rocket, the immediate task of the rest of Krause's Rocket-Sonde Research Section was to compete for berths on V-2s, gain access to APL's Aerobees, and plan for specific experiments. Krause chose cosmic-ray physics for himself, but left the bulk of the work to Serge Golian and Gilbert Perlow, who had stronger backgrounds in nuclear physics and cosmic-ray research.[37] Homer Newell led a small mathematical analysis group that included Eleanor Pressly, who performed tasks ranging from atmospheric modeling and trajectory analysis to electronic circuit design.[38] Ralph Havens and Herman E. La Gow led a larger group assigned to develop instruments for measuring temperature and pressure. The highest priority and the most manpower were given over to studying the radio propagation characteristics of the ionosphere, an NRL tradition. Headed at various times by J. F. Clark and T. Robert Burnight, the group included, among many others, C. Y. Johnson and J. C. Seddon, all of whom were adept at radio technology and electronics.[39] Throughout 1946, these were the principal areas of upper atmospheric research in the branches of the Rocket-Sonde Research Section.[40]

## Richard Tousey Enters the Picture

E. O. Hulburt, who had been present when Army Ordnance made its offer on 16 January 1946, moved quickly to ensure a place in the upper atmospheric research program for his Physical Optics Division. His most significant asset, Richard Tousey (figure p. 80), was highly adept in vacuum ultraviolet spectroscopy and would be the best person to build ultraviolet solar spectrographs. Hulburt duly made the suggestion, and Tousey, head of the

Richard Tousey at White Sands for the launch of V-2 54 on 18 January 1951. Richard Tousey collection. NASM SI 90–2337.

Micron Waves Section, immediately made the project his own.[41]

Hulburt and Krause arrived at an administrative compromise for the spectroscopic work; it was to be under the "cooperative direction" of Tousey and C. V. Strain, the former remaining in the Physical Optics Division with his staff, and the latter reporting to Krause. Eric Durand was also assigned to this group, because he had experience in optics and had in fact wanted to perform the spectroscopic studies himself.[42]

Tousey brought in Francis Johnson, a physicist with training in laboratory vacuum ultraviolet techniques and experience in meteorology and weather forecasting, and William Baum, a physics graduate student from Caltech who had enlisted in the Navy, was stationed at Caltech, and then at NRL in the Optics Division at the end of the war. Baum had training under physicist Ira S. Bowen at Caltech, was particularly interested in optics, and was familiar with a wide range of design and drafting techniques; in the last months of the war, Baum was studying the effects of loading in electrical elements in fire-control systems. He knew how to "build an instrument with optics in it."[43]

Johnson and Baum were brought into the V-2 program in early February 1946. Johnson recalls that the co-direction between Tousey and Strain was something of an "administrative misfit" and that it "didn't work out too smoothly on the whole." For practical purposes, Tousey was the leader of the group. Johnson felt that Tousey "by all odds, had the greatest competence in this area," and even though the Rocket-Sonde Research Section would have wanted greater control, "Tousey was unwilling to grant it."[44]

Tousey was able to resist because, in addition to having Hulburt's support, which was exerted in a constant quiet, gentlemanly manner, he and Krause were at the same administrative level; both were section heads, even though Tousey's staff was far smaller than Krause's.[45] Krause, of course, controlled access to the rockets and the V-2 Panel—he more than anyone else had made it all possible—but he always respected Tousey's primacy in the field of solar research with the V-2.

A childhood fascination with radio and machine-shop experience brought Tousey into physics at Tufts, and then into graduate work with Theodore Lyman at Harvard, where he began his long career exploring the vacuum ultraviolet.[46] Opting for an experimental thesis on the reflectance of metals in the extreme ultraviolet, Tousey's professional life became dedicated to detecting vacuum leaks, devising efficient pumping and contaminant purging techniques, and preparing and developing fluorescing oil-emulsion photographic plates that could detect extreme-ultraviolet radiation.[47]

After graduation, Tousey taught at Tufts and continued in ultraviolet research until June 1941, when he moved to the Naval Research Laboratory, choosing NRL because he had known Hulburt for some time through a fortuitous meeting on a sailboat dock in Maine.[48] Sensing in Hulburt a research manager who knew how to avoid the "cut-throat competition" typical of Tufts and Harvard, Tousey happily transferred his life and passion for birds, flowers, music, and the vacuum ultraviolet from New England to Washington.[49]

War work on atmospheric transparency, scattering phenomena, and physiological optics, all directed to problems of daytime navigation using stars, brought Tousey into contact with astronomers Donald Menzel and Walter Orr Roberts, who were developing Harvard's High Altitude Observatory at Climax, Colorado. This was Tousey's first extended contact with astronomers, and growing out of this activity came an interest in sky brightness and polarization at high altitudes. Thus, with training and experience in the vacuum ultraviolet and wartime experience in problems of physiological optics, atmospheric scattering, night vision, optical reconnaissance, and the infrared, Tousey was well placed to design instruments to detect the ultraviolet spectrum of the sun.

## NRL As Provider

Krause not only had to develop a scientific program and build a sounding rocket, but also had to provide replacement warheads and telemetry services to the other member of the V-2 Panel. After numerous meetings and countless phone conversations, Krause pressured the Naval Gun Factory to produce 25 cast and machined warheads to specifications set by the Army and General Electric (figure p. 82). At the time of the panel meeting on February 27, Krause announced that the Gun Factory could provide the warhead shells at a unit cost of $1,000, which NRL would bear.

Telemetry became NRL's largest contributed effort and over the first two years involved more than 20 staff in the division and outside contracts in the hundreds of thousands of dollars. During this period, the NRL pulse-time telemetry system was continually refined to improve reliability, signal strength,

An early example of the replacement V-2 warheads built by the Naval Gun Factory. U.S. Navy photograph. NASM SI 82–4615.

and number of channels; at the same time its weight and dimensions were reduced.[50]

Through Krause's efforts, NRL became a primary provider, second only to Army Ordnance. NRL warheads would be shipped to each panel member for instrument integration, and telemetry systems provided by NRL would be installed and operated by NRL crews at White Sands, along with cooperative teams to man the field stations. APL, a substantially smaller institution, could not compete as primary provider, but did wish to develop its own telemetry capability, mainly because it already had considerable expertise in the field during the war. APL did compete in developing the Aerobee, however, and also acted as provider for several academic subcontrac-

tors who were drawn into upper air research at the close of the war.

## The Applied Physics Laboratory

The Applied Physics Laboratory was formed by Merle Tuve in March 1942 to develop a special miniature radio proximity fuze, a device that could cause an artillery shell to explode at a predetermined distance from its target. Created as Section T of the wartime Office of Scientific Research and Development (OSRD), APL was composed of DTM staff from the Carnegie Institution of Washington and was housed behind a used-car dealership on Georgia Avenue in Silver Spring,

Maryland. APL's charter members included Tuve and his senior physicist colleagues such as Robert B. Roberts and Lawrence Hafstad, and junior postdocs like James A. Van Allen, among many others. By the end of the war, APL had built a successful fuze, managed its production, and introduced it into combat in the Pacific.[51]

With the end of the war in sight, Tuve initiated plans to make APL permanent, arguing in October 1944 that complete demobilization would create national instability, and that the Navy should seize the advantage and hold some of its best civilian groups together.[52] An architect of the concept of the permanent mobilization of science, Tuve believed that in the new world, "Defense is henceforth more of a civil job than a military one."[53] In October 1944 the actual projects APL would administer were not defined; by January 1945 they became sealed when APL became a central Navy Bureau of Ordnance (BuOrd) contract laboratory for guided missile research under the code name "Bumblebee."[54] APL thus emerged from the war deeply involved in guided missile development with its Bumblebee and associated Lark missile programs, and its senior staff were prominent members of the Guided Missile Committee of the Joint Committee on New Weapons and Equipment of the Joint Chiefs of Staff.[55]

APL's Bumblebee was a broadly based research and development program that would lead to a "supersonic, rocket-launched, ramjet-propelled, radar-guided [antiaircraft] missile system."[56]It included research on propulsion, guidance, and aerodynamics characteristics for short-range antiaircraft and long-range cruise-type missiles. APL managed Bumblebee through subcontracts with universities and industry, including Princeton and the University of Virginia.[57]Even though the staff's highest priority was "fundamental military research," or basic research informed by military need, APL also looked for ways to diversify.[58]

In January 1946, Robert B. Roberts believed that Tuve should trim Bumblebee staff in anticipation of a postwar downturn. To retain the best staff, however, Roberts sug-

gested that APL take on one or more new projects. Roberts suggested either "V-2 tests for upper air research including cosmic rays" or the "instrumentation of atomic bomb tests."[59] Roberts preferred the V-2 work, because it was less expensive, would not seriously affect the Bumblebee project, and because their Princeton subcontractor was "already active in this program."

## Princeton and APL

Princeton indeed was active. As noted in Chapter 5, Rudolf Ladenburg represented both APL and Princeton in the early meetings with Army Ordnance. He was also the architect of Princeton's program, urging Myron H. Nichols in November 1945 to prepare a detailed proposal to Merle Tuve recommending that a portion of Princeton's contract research group engage in research on the physics of the upper atmosphere as part of Project Bumblebee. Nichols was technical liaison to APL for Princeton's telemetry subcontract, and with the end of the war he was looking for a continued relationship that would maintain his group.[60]

Nichols had little trouble linking upper atmospheric research and Bumblebee. Beyond providing improved knowledge of the upper atmosphere, especially the ionosphere, which would benefit Bumblebee directly, basic research of this kind would make Bumblebee more appropriate to a university campus, would help attract good physicists to Princeton, and would endow APL with a prestige unavailable through missile research alone. Nichols felt that a university was the best place to do this work: "The government and service laboratories are tied up with civil service and military red tape—some of the most creative men will not work under these conditions."[61] Princeton's telemetry contract work for APL's Bumblebee Program was only one of its many war-related activities. Princeton faculty members were leaders in nuclear physics and had been prominent in the Manhattan Project.[62] Accordingly, Nichols argued that Princeton could gather up all the expertise required to perform every conceivable type of upper atmospheric research.[63]

Pulling out all stops, Nichols claimed that members of Princeton's acclaimed astronomy and electrical engineering departments, as well as members of the Institute for Advanced Study and nearby RCA Laboratories, would likely be interested and available to lend expertise, staff, and graduate students. Nichols also predicted that Princeton's research would meet the needs of guided missile development. In particular, Princeton's cosmic-ray research with V-2s would help to improve the telemetry systems they were still developing for APL and the Navy.[64] The combination was as unbeatable as it was optimistic.[65]

Merle Tuve approved Nichols' proposal, and Nichols ultimately became Princeton's representative on the V-2 Panel. Nichols originally proposed that Princeton represent APL's interests in upper atmospheric research as well, but in February, James Van Allen was given that job. According to an NRL observer, APL wanted to be an independent member of the V-2 Panel. They were "not satisfied to work through Princeton."[66]

### The APL High-Altitude Research Group

With prominent university professors advocating upper atmospheric research and its easy association with the central postwar APL mission, Tuve saw this activity as a means of achieving one of his original goals: establishing stronger ties with "the regular work of the Johns Hopkins University.[67] Thus, well before Roberts made his suggestion, Tuve was seriously pondering the possibility of having James Van Allen manage a V-2 program.[68] He therefore sent Van Allen, still in Navy uniform, to the 16 January meeting at NRL with Army Ordnance. Van Allen was excited by the prospect of V-2 research, and by the end of January both Roberts and Tuve decided that Van Allen would develop an APL upper air research program, to be staffed by not more than 15 people (figure p. 85).[69]

Roberts carved out a small group with dual goals identical to those of the NRL program: to develop a launch vehicle and use it to conduct high-altitude research. The latter activity, lumped under "discretionary research," was subject to Tuve's control, and was not to take more than 5 percent of the laboratory's budget. Thus Van Allen's group and a few others at APL working in basic research came to be known internally as the "5 percenters."[70]

A significant portion of Van Allen's administrative time and technical attention was given over to the design of the launch vehicle, later called the Aerobee, supported by the Bureau of Ordnance and contracted to the Aerojet Engineering Corporation and the Douglas Aircraft Company. The Aerobee was very much in consonance with APL's postwar mission, to act as a central laboratory engaged in "fundamental military research." It was justified by a Bureau of Ordnance requirement that the APL acquire experience with liquid-fueled rockets in addition to the air-breathing ram jets that powered Bumblebee. Thus Aerobee was a part of Bumblebee's search for liquid-fueled antiaircraft and surface-to-surface missile systems.[72] Accordingly, its dimensions were closer to those of an antiaircraft missile rather than the V-2. Aerobee would also establish the APL as a provider; certainly, once the V-2 rockets ran out, the existence of a proven, economical vehicle like Aerobee would become vital. Indeed, as we have seen, the Aerobee became APL's calling card for membership on the V-2 Panel.

### James Van Allen's Staff Building

In late January, Van Allen had six slots to fill.[73] By mid-February, Van Allen added six more and brought in Howard A. Tatel, R. B. Baldwin, S. A. Buckingham, and John J. Hopfield. Soon Baldwin and a few others were shifted to other tasks. This shifting caused Van Allen constant manpower problems. Even so, he was confident that his group would have cosmic-ray counters ready by June, along with a solar spectrograph, ozone absorption spectrograph, gas-sampling devices, and a temperature monitor for the warhead. The group would also build wire tape recorders and parachute recovery systems. In late February, Van Allen boasted to Nichols, "This group represents considerable experience in nuclear physics, spectroscopy, and geophysics."[74]

James Van Allen and APL's High-Altitude Research Group, circa December 1946, and the warhead destined for the first night flight of a V-2. Back row (l to r): Howard Tatel, Arthur E. Coyne, Lorence W. Fraser, S. Fred Singer, J. V. Smith, J. W. Barghausen, L. D. Moritz, J. E. Bilderback, USN, B. J. Chafee, A. V. Gangnes, and James A. Van Allen. Front row (l to r): R. S. Ostrander, F. W. Loomis, Shirley R. Mc-Collum, D. G. Klampe, and James F. Jenkins. The Johns Hopkins University Applied Physics Laboratory, neg. no. 225685.

Van Allen talked up the new work among his colleagues around the laboratory, and found many takers. Lorence W. Fraser was an electronics and mechanical engineer, an "all around capable person."[75] Fraser had field experience testing rockets, and became the systems engineer for the group. He and Russell Ostrander took care of the technical details of much of the equipment Van Allen designed.[76] Howard Tatel had a background in nuclear physics at DTM and so became Van Allen's principal collaborator in the cosmic-ray work.[77] Clyde Holliday had experience in the photographic tracking of missiles and aircraft and diagnostic high-speed photography

for proximity fuze testing. With Van Allen, he planned to place aerial cameras in V-2 rockets to photograph the earth and record rocket aspect.

John J. Hopfield, along with Harold Clearman and J. Allen Hynek, designed several ultraviolet solar spectrographs to do the types of research suggested by Ladenburg and pursued at NRL. To APL engineers, Hopfield was an "old line experimenter" who "was the bane of . . . practical people."[78] But, like Tousey, he knew ultraviolet laboratory research, and was very interested in getting a spectrum of the sun. Hynek had been an administrative assistant for Tuve during the war but was a trained and well-connected astronomer. In addition to collaborating with Hopfield, Hynek worked to get other astronomers interested, succeeding with Jesse Greenstein of the Yerkes Observatory.

As of June 1946, Van Allen's group included 16 people divided into research and engineering staff. In contrast to the NRL group, APL's High-Altitude Research Group was self-contained, although a few members had duties in other sections. Clyde Holliday, for instance, was connected primarily with electronics and photography in a Bumblebee aerodynamics group.[79] All phases of the work were still managed within one section, however, so that Van Allen could be directly involved with all aspects of scientific instrument development.

James A. Van Allen was an appropriate person to head the group. Young and energetic, Van Allen had gained valuable practical and organizational experience during the war making the proximity fuze operational.[80] He was an excellent bet for heading a successful instrument design effort.[81] As he recently pointed out:

> I was never cut out to be a theoretical physicist. I appreciated theory, but I never felt I could flourish in trying to do theory. I was more experimentally inclined and phenomenologically inclined. That has definitely been my career all along. . . [I] liked to do things I understood thoroughly and could visualize in more concrete terms.[82]

Trained in nuclear physics at the State University of Iowa (now the University of Iowa), Van Allen became Merle Tuve's postdoctoral fellow in September 1939, working in nuclear physics. But at DTM, Van Allen also came to know geophysicists and their work, such as Harry Vestine and his studies of the earth's magnetic field, and Scott Forbush and his efforts in correlating cosmic-ray phenomena with solar activity.[83] But before Van Allen had time to develop any personal research momentum, DTM entered the war and he was brought into Tuve's program to develop photoelectric and radio proximity fuzes

for all forms of military ordnance. Among his other accomplishments, Van Allen had created a ruggedized electronic circuit for a photoelectric fuze containing miniaturized Raytheon vacuum tubes. The tiny system could withstand an unbelievable acceleration of 20,000 times the force of gravity.[84] Van Allen recalls that his experience designing and testing this ruggedized electronic system was of value in his postwar rocket research; in fact, he used the very tubes that were developed at Raytheon and later at Sylvania in his cosmic-ray instruments.[85]

After an eight-month tour in the Pacific, Van Allen was assigned to the Bureau of Ordnance as liaison between APL and BuOrd for the continued refinement of the fuze. He later took a second tour in the Pacific at the end of the war and then went back to BuOrd as a liaison officer before returning to civilian life in February 1946. It was in this position that Van Allen attended the formative meetings of the V-2 Panel and then worked to establish the High-Altitude Research Group at APL, which gained formal BuOrd status on 21 February 1946.[86]

Princeton and APL found that they could coexist on the V-2 Panel, since a precedent for duplicating scientific objectives had already been established at the first meetings. Nichols did, however, narrow Princeton's focus, partly in deference to APL's interests, whereas Van Allen pushed for a program that matched NRL's in all areas save for ionospheric radio propagation. Attending the March 1946 JPL symposium, Van Allen reported that APL was interested in cosmic-ray research, solar spectroscopy, and atmospheric physics: a full agenda.[87]

## The Applicability of Basic Research

As with Krause at NRL, Van Allen at APL was keenly aware of military interests and the general value of the data he hoped to obtain to better understand the behavior of missiles. In late March he wrote to L. R. Hafstad, APL director for research, answering questions from BuOrd about how cosmic-ray counters could be used as altitude-measuring devices on missiles. Van Allen believed that he had an even better mechanism: he proposed using the transmission of the ultraviolet solar spectrum by the earth's atmosphere to gauge altitude.[88]

Of course, both Van Allen, as a member of a civilian contract laboratory, and Krause, at a military-based center such as NRL, were expected to let military interests and needs guide their research. Krause argued that "basic" research need not have immediate military application to ensure military preparedness, yet he never hesitated to point out military benefits; in April 1947, in a report to the Joint Research and Development Board, he predicted that "although the specific applications [of basic research] are to a large extent unpredictable, the long range value of such research to the National Defense is certain to be great."[89]

Both Krause and Van Allen aligned their concepts of basic research with the interests of their military patrons. They worked within institutions that assumed such thinking. Both were ready to build rockets, provide considerable manpower support at White Sands, and produce complete telemetry systems, as well as warheads in NRL's case. Both realized that the infrastructure needed to conduct upper atmospheric research for themselves and for other members of the panel was the same infrastructure desired by their patrons, the new Office of Research and Inventions as well as the older Bureaus of Ordnance and Aeronautics, which were ready to support any activities that aided the development of guided missiles.

## Contracting to Catch Up

The Navy, through NRL and APL, was the first to be ready to instrument V-2 missiles. Army Ordnance also made sure that its own Aberdeen Proving Grounds as well as the Signal Corps were represented and active on the V-2 Panel and in the program. Running a distant third, but closing fast, were the Army

Air Forces Air Technical Services Command at Wright Field, the Watson Laboratories in New Jersey, and Watson's newly formed Cambridge Field Station, in July 1949 renamed the Air Force Cambridge Research Laboratories (AFCRL), later known as the Air Force Cambridge Research Center (AFCRC), and now called the Geophysics Research Directorate's (GRD) Geophysics Laboratory.[90]

The Cambridge Field Station (CFS) was created at the close of the war as a site where the military might attract scientific personnel from the Radio Research Laboratory at Harvard and from MIT's Radiation Laboratory. Specialists in radio and radar research were first recruited under the jurisdiction of the Watson Laboratories at Red Bank, New Jersey.[91] One of the most active recruiters was Marcus O'Day, who worked to build a special navigation laboratory.

### Marcus O'Day and the Army Air Forces' Cambridge Field Station

Marcus O'Day (figure p. 88) received a doctorate in physics from Berkeley in 1923, taught radio physics and electronics at Reed College before the war, and had been at Harvard's Radio Research Laboratory since 1941 in charge of the Special Beacon Group, specializing in radio navigation and guidance systems, and became the first group chief to sign up with the Air Technical Services Command (soon to be reorganized as the Air Materiel Command). O'Day's Navigation Laboratory at the Cambridge Field Station site near MIT was created to continue research on radar beacons for the tracking, guidance, and control of missiles.[92]

O'Day wanted to design precise tracking systems to triangulate the positions and trajectories of missiles, using pulse radar techniques at first, and then automatic cycle-matching schemes that had to be tested on missiles. Thus, he was naturally attracted to the V-2 program because he wanted to see how the passage of a radio frequency transmitter through the ionospheric F region would affect the intensity of natural ion frequencies in that region. O'Day's interests in upper atmospheric research were therefore related to

The Geophysics Research Directorate at the Air Force Cambridge Research Laboratory, circa 1949–50. First row (l to r): Marcus O'Day, Peter Wyckoff, and Nathaniel Gerson. Far left back row: Milton Greenberg. Air Force Geophysics Laboratory.

practical problems in radio communication and navigation, as well as to the development of electronics systems.[93]

O'Day attended the third V-2 Panel meeting at NRL on 24 April 1946, and from then until the next meeting at APL on 3 June he petitioned for membership as a representative of the Air Forces. Krause at first resisted O'Day's bid, because the Air Forces were already represented by the Michigan contractor, William Dow. O'Day's appeals led the V-2 Panel to split the Air Forces' representation: Dow would represent Michigan and O'Day the interests at Watson Labs and Wright Field. This suited Dow as he did not want to be responsible for the Air Forces' larger interests, which included an effort to use the V-2 to develop ballistic reentry techniques.[94]

Starting small in late 1945, O'Day's Navigation Laboratory rapidly gained momentum and staff within the CFS, which itself had grown to some 350 technical and 450 nontechnical personnel by June 1946.[95] O'Day, like Van Allen and Krause, had a strong personal interest in new technology. In the words of an early staffer, O'Day "was not only a lab director but . . . was very much involved

personally in experiments or getting right into the actual lab work."[96]

The interests of the Air Forces could be addressed by all types of upper air research, especially atmospheric composition and ionospheric radio propagation: anything that might enhance its expertise in electronic detection systems and improve knowledge of its medium of operations. These interests became concentrated at CFS as various projects and contracts initiated at the Watson Labs and at the Air Materiel Command were transferred to the CFS, including William Dow's project at Michigan. O'Day thus became coordinator for all Air Forces work on the properties of the upper atmosphere, as well as "those [studies with missiles] whose primary objective was of an engineering nature."[97]

O'Day, acting as contract agent, served Army Air Forces interests and policy, actively soliciting university groups interested in high-altitude research and engineering development. In most cases, O'Day brought in new groups that had not concentrated in research before. Army Air Forces contracts, O'Day later argued, "made it possible to build up good research teams, acquire high-grade equipment and better facilities."[98] There were good arguments in favor of university contracting, such as the need for geographical diversification of technical centers, and the need to support university physics departments to generate and maintain a pool of talent "cognizant of, and thinking about, Air Force problems."[99]

## Air Forces Contractors and Research Areas

Like the Army and Navy, the Army Air Forces hoped to attract the best personnel it could find. If it couldn't hire them, it would contract with them, or with their institutions. Coming in late, without a major captive military laboratory as a base of operations, the Army Air Forces at first contracted with external groups in universities and industry, while it developed its own research facilities.[100] Watson Labs recruitment teams therefore sponsored a series of informal meetings and conferences in late 1945 and early 1946 that attracted many university scientists and potential contractors. Attending one of these conferences, JPL's William Pickering reported that the projects the AAF wished to contract for "are visualized as being of a fundamental research nature resulting in reports." With the Air Forces providing equipment and funding, the work could be published in the open literature "except where military secrecy is required." "The projects must have some bearing on air corps problems," Pickering added but advised that "this is worth following up as being a good way to finance electronic research."[101]

Air Forces policy gave O'Day something less than a free hand, not in choosing contractors, but in choosing research technique. Since the Cambridge Field Station was not allowed to "duplicate the work of any other agency," he had to give up the idea of measuring the earth's magnetic field from a rocket (NRL's domain), of investigating cosmic rays (they were being studied by APL and NRL), or of employing mechanical types of air sampling for composition studies (the Signal Corps' specialty). But this proved to be of little concern, since William Dow's expertise was wholly within electronics, and O'Day could develop pointing controls for solar research, since similar plans had not yet emerged in the Navy.

O'Day scoured the country and exploited university contacts to ferret out new sites for research. Typically, he and his staff would contact people who had demonstrated through publications or reputation the type of expertise they desired to capture.[102] O'Day helped Harvard astronomers Donald Menzel and Walter Orr Roberts develop the Sacramento Peak Observatory, first as a tracking station for planned Holloman Air Force Base missile firings, then as a full coronographic facility.[103] Sac Peak's coronal observations could alert Holloman to launch a probe in order to capture short-lived solar bursts and flares that were the suspected cause of ionospheric disruptions and radio fadeouts. The rocket-borne detectors to be fired from Holloman came out of another related O'Day initiative, this time at the University of Colorado, where contracts were placed to work on solar spectroscopy and to develop solar pointing controls for rockets.

A third major effort O'Day mounted, in concert with the interests of the Air Materiel Command (AMC), was centered at the Franklin Institute in Philadelphia and at other contract sites. It was a plan to extend the V-2 frame by one caliber to improve payload capabilities and to test AMC parachute recovery systems.[104]

Although AAF research policy emphasized contracting, the Cambridge Field Station and O'Day's Navigation Laboratory were not merely technical contracting and monitoring entities. O'Day's own staff was eager to prepare radio propagation experiments itself, and to play an active role in upper air research. Accordingly, a member of his staff reported in June 1946 that even though the first Michigan instruments were beyond CFS's control, subsequent "launchings made in conjunction with the University of Michigan will carry beacon equipment designed and built at Cambridge Field Station."[105]

Internal projects at the Navigation Laboratory were directed toward practical problems in communication and navigation, but they were also designed to complement projects undertaken by its academic and industrial contractors. Typically, the physicists on O'Day's staff understood the technology but needed guidance on the previous studies to consult and observational methodology. For example, staff physicist Henry A. Miley planned to measure sky brightness as a function of altitude using photoelectric sensors and so turned to Harry Wexler, chief of the Special Scientific Services Branch of the Weather Bureau, for suggestions on the literature to read, specific methods, and technique.[106] Other staff members collaborated with William Dow to prepare ambient pressure and temperature instrumentation, and with Wright Field's Personal Equipment Laboratory to develop parachute systems for equipment and data recovery.[107]

Not only did the Air Force come late into the program, but it failed to establish an adequate presence at White Sands or even a sympathetic infrastructure within the service. O'Day and the reorganized U.S. Air Force also remained somewhat outside the collegial V-2 Panel. Although he was a member in good standing, O'Day argued frequently that his programs were not receiving the attention and priorities they deserved in the first years of the panel's deliberations.[108] His contractors, however, received fairer treatment from the panel. Here we confine our attention to the largest early contract O'Day managed in upper atmospheric research, that to the University of Michigan and William Dow. We leave for later chapters other contract groups that became active later in the 1940s, such as the Physics Department of the University of Colorado.

## The University of Michigan and Project Wizard

William Dow (figure p. 90) was not looking for work at the end of the war, nor was he remotely interested in the physics of the upper atmosphere. His involvement came directly as a result of his university's involvement in an AAF antiballistic missile program

William and Edna Dow, summer 1949. William Gould Dow collection. NASM SI 89–21506.

called Project Wizard, created and headed by Emerson W. Conlon, chairman of the Aeronautical Engineering Department, who enjoyed close contacts within the Air Technical Services Command at Wright Field.

Both Conlon and Dow, in Michigan's College of Engineering, took advantage of their university's "effective pattern for sponsored research involvement" developed by its Department of Engineering Research.[109] It could organize large-scale engineering efforts that coordinated military contracts; in physics, it meant that by 1951 Michigan would have the largest number of faculty researchers in the United States and that more than 90 percent of them were sponsored by defense contracts.[110]

In December 1945, ATSC asked Conlon to develop pilotless aircraft "capable of intercepting the V-2 type of missile."[111] Through the Department of Engineering Research, Conlon found campus colleagues with war experience in propulsion, remote-controlled guidance systems, and proximity fuzes who thought that they could build an antiballistic missile system "more quickly and with less expense than a commercial organization."[112]

One of these faculty was William G. Dow, professor of electrical engineering.

On 7 January 1946, Dow and others accompanied Conlon to Wright Field to suggest a long-term relationship.[113] Wright Field agreed and provided one million dollars for a pilot study to determine the performance characteristics of such a missile system and to begin developing instrumentation for obtaining data from the upper atmosphere and for the testing of pilotless aircraft.[114] Conlon was the architect of the overall plan, which included upper atmospheric research. Preparing for a second visit to Wright Field on 21 January, Conlon invited Princeton's Myron H. Nichols to suggest which upper atmospheric studies would be needed to build Michigan's antiballistic missile system.[115] Conlon's notes in preparation for this meeting outlined his rationale:

> *The region above 100,000 feet is essentially uninvestigated and in order to determine accurately the flight characteristics of long-range rockets and their interceptors, densities, temperatures, and composition need to be known. In order to shed more light on radio propagation through and reflections from the ionosphere, more data on the ionization in the upper atmosphere are required.[116]*

Thus like Krause at NRL and Van Allen at APL, Conlon at Michigan became interested in upper atmospheric research as an essential ingredient in its antiballistic missile development program, Project Wizard. Specifically, Wizard would benefit from research on radio frequency propagation in the E and F layers and on the vertical distribution of ions. Michigan would design both the flight equipment, and the necessary ground equipment.[117]

From the beginning, both Conlon and Dow hoped to hire Nichols away from Princeton to head up Wizard's upper atmospheric research effort. They were suitably impressed with his performance at Princeton's telemetry conferences, and Conlon was able to offer him a permanent position. Nichols soon accepted, but he had to finish the spring semester at Princeton, even though he advised Conlon in late January that Michigan had better not wait for him to get started: "the firing of the A-4 rockets is scheduled to start in April and [is expected] to be completed within the next eight months."[118]

At first Dow saw upper atmospheric research as a fact-finding and report-writing activity that was distant from his preferred line of building instruments. Even so, Conlon convinced Dow to step in to get the research organized, finding $60,000 for the first year from a Wright Field contracting officer; this was all, Conlon felt, the university could "intelligently" spend on upper atmospheric research.[119] Dow agreed, admitting that radio frequency propagation instruments were logically the responsibility of the Electrical Engineering Department, and Dow was the campus vacuum-tube expert. He therefore warmed up to the instrument-building side of upper atmospheric research. His patriotism also augured well for compliance: "continued military competence is important to all of us, and it is part of the University's job, where it can, to serve the public interest by technical guidance as well as in education; also, military vacuum tubes offer special technical problems that are interesting professionally."[120]

Patriotism blended well with patronage. If Dow agreed to take on the project, he could hire back from industry some of his best students.[121] Thus as Conlon built Project Wizard, Dow, following Nichols' ideas, gathered together a small instrument group.

### Dow and "the behavior of charged particles in a vacuum"

After Dow attended his first V-2 Panel meetings, he knew that O'Day and the Air Forces were a thorn in the sides of Krause and Van Allen, but he felt personally welcome by the panel, especially by Nichols: "No doubt this was largely because I brought with me money, skilled hands, and know-how in a needed area."[122]

William Gould Dow did indeed bring special skills and a university affiliation. He also brought a competitive spirit and drive that established Michigan as a leading player on the V-2 Panel. As an expert experimentalist "in the behavior of charged particles in a vacuum," Dow later rationalized placing his laboratory on a rocket as a study of an "outer-atmosphere vacuum."[123] William Dow was always more interested in experimentation with vacuum tubes than collecting data about the upper atmosphere. Nelson W. Spencer recalls that his old teacher regarded the upper atmosphere as "the greatest vacuum of all."[124]

Dow, a licensed engineer with a master's in electrical engineering, had been teaching electrical engineering and electronics at Michigan since 1926 with interludes of working in the electrical power industry.[125] In 1943 Dow joined Harvard's Radio Research Laboratory to develop transmitters for radar countermeasures. He soon had about 35 physicists and engineers under him and was responsible for improving high-power vacuum tubes. Dow also became a member of the National Defense Research Committee (NDRC) Division 15 Vacuum Tube (VT) Development Committee, consulted for the Signal Corps, and for three months was liaison for the VT Committee in England. By early October 1945, as he prepared to return to Michigan, Dow was awash in offers, mainly from the Special Projects Laboratory at Wright Field, to act as a consultant and also had contract offers from Harry Diamond at the National Bureau of Standards.[126] But Conlon's initiative drew Dow's attention back to campus and toward Wizard.

## Forming Michigan's Upper Atmosphere Group

By early February 1946, Dow had agreed to take on upper atmospheric research in Nichols' absence. Not willing to be distracted from his primary interests, however, Dow was happy to follow Nichols' advice on how to get started.[127] Nichols shared his ideas on how to probe the ionosphere, but he also wanted to alert Dow to just how far behind the Army Air Forces were and how important it was that he become an activist member on the V-2 Panel.[128] Following Nichols' advice, Dow took his first opportunity at the 27 February

meeting of the panel to identify Michigan's program: to study the radio propagation properties of the upper atmosphere using ground-based microwave radar and low-frequency transmitter beacons in the missile.[129] Since Michigan's first flight was scheduled for June 1946, they had to get to work.

One of the values of the monthly V-2 Panel meetings was to put scientists in contact with patrons. At the 27 February meeting, Dow met an old wartime friend, Lt. Col. Harold A. Zahl, the new chief scientist and director of research at the Evans Signal Laboratories (SCEL), who had just won approval for Army Signal Corps representation on the panel. The two went to lunch to talk over mutual interests, and Dow emerged not only well fed, but closer to his dream of a microwave-frequency vacuum-tube laboratory at Michigan.[130] The hitch was that Zahl wanted Dow to contract with the Signal Corps for upper atmospheric research, too, which Dow agreeably seized as a means of funding Nichols' group when he arrived at Michigan. As he recently described himself, Dow always "wanted to do research on something that somebody wants."[131]

The people Dow brought together to perform upper atmospheric research were, like himself, trained in electrical engineering and were mainly his former students. John Strand returned from Bell Laboratories in February, and Nelson Spencer returned in March from the Cornell Aeronautical Laboratories in Buffalo, where he was working on a Bumblebee subcontract. Joining the others, including Robert Buritz, Alfred Reifman, and F. V. Schultz, Spencer eventually led the small group when Strand was given the larger responsibility of managing Dow's overall involvement in Project Wizard at Michigan's off-campus Willow Run airport facility.[132]

Dow's group worked along two lines: their first responsibility was to carry out Nichols' plan to measure electron and ion densities in the ionospheric E layer using the time delay of a radio-frequency pulse that had a frequency slightly higher than the critical frequency (threshold transmission frequency) of the E layer. Dow thought he had been assured that Wright Field would provide the ground facilities needed to make this work, but they never materialized. By July 1946, with their first launch delayed until August, Dow threatened to bolt from the project unless Wright Field met its obligations.[133] But by this time, Dow's contract had been transferred from Wright Field to Marcus O'Day's Cambridge Field Station, and O'Day had other plans.[134] O'Day felt that Michigan was ill-equipped to take on such a large project, and so he took on the radio propagation experiment for himself, to justify his V-2 Panel membership. O'Day, in fact, knew that Dow preferred to work from his strengths in vacuum-tube development, which had been his second responsibility on his Wright Field contract.

O'Day's reorganization of the contract responsibilities suited Dow just fine. From the summer of 1946 on, his group would build vacuum ion gauges to make in situ density and temperature measurements. This was an activity, Dow recalls, that made "good use of his long-time expertise in measuring attributes of a partial vacuum."[135] It also allowed Dow to turn to another familiar technique to determine positive ion concentrations from measuring profiles of electron temperatures in the ionosphere with a Langmuir probe, a device he had been using since the 1930s.[136]

When Spencer returned, Strand, as group leader and instrument specialist, and Reifman, as theorist, were already deeply involved in designing thermionic ionization gauges and figuring out how their in situ measurements could be converted into useful electron temperatures, pressures, and densities.[137] Everyone knew that "Dow was the boss," as Spencer recalls, but "he was not a directing-type boss," and encouraged cooperation as equals.[138] "We knew each other from the past, and there was plenty to do. And we never had any formal meetings or any formal organization or anything. We just did it." Dow chose staff on the basis of who could get the job done. As an undergraduate, Dow recalls, Spencer could "make anything work. He had helped my efforts in some consulting work involving very complex instrumentation in a client's plant."[139]

In the following two years, Dow built up both his new Microwave Vacuum-Tube Engineering Laboratory and his upper atmosphere group. When Strand moved to Willow Run, he was replaced by Gunnar Hok in 1948.[140] Hok was a sophisticated electronics specialist and with Dow designed Langmuir probes for V-2 flights that Spencer's technicians then built.[141] Although Dow never lost interest in his upper atmosphere group and continued to attend V-2 Panel meetings, he quickly placed others in charge of the detailed design of the instruments and the analysis of data. He was, according to Nelson Spencer, "a restless sort and usually turned over the direction of each lab a few years later."[142]

## Nichols and Dow Meet at 80 Kilometers Altitude

By the time that Myron Nichols arrived at Michigan, in July 1946, Dow's group was firmly established and not about to evaporate, and Nichols' ideas for radio probing had been assumed by O'Day. In the interim, however, Dow had secured support for Nichols from the Signal Corps, and the parties had orchestrated a plan to avoid duplication and conflict. Dow's Air Force contract provided access to missiles assigned to the Air Materiel Command, and his electronic devices worked best at altitudes above 80 kilometers. Nichols' project in the Aeronautical Engineering Department gained access to missiles through the Signal Corps (SCEL) and was devoted to studying the atmosphere below 80 kilometers. Both projects were related to Wizard, and Michigan's Department of Engineering Research was responsible for keeping all the contracting clean and straightforward, but Spencer recalls that this administrative structure created natural competition "across the board" with Nichols' group.[143]

Nichols brought engineers and physicists from Princeton's Palmer Physical Laboratory to seed his group under Conlon in the Aeronautical Engineering Department. They had been working on the Lark missile program at Princeton and in other guided-missile projects under APL's Bumblebee umbrella. Mainly electronics specialists in telemetry, they now started a meteorology program at Willow Run Airport, the site of Project Wizard, to take air samples from rockets and through them study the nature of the 30 to 80 kilometer region, the seasonably variable transition layer between the stratosphere and the mesosphere.[144] Since Nichols was also responsible for creating a graduate curriculum in aeronautical instrumentation, he depended on his staff, led by Leslie M. Jones, to design and build the rocket instruments.[145]

Nichols' choice of mechanical sampling not only helped to avoid direct competition with Dow, but was also considered the best means of sampling the lower and denser regions of the atmosphere to determine its composition. In the quiescent stratosphere, he argued, diffusive separation should promote a composition gradient where light gases dominated at higher altitudes. "Our task was to find out where and how much," Jones recalls, in order to find out where diffusive separation ended, and the mesosphere began. "The deceptively simple approach of capturing samples in rocket-borne bottles was undertaken and constituted the major effort of the Laboratory for over ten years."[146] Air sampling was indeed an important technique for Nichols' group, but it was not the only one they tried in the first years; when most of their first ideas failed, they would eventually return to radio Doppler techniques.[147]

## Harvard College Observatory

Despite his enthusiasm at the January 1946 Washington meetings, Harlow Shapley, director of Harvard College Observatory, did not want any of his staff to become heavily involved in rocketry: such a nontraditional area was too dependent on military funding. They could advise, however, if they wished.[148] He knew that someone from Washington would be visiting Harvard soon to publicize the program, so he advised Menzel that although "the whole thing down there was very attractive," he had serious reservations:

> *For us there is hardly time to do anything about it because they expect the cooperating universities to devise and build up their equipment, electronics & all, and have it ready by early summer. We are all too much tied up; and there will later be American bombs [sic] and we need not depend wholly on the V-2's.*[149]

Without Shapley's support, it was difficult for Harvard astronomers to be involved, even though both Menzel and Fred Whipple, through their wartime military contacts, were gaining some autonomy over their careers. Whipple, although a rapidly rising star at Harvard, still did not have a permanent faculty appointment; he therefore had to be careful and limit his activities on the V-2 Panel.[150]

Shapley was a complex fellow. Facing daunting institutional demands, his first priority was to rebuild Harvard's traditional research programs that depended on its three off-campus observing stations.[151] But he could also play up patriotism, reminding his Observatory Visiting Committee that his staff had performed admirably during the war, at great cost to the Observatory's programs. And in the emerging Cold War, Shapley wondered if "our experts in ballistics and optics [should] turn entirely from their war work?"[152] Shapley looked on research with rockets as war work.

Shapley was, indeed, worried about losing his best staff to military research, and his search for support was constantly compromised by this fear.[153] But for all his concern over a growing military presence on campus, Shapley could not prevent Menzel or Whipple from pursuing, within limits, personal research programs with Army and Navy patrons. Of the two, Whipple had interests of more direct military concern and so would become Harvard's representative to the V-2 Panel—a service Shapley saw as politic. Shapley was simply unable to offer viable alternatives, later rationalizing to the Visiting Committee that the promising results of Whipple's photographic techniques were made possible by the Navy, "with gratifying freedom to the Observatory for publication of results."[154] Shapley did not mention that Whipple's *data* were initially classified by Naval Ordnance.[155]

### Fred Whipple and Upper Atmospheric Research

Fred Lawrence Whipple (figure p. 95) attended the first V-2 Panel meeting at Princeton on 27 February 1946 as Harvard's representative. Unlike the other charter members of the panel, Harvard did not represent any of the services, nor did Whipple, in deference

to Shapley's wishes, intend to be a "working member" actively pursuing experiments to fly on rockets. Thus, Harvard violated the primary criteria for membership on the panel.

But Whipple's presence was important to the panel; it demonstrated that at least one recognized scientist felt that the panel was engaging in a worthwhile scientific activity. From the 1930s, Whipple had been a student of meteor orbits and origins, but he also was aware, partly as a result of following the progress of scientific ballooning, that there was a serious lack of knowledge of conditions in the upper atmosphere beyond bal-

**Fred L. Whipple circa 1952 with the Baker Super Schmidt Meteor Camera. American Institute of Physics Niels Bohr Library.**

loon heights. In 1938 he pointed out, as had many before him, that meteor photography could yield "direct information concerning the physics of the atmosphere."[156] With his peripatetic Estonian mentor Ernst Öpik, as well as colleagues Peter M. Millman, J. A. Pierce, and others at Harvard, Whipple was in the midst of an array of meteor researchers who employed spectroscopic, visual, and radio techniques to study both meteor phenomena and their effect on the atmosphere and ionosphere.[157] Whipple thus was in a position to appreciate both the astronomical and the geophysical aspects of meteor research.

During the war, Whipple developed radar countermeasure techniques as a member of Harvard's Radio Research Laboratory. Whenever possible, he continued his astronomical interests; in 1942 he prepared an invited paper explaining how meteors could be used to probe the upper atmosphere, which later was cited by V-2 Panel members as the standard work on conditions in the upper atmosphere, against which their in situ studies had to be judged.[158] Whipple's 1943 study, "Meteors and the Earth's Upper Atmosphere," was also an important vehicle through which he gained patronage after the war. Thus, Harlow Shapley sarcastically observed to his staff in August 1946 that the Navy's Bureau of Ordnance was willing to fund Whipple's meteor research "because the shooting stars perform in that same level of the earth's atmosphere where the shooting rockets and the rocket ships of the future are planning to operate."[159]

At the February 1946 V-2 Panel meeting, Whipple announced that Harvard and MIT would conduct ground-based meteor observations to determine atmospheric densities and pressures. He also hinted that Harvard might provide small packages for biological and solar studies, all passive, but by the end of April, these were clearly secondary interests.[160] Whipple's attention was focused primarily on his growing meteor program.

Through astronomers he knew in the Bureau of Ordnance, Whipple found funding for his meteor research, even though one of them, P. C. Keenan, warned that "backing by one of the services means that it will be hedged about with classifications and restrictions."[161] But Whipple was in a hurry and could not afford to wait for civilian support. He had to find funds to maintain his growing program, which included the services of astronomers Zdenek Kopal, based then in the Electrical Engineering Department at MIT in its Center of Analysis, and Luigi Jacchia, in Harvard's Astronomy Department. The differential analyzers available to Kopal were critical to Whipple's meteor project.[162]

The Ordnance Department was interested in high-speed aerodynamics and exterior ballistics. They were thus interested in what high-velocity meteor trails would reveal, and hence they were interested in Whipple. By the end of January 1946, Whipple, Shapley, and their Harvard deans had negotiated a two-year $60,000 contract with BuOrd for meteor observations and reductions and started talking about a long-range program at one of the Naval research stations.[163] It was in this context alone, that of studying conditions in the high atmosphere, that Whipple remained active on the V-2 Panel, even though his contacts with other governmental and scientific bodies would prove vital to the health of the panel.

Whipple became more a critical networking agent than a working activist in maintaining support for upper atmosphere research, in his multiple capacity as panel member, member of the Joint Research and Development Board (JRDB), and his memberships on many geophysical panels. Whipple also consulted on many technical problems the V-2 panel faced, and his own research continued to inform the panel's activities, including its detailed estimates of upper atmospheric conditions, seasonal variations in the density of the upper atmosphere, aeroballistic heat transfer, and winds in the upper atmosphere.

## Alignments and Goals

We have now been introduced to the key players and institutions represented by the V-2 Panel who were active in the first years of rocket research, learning something of the arguments both the scientists and their military patrons used to justify upper atmospheric research with the V-2. Those arguments, as we have found, were closely aligned. Each group saw upper air research as a complement to the primary goal of developing ballistic missile systems, radio and radar detection systems, and ballistic missile countermeasures systems.

The members of the V-2 Panel had at least one goal in common: to build analytic tools that could solve specific problems of interest to their patrons. Save for Whipple, none of the organizers of upper atmospheric research were professional geophysicists, meteorologists, or astronomers in the traditional disciplines served by rocket research. Some, such as James Van Allen, Serge Golian, or Gilbert Perlow, had training in these fields and specific research interests, but only Whipple had actually established a research agenda that could be addressed by placing scientific instruments on rockets. Since the problems they would now address were by and large new to them, most panel members organized their research around the vehicles the military provided and the tools they could build, with a view to uncovering the information the military needed to build better missile systems.[164] In so doing, they aligned scientific research with national interests, particularly those of national security.[165]

The first practitioners of scientific rocketry soon found that their world had expanded to include a new site for research, one far removed from the traditional campus laboratory. Nowhere was the context for such research more vividly defined than at Army Ordnance's White Sands Proving Ground in New Mexico.

# Notes to Chapter 6

1. Names come from rosters in the technical reports of each of the groups, as well as from authors of, and contract groups identified in, relevant bibliographies and in the contractor files of the Air Force Cambridge Research Center. See Benton (1959); and Library files, CF/AFGL.

2. See, for instance, J. J. Bartko to Frank J. Malina, 21 February 1946. NRL/NARA.

3. See Dennis (1990, 1991) for an examination of the origins of APL's style and managerial practices that created the fuze and the wherewithal to move it into production.

4. Information on the backgrounds of AMC contractors is from contract reports listing vitae. CF/AFGL.

5. Leak (1954), p. 1. To the extent known, these seem to be internal NRL classifications, and not Civil Service definitions. The degree to which the scientists classified themselves is unknown.

6. Newell (1980), p. 33.

7. At the time Krause was completing his doctoral dissertation, A. Hoyt Taylor, a former Wisconsin professor and now chief radio scientist at NRL, informed the Wisconsin physics department that NRL needed physicists. Krause responded quickly to a contract position, and in September 1938 moved to NRL. He recalls an immediate fascination with the radar program at NRL under R. M. Page and Arthur A. Varela. See Krause OHI, pp. 9—13rd; Allison (1981), pp. 41—43.

8. Krause OHI, p. 27rd.

9. See Rosen, Kuder, and Pettitt (1945).

10. Rosen to the author, 10 October 1990, p. 2.

11. As historians Bruce Hevly and David Allison have shown, basic scientific research was not NRL's strength before the war. The laboratory performed tasks delegated by Navy bureaus, such as the testing and evaluation of equipment. Still, under the leadership of A. Hoyt Taylor, physicists such as E. O. Hulburt performed first-rate radio propagation studies, and it was to Hulburt's legacy of ionospheric research that Krause pointed as he argued that basic research be established at NRL as a component of a new initiative in ballistic missile development. See Allison (1981), and especially Hevly (1987), who contends that Hulburt was to a large extent responsible for demonstrating that basic research was possible at NRL.

12. Laboratory Order 28—45, June 1945 and Laboratory Memorandum 78—45, 3 July 1945, A3—2(B) memo 270—81/45. HONRL. See also Hevly (1987), pp. 57, 113.

13. Jerry Kluttz, "New Hopes to Expand Research Laboratory," in "The Federal Diary," *Washington Post* 10 September 1945. Item 18 in Krause clipping file, SAOHP.

14. Ernst Krause "Manuscript Biography," p. 6. EHK/NASM.

15. J. J. Bartko, "Minutes of Committee on Guided Missiles Conference," Record of Consultative Services, 16 October 1945, p. 1. Meeting date, 3 October 1945. Box 13, S-F42–1/84, NRL/NARA. Members included representatives from metallurgy, chemistry, physical optics, sound, and Navy consultants on aerodynamics, along with a U.S. Army liaison officer. On the structure of these committees, see Hevly (1987), pp. 82–84.

16. See, for instance, Laboratory Order 41–46, 26 November 1946, A3–2(8), Order 176 "Reorganization of Radio Divisions," reference a: "Research and Development Program of the Naval Research Laboratory, October 1946." HONRL. On the origins of ORI, which soon became the Office of Naval Research, see Sapolsky (1990).

17. Excerpted from J. H. Garvin to "Distribution," 13 December 1945, "NRL Policy" A3–2(8) (200–106/45). HONRL.

18. The only hard evidence of these deliberations comes from a mention in Ernst Krause, "Report on a Proposed Guided Missile Program for the Naval Research Laboratory," 3 December 1945. HONRL, copies in EHK/NASM; and in Serge Golian, "Registered Laboratory Notebook no. 1," entry for 12 December 1945. HONRL. Here Golian lists the types of observations possible from rockets. The next several entries in his notebook record his reading and analysis of reviews of indirect estimates of the composition of the upper atmosphere, to 140 kilometers.

19. Rosen (1955), p. 19; Rosen to the author, 10 October 1990. Rosen OHI, pp. 48–67; and conversations with the author, November and December 1990; Krause OHI. See also Newell (1980), pp. 33–34.

20. Rosen OHI, p. 55rd. Bruce Hevly (1987) has explored how Hulburt's research on the ionosphere at NRL before the war influenced the laboratory's postwar agenda. Rosen's recitation however, was more political than intellectual.

21. Rosen (1955), p. 20; Rosen OHI, p. 60rd.

22. Ernst Krause "Manuscript Biography," page 6. EHK/NASM. Thor Bergstralh, who had been a technical liaison specialist at MIT's Radiation Laboratory during the war, remembers that during one of Krause's recruiting tours for NRL, with Homer Newell and Gilbert Perlow, they talked of continuing research on radar and short-range infrared guided missile systems. But by the time Bergstralh switched to NRL to work on infrared detection and guidance systems, Krause's group had just decided to move into upper atmospheric research, "which as a matter of fact I was quite happy with. I was delighted to be able to work in that rather than in guided missiles." Thor Bergstralh OHI, p. 18, see also pp. 16–20.

23. See Ernst Krause, "Report on a Proposed Guided Missile Program for the Naval Research Laboratory," 3 December 1945, p. 3. HONRL; copies in EHK/NASM. Krause defined "basic" and "applied" research at the outset of his report as undirected and directed research; the former was research into "the fundamental composition and properties of the materials which constitute the universe and the study of the phenomena which surrounds and relates these materials." Krause felt strongly that although both were needed, they had to be completely independent of one another. Otherwise, applied research would dominate in an institution such as NRL. See Hevly (1987), chap. 2 for a discussion of basic and applied research, as defined within the institutional context of NRL. Krause's contention that other parts of the Navy were planning stems from his knowledge of Comdr. Harvey Hall's efforts. By October 1945 Hall had established a committee to evaluate the feasibility of putting a satellite into space. See Hall (1963), p. 414; Stares (1985), p. 25.

24. Krause, "Report . . ." 3 December 1945. Ibid., p. 10.

25. Ibid.

26. Ibid., p. 7.

27. Ibid., p. 8.

28. Memorandum, Special Committee on Guided Missiles to Director's Council, undated (document is bracketed by Krause's presentation on 3 December and the formal laboratory orders that created the missile program, dated 17 December 1945). HONRL, copies in EHK/NASM.

29. Two of the preexisting divisions were renamed and restructured. The Fire Control Division, headed by R. M. Page, became the Fire, Missile, and Pilot-less Aircraft (F,M&PAC) Control Division. The Special Electronics Research and Development Division, headed by J. M. Miller, became the Electronic Special Research Division, and Krause's Guided Missiles Subdivision contained within it became the Rocket-Sonde Research Subdivision. See "Changes in Organization of Guided Missiles Research Program," Laboratory Order no. 47–45 (17 December 1945). HONRL, copies in EHK/NASM.

30. "Function of the Rocket-Sonde Research Subdivision," Laboratory Order no. 46–45 (17 December 1945). HONRL, copies in EHK/NASM.

31. H. A. Schade, "V-2 Works for Science," enclosure A to G. L. Webb to L. M. Slack, ORI, 31 May 1946. Box 98, folder 2, NRL/NARA.

32. Krause OHI, pp. 32–33; Krause OHI added commentary, pp. 46–47. Milton Rosen, however, had no recollection of Krause ever saying this. Rosen OHI, pp. 65–67rd.

33. Ernst Krause, "Report on a Proposed Guided Missile Program for the Naval Research Laboratory," 3 December 1945, p. 5. HONRL; copy in EHK/NASM.

34. Krause "Report . . ." pp. 5–6. Newell (1959b), pp. 235–36.

35. On the Corporal program, see Koppes (1982). Rosen, a product of the University of Pennsylvania's Moore School of Electrical Engineering, was hired at NRL one year after Krause to work on radio pulse techniques, and like his NRL colleagues, he had experience with missile research but it was confined to radio guidance and countermeasures.

36. These negotiations took place during the March 1946 "Guided Missiles and Upper Atmosphere" symposium at JPL. See M. W. Rosen "Record of Consultative Services," 25 March 1946. NRL/NARA. See also Rosen (1955). Rosen OHI, p. 76rd; and letter, Rosen to the author, 10 October 1990, p. 3.

37. Golian had worked under Marcel Schein in high-energy physics at Chicago, and Perlow, who also had been a graduate student at Chicago working in nuclear physics under Sam Allison, had interests that naturally drew him to cosmic-ray studies. Gilbert J. Perlow to the author, 23 February 1987; and Thor Bergstralh OHI, p. 20rd.

38. See Elanor Pressly registered notebook no. 6322, entries between 18 March and late May 1946. HONRL.

39. See J. C. Seddon, registered laboratory notebook no. 304, p. 92, entry for 28 August 1952. HONRL.

40. Schade granted Krause's group section status in late January to give it a higher staff ceiling (technically 66) to cope with the complex liaison and service tasks Krause required if he was to maintain provider status on the V-2 Panel, and to match the status enjoyed by Tousey's section in the Physical Optics Division. Krause to Code 101, 28 March 1946. NRL/NARA.

41. Tousey OHI, pp. 75–76rd. John Sanderson accompanied Hulburt to the 16 January 1946 meeting, and recalls that as they returned to their offices, Hulburt suggested that Tousey perform ultraviolet research with the V-2. Sanderson to the author, 21 December 1990.

42. Tousey OHI, p. 109rd. Krause recalls only that Hulburt was "most enthusiastic" for his proposal. Krause OHI, p. 54rd.

43. Baum OHI, p. 9rd; and William A. Baum registered notebook no. 5401, March/April, 1945. HONRL.

44. F. S. Johnson OHI, p. 22rd; see also Tousey OHI, p. 79rd.

45. Hulburt's ability to bring Tousey and hence part of his Optics Division into the rocket work was a result of his senior position at NRL, as well as the fact that much of what Krause hoped to accomplish in scientific research had been Hulburt's territory in the prewar era. Hulburt's seniority, the respect Krause, Tousey, and others held for him, and his quiet diplomatic style earned for him continuing influence; in 1949 he became NRL's first civilian director of research. See Hevly (1987), pp. 26–27, 45–46. Tousey OHI, pp. 79–80rd; no. 2, p. 104rd.

46. The vacuum ultraviolet is the spectral region blocked primarily by continuum absorption due to atomic oxygen and ozone roughly between 300 and 10 nanometers. Tousey was aware of Lyman's interest in detecting the far ultraviolet spectrum of the sun beyond the atmospheric cutoff, but he never participated in that interest. Tousey OHI, pp. 7–11rd. See also [Tousey] (1961).

47. Tousey OHI, pp. 37–43rd. In particular, Tousey sought a better understanding of the oil emulsion process and helped to improve methods of calibrating intensities derived from these types of photographic plates by examining how they deviated from the known laws of photographic reciprocity.

48. Tousey recalls meeting the Hulburt family on several occasions in the summers before he knew who Hulburt was. Tousey OHI, p. 54rd. See Tousey (1936, 1939).

49. Tousey OHI, p. 57.

50. E. B. Stephenson, "Request for Assignment of Problem," 27 February 1947, enclosure A: Supporting Data, p. 1. Box 99, folder 5, NRL/NARA. The first 10-channel system was based on one they had developed for the JB-2 Loon missile during the war and is described in Garstens, Newell, and Siry (1946), chapter IIC. NRL telemetry failed frequently in the first flights in the summer of 1946, which caused Krause to redouble efforts to improve the system. Krause's crusade won the confidence of other members of the panel, such as Van Allen, whose APL group was developing a competing system. James Van Allen to Jesse Greenstein, 21 June 1946. JGP/CIT. The NRL 23-channel system proved to be highly reliable. See C. H. Smith to Code 120A, 7 January 1947. Box 99, folder 5, NRL/NARA. By early 1947, an improved 30-channel system was undergoing laboratory tests, and plans were progressing to increase the number of channels to 450. See V-2 Panel reports during the summer of 1946 and through 1947.

51. The APL history presented here follows Dennis (1990, 1991). See also Gibson (1976); and [APL] (1983). For a general survey of other major participants in developing proximity fuzes, see [National Bureau of Standards] (1945).

52. M. A. Tuve to C. L. Tyler, 10 October 1944, pp. 1–2. Box 101, "Notes on Post War Research" file, MAT/LC.

53. Merle Tuve to Director's files, 5 September 1945. Box 116, "APL Administrative Memoranda," MAT/LC.

54. G. F. Hussey, Chief BuOrd to Coordinator, Research and Development, 11 January 1945. JVA/APL.

55. The Guided Missile Committee of the Joint Committee on New Weapons and Equipment of the Joint Chiefs of Staff was developing a series of technical panels in late 1945 similar in character to the activist V-2 Panel. Tuve was chairman of the Committee's Propulsion Panel; others were headed by Clark B. Millikan of JPL/Caltech (Aerodynamics and Ballistics); W. A. MacNair of Bell Laboratories (Guiding and Control); R. E. Gibson of APL (Launching); and R. H. Kent of Aberdeen (Warheads and Fuses). They were formed to "provide the Guided Missiles Committee with advice and recommendations concerning the technical policies and scientific aspects of the national program" and to "foster liaison and to increase the effectiveness of interchange of information and ideas between specialists engaged in guided missiles research and development." See "Announcements," Guided Missiles Technical Panels. Announcements of 8 December 1945 and 5 January 1946. JCNWE, 911 J859, OCO/NARA. On Tuve's prominence in postwar planning, see also Kevles (1975).

56. On Bumblebee, see Gibson (1976), pp. 8–14. See also Dennis (1991).

57. APL also entered the visible world of civilian research through this shift to university-based contracting and thereby aligned itself with the type of research conducted on a university campus. M. A. Tuve to Director's Office Files, "The Postwar Situation and the Policy which it Dictates for the APL," 5 September 1945 (transcript of dictation), p. 12. Box 116, "APL Administrative Memoranda," MAT/LC.

58. APL's staff defined research as "normal peace-time research whose function is to explore and to discover qualitatively new fields . . ." within the sphere of looking for "new methods, techniques, and weapons which will drastically alter the course of a war" without consideration of their applicability. Quotes taken from "Young Turks" Committee to M. A. Tuve, 7 September 1945, pp. 4–5. Box 114, "APL Administrative Files, September 1945," MAT/LC. Their sense of "fundamental military research" is similar to the concept of basic research being conducted within a military laboratory: it is open-ended, but informed by the mission of the institution.

59. Robert B. Roberts to Tuve, 28 January 1946, p. 1. Box 119, "Blue Notebook," MAT/LC.

60. Nichols had produced a successful telemetry system based on the technique of frequency-

deviation multiplexing, in distinction to NRL's pulse-time modulated telemetry design. On Princeton's telemetry system, see "Papers Presented at: I. Princeton Telemetering Symposium; II. APL Telemetering Conference," *Bumblebee Report no. 42* (December 1946). JPLHF5–449, JPL/L.

61. Nichols to Tuve, 26 November 1945, p. 2. Within this letter, Nichols also noted that he had made this proposal originally to H. D. Smyth, a senior member of the department. Box 119, "Unmarked Black Folder," MAT/LC.

62. Leitch (1978), p. 365. The Palmer Physics Laboratory, since the days of Owen W. Richardson, had been a leading site for experimental research in electricity and magnetic phenomena, and later in electronics. Both nuclear and high-energy physics had been central programs at Princeton since they decided to build a cyclotron in the 1930s. See "Owen Willams Richardson," *Dictionary of Scientific Biography*, pp. 419–22; Thorp, Myers, and Finch (1978), pp. 181–82.

63. These included, with names attached: (1) spectroscopy and physics of the upper atmosphere (R. Ladenburg, A. Shenstone, H. D. Smyth, and L. A. Turner); (2) nuclear physics and cosmic rays (R. Ladenburg, H. D. Smyth, L. A. Turner, J. A. Wheeler, M. G. White, and E. P. Wigner); (3) electronics, microwaves, telemetering (W. B. Roberts, L. A. Turner, and M. G. White); (4) graduate students who would be attracted to study with the above men; (5) a large facility already organized to handle electronics and telemetering problems. Nichols to Tuve, 26 November 1945, p. 1. Box 119, "Unmarked Black Folder," MAT/LC.

64. Noted by J. A. Wheeler, "General Survey of the Princeton Project Program—Cosmic Rays and Telemetering," 28 August 1946. Appendix IV to H. D. Smyth, "Annual Report NOrd 7920," n.d. PP/P.

65. Nichols appended a 1 November 1945 memorandum prepared by Ladenburg that elaborated on these research areas and advocated the direct involvement of prominent scientists experienced in problems of the upper atmosphere: "It is, therefore, suggested that experts of the fields concerned should be asked for cooperation, either by giving advice about the best methods to be used or

even by supplying the necessary instruments." R. Ladenburg, "Memorandum Concerning Study of the Upper Atmosphere," 1 November 1945. Appended to Nichols to Tuve, 23 November 1945, p. 3. MAT/LC. Ladenburg enjoyed an international reputation in many fields; at the time he was a member of the International Subcommittee on Ozone for the Royal Society's Gassiot Committee, and he claimed that his colleagues, including Sydney Chapman of Imperial College, G. M. B. Dobson of Oxford, and others would be "most interested . . . in the contemplated high altitude studies by use of rockets." Ibid., p. 3.

66. C. H. Smith, "Record of Consultative Services," 20 February 1946, p. 1. Meeting date: 18 February 1946. Box 32, NRL/NARA.

67. On Tuve's original goals for postwar APL, see M. A. Tuve to Director's Office Files, "The Postwar Situation and the Policy Which It Dictates for the APL," 5 September 1945 (transcript of dictation), p. 12. Box 116, "APL Administrative Memoranda," MAT/LC. I am indebted to Michael Dennis for providing this material.

68. See M. A. Tuve, "Contracts Division Plan" fragment notes, 2 January 1946. Box 114, "TC Technical Planning and Programs," MAT/LC.

69. Robert B. Roberts to Tuve, 28 January 1946, p. 1. Box 119, "Blue Notebook," MAT/LC.

70. Van Allen recalls that they suffered mild derision for "getting a free ride" from the weapons research that continued as the staple of the laboratory. Van Allen OHI, p. 157rd.

71. C. H. Smith, "Post Conference Notes," Record of Consultative Services, 13 February 1945, p. 4. Box 32, NRL/NARA.

72. See R. E. Gibson to files, 17 and 18 April 1946. "R. E. Gibson Chron Files, 1946." APL Archives, noted in Dennis (1991), p. 10 n. 31.

73. Robert B. Roberts to Tuve, 28 January 1946, p. 1. Box 119, "Blue Notebook," MAT/LC.

74. Van Allen to Nichols, 25 February 1946. NRL/NARA.

75. Van Allen (1983), p. 17. Fraser had left APL for Kodak in late 1945 but was encouraged to return in April 1946 to the high-altitude group. Lorence Fraser OHI, pp. 65–67rd.

76. Van Allen OHI, p. 174rd.

77. "The Johns Hopkins University Applied Physics Laboratory," 7 June 1946, p. 20. Box 116 "Section T/OSRD&JHU," MAT/LC.

78. Fraser OHI, pp. 95–96rd.

79. On early identifications of staff, see Van Allen to Nichols, 25 February 1946. NRL/NARA; "Record of Consultative Services," 21 February 1946. Box 32, NRL/NARA.

80. Van Allen OHI, p. 109rd.

81. Van Allen has been the subject of several biographical profiles. See, for example, Thomas (1960), pp. 111–32; and "James A. Van Allen," *Bulletin of the Atomic Scientists* (December, 1973). Van Allen (1983, 1990) has also written personal memoirs on his scientific career.

82. Van Allen OHI, pp. 59–60rd.

83. Forbush in 1939 discovered a correlation of cosmic-ray flux with the 11-year solar cycle.

84. "James Alfred Van Allen." Biographical statement in SAOHP/NASM, p. 10; Van Allen OHI, p. 87rdff.; [APL] (1983), pp. 4ff.

85. Van Allen OHI, p. 124rd.

86. See "High Altitude Research at the Applied Physics Laboratory of the Johns Hopkins University—A Brief Summary," 10 April 1947, p. 2. "V-2 Survey" folder, HT/DTM.

87. Van Allen's first thought was to mount a Geiger counter in a V-2 to detect cosmic-ray primaries. He would also record cosmic-ray events as a function of altitude with single counters and with coincidence counters and would use thick emulsion plate stacks and Wilson cloud chambers to record the nature of the rays. Solar physics studies would record the ultraviolet spectrum of the sun to 60 nanometers as a function of altitude; the group was also interested in measuring the absolute value of the solar constant "at extreme altitude." Atmospheric physics experiments included air sampling, in situ composition studies using a spectrometer and capillary spark, and skin temperature measurements. James A. Van Allen, "Abstract of Paper Presented at Aeronautical Symposium—Experiments Planned by the Applied Physics Laboratory, Johns Hopkins University," 16 March 1946, n.d. Box 311, MAT/LC; and "Abstracts of Papers Presented at the Guided Missiles and Upper Atmosphere Symposium" held 13–16 March 1946. JPL-GALCIT Publication no. 3 (February-March 1946), pp. 144–45.

88. Van Allen to L. R. Hafstad, 29 March 1946, p. 1. See also J. A. Van Allen "Rough Quantitative Considerations in the Use of Cosmic-Ray and Ultra-Violet Altitude Meters," 2

April 1946. Box 118, "Bumblebee Series of Missiles," MAT/LC.

89. E. H. Krause, "The Accomplishments of the First Year of Rocket-Sonde Research in the Upper Atmosphere," chap. 1, p. 1, in Newell and Siry (1947a).

90. The many name changes in large part reflect the restructuring of Air Force research and development policy, especially by factions within the Air Force that wished to distinguish themselves from the Air Materiel Command. See Malcolm R. Fossett, Jr., "Demise of the Air Force In-House Basic Research Laboratories," n.d., Air War College Research Support Summary no. 5911, pp. 10–11.

91. On the history of the Cambridge Field Station, see Julius King, Jr., *History of Air Force Cambridge Research Center* (Air Research and Development Command, USAF, n.d. AUD5); Sigethy (1980), p. 27ff.; Liebowitz (1985), p. 1ff. In addition to personnel, tons of equipment surplus to the MIT Radiation Laboratory effort when it closed were divided between the Army and Navy, the former's portion coming to CFS, which was, in effect, the AAF's expressed desire to continue what had been started at the Radiation Laboratory during wartime.

92. Biographical data on O'Day provided by Ruth Liebowitz, AFGL historian, compiled from a biographical database for AFCRL scientific personnel, 7 December 1960. The primary assignment was for radar navigation aids, O'Day personal statement, "Exhibit 1" in ibid. See also Jules Aarons OHI, p. 1.

93. Jules Aarons OHI, p. 2.

94. Megerian to Bain, "V-2 Report no. 4," 6 June 1946, p. 3. V2/NASM. Dow's contract with the Signal Corps was on behalf of Nichols, who was being courted for Michigan at that time. O'Day felt that Dow's willingness to be replaced as liaison to the V-2 Panel for Air Materiel Command interests was based upon his knowledge of Project Blossom. The Army Air Force wanted to replace the warhead section of a series of V-2s with the front section of an experimental supersonic aircraft under development. The craft would be ejected from the high-velocity V-2 and would then be retrieved by parachute. O'Day pointed out in 1954 that "a great portion of the Air Force effort in rocket-borne research

was concerned with parachute studies at supersonic speeds." O'Day recalls that Dow realized that his liaison position was incompatible with University duties: "The knowledge of the performance of the parachute seemed to be of equal or greater importance than the successful lowering of the scientific equipment." O'Day (1954), p. 2.

95. Liebowitz (1985), p. 2.

96. Adolph Jursa OHI, p. 8.

97. O'Day (1954), p. 2. Specific engineering projects O'Day later identified that were given high-priority status at Wright Field were (a) parachute studies; (b) measurements of the rise in skin temperature at many points on the rocket; (c) development of special gyros for aspect measurements; and (d) studies of the behavior of animals in a space vehicle. Kennedy (1983), pp. 63–64, outlines some of the physiological and biological experiments performed on V-2 fights, which ranged from sending spores, seeds, and fruit flies into space for mutation studies, to flights of monkeys and a mouse.

98. O'Day (1954), p. 3.

99. Ibid., pp. 2–3.

100. See Sigethy (1980), pp. 22, 25–27; and on the research policy of the Air Forces, see Malcolm R. Fossett, Jr., "Demise of the Air Force In-House Basic Research Laboratories," n.d., Air War College Research Support Summary no. 5911.

101. William Pickering to H. Victor Neher, n.d. [circa February 1946]. HVN/CIT.

102. Adolph Jursa, a later member of O'Day's staff, gained early experience as a student working for one of the CFS contractors at the University of Rhode Island. He recalled that his professor, John Albright, was contacted by O'Day's people "to build some early instrumentation on rockets for measuring upper atmosphere conditions, based on what they read from a book that he authored." Adolph Jursa OHI, p. 2.

103. Aarons OHI, p. 2; see also Walter Orr Roberts OHI, pp. 98–118rd; Menzel—Roberts correspondence, 1947–50. WOR/UC; and Bushnell (1962).

104. This was a top priority program in support of the Air Force's interest in using the V-2 as a booster for propelling experimental supersonic aircraft mockups into the high atmosphere, to release them and study their aerodynamic characteristics upon reentry.

The first stage in this experimental program was the Blossom Project, which grew to involve many university and industry contractors, including the Cook Research Laboratories in Chicago, Boston University, the Bell Aircraft Corporation, and Temple University, working in conjunction with AMC specialist laboratories at Wright Field and elsewhere. See contractor reports, Franklin Institute Laboratories for Research and Development, Mechanical and Civil Engineering Division, Contract AF-19(122)-33. AFGL Library.

105. Gilbert O. Hall, "V-2 Program of the Navigation Laboratory," 12 June 1946 Status Report WLCES-2A. MOD/AFGL. Hall was identified as liaison engineer in the Navigation Laboratory for V-2 activities.

106. Henry A. Miley to Harry Wexler, 18 July 1946; Wexler to Miley, 30 July 1946. HW/LC.

107. [Navigation Laboratory], "Report To V-2 Panel Meeting of 25 March 1947," 21 March 1947, pp. 1–3. WLCAL–1–21, MOD/AFGL. Other planned activities included ionospheric wave propagation effects; composition profile measurements of the upper atmosphere using on-board spectroscopic observations of high-power radio frequency arcing; upper air turbulence investigations using gaseous ejection from the missile; acoustic measurements of the interior of the V-2; and a study of the acoustic noise spectrum in the atmosphere by means of an ejected microphone system.

108. See Homer Newell, "Record of Consultative Services," 12 September 1947, routing slip comments. Box 99, folder 9, NRL/NARA. See Marcus O'Day to NRL, 24 February 1950 and accompanying routing slip commentary; J. W. Goodwin to NRL, 2 February 1950, and accompanying routing slip comments; NRL to Members UARRP, 3 January 1950, and accompanying routing slip commentary. Box 101, folder 26, NRL/NARA. See also Ernst Krause OHI, p. 53.

109. Quotation from William G. Dow to the author, 25 October 1986, p. 2. See also Robert P. Weeks, "The First Fifty Years" (typescript draft of history of Department of Aeronautical and Astronautical Engineering, October 1964), p. 19. Box 15, Michigan, College of Engineering Collection, BHLUM.

110. See Forman (1987), table 1, p. 195. In 1969 Michigan was second only to MIT in re-

ceiving funding from DoD. See Nelkin (1972), p. 27.

111. Emerson Conlon to Commanding General, ATSC, "Proposal for Ground-to-Air Pilotless Aircraft Research Program," 26 January 1946. WGD/BHLUM.

112. Ibid.

113. [E. W. Conlon] "Pilotless Aircraft Research Program—Conference at Wright Field January 7—Memorandum," 8 January 1946. WGD/BHLUM.

114. Michigan would also study supersonic aerodynamics, radio, and radar methods of target location and infrared detection devices. Ibid.

115. Nichols was already known to Conlon, primarily through the Princeton telemetry conferences that were being sponsored by APL/Bumblebee. Nichols had sent out an announcement for the February conferences, and both Dow and Conlon attended. Dow met Nichols there and discussed the possibility that he might come to Michigan with part of his Bumblebee group. Dow to the author, 25 October 1986.

116. Conlon to Commanding General ATSC, Wright Field: re "Proposed Contract for Development of Instrumentation for Pilotless Aircraft. Draft proposal to be used for discussion during meetings at Wright Field on 21–22 January," 19 January 1946. WGD/BHLUM.

117. Fragment titled "Exhibit A," 10 January 1946, WGD/BHLUM.

118. E. W. Conlon, "Memorandum: Research Program for Wright Field in Pilotless Aircraft," 25 January 1946. WGD/BHLUM.

119. See, for instance, Dow to Conlon, 16 January 1946. WGD/BHLUM.

120. Dow to Maj. Walter N. Brown (Special Projects Lab., Wright Field), 11 October 1945. Box 4, folder 2, WGD/BHLUM.

121. E. W. Conlon, "Memorandum: Research Program for Wright Field in Pilotless Aircraft," 25 January 1946. WGD/BHLUM.

122. Dow to the author, 25 October 1986.

123. Ibid.

124. Nelson Spencer OHI, pp. 35–36. In 1953, when planning to attend summer meetings at Oxford on the upper atmosphere, Dow spent most of his time and energies planning for tours of British vacuum-tube facilities. See Dow to Guy Suits (GE), 18 May 1953, and Dow to H. C. Steiner (GE), 18 May 1953; Dow to A. L. Samuel (RDB Panel On Electron Tubes), 25 February 1953, A. L. Samuel to Dow, 7 April 1953; and Dow to A. W. Schrader, 30 June 1953. Box 4, WGD/BHLUM.

125. William Gould Dow, "Application for Federal Employment," 8 October 1945. WGD/BHLUM; and W. G. Dow to the author, 8 August 1990, pp. 18–31.

126. Dow preferred the NBS contact as it built on his RRL responsibilities: "It represents a continuing contact with the military needs for special tubes, which is professionally interesting, and offers an opportunity to keep in active touch with ultra-high-frequency tube development techniques . . ." Dow to A. H. Lovell, Chairman, Department of Electrical Engineering, University of Michigan, 7 October 1945. Box 3, NBS Correspondence, WGD/BHLUM. On Dow's reluctance to jump prematurely into Wright Field activities, see Dow to Maj. Walter N. Brown (Special Projects Lab., Wright Field), 11 October 1945. Box 4, folder 2, WGD/BHLUM.

127. On Dow's preoccupation with electronics and tube development, see Dow to N. R. Scott (Special Projects Laboratory), 15 February 1946. Box 4, WGD/BHLUM. Dow's early visits to Wright Field with Conlon also revealed to him their need for vacuum-tube development. He suggested studies of amplifier tubes at Michigan for the improvement of receiver sensitivity. The project, Dow argued, would be "of a sufficiently basic and analytical nature to be suitable for" university involvement, adding: "Our approach to it would probably be of a fundamental and analytical nature rather than an effort to invent at the earliest possible date a new type of tube, and I believe that just such fundamental work is what this particular problem needs now." Dow felt that the tube project should be part of the guided missile project, not a special separate contract.

128. Dow acknowledged Nichols in Schultz, Spencer, and Reifman (1948), p. 13. See also M. H. Nichols, "Measurement of Ionization Density by Radio Methods," 29 January 1946. Box 3, folder 1, M668 "Notes and Reports," WGD/BHLUM.

129. See Megerian to Bain, "V-2 Panel Report," 7 March 1946, pp. 7–8; and its "enclosure A," Brig. Gen Alden Crawford to Commanding General, ATSC, n.d., V2/NASM.

130. Dow to the author, 25 October 1986, p. 2; 8 August 1990, pp. 35–38. Harold A. Zahl was an important patron throughout the early years of upper air activities. Director of research since 1946 at SCEL at Fort Monmouth, N. J., Zahl had been in the Signal Corps for 15 years, entering directly upon obtaining his Ph.D. in physics from Iowa in 1931. See Zahl (1952, 1968); Stewart and Zahl (1947).

131. Eventually, the Signal Corps did fund Nichols, through Michigan's Department of Engineering Research. Dow to the author, 8 August 1990, pp. 36–38; quotations from p. 38. Zahl's proprietary attitude was representative of military monitors of sponsored research. See Forman (1987), pp. 205–06.

132. Lin and Bunch (1984), p. 12; Dow to the author, 11 August 1990, p. 40. Dow was always deeply involved with Wizard and remained concerned that it might get out of hand. In September 1946, after a particularly lengthy and confusing Project Wizard review, where Dow felt compelled to state what the boundaries of his research were, he reminded John Strand that their "customer is the technical branch, not [the] pilotless aircraft" branch at Wright Field, and that as such, they were to confine their activities appropriately: "I called attention of the group to our feelings that it was not generally part of our responsibility to attempt to develop a complete detection system [for Wizard]. This rather shocked certain people, which merely indicates that they completely fail to understand the magnitude of the detection problem." On the other hand, Dow remained just as tenacious that his group maintain accountability with their patrons; both Strand and Floyd Schultz had to take care to complete full trip reports to White Sands, where they had examined tracking facilities. "That is part of keeping the customer happy," Dow cautioned. W. Dow, "Memorandum to John Strand." 23 September 1946. WGD/BHLUM.

133. Dow to R. F. May, 2 July 1946. WGD/BHLUM.

134. See Liebowitz (1985); and O'Day (1954), p. 2.

135. Dow to the author, 7 September 1990, p. 43. On the evolution of the group's work, see also Dow, Reifman, and Schultz (1947), p. 2; Schultz, Spencer, and Reifman (1948), p. 13. On the radio propagation technique, see M. H. Nichols, "Measurement of Ionization Density by Radio Methods," 29 January 1946. Box 3, folder 1, M668 "Notes and Reports," WGD/BHLUM.

136. Dow to the author, 7 September 1990, p. 43.

137. Spencer OHI, p. 34rd.

138. Spencer OHI, p. 37rd.

139. Dow to the author, 25 October 1986, p. 3.

140. Dow to Landsberg, 4 April 1948. Oxford folder, WGD/BHLUM.

141. Spencer OHI, pp. 46–47rd. Although initially intended to provide ion densities, their greatest usefulness ultimately was to measure the profile of the electron temperature in the ionosphere, as we shall see in Chap. 16.

142. Lin and Bunch (1984), p. 12.

143. Spencer OHI, pp. 24–25rd, 38–39rd. See also Lin and Bunch (1984); Dow to the author, 25 October 1986, p. 4; and Dow to R. F. May, 2 July 1946. WGD/BHLUM.

144. Robert P. Weeks, "The First Fifty Years" (typescript draft of history of Department of Aeronautical and Astronautical Engineering, October 1964), p. 21ff. Box 15, Michigan, College of Engineering Collection, BHLUM.

145. These also included instruments to measure stagnation pressure and temperature at the tip and sides of the warhead. See Megerian to Bain, 10 September 1946 "V-2 Panel Report no. 6," p. 13. V2/NASM. As a new associate professor at Michigan, Nichols was responsible for setting up a series of new graduate-level courses using his expertise in telemetry. Through the fall of 1946 and into 1947, he therefore concentrated on developing new courses of study in electronic instrumentation beyond telemetry systems. These included the "dynamical and random response of instruments, wind tunnel and flight test instrumentation, automatic control, and engineering applications of the electronic differential analyzer." Weeks, pp. 21, 25.

146. Ibid., Weeks, pp. 22–23.

147. Nichols' group quickly grew to more than 20 people as they searched for diverse methods to determine air density, pressure, and temperature using photographic shadowgraph devices to measure the angle of

shock waves from the nosecone during flight. They also built tiny pressure-sensitive moving probes for the slip stream of the rocket, and a wide array of other pressure sensors around the tip of the rocket. All these instruments eventually failed during the V-2 period because the designs were far too complex. See Bartman (1954); and Jones (1954).

148. On Shapley's fear of military support, see DeVorkin (1991) and Kidwell (1991); and Shapley to Walter Orr Roberts, 25 July 1947, p. 1; see especially Shapley to Roberts, 8 August 1947; Roberts to Shapley, 13 August 1947. WOR/UC. For recollections, see Bok, Hoffleit, Roberts, and Whipple oral histories, SHMA/AIP and SAOHP/NASM.

149. Shapley to Menzel, 9 January 1946. HUG 4567.5.2, DHM/HUA. Shapley's terminology was often sardonic; he thought of missiles in terms of their payloads, bombs.

150. He was given tenure in August 1946. On Whipple's status, see Harlow Shapley, "Tenth Informal Memorandum from Harlow Shapley," August 1946. FLW/SIA.

151. On Harvard College Observatory before the Shapley years, see Jones and Boyd (1971). As yet there is no coherent history of Harvard College Observatory during the Shapley years. For a sense of it, see Shapley (1969); DeVorkin (1984); and Kidwell (1990, 1991).

152. Harlow Shapley, commentary in minutes, "Meeting of the Visiting Committee of the Harvard Observatory with the Observatory Council," 15 October 1946. HCO Council folder, 1946–1948, FLW/SIA.

153. Shapley's rhetoric for support belied his continuing fears, expressed to his own staff, that wartime pressures remained on trained manpower in a world presumably at peace. Funds available from the Army and Navy in late 1946 for research and development, amounting in Shapley's estimate to almost $100 million, represented a "core of extreme pressure on universities and private research and results in [the] ionization of scientists." See Harlow Shapley, "Minutes," Harvard College Observatory Council, 10 December 1946. HCO Council folder, 1946–48, FLW/SIA. Shapley reported here that the fledgling NSF proposal had just been "canceled by Congress. Sad" and even if it had lived, it was worth at most only $5 million.

He added in his distinctive prose: "Armavy theft of American scientists noted even by armavy as serious." In Shapley's terms, "Armavy" was a conflation of Army and Navy. Shapley's prominent role in the early deliberations over the existence of an NSF are identified by J. Merton England (1982).

154. Harlow Shapley, commentary in minutes, "Meeting of the Visiting Committee of the Harvard Observatory with the Observatory Council," 15 October 1946. HCO Council folder, 1946–48, FLW/SIA.

155. It was possible, however, to receive the data by direct request, presumably if one had clearance. See F. Whipple to J. Kaplan, 21 October 1946. FLW/HUA.

156. Whipple listed this application third, however, behind his astronomical interests. Whipple (1938), p. 499. On early meteor studies and their importance for determining the conditions in the upper atmosphere, see Doel (1990a), chaps. 1 and 2, esp. p. 55; McKinley (1961); Satterly (1923); and Hoffleit (1988). On the stratospheric balloon flights of the 1930s, see DeVorkin (1989a); Whipple recollections, May 1989. Physicists and meteorologists had long considered how meteors could provide information on temperature and density in the atmosphere. See Lindemann and Dobson (1923), who examined "the theory of what happens when a meteor appears."

157. See McKinley (1961). Ernst Öpik, along with Fritz Zwicky, deserves careful historical attention. Both were extremely fertile, brilliant scientists, yet because of their personalities they never enjoyed the recognition many feel they deserved.

158. See, for instance, Mitra (1948), p. 89, referencing Whipple (1943). Whipple prepared his paper for an interdisciplinary spectroscopic conference at the Yerkes Observatory. See letters between Otto Struve and Fred Whipple, 27 March 1942, 6 April 1942, and 12 June 1942. Roll 3, Section 6, OS/AIP. The paper in question is Whipple (1943); Ron Doel (1990a), chap. 2, p. 66, note 86, has argued that Whipple reversed his research priorities during the war, emphasizing atmospheric research. On Whipple's war work, see Whipple OHI 1977. SHMA/AIP. See also Peggy Kidwell (1990, 1991).

159. Harlow Shapley "Tenth Informal Memorandum from Harlow Shapley," August

1946, p. 6. Box 4, HCO Council folder 1946–48, FLW/SIA.

160. By the second panel meeting, Whipple talked only of sending small packages of eggs, seeds, and grains prepared by Harvard biologists to study their resistance to the conditions of space, and to look for mutations caused by high-energy cosmic radiation. He also proposed flying packages of X-ray film to gain information about the amount of X-ray radiation emitted by the sun. By the end of April, Harvard's participation reduced to the biological package and the X-ray film package, very simple passive experiments that needed only to be placed into containers that were rugged enough to survive the flight and impact. See V-2 Panel reports no. 1, no. 2, and no. 3. V2/NASM.

161. P. C. Keenan, an Ohio astronomer, and Raymond J. Seeger, both within the Naval Ordnance Laboratory in Silver Spring, Maryland, advised Whipple to apply directly to BuOrd, with their NOL support. Neither NOL nor the Bureau's Naval Ordnance Test Station, in Inyokern, California, had yet developed mechanisms for external contracts. See Keenan to Whipple, 1 October 1945. FLW/HUA. Keenan, wizened by his wartime experience, kindly advised his younger colleague that it might be better to "have the meteor photography supported by a civilian government organization, such as [Vannevar] Bush proposes." On Bush's arguments for a civilian science agency, see England (1982). Whipple probably well knew that a civilian agency was not soon to appear. See Kevles (1975).

162. Kopal (1986), pp. 191–203 recalls his war years in the Center of Analysis (called then the Bureau of Analysis) as years away from astronomy performing tasks in developing firing tables for the Navy using differential analyzers, which, he argues, required the mathematical skills of an astronomer. After the war, Kopal remained in the Center as its director of special computing, heading a large group that performed computational studies of supersonic aerodynamics and ballistic reentry for the Navy. Ibid., p. 204. Kopal rationalizes his participation with Jacchia in the meteor analyses in terms of the existence of the V-2: "it suddenly became of operational importance to learn more about the structure of the upper atmosphere of the Earth." Curiously, Kopal fails to identify Whipple's role in the meteor research program, identifying Jacchia as its head. Ibid., p. 205.

163. See Harlow Shapley to Vice Admiral G. F. Hussey, Jr., 30 January 1946, with attachments; and Peter Millman, "Visit Report— M.I.T. and Harvard University," 20 March 1946. FLW/HUA.

164. In creating the panel, Army Ordnance wanted it to coordinate and collect all data pertaining to the behavior of the missile and its aerodynamic and propulsion characteristics. Army Ordnance's Ballistic Research Laboratory searched for ways to improve its ability to analyze telemetry data, whereas the Signal Corps and the Air Force searched for ways to improve optical and radar tracking and control systems; the latter most concerned with its mission to develop missile countermeasures. See, for instance, Gilbert O. Hall, "V-2 Program of the Navigation Laboratory," 12 June 1946 Status Report WLCES-2A, p. 1. MOD/AFGL.

165. This perspective owes much to Forman (1987), as well as a remark by Nelkin (1990), p. 18, in a review of Mukerji (1989).

# 7

# A New Site for Research: White Sands Proving Ground

When Richard Tousey retired from NRL, several generations of colleagues and associates presented him with three bound volumes of his papers. The endpapers were maps of the White Sands area, not only a fitting tribute to a man whose talent and mission in life were tested in the sands of the Tularosa Basin but also a testimonial to the central position the White Sands Proving Ground had in the origins of the space sciences.

The White Sands Proving Ground, now known as the White Sands Missile Range, occupies much of the Tularosa Basin between the San Andreas and Organ Mountains on the west and the Sacramento Mountains on the east. Extending about 160 kilometers from north to south and 40 to 60 kilometers from east to west, White Sands is a beautiful and desolate place where sand dunes, alkali flats, dry lakes, and lava flows are broken only by outcrops of serpentine, mottled with scrub, mesquite, yucca, and cactus. Few roads cross the region; there are occasional instrumentation outposts and airstrips. The city of Alamogordo and the Holloman Air Force Base (then known as the Alamogordo Army Air Field) lie due east of the Ground's midrange and about 80 kilometers northeast of the main post area, which is just north of the Army's Fort Bliss Military Reservation, Texas, where Von Braun's Germans were housed.

White Sands became more than the site for testing missiles and their payloads. It became a severe testing and training ground for upper atmosphere researchers and their military patrons working together in the postwar era. It would be the site for a completely new type of scientific activity—one whose broad out-lines were defined by the landscape of the national security state.

Our intent here is to gain a sense of the scale of the infrastructure within which those who used rockets for scientific research had to live. Although dependent on this structure, the research groups were not entirely comfortable within it at first.

## Origins of the Proving Ground

In the late fall of 1945, as the first convoys of V-2 parts were arriving from Europe via ship, train, and truck (some 300 railroad carloads in all) and as von Braun's team was heading for internment and sanctuary at Fort Bliss, crews from the Army Corps of Engineers were putting the finishing touches on permanent staff quarters at White Sands and were refurbishing old hangars for the assembly and testing of the captured weapons. The Army had established the White Sands facility in February 1945 after many months of searching for a suitable missile and bombing test range.[1] Construction of the post began in June 1945, with the importation of surplus Civilian Conservation Corps buildings from the Sandia base outside Albuquerque.[2] As a multifunctional testing range for all types of rocket ordnance, including JPL's Corporal series, GE's Hermes, and Bell Laboratories' Nike, as well as a multitude of smaller ordnance systems, White Sands was a multimillion dollar operation, yet it was far from Army Ordnance's top priority.[3]

During the summer of 1945, launch pads and a blockhouse were constructed about 10 kilometers from the main post. The block-house had concrete walls 3 meters thick, and a pyramidal roof of reinforced concrete with a maximum thickness of 8 meters to withstand a direct hit from a rocket in free-fall (figure, p. 112). The earliest firings at White Sands were of tower-launched Privates and WAC Corporals, developmental rockets that paved the way for the Army's planned Corporal missile. The first WAC was fired on 11 October 1945 with a Tiny-Tim booster. The 7.6 meter WAC rose more than 70 kilometers and crashed to earth only 1 kilometer from the launch pad.[4]

## The Navy Moves to White Sands

Abiding by Joint Chiefs of Staff wishes, Army Ordnance encouraged as much activity by outside contractors as possible; it recognized the value of close working alliances "with the best brains and best resources of American science and industry."[5] Since White Sands operations were expensive, and not of top priority, Army Ordnance was happy to invite the Navy to share the space in late 1945. By the spring of 1946, the Army agreed to give the Navy access to these facilities at least 25 percent of the time, which fit the Navy's predicted launch facility needs.[6]

To accommodate the diverse rocketry interests within the Navy's many institutions, including the Bureaus of Ordnance and Aeronautics, the Office of Research and Invention, the Naval Research Laboratory and contractors such as the Applied Physics Laboratory, a newly formed Navy Department Guided Missile Section, established by the Office of the Chief of Naval Operations (CNO), coordinated the construction of a Naval facility at White Sands in the summer of 1946, which was completed early in 1947.[7] The process of establishing this Navy pres-

**Main post area at White Sands Proving Ground, circa 1946. Army Ordnance. NASM SI 80–12318.**

ence was, however, far from simple; its success in part was due to lobbying by the V-2 Panel and the persistence of Ernst Krause.

Krause and Rosen at first pressed for an independent launch facility at White Sands, predicting that they would be firing at least 13 V-2s, 5 Aerobees, and 10 Vikings (then called Neptune) through the end of 1948. But the U.S. Naval Unit established by the Bureau of Ordnance on 14 June 1946 was a cooperative scheme.[8] CNO did not want to commit some half million dollars for a launch system that was not yet needed.[9] Although CNO's decision was a prudent step, shared responsibilities with Army Ordnance did create many procurement bottlenecks and hassles, such as the considerable paperwork required just to get a jeep painted proper Navy colors.[10] And yet, in the summer of 1946, the Navy was well served in comparison with the AAF contractors. The Navy would eventually build its own extensive firing facilities for the Aerobee and Viking (Chap. 10), and the AAF ultimately established a completely independent firing facility at nearby Holloman. In the interim, shared use meant shared responsibility, but not authority.

## Shared Responsibilities but Reserved Authority

Army Ordnance, through the General Electric Company and its Ordnance Research and Development Service Suboffice (Rockets) at Fort Bliss, controlled the missiles. The Navy, of course, had technical control of its own instrument packages and eventually of its own range facilities for monitoring and tracking each flight. But it had to comply with all Army regulations and procedures.[11]

The Ballistic Research Laboratory (BRL) at Aberdeen was in charge of coordinating tracking and analysis, while NRL and GE shared responsibility for determining missile behavior, the former providing telemetry data services and the latter overall coordination. The Signal Corps, Army Ground Forces, and the Army Air Forces took care of missile detection, but range safety was the sole responsibility of Army Ordnance, delegated by Col. Harold R. Turner, commanding officer at White Sands. Turner also controlled all radio and radar frequencies. Studies of the "physics of the upper atmosphere," as Toftoy called them, were the shared responsibilities of NRL, APL, Princeton, Michigan, the Signal Corps, and the Air Materiel Command, with NRL as coordinator, since NRL's Krause was chairman of the V-2 Panel.[12]

## Tracking and Telemetry Services

The two services most critical to assessing the performance of the rockets and retrieving scientific data from them were tracking and telemetry. The Ballistic Research Laboratory had primary responsibility for the former, whereas the latter was a combined effort of General Electric and NRL. BRL was Army Ordnance's key trajectory calculator and predictor, and its most technically diverse in-house research and development agency.[13] BRL provided radar beacons, Doppler techniques, and optical tracking systems composed of high-speed cinetheodolites captured in Germany, as well as domestic varieties.

BRL was also responsible for providing all agencies with detailed tracking data, but soon after firings commenced at White Sands, the system bogged down, because priorities had not been properly set by Army Ordnance. In June 1946, V-2 Panel members complained that they were not getting adequate tracking data from BRL from the first flights and that delays were already close to 30 days. They wanted data from flight to flight, for planning and evaluation. Krause later reported privately to his NRL section that BRL's "attitude on the program was one of developing instruments for the accurate detection and tracking of the V-2 type and that actual data on the V-2 was secondary [to them]."[14] By October, Krause had taken the matter to the Pentagon and, with the support of panel members, secured Army Ordnance's commitment for better communication between NRL, BRL, White Sands, and the Signal Corps and a promise that tracking data would be

Interior of the block-
house showing the firing
panel circa August 1946.
Much of the compo-
nents were still German.
Nelson Spencer collec-
tion. NASM SI 89–1105.

provided quickly. To meet the demands of the
other panel members and services, BRL
promptly contracted with a local college for
range personnel and preliminary data analysis.[15]

The optical work was coordinated by Louis
Delsasso, BRL's representative on the V-2
Panel. In addition to his own staff in Maryland, Delsasso, a physicist from Princeton who
had been recruited in 1943 as BRL's chief
physicist, hired local specialists to augment
his staff.[16] One was the veteran astronomer
and discoverer of Pluto, Clyde Tombaugh,
who developed an optical system to detect
parts of V-2 rockets from distances up to 160
kilometers.[17]

Tombaugh, as a supervisor in the optical
section, thus became part of a large team under Dirk Reuyl, who was trained in astronomy and was the chief of BRL's optical measurements branch. Tombaugh, a life-long
amateur telescope maker, had practical ex-

perience in using high-powered telescopes.
Collaborating with his brother-in-law, James
Edson, Tombaugh applied what he knew to
long-range optical reconnaissance of rockets
in flight.[18] Tombaugh's unorthodox instruments were at first criticized by Ordnance
staff, but they showed that V-2 rockets gy-
rated wildly during flight. This happened
within a year after the firings started and gave
Tombaugh wider responsibility for all optical
range instrumentation. By the late 1940s, he
was running upwards of 50 stations and 80
people.[19]

Optical tracking and data reduction of ballistic missile trajectories were activities that
had long attracted astronomers. In addition
to Tombaugh and Reuyl, Harvard's Dorrit
Hoffleit and Theodore Sterne were involved
in the analysis of the optical ballistic data.[20]
In the nine years Tombaugh spent at White
Sands, his teams tracked Aerobee, Nike, and
Honest John missiles, as well as the larger

Corporals, Vikings, and V-2s. Tombaugh was well suited for White Sands duty. He came from the plains and loved the West. To NRL's Milton Rosen, Tombaugh "looked like an old-time prospector . . . wandering around in the hills."[21]

If Tombaugh felt he had to prove the veracity of his optical systems over the others, Dorrit Hoffleit noted soon after that "No one system has yet proved 'best' for all the information that is wanted on a missile's entire flight path and behavior."[22] Determining the position and velocity of the missile, its accelerations, spin, yaw, and roll required a complex of different devices. Her experience with all of them, primarily in data reduction and analysis at BRL where she pioneered a number of punch-card techniques, brought her to a special appreciation of the DOppler Velocity and Position method, or DOVAP.[23] DOVAP included a continuous wave transmitter, located about 2.5 kilometers behind the launch site, and four receiving stations spaced at wide intervals on the range. The transmitter sent out a signal along the missile trajectory, which was received by the four stations and by a transponder on the missile. The signal received on the missile was doubled in frequency and then retransmitted. Each ground station also doubled the original frequency and mixed it with the retransmitted doubled frequency from the missile. The doubled frequency received from the missile, however, would be different from the doubled ground signal because of the Doppler effect caused by the missile's motion. The mixed signals would therefore beat against one another with a frequency that could reveal the motion of the missile. Each beat signal was transmitted over wires to a central photographic recording system, and the resulting data tapes were sent back to BRL for analysis.[24]

In the field, the DOVAP systems were apparently straightforward to operate. Complete data reduction, however, required electronic computers.[25] The ENIAC computer at Aberdeen, as is well known, could provide a trajectory analysis in "a time comparable to the time of flight of the missile itself." But it took two full days "to get the machine set up for the specific problem."[26] Even though the ENIAC was available after July 1947, getting time on it was not always possible. For practical reasons, a complete reduction, even in 1949, was apt to take "several weeks of film reading, interpretation, and computation."[27] Without such high-speed electronic assistance during the first launch season in 1946, the complete analysis (for each half-second of a typical five-to-eight-minute trajectory) by a "skilled computer using desk machines" would have taken 10 weeks of working time.[28] Given the amount of effort required, it is not surprising that BRL could not supply the V-2 Panel with its trajectory data in a timely manner.

There were sets of Signal Corps S- and X-band reflective radar tracking sets at White Sands for range, azimuth, and elevation information. These radar systems yielded first-order trajectory analysis and identification of the impact point. Their application began during the first firing season and was undertaken not only to provide range control and data, but to refine the radar systems themselves at the Evans Signal Laboratory.[29]

The extensive tracking network at White Sands served a number of purposes. Certainly trajectory analysis and fire control were immediate range needs, but the time analysis of the ballistic path and the establishment of field techniques for firing missiles were also of interest to a wider audience. Army Ordnance wanted improved tracking and control systems and faster and more efficient means of data reduction. They wanted, in sum, to explore any means of incorporating radar technology into ballistic missile systems.

Much the same can be said for telemetry, but here, many systems were tried out after General Electric found it could not satisfy range needs in February 1946. NRL furnished all the initial on-board telemetry, ground stations, and manpower for the first V-2 firings. The hardware was easily adapted from wartime designs familiar to Krause's old Communications Security Section.[30] Providing hardware fit NRL's new image as the Navy's source of critical technology. Krause knew that the sets on hand could not accommodate all of the housekeeping and data telemetry that would be needed, so, with ORI's backing ensured, he promised to have 23-channel and 30-channel sets available soon.[31] The war had given rise to several new sophisticated multichannel telemetry designs that were based upon miniature vacuum-tube circuitry (figure, p. 114). Although a choice was now available, both NRL and GE insisted that all instruments had to be standardized to one system. NRL preferred its pulse-time system and rushed to make the hardware available. Even so, competing telemetry systems were allowed by Army Ordnance for White Sands firings, but only after considerable pressure had been applied; both Princeton and APL pushed to demonstrate the capabilities of their competing Bumblebee designs, which used the

technique of frequency-deviation multiplexing.[32]

Telemetry was not, however, as important to Van Allen as it was to Krause. When asked how it all worked, Van Allen replied that the signals were somehow "unscrambled" on the ground, and admitted: "Don't ask me how."[33] What NRL provided, therefore, was an expensive specialized service far beyond the financial and technical commitment that others could handle. The eventual system, including an emergency cutoff circuit, cost NRL some $460,000 in 1947 for equipment and contract manpower for six months of operations, not including Army Ordnance manpower at White Sands.[34]

The first 10-channel telemetry systems failed frequently, prompting Krause to launch a crusade not only to improve their reliability,

NRL 23-channel airborne telemetry unit showing pulse-time modulator, with some of the plug-in circuits removed revealing the miniature vacuum tubes developed, according to NRL Report R-2955 and Krause (1949), for hearing aids by Raytheon, who manufactured the telemetry units for NRL. U.S. Navy photograph, Ernst Krause collection. NASM SI 83–13902.

ruggedness, and signal strength, but also to reduce their weight and increase the number of channels. With 21 NRL staff at work full-time on the problem, 23-channel sets were operational by the summer of 1946, and more improvements were promised. Although Krause was always willing to provide the hardware without complaint to other members of the V-2 Panel, he wanted help with supplying the manpower required to staff NRL's telemetry units at White Sands.

Through the first summer of firings in 1946, NRL found that it needed at least eight technical personnel stationed at White Sands to operate and maintain both base and range services. The National Bureau of Standards (NBS) provided three technicians for one range station, but they were there primarily to conduct ionospheric propagation studies. In September, however, Newbern Smith informed Krause that the NBS could no longer participate for lack of travel funds.[35] An "economy wave in the military services is now beginning to pinch the V-2 program," one V-2 Panel observer reported later, and the first casualties were transportation services and manpower.[36]

Krause did not want to lose the NBS range station, both because of its telemetry services and because its ionospheric measurements served many members of the V-2 Panel. He appealed to V-2 Panel members to donate manpower equally, but they resisted, arguing that Krause should instead appeal to the Joint Research and Development Board. In November, however, the NBS agreed to delay its pull-out, but Krause still needed more manpower, warning that "because of the present [turnover] in personnel and lack of men willing to stay in the desert," telemetry services would continue to deteriorate.[37] By early 1947, however, his protestations yielded sufficient funds from the Navy to hire local contractors who would be happy to sit in the desert. BRL's success with the New Mexico College of Mechanic Arts and Sciences soon brought an NRL contract for telemetry field services.[38]

## Living and Working in the Desert: Policies and Pressures

When Marcus O'Day visited White Sands in June 1946, he was appalled by the messing and housing, which he described as "intolerable." He had agreed originally to assign several AFCRC men to White Sands, to be attached to Krause's unit there, but now was "unwilling to order any of his group to White Sands until conditions are improved."[39] C. H. Smith and Krause had repeatedly brought the inadequacies of the messing and housing to the attention of Colonel Turner as well as to Bain (figure, p. 116) and Toftoy, and tried to assure O'Day that once the Navy facilities were completed, conditions would improve. Their most persistent fear was that the lack of adequate air-conditioning facilities and protection against the drifting sands would make the instrument preparation areas unusable.[40]

Panel members were loath to donate junior staff to extended tours in the desert sands because the living and working conditions were so poor. Through the summer and fall of 1946, construction delays, personnel problems, equipment shortages, and general coordination headaches made White Sands a tough duty station. Holger Toftoy's staff man at Fort Bliss, Maj. James Hamill, constantly sent reports of problems back to the Pentagon. Toftoy was happy to hear in October 1946 that housing facilities at White Sands were starting to improve, but one month later, in an economy drive, the Army Air Forces terminated housing facilities for 70 families of White Sands personnel on the Alamogordo Air Base.[41]

Lorence Fraser, Van Allen's assistant from APL, recalled that he tried to drink everything in sight when the temperature was 110 and the humidity 2 percent; and without air conditioning, or even swamp coolers, it was impossible to get hydrated and concentrate on the work at hand.[42] The poor living conditions reduced morale and affected services, but the latter were also impaired by a lack of clear coordinating mechanisms for getting things done. In a base where the Army was

Lt. Col. Harold R. Turner (l) and Col. James B. Bain in the White Sands blockhouse. NASM SI 80–4099.

in charge and the Navy had established a presence, working under an Army Air Forces contract could be frustrating. One early experience of the Michigan group demonstrates the depth of the problem.

When William Dow sent his University of Michigan team and subcontractors south in June 1946 to prepare for their 22 August launch, using the radio system suggested by Nichols, the first thing they discovered was that local AAF officers based at Alamogordo were not properly briefed on the V-2 program, its contractors, or their needs and so refused to provide facilities and equipment. There was also an embarrassing conflict within the Wright Field command: a technical officer representing Wright Field's Electronics Subdivision at Alamogordo tried to make arrangements for the V-2 work but was over-ruled by a higher-ranking liaison officer.[43] Equipment promised by Wright Field failed to arrive on time, and when it did show up, the equipment was untested, and there was more red tape to cut through before it could be tested and fixed at White Sands. The facilities for testing equipment were inadequate, and the Michigan group had to go hat-in-hand to anyone who seemed to have control of a screwdriver to try to get things done before launch. Most frustrating, and nearly fatal to the program, was the group's inability to secure transportation on the range. The team's work sites were spread between 9 and 90 kilometers apart across the desert, and Alamogordo Air Base officials "did not recognize that civilian engineers, of any of the three A.A.F. technical agencies, were sufficiently responsible individuals to be trusted to sign [for] vehicles in or out of the car pool."[44]

In a closely typed 11-page complaint to Maj. Gen. Curtis E. LeMay, who was then deputy chief of air staff for research and development, Dow blamed "the failure to get an early start [in the V-2 program, which] severely handicapped A.A.F. participation." Dow wished to call LeMay's attention to "various serious organizational handicaps" that stemmed from that failure and demanded action from LeMay.[45] Among Dow's many charges, one that probably pleased LeMay was that the AAF lacked a suitable infrastructure to manage scientific research, especially in the field: "There was no established, dependable, A.A.F. research contractor with a flexible, well-staffed organization ready and able to pick up the overall load immediately" when the V-2 work began.[46] In a follow-up letter, Dow argued that "present army practices for supply and procurement are often completely hopeless as applied to a research and development unit" and that military classification seriously impedes scientific progress, draining military research and development centers of the best minds: "*If* we are to fight scientific wars, we must re-organize military establishments accordingly; perhaps the AAF can lead the way."[47]

LeMay's staff was happy to receive Dow's complaints; in fact, they encouraged him. At the time, LeMay was not in control of research and development, as it was still in the hands of the Materiel Division. His staff therefore saw complaints like Dow's as useful ammunition in their efforts to wrest control of this responsibility.[48] LeMay replied in November, saying that Dow's grievance was similar to other reports he had received and that "I am hopeful that you will see some improvement in our position before long."[49]

What saved Dow's group, as it turned out, were his good relations with the Signal Corps at White Sands. The Signal Corps had an "extremely helpful, cooperative, and technically sound attitude." They allowed the Michigan group cooperative use of their tracking facilities.[50] Dow assumed that a high degree of cooperativeness and comradeship existed in the other groups, and in their dealings with the Army. But another member of the Michigan party soon informed him that this was not so.

Melvin Gottlieb, of the Ryerson Physical Laboratory, University of Chicago, had been a member of Dow's field party to White Sands, and claimed that the treatment they experienced extended to others. He complained to Dow about the inequities and inefficiencies, the blatant hostilities, and the insensitivity at Alamogordo and White Sands, recounting tale after tale. But, Gottlieb reported, the worst part of the experience was the treatment they themselves received and the AMC's incompetence.[51]

The failure of the 22 August AMC V-2 did not help matters, either. It reached an altitude of 100 meters and then leveled off into a flat spin (figure, p. 118). After six seconds, its fuel was cut off by radio command from the blockhouse, and it crashed after 10 seconds of flight. Yet, Dow took the failure of the vehicle in stride. He told LeMay that the experience provided useful lessons, adding that "the work has given many A.A.F. agency staff members a valuable opportunity to study problems involved in rocket firing, guidance, control, and instrumentation generally."[52]

If Dow's experience represented those of a civilian user on the periphery, Ernst Krause's represented those of an insider and provider. He enjoyed the comparative luxury of a staff presence at White Sands, but he also faced ever-increasing coordination problems in trying to meet his commitment to supply technical services at White Sands, including warheads and telemetry. C. H. Smith and then Thor Bergstralh were both responsible for coordinating with NRL's White Sands staff and had to keep track of all deliveries and manpower needs, as well as every milestone for each launch, including housing for 50 additional NRL personnel who were to arrive from Washington for the first NRL launch in June 1946 (table 7.1).

In May, Smith drafted regulations for the use of Navy facilities, from shop equipment to the motor pool, and also developed schedules for the delivery and installation of in-

A V-2 out of control seconds after launch. This is either the fateful 22 August flight of Michigan's V-2, which reached just over 100 meters altitude before spin out, or Princeton's V-2 which leveled off at about 360 meters on 14 November 1946, according to Smith (1954). The V-2 gantry was being constructed during August 1946, so the identification is not certain. But the behavior observed here is fully reminiscent of both flights. U.S. Navy photograph, Ernst Krause collection. NASM SI 83–13888.

strumentation at White Sands.[53] Many of these regulations depended on services from the Army, which did not materialize on the Navy's schedule. An ever-present problem, encountered as well by Dow's team, was the lack of reliable transportation to Las Cruces for personnel, either on business or on leave. Before the Army established a bus service, the Navy personnel scrounged what they could, often breaking regulations.[54]

NRL also had to establish and maintain effective working relations with BuOrd and with the Naval Ordnance Test Station (NOTS/ NOTU) at Inyokern, California, which was thinking about setting up its own Naval Ordnance Test Unit at White Sands. Policy matters, station regulations, responsibilities for shared facilities, and general communications and governance of naval personnel all posed problems. Krause hoped that he could secure shared technical services from NOTS, but ORI and BuOrd did not have an officer available. BuOrd suggested that one be assigned from NRL, but this was not possible since Marine officers stationed at NRL were there to gain technical and not administrative experience.[55]

NRL and ORI expected no major problems to develop internally, but the situation grew far more complex when APL entered the picture, to say nothing of the Air Materiel Command, the Ballistic Research Laboratory, the Signal Corps, and possibly Inyokern (NOTS/NOTU). Joint responsibilities for range instrumentation between these groups, as well as shared housing and laboratory facilities, were all decided by ad hoc committees in the summer and fall of 1946, cobbled together by Toftoy's office. The trick was to be able to have technical people on these committees who knew the needs of each rocket group and who also had the authority to commit resources. In effect, by the end of 1946, Krause, even as an insider, was experiencing many of the same bureaucratic frustrations with interservice coordination that Dow had felt as an outsider.[56]

TABLE 7.1   Countdown for 27 June 1946 NRL Flight

At the Naval Research Laboratory:

| | |
|---|---|
| 1 April: | NRL personnel begin leaving for White Sands, leave dates shift dependent upon responsibilities, and slips due to late arrival of warhead to NRL:<br>—Rocketry personnel (9 people)<br>—Ionosphere instrument specialists (11 people)<br>—Astrophysics (2 persons from RSRS; 2 from Optics)<br>—Electronic Instrumentation (11 people)<br>—Cosmic Rays (3 people)<br>—Atmospheric Physics (2 people)<br>—NRL Shop Personnel (4 people)<br>—Physical Optics (2 people)<br>—Bureau of Standards liaison personnel (6 people). |
| 3 May: | Two railroad flatcars leave Washington carrying 1.5 ton panel truck, 3 power trailers, one passenger sedan, one box of wave guides, and one Bureau of Standards truck filled with electronics. |
| 8 May: | Two flatcars leave Washington with more trailers, passenger cars, another panel truck, a power mast, all filled with electronics and power equipment, and one K65 Electronics trailer fitted for fieldwork. |
| 8–14 May: | Final production of warhead and fitting of pressurized doors at Naval Gun Factory. |
| 9 May: | Three flatcars leave Washington carrying another panel truck, another car, and two K65 trailers, one filled with electronics and the other fitted out as a field machine shop. |
| 15 May: | The Naval Gun Factory delivers the first warhead in what it called "Project Stardust" to NRL, and all laboratory equipment was ready by that time for insertion and testing. It was due originally on 25 April and was to be flown to White Sands by 15 May. Its delay was caused by fabrication problems with hand-fitting the access doors. |
| 15 May: | Warhead mounting hardware and warhead balancing machine, counterweights, and pressurized chambers delivered to NRL. |
| 15 May–10 June: | Installation of all equipment in warhead at NRL. |
| 16–17 May: | Cutting of all holes and fitting of all electrical connections in warhead preparatory to instrumental integration. |
| 18 May: | Transport plane carries telemetry equipment. |
| 20–28 May: | Installation of all scientific equipment, including batteries. |
| 29–31 May | Assembly of complete nose section, weighing and balancing. |
| 3–10 June: | Testing of all systems, and sealing of warhead. |
| 5 June: | 40 visitors to NRL from BuAer, BuOrd, BuShips, ORI, and Army Ordnance attend demonstration of V-2 warhead installation. |
| 10 June: | Warhead air-shipped from NRL to White Sands, carrying cosmic-ray ground station, cosmic-ray test sets, and four personnel. |

At White Sands:

| | |
|---|---|
| 22–24 May: | Installation of telemetry antennae and airborne telemetry units. |
| 27–31 May: | Installation of wave guides in missile, and completion of antenna cabling. |
| 3–11 June: | Completion of all wiring for telemetry system at White Sands. |
| 12 June: | Warhead arrives at White Sands. |
| 13–14 June: | Preliminary work on warhead, instrument chamber, and tail section in instrument assembly building. |
| 17 June: | Check cosmic-ray instrumentation, and test all pull-away plugs and ground control cables. |
| 18 June: | Mount warhead and make all connections to missile. |

TABLE 7.1    Countdown for 27 June 1946 NRL Flight (continued)

| | |
|---|---|
| 19 June: | Wiring check. |
| 20–21 June: | The missile, still in the assembly building, is under the complete control of NRL. All electrical links, electronics interference, and operating charactristics are checked. |
| 22 June: | Warhead buttoned up. Pressure test at 1/2 atmospheres. |
| 24 June: | Painting (no testing). |
| 26 June: | Missile transportd to launch pad and set vertical. |
| 26 June: | 2 hours of testing during the day: run complete system 15 minutes on external battery, check for internal consistency; then run 15 minutes of testing internal battery. Check all ground stations vs. radars. Checkout all systems on stand. Run telemetry for 5 minutes and check all channels. |
| 27 June: | 5 a.m.: Recheck that all systems are operating, turn on Doppler beacon. All instrument checks completed by 8 a.m. Fuel and liquid oxygen filling commences. Problems arise with oxygen fill line. By 3 p.m., flight called off until 12 noon on 28 June. |
| 28 June: | 9 a.m.: All equipment rechecked. Fuel problem fixed, and tanks are filled, followed in one hour by hydrogen peroxide and permanganate fueling of turbopumps. Firing time set at 12:30 p.m. With X = 12:30 p.m.: |

| | |
|---|---|
| X − 1 hour: | All fueling to be completed. |
| X − 20 minutes: | Red smoke bomb released indicating standby for firing. All personnel must leave launch site for safety areas. |
| X − 10 minutes: | Telemetry systems on. |
| X − 4 minutes: | All NRL equipment on. |
| (add 3 minutes due to firing delay) | |
| X − 400 seconds to X − 290 seconds: | Telemetry synchronized. |
| X − 290 seconds to X − 180 seconds: | Launch crew sealed in blockhouse. |
| X − 180 seconds: | Blockhouse takes full control. |
| X − 120 seconds: | Red star shell fired. Start of firing operations. |
| X − 100 seconds: | Telemetry operating. |
| X − 18 seconds: | Preliminary stage on. |
| X − 6 seconds: | Ignition clear. |
| X − 2 seconds: | Main stage ignited. |
| X − 100 seconds to X + 90 seconds: | Telemetry blanked out by 30 cycle interference. Interference cleared up after 90 seconds. The plan had been that if the telemetry failed, the launch would be delayed. But this failure occurred at the "worst possible time." |
| X + 90 seconds to X + 270 seconds: | Telemetry operating properly. |

*Source:* Chronicle of nearly three months of effort for NRL's first full flight. Countdown taken from: "Record of Consultative Services," 5 June 1946. Box 98 Folder V2 Rockets No. 4. NRL/NARA; and Megerian to Bain, 15 July 1946, "V-2 Report no. 5," p. 5.

By early 1947, at Toftoy's urging, Turner improved White Sands policies and procedures for the preparation of warheads and clarified what services civilian groups could expect. He took care to provide a long and helpful memorandum to all civilian and military users of the facility discussing safety, security, and operating procedures. It included the names of officers responsible for each service, their addresses, and phone numbers, and announced the Army would now provide both air and ground transportation.[57]

Early popular accounts of working and living conditions at White Sands romanticized the place. GIs in jeeps bounding over the desert dunes to the crash sites were likened to cowboys rounding up cattle (figure, p. 121).[58] Milton Rosen, NRL's chief scientist for Viking, recalls the romance of the rocket and of White Sands in his popular 1955 book, *The Viking Rocket Story*. The harsh conditions and the rigors of preparing missiles for launch at White Sands pitted people against one another, but "there was, withal, a spirit of fraternity that men find when they live and work together and share their gripes and annoyances."[59]

Rosen's depiction of comraderie in the desert is hard to find in the daily accounts, memoranda, and records; it survives mainly in popular accounts of the day.[60] Certainly there was excitement and an immediacy living and working with comrades under trying conditions, but the collegiality was often overshadowed by the lines drawn by the military services. One who did something about it was Comdr. R. B. McLaughlin, the first head of the Naval Unit, who built a swimming pool on the post. It was never approved or explicitly funded—-but was deeply appreciated by everyone who used it.[61]

The Applied Physics Laboratory search team for a spring or summer 1947 flight. (l to r), front: R. S. Ostrander, Lorence Fraser (sitting), and J. J. Hopfield. Lorence W. Fraser collection. NASM SI 90–7785–1.

## Conflicting Personalities and Perceptions

Melvin Gottlieb was right; the experience of the Michigan team in August 1946 was not unique. Under difficult and unfamiliar living conditions and under the constant pressure of maintaining launch schedules and making the whole system work, sensitive personalities often clashed among the different groups. Enlisted men, contract civilian workers, technical staff, Germans, and visiting scientific and technical personnel all began to chafe, especially after the failed launches during the summer of 1946.

"The operating personnel at White Sands," the Weather Bureau's Harry Wexler reported in October, "because of the high incidence of failures, is getting quite jittery and the scientist finds he has to work under more and more restrictive regulations."[62] In January 1947, from his office in the Pentagon, Holger Toftoy hoped that Turner, Hamill, and White Sands technical director Maj. Herbert L. Karsch could work with Ernst Krause to restore "harmonious relations between the scientific world and the Ordnance Department."[63]

Toftoy was then trying to quiet a flap that was fulminating after the first night-firing of a V-2 in December 1946. The source of the disturbance was a *New York Times* reporter's claim that scientists preparing their experiments at White Sands, such as Caltech physicist Fritz Zwicky and Harvard astronomer Fred Whipple, were being "treated as trespassers or people there under sufferance."[64] According to the *Times* reporter, Colonel Turner and his staff barely tolerated the presence of scientists and considered their presence a great concession to their primary mission of "firing off these rockets month after month as a routine operation, regardless of anyone else."

The trouble started during a press briefing at the launch site, when Zwicky attempted to distribute a flyer on his work to the assembled reporters. Turner apparently blocked Zwicky from doing so, and Zwicky exploded. Zwicky, the *Times* reporter claimed, was being abused because he wouldn't play the Army's game. Even those who did play by the rules, the reporter continued, "indicated that they had had this protocol rubbed into them by paying over-elaborate verbal obeisance to the Army, one after another, to the point where it was embarrassing to auditors—palpably somewhat with tongue in cheek and gun in back."[65]

Toftoy first saw the draft text of the *Times* reporter's story in the hands of his boss at Ordnance, who was being asked to issue a statement to the press. They both agreed that the best defense was a rebuttal from Krause, representing the V-2 Panel scientists. Krause rallied Whipple and Van Allen to the defense and dutifully responded, "Never have I heard such criticism nor the need for such criticism as brought out in the unsigned letter."[66]

Krause defended Turner, pointing to counterevidence that the Army respected the scientific presence at White Sands. He recounted instances when Turner went out of his way to accommodate scientists, such as delaying a launch for their convenience.[67] Krause discreetly avoided giving examples that would support the *Times* reporter's allegations. But he had them: In March 1946, before any V-2 missiles were fired, Krause and Milton Rosen stopped at White Sands en route to JPL, and were led into a spare office where, sitting behind a cluttered desk "was a large, rugged man in a khaki shirt open at the neck . . . . He motioned us to sit down, rocked back in his chair and bellowed out, 'No one from Washington comes out here to do anything. They just come to snoop around. That what you're here for?' "[68]

This image of Turner was shared by many of those visiting White Sands. Turner may have appeared to be gruff and quick tempered, posing as a stereotypical red-faced field officer, but he was in fact a dedicated engineer and organizer.[69] The problem was, Turner and people like Rosen did not speak the same language. Even James Van Allen, who agreed with Army Ordnance policy and who certainly supported good relations with White Sands personnel, recalled that relations were poor. In sum "the [White Sands] rocket engineers thought we were a terrible pain in the neck to have to cope with."[70] Krause's testimony was countered in private by many at NRL, including Richard Tousey, who was frequently frustrated by his lack of control over when a rocket would actually be launched on a particular day.[71] Thor Bergstralh also found that if an experiment was not ready at launch time, the missile would probably still be fired because it contained other experiments that were primed and waiting to fly. On at least one occasion a camera was launched in the full knowledge that a lens cap was still in place.[72]

Conflicting personalities and perceptions did indeed often get in the way at White Sands. According to Julius Braun, then an Army Infantry captain assigned to White Sands in June 1947 to learn about firing procedures and optical and electronics tracking units, "Quite a few of the missileers looked on the scientists as overly educated prima donnas totally lacking in common sense or mechanical ability."[73] Some scientists, he added, could be condescending, demanding of attention, and "often unreasonable in expressing their pressing needs." Some would not appreciate the practical problems of making the rockets work reliably, Braun observed, and "most of the scientists felt that WSPG was merely an extension of their laboratories." Some were even of the opinion that the rocket program "could only be justified as a supporting activity to their investigations, when in reality all their payloads were secondary (in the eyes of the missileers) to the more important task of developing future weapons and training people to use them."[74]

Not all scientists fit this stereotype. Some gradually learned that the weeks and months spent preparing instruments in their laboratories did not give them any special priority. To come to this enlightened perspective, however, required time in the desert, in order to learn when it was time to launch.

## Preparing for Flight: Learning When to Launch

Krause's defense of Turner made political good sense as much as it was a testimony to Toftoy's sympathy for the scientific aspects of the firings. From the start, Toftoy equated the scientific work with ballistic missile development. His interpretation of Ordnance policy calling on staff to make the most effective use of each firing was to cancel firings "whenever a delay would result in more complete evaluation of data from preceding missile or to gain better instrumentation."[75] He found Richard Porter and Ernst Krause delighted to follow this policy, but in May 1946, when Toftoy told Turner to cancel the next "shoot" and to slow things down in general at White Sands, "Turner wished to let the schedule [remain] unchanged."[76] Turner stuck to his guns until Herb Karsch and GE contract personnel complained that they could not keep pace with the Army's original schedule.[77] Only then was Toftoy able to slow down the firing schedule to one launch every two weeks. Even so, Turner wanted to keep to a schedule. In the face of this resistance, Toftoy restated his policy in November 1946 in no uncertain terms: "Every effort will be made to adhere to [the V-2 firing] schedule but the firing of any rocket may be postponed or cancelled if the rocket itself or any of the scientific or ballistic instrumentation is not ready on the scheduled date."[78]

No policy, however, could cover all the complexities of preparing for a launch, nor could it make everyone conform. Indeed, many factors determined when and when not to fly, and they were a large part of the acculturation process the scientific groups had to experience.

Pressures influencing launch dates were best appreciated by those closest to the actual work in the desert. Thor Bergstralh recalled his relief when, "once in a while, the Army and G.E. would have a problem and have to delay a firing. . . . I think everyone felt, I'm sure we did, that the schedule was pretty high pressure."[79] In spite of the stated policies Toftoy advocated, and in contrast to Krause's recollections, Bergstralh states that Turner's locally enforced unwritten policy was that if a rocket was fueled and ready for launch and a majority of systems were working properly, the failure of one lone system or experiment would not stop the launch, as long as it did

The V-2 gantry was constructed between August and November 1946 to allow easier access to all parts of the missile. U.S. Navy photograph, Ernst Krause collection. NASM SI 83–13874.

not jeopardize the launch itself.[80] Bergstralh appreciated that Turner had his own priorities, and that they were based on practical issues. If an instrument did not check out in the blockhouse, even though the rocket was fueled and ready to go, the experimenter still could, and sometimes would, entertain the dangerous option of going out to the rocket to try a quick fix:

*You had to realize that there [was] no way [to] hold the whole thing, because of the hazard of holding a rocket too long, unless you could get up and fix it in a hurry. If you could do that, that opportunity would be allowed. . . . But if it were something that would require quite a bit of time, the idea of holding that rocket full of oxygen and alcohol, in that condition, we felt was more hazardous to the other experimentation than taking the chance of going [out to the launch pad].*[81]

Balancing the costs of delays with the need to have all systems ready created a confusing, ever-changing schedule for firings that was often modified without the full knowledge and understanding of every participant, especially those who worked on the scientific teams. Krause knew this, and he knew that his endorsement of Toftoy's policies, in answer to the Zwicky incident, was an essential part of reality at White Sands. But he also told Toftoy in February 1947 that he did not want to give the "impression that we feel the White Sands Proving Ground is a little bit of heaven. We definitely do not feel this way."[82]

## Tightening Controls

White Sands moved a bit farther away from heaven and closer to a functional reality when Brig. Gen. Philip G. Blackmore replaced Turner as commanding officer of White Sands on 4 August 1947 and instituted immediate changes in firing procedures. Instrument teams would no longer be allowed to monitor and control their devices from the blockhouse. They had to devise automatic means of turning on their instruments, using switches linked electrically to the countdown routine or using on-board accelerometers. Safety restrictions were tightened considerably, and all "installation times, testing times, completion dates, etc., [were] to be strictly adhered to."[83]

Some of these procedural changes came as welcome news to the trench workers like Bergstralh, Nelson Spencer, DeWitt Purcell, and Lorence Fraser. Instrument teams no longer had free access to the missile the night before a launch, but now, with a full gantry available (figure, p. 124), equipment could be installed after the missile was erected on the launch pad. Previously, GE had first access to the missile after it was erected and did not allow the scientists in until it had finished its preparations. As a result, there was usually a scramble for access at the last moment and confusion over priorities. Now, the instrument people went first, before the GE crew made its final preparations.

Michigan's Nelson Spencer did not mind a general tightening of security as the firings continued through 1947, but his experience thus far alerted him that his group had to be "prepared to install . . . under someone else's schedule."[84] Unlike elite academics like Zwicky, Whipple, or Greenstein or instrument specialists like Tousey, or even aggressive leaders like Van Allen and Krause, Spen-

cer understood the situation at White Sands; he was ready to take whatever steps were necessary to minimize conflict. Bergstralh was equally ready to cooperate, knowing that Krause and Toftoy were doing all they could to make the disparate groups work together. The leaders of the field teams, such as Fraser from APL, also appreciated the demands of the range, were more tolerant of the restrictions imposed by the military, and were less driven by demanding personal scientific agendas during their visits to the range.

In the first decade, White Sands activities increased by a factor of 50, based on the number of firings alone.[85] As the number and variety of missile and ordnance firings increased and as the base acquired more personnel, White Sands grew into a small city. Increased size and activity brought increased bureaucracy, especially in matters pertaining to range safety and security.[86] On 1 April 1948, the Naval Unit officer-in-charge issued an 80-page *Organization and Regulations* manual for all civilian and Naval personnel. In addition, as the manual made clear, Naval personnel were still subject to Army regulations.[87]

Eventually, people like Bergstralh, Spencer, Purcell, and Fraser learned to live with these rules, and even used them to advantage. They knew that life in the desert was certainly a tradeoff. They knew that Toftoy and his staff, unless pressured, wanted to accommodate the scientific experiments, but they also knew that the longer the V-2s remained in the desert sand, the more unreliable they became, and the more costly they were to fire.[88] Delays meant increased costs, and the V-2 was, by far, the largest and most costly piece of ordnance being fired in the first few years of activity at White Sands.

## Drama in the Desert

Preparing for a V-2 launch was comparable to fighting a war. The enemy was time, procurement red tape, aging V-2s, balky equipment, and an infrastructure unable to cope, unsure of its mission. For Harold Turner, a

**Stages in preparing a V-2 for flight. Scenes are not from the same flight, but all are from the spring and summer of 1946.**

Staging area in front of the V-2 assembly building, White Sands Proving Ground. Applied Physics Laboratory, JHU/APL archives.

Telemetry checkout for an NRL launch. Ralph Havens is at upper right, Thor Bergstralh is wearing a white cap standing at left next to Ernst Krause who is peering into the telemetry chamber. F. S. Johnson is behind Bergstralh. Ernst Krause collection. NASM SI 83–13873.

After assembly the V-2 is transported to the firing area, some 10 kilometers from the assembly area. Project Hermes. NASM SI 79–13150.

Before the gantry was completed, vehicle and final instrument checkout were accomplished from extension ladders. NRL launch, June 1946. The WAC Corporal launch tower is at right. William Baum collection. NASM SI 90–6481.

successful battle won was a V-2 flying straight and true into space, and on time. For Herb Karsch, or General Electric's field crew, victory included complete diagnostics about the missile in flight. For the scientists, a win required not only good diagnostics, but a data stream that revealed something new about the high atmosphere.

Whatever their goals and frustrations, however, those who built, tested, and flew scientific instruments at White Sands were all learning how to perform research in an environment unlike any that had supported science in the past. Somewhat similar to the experiences of 19th-century naturalists who depended on, but often were compromised by, the direction of naval commanders whose missions were different from theirs, scientists preparing for flights at White Sands had to organize their tasks around a vehicle not built for research; yet many were captured by the fascination of it.[89] Even those who left rocket research quickly, like Jesse Greenstein from Yerkes, whom we will meet in Chap. 11, were taken by the drama capping their weeks and months of frustrating effort. At a 1947 Yerkes symposium, Greenstein talked about this new way to conduct scientific research:

> *The takeoff and the flight are majestic events. After a breathtaking, slow initial rise the rocket and its jet disappear from sight and the track is marked only by the white vapor-condensation trail formed in the wake. . . . The flight is silent except for the few seconds at takeoff, and the landing is marked only by the reverberations of the shock waves from the distant mountains.[90]*

Greenstein's academic contemporaries, like William Baum, who left Tousey's group in late 1946 to resume graduate studies at Caltech, shared his fascination for White Sands and the experience of sending a scientific instrument into space. For them, as well as their compatriots at NRL and APL, White Sands became a testing ground for personal choices. As we will find in later chapters, the fate of scientific payloads at White Sands more often than not determined the future course of par-

ticipating individuals and groups. Many would find at White Sands that, in order to persevere in the activity, they had to define their scientific goals in relation to rocketry. For those who thought this an inappropriate personal choice, White Sands became only an interlude. But for those who thought it appropriate, White Sands became a lifestyle.

## Notes to Chapter 7

1. Brown, et al. (1959), pp. 14—18. A series of real estate directives from early 1945 through late 1947 established the range. At first, land use permits were issued so that ranchers and miners could still graze their herds and exploit the mineral deposits, but after 1947 the Department of Defense restricted use of the area completely.
2. Ibid., pp. 20–21.
3. The first annual budgets for White Sands were in the range of $1 million, with some $3 million provided for construction, and another $2 million earmarked for general rocket development, aside from long-range missile testing. Army Ordnance provided less than one-fifth the support given over to armored tank development and less than half as much given over to improving traditional small-rocket ordnance. L. H. Campbell, Jr. [Lt. Gen., Chief of Ordnance, OCO], "Plan for Ordnance Department Post War Research and Development," 15 June 1945. Box A763, entry 646A, OCO/NARA.
4. On JPL and the Corporal series, see Koppes (1982), p. 23ff.
5. Lt. Gen. L. H. Campbell, Jr. [Chief of Ordnance, OCO], "Plan for Ordnance Department Post War Research and Development," 15 June 1945, pp. 17–18. Box A763, entry 646A, OCO/NARA.
6. After a Navy Intermediate Land Test Range Investigation Board evaluation endorsed the use of White Sands, and conferences were held between Army Ordnance, ORI, BuAer, BuOrd, NRL, and the CNO in April 1946, NRL was asked to define what the Navy needed at White Sands. See "Joint Army-Navy Meeting on Army Ordnance Research and Development, the Pentagon, 26 June 1946." 26 June 1946, pp. 41–44. Box 767A, OCO/NARA; and "NRL Requirements at White Sands Proving Ground," Enclosure A to F. W. MacDonald

to CNO (Code OP-06, Captain S. Teller), 22 April 1946, p. 1. NRL/NARA.

7. "Joint War Department—Navy Department Press Release," attached to ibid., "NRL Requirements . . ."

8. Admiral Forrestal to "All Ships and Stations," 14 June 1946; "NRL Requirements at White Sands Proving Ground," p. 3 of cover letter, enclosure A to F. W. MacDonald to CNO (Code OP-06, Captain S. Teller), 22 April 1946, p. 1. NRL/NARA.

9. At least 500 personnel would be required to build and maintain the separate facility and take care of the combined needs of all Navy activities. M. W. Rosen, "Record of Consultative Services," 24 April 1946. NRL/NARA.

10. In October 1946, for example, Navy officers at White Sands wanted to paint Navy jeeps special colors to distinguish them from Army jeeps. The surviving documentation file contained several linear inches of onion skin text representing many weeks of haggling. See, for instance, J. A. McNally to Commander (DATD) 11th Naval District, 11 October 1946. Box 98, folder 4, NRL/NARA.

11. Van Allen, Townsend, and Pressly (1959), p. 58.

12. H. N. Toftoy to Commanding Officer, WSPG, 10 June 1946. Box 98, folder 2, NRL/NARA.

13. See Brown, et al., (1959), pp. 32–33; and Scientific Advisory Committee, Aberdeen Proving Ground folder, 1944/1945, in "SAC" folder, OCO/NARA.

14. E. H. Krause, "Record of Consultative Services," 29 October 1946, record of meeting of October 25 at the Pentagon. Box 98, folder 4, NRL/NARA.

15. It contracted with the New Mexico College of Agriculture and Mechanical Arts (now New Mexico State University) for optical tracking data reductions, and the Navy also contracted with the college electrical engineering department for radio cutoff services. E. R. Toporeck, G. R. Carley, and G. R. Sutherian (NOTS) to N. A. Renzetti, "Report on Trip to White Sands . . . ." 26 August 1948. Copy in JB/NASM. I am indebted to Gen. Julius Braun for providing this and other material relating to tracking facilities at White Sands.

16. *American Men and Women of Science.* 10th ed., p. 911.

17. A popular well-illustrated review of the optical tracking facilities is Mann (1948); technical reviews are the *White Sands Missile Range Technical Catalog* (Army Ordnance, n.d.), especially vol. 2, sect. 1: "Optical Flight Measurement." See also Delsasso, de Bey, and Reuyl (1947). Tombaugh recently recalled that, after World War II, he hoped to return to Lowell Observatory, but budget cuts and politics there denied him a position. His brother-in-law, James Edson, had worked at Aberdeen during the war and drew Tombaugh to White Sands in 1946 to collaborate in developing long focal-length optical tracking systems set on modified gun mounts. Clyde Tombaugh OHI, pp. 14–16rd.

18. Tombaugh's high-powered reflecting telescopes were able to yield not only trajectory data, but could record rocket aspect, pitch, and roll once high-contrast black and white markings were added to the vehicles, an addition Tombaugh takes some credit for developing. Tombaugh OHI, p. 62rd; confirmed in Rosen OHI, p. 127rd; and in Julius Braun to the author, 4 February 1989. Tombaugh's augmentation of the Askania systems is confirmed in E. R. Toporeck, G. R. Carley, and G. R. Sutherian (NOTS) to N. A. Renzetti, "Report on Trip to White Sands . . .," 26 August 1948. Copy in JB/NASM.

19. These included the tracking cinetheodolite Askania network, a set of trailer-mounted high-speed Bowen-Knapp fixed-motion cameras trained on the launch scene, several tracking Mitchell photo-theodolites, a set of fixed ballistic cameras employing high-speed rotating shutters that could provide information on positions and velocities, and Tombaugh's large camera/telescopes. Brown, et al. (1959); [Naval Unit, WSPG] "Facilities of the White Sands Proving Ground," 15 June 1948, pp. 16–20. JB/NASM. Tombaugh OHI, pp. 20–21rd.

20. During the war, they, along with about a dozen other astronomers, engaged in a wide range of observational and theoretical work on interior and exterior ballistics. Hoffleit, Sterne, Reuyl, Robert Atkinson, Edwin Hubble, Martin Schwarzschild, among others, were full-time staff members there, and a few astronomers acted as advisers. For a partial listing of mathematicians, physicists and astronomers, see Goldstine (1973), p. 131. See also Hoffleit (1949), p. 172; and D. Hoffleit OHI; Subrahmanyan Chandrasekhar OHI, pp. 89–90; Martin Schwarzschild OHI. AIP/SHMA.

21. Rosen OHI, pp. 127–28rd.

22. Hoffleit (1949), p. 172.

23. The origin of the DOVAP system using multiple ground stations merits further comment. Some argue that DOVAP was developed independently at BRL during the war by Frank S. Hemingway working under Delsasso, but that the Germans had also developed a similar system, which saw first use in June 1944. No attempt has been made here to study the detailed characteristics of these two systems, but it is clear that the Germans developed many Doppler systems for both tracking and jamming, as E. H. Krause and others found during interrogations in May 1945. See R. H. Block and E. H. Krause, "Further Interrogation of Steinhof, Kirchstein, and Mulner," 18 May 1945, pp. 219–28, in [Army Ordnance] *Peenemünde East, through the Eyes of 500 Detained at Garmish,* known also as *The Story of Peenemünde, or What Might Have Been.* Copies from AAF Historical Office, and from the JPL Library file 519.652–1, 4055–6A (1945). Navy interrogation reports on the use of Doppler techniques in Germany can also be found in "Structural Flight Test Equipment Developed or Used by the Peenemunde Group," U.S. Nav. Tech. Mission, Europe, 180–45 (27 July 1945); NRL Report 4362, Bulletin 125 "Radio-Control System of V-2," as noted in Serge Golian, "Registered Laboratory Notebook no. 1" (3 December 1945). HONRL. For a description of DOVAP operations and theory, see Hoffleit (1949); Berning (1954); and Newell (1953), pp. 217–21. See also L. G. de Bey and Dorrit Hoffleit, *Ballistic Research Laboratory Report no. 677* (BRL, 1948), from which Hoffleit (1949) was derived. Both Julius Braun (letter to the author, 20 July 1990) and Tom Starkweather (letter to the author, 10 July 1990) discuss Hemingway's contributions, while Warren Berning (letter to the author, 25 June 1990) identifies elements of German priority, and what the BRL design borrowed. Braun adds that DOVAP meant to some "*Determination Of Velocity And Position.*"

24. See, for instance, Delsasso, de Bey, and Reuyl (1947), pp. 37–56; [Naval Unit], "Facilities of the White Sands Proving Ground," 15 June 1948, pp. 22–23; E. R. Toporeck, G. R. Carley, and G. R. Sutherian (NOTS) to N. A. Renzetti, "Report on Trip to White Sands . . .," 26 August 1948. JB/NASM. See also Hoffleit (1949).

25. See Julius Braun to the author, 4 February 1989, with accompanying documentation. JB/NASM. On the ENIAC, see Goldstine (1973); and Stern (1981).

26. Hoffleit (1949), p. 175. Even though ENIAC was built for trajectory calculations, the critical application at first was to solve design problems for the superbomb at Los Alamos. Goldstine (1973), p. 234; and Stern (1981), p. 33.

27. Hoffleit (1949), p. 174. This was for a complete analysis. She also noted a "quick method" taking only a few minutes' time for determining a few points along the trajectory for preliminary needs.

28. Computing time for an 800-point trajectory analysis was reduced to about four weeks when newly introduced IBM relay multipliers were brought to the task. These machines failed quite often and required constant supervision. By 1948, two of these machines were wired in tandem and provided a completely checked analysis in about two weeks. BRL also used Bell relay machines designed by Stibitz that could run unattended continuously and produce a full analysis in about three days. Hoffleit, (1949), p. 174.

29. Delsasso, de Bey, and Reuyl (1947), p. 36.

30. J. N. Davis, "The NRL Pulse Telemetering System," in *Bumblebee Report* no. 42, "Princeton Telemetry Symposium" (December 1946), pp. 36–37. NRL had available two dozen 10-channel pulse-time modulated systems from their JB-2 pilotless aircraft program. Through the spring of 1946, General Electric was still claiming that a worker's strike was preventing it from providing the telemetry systems.

31. He directed S. W. Lichtman, P. R. Shifflett, J. M. Bridger, and several others to prepare specifications and requests for proposals from industries such as Sylvania Electric Products, Inc., and Raytheon Manufacturing Co. By the time of the 27 February 1946 panel meeting, no contracts for telemetry systems had been let, but ORI had approved the appropriation of two dozen sets at $3,000 each.

32. The NRL system sampled in rapid sequence in either the 10, 23, or 30 channels available, for between 50 and 200 microseconds each, and produced a linear pulse proportional to the voltage present on that channel. The sampling was produced electronically by controlled multivibrators. See "Papers Presented

at: I. Princeton Telemetering Symposium; II. APL Telemetering Conference," *Bumblebee Report* No. 42 (December 1946). JPL/L JPLHF5–449.

33. Van Allen to Greenstein, 6 December 1946. JGP/CIT.

34. The telemetry signal received at White Sands was handled by a facility set up later on a combined ONR and Army contract to the New Mexico College of Mechanic Arts and Sciences (now New Mexico State University). E. B. Stephenson, "Request for Assignment of Problem," 27 February 1947, enclosure A: Supporting Data, p. 2. Box 99, folder 5, NRL/NARA. The first NRL telemetry systems are described in Garstens, Newell, and Siry (1946), chapter IIC; Heeren, et al. (1947); and Homer E. Newell, "Consultative Services Record," 27 September 1948, Serial Number 3420–391/48(3420). Date of Conference: 24 September 1948. Box 100, folder 17, NRL/NARA.

35. "V-2 Panel Report No. 6," 10 September 1946, p. 4. V2/NASM.

36. Harry Wexler to Chief of Bureau, 4 October 1946. HW/LC.

37. "V-2 Panel Report No. 7," 12 November 1946, p. 5. V2/NASM.

38. See E. B. Stephenson, "Request for Assignment of Problem," 27 February 1947, enclosure A: Supporting Data, p. 2. Box 99, folder 5, NRL/NARA.

39. C. H. Smith, "Record of Consultative Services," 20 July 1946. Box 98, folder 3, NRL/NARA.

40. Ibid.

41. Holger Toftoy Diary, entries for 2 October and 4 November 1946. HT/ASRC.

42. Fraser OHI, p. 42.

43. W. G. Dow, "Review of A. A. F. Participation in the V-2 Firing of August 22, 1946," typescript, r.d. and second draft, 9 September 1946. Box 3, folder 3, M669, WGD/BHLUM.

44. Ibid., pp. 7–8.

45. Ibid., p. 2.

46. Ibid.

47. Dow to Maj. Gen. Curtis LeMay, deputy chief of staff, AAF R&D, 15 October 1946, pp. 3–5; quotations on pp. 3, 5. Box 5, Engineering Research Institute Corresp. 1946–49, WGD/BHLUM.

48. Dow did not send his critique directly to LeMay at first, but to Ralph P. Johnson, a civilian scientist in LeMay's office. Johnson felt that Dow's criticisms and observations should go directly to LeMay quickly, adding: "We are agreed that the time is here for some action, and are hopeful that something effective can be done." Ralph P. Johnson to Dow, 27 September 1946. WGD/BHLUM. This perspective was provided by Jack Neufeld.

49. LeMay to Dow, 8 November 1946. WGD/BHLUM.

50. W. G. Dow, "Review of A. A. F. Participation in the V-2 Firing of August 22 1946," typescript, r.d. and second draft, 9 September 1946, p. 11. Box 3, folder 3, M669, WGD/BHLUM.

51. On several occasions, Maj. Herbert L. Karsch, White Sands technical director, worried aloud that the Michigan warhead would not be ready in time because of these problems. Their flight was almost canceled because AMC failed to provide the necessary drawings or plans for integrating different experiments into the warhead. Worse yet, even after the warhead arrived, it was incomplete, so the system could not be tested properly. Gottlieb agreed with Dow that Signal Corps people were more forthcoming than "our own Air Corps." Melvin Gottlieb to William Dow, "Report on field party activities in connection with V-2 firing of Aug. 22, 1946," 3 April 1947, pp. 3–4, 8. WGD/BHLUM.

52. W. G. Dow, "Review of A. A. F. Participation . . . August 22 1946," 9 September 1946, p. 1. Box 3, folder 3, M669, WGD/BHLUM.

53. F. W. MacDonald to Army Ordnance Test Station, WSPG, 9 May 1946; Enclosure A: "Rocket Sonde Research Section White Sands Bulletin No. 2 April 30, 1946." Box 98, folder 2, NRL/NARA.

54. Typically, they incorrectly signed out jeeps and automobiles from their motor pool. This caused a heavy strain on the few vehicles available, and the men neglected normal maintenance. As a result, the vehicles were in poor shape, and Hanks suspended the use of Navy vehicles for personal leaves to Las Cruces, which caused a "large howl" among his men. H. C. Hanks to C. H. Smith, 20 July 1946. Box 98, folder 3, NRL/NARA.

55. C. H. Smith, "Record of Consultative Services," with Enclosure A: "Agenda for Meeting," 7 August 1946. Box 98, folder 3, NRL/NARA.

56. Nolan R. Best, "Record of Consultative Services," 25 November 1946, p. 2. Box 98, folder 4, NRL/NARA.

57. [Harold R. Turner] to all users, "Facilities at White Sands Proving Ground," 1 January 1947. WSPG folder, WGD/BHLUM.

58. Lang (1948), p. 155.

59. Rosen (1955), p. 43.

60. See Lang (1948).

61. Rosen (1955), p. 42. McLaughlin, along with Tombaugh and a few others, is also remembered for reporting the earliest sightings of unidentified flying objects around White Sands, for which they were roundly chided. Ibid., and McLaughlin to E. Krause, 29 June 1947; Newell to Krause, 9 July 1947. EHK/NASM; Warren Berning to the author, n.d. 1990; and Homer Newell, "Rocket-Sonde," text of talk to AIAA Sounding Rocket Vehicle Technology Specialist Conference, Williamsburg, VA, 28 February 1967, pp. 5–6. JPL/L 3–41. Jacobs (1975), pp. 57–58 provides context for McLaughlin's UFO reports.

62. Harry Wexler to Chief of Bureau, 4 October 1946. HW/LC.

63. H. Toftoy to E. H. Krause, 24 January 1947. EHK/NASM.

64. H. Toftoy to E. H. Krause, 2 and 24 January 1947. EHK/NASM.

65. Ibid.

66. E. H. Krause to Colonel Holger Toftoy, 11 February 1947. EHK/NASM.

67. At one of the first NRL launches on 27 June 1946, Krause reminded Toftoy, the Army and GE were having trouble readying the V-2, which delayed the launch until late in the day. Herb Karsch asked if this delay would affect the scientific experiments, especially the solar observations. Krause responded that it would not, as long as it was a short delay. But when the delay continued, Krause asked that the firing be postponed until the next day, and Turner approved it. Krause believed this was significant evidence that the Army was bending over backwards to be agreeable, because: "It must be remembered that this was one of the earlier V-2 firings and the number of visitors present, including numerous 'star' officers, was very large." Krause was especially impressed because some of the top brass and civilian VIP visitors could not stay over for a firing on the 28th. Although they both knew the great "prestige and publicity value of firing that day." Still, Krause testified, "without a moment's hesitation, [Turner] asked me what time I would like to fire the next day." E. H. Krause to Col. Holger Toftoy, 11 February 1947. EHK/NASM.

68. Rosen (1955), p. 38; see also Rosen OHI, p. 117–18rd.

69. Turner had been an engineering instructor and a member of the Ordnance Reserve in the 1920s. From 1922 until 1941 he was a designer and commercial engineer at General Electric's Pittsfield, Massachusetts, facility. He went on active duty in that year and became an industrial engineering officer within OCO before being assigned to identify and build White Sands. Brown, et al. (1959), pp. 5–7.

70. Van Allen OHI, p. 142. Van Allen added that Ernst Stuhlinger, a member of Von Braun's team, helped to smooth the way between scientists and White Sands personnel.

71. He knew that exact launch times were specified and were made largely for his convenience, but he also knew that delays were often caused by purely technical problems, rather than Army policy. Still, it was a source of frustration, and something to grouse about to colleagues. Richard Tousey to W. O. Roberts, 21 May 1947. WOR/UC.

72. Anonymous source; and T. Bergstralh OHI, p. 33rd.

73. Julius Braun to the author, 4 February 1989. JB/NASM.

74. Ibid.

75. Holger Toftoy Diary, entry for 15 May 1946. HT/ASRC.

76. Ibid.

77. According to William Dow's notes from a V-2 Panel meeting. See notebook entry 3 June 1946. WGD/BHLUM.

78. Holger Toftoy, "All Interested Agencies," 8 November 1946. Box 98, folder 4, NRL/NARA.

79. Bergstralh accompanied most of the NRL experiments to White Sands between 1946 and 1950 and was responsible for the warhead payload once there. After he and his staff checked all warhead systems in the hanger and then on the gantry, which upon its completion in November 1946 greatly aided final installation and checkout, they performed additional tests in the blockhouse. But Bergstralh could not predict that all payload instrument systems could be checked from there. It was not unusual to find that the wiring diagrams prepared at NRL did not match the devices that showed up at White Sands. T. Bergstralh OHI, p. 36rd.

80. Bergstralh OHI, p. 33rd.

81. Bergstralh OHI, pp. 35–6rd. Obviously, recollections of unrecorded range procedures de-

pend on the experiences and perceptions of the parties involved. It seems that some missile launches were delayed and others were not, and each was handled on an individual basis. Although the original weekly firing rate was relaxed in June to one every two weeks, the original rapid preparation schedule was maintained until later in the year, when firing failures exceeded successes. Only then did the Army agree that if a missile package or the missile were not ready, the launch would be canceled and that particular missile would be moved to the end of the roster, so that the original schedule might remain intact. This helps explain why the numerical order of the missiles was not chronological in tabulated firing histories. See, for instance, Newell (1980).

82. E. H. Krause to Col. Holger Toftoy, 11 February 1947, p. 3. EHK/NASM.

83. Although these field rules were eventually publicized and implemented, scientific teams were at times allowed into the blockhouse for special needs, on an ad hoc basis. Ibid., p. 2. Turner remained at White Sands as Executive Officer until October 1947, returned to Washington D.C. for a few months, and then was assigned to the Research and Development Division, OCO until he assumed command of the Advance Headquarters, Joint Long-Range Proving Ground, Cocoa, Florida, and helped to build that facility as he had White Sands. Brown, et al. (1959), p. 7.

84. Spencer added: "We should also design our devices so that they are as completely independent as possible of the services of others such as switching, rocket wiring, and power supply." Nelson W. Spencer to Dow, "Visit to White Sands Proving Grounds, October 14, 1947," 20 October 1947, p. 3. Box 3, folder 2, M669, WGD/BHLUM.

85. The greatest jump came between 1954 and 1955 when the rate doubled, with the Corporal, Nike, and Honest John leading the way. On statistics of firings, see Brown et al. (1959), pp. 72, 76–7; Newell (1959a), pp. 167–70, 242.

86. The most celebrated event took place when a wayward V-2 landed near Juarez; but other missiles were known to threaten El Paso or Las Cruces. Brown, et al. (1959), pp. 82–3; Ordway and Sharpe, (1979), pp. 354–55. One of the most dangerous but not unusual problems was to be fired on when down range. NRL's ionospheric field antennae were inadvertently used as targets for Honest John missile tests when the NRL crew was present. See Seddon to the author, 24 August 1990. A major safety issue arose in 1947 when Turner delayed firings of all guide-rail launched arrow-stabilized sounding rockets such as the Aerobee until the V-2 Panel satisfied Ordnance that windage data and launch tower tilting capabilities were adequate to ensure safe firings. Van Allen, Ernst Krause, and Michael Ference developed an encompassing safety plan which satisfied the Army that the Aerobees could be launched with a sufficient margin of safety. General Saylor, who had replaced Barnes, did what he could to help out as well; he gave Aberdeen a high-priority task to develop adequate windage tables. "V-2 Panel Report No. 11," pp. 12–14. On these and other safety problems, see Homer Newell, "Rocket-Sonde," text of talk to AIAA Sounding Rocket Vehicle Technology Specialist Conference, Williamsburg, VA, 28 February 1967, p. 6. JPL Library Historical Files source list 3–41; letters, Krause to Newell, 7 July 1947; Newell to Krause, 9 July 1947, Krause Papers, copies at NASM; V-2 Panel Report No. 11, 15 July 1947, p. 5.

87. [W. A. Gorry]. *Organization and Regulations*, U.S. Naval Unit White Sands Proving Ground, Las Cruces, New Mexico, 1 April 1948. Box 100, folder 13, NRL/NARA.

88. See, for instance, Toftoy diary entry, 15 May 1946. HT/ASRC; and consistent references to this fear in the V-2 Panel minutes throughout the 1940s.

89. We will develop this point in Chap. 11–17. Both Knight (1986) and Miller (1968), pp. 173–74, for instance, discuss episodes in the history of scientific exploration that were compromised by clashes between naval commanders and their scientific complement. DeVorkin (1989a) has made these connections for manned scientific ballooning in the 1930s.

90. J. Greenstein, in Kuiper (1947), p. 116.

# 8

# The First Launch Season: Technical Problems

Nothing was routine during the first year V-2 rockets were launched from White Sands. Technical problems, procedural foul-ups, and policy conflicts dominated life in the laboratory and on the proving ground. During what Army Ordnance called the first launch season, which started in April 1946 and lasted until the end of the year, the members of the V-2 Panel came face to face with the realities of conducting research with rockets. Among the many obstacles that lay before them, the most disturbing was poor data retrieval. Here we examine how these groups dealt with this and other problems as they and their patrons struggled to acquire a working knowledge of the V-2.

## Bits and Pieces, The Recovery of Data

Even though NRL and APL's scheduled flights started in June 1946, both Krause and Van Allen managed to place small packages on the first flights managed by General Electric. Krause's were purely diagnostic, designed to assess the possibilities for physical retrieval of data, whereas Van Allen hoped to get a jump start on detecting cosmic-ray primaries. Van Allen's wartime experience with the proximity fuze at APL taught him how to design simple and reliable devices that could stand the rigors of violent acceleration, and he was ready to demonstrate his acquired skill.

Van Allen had less than a month to meet the first launch of a V-2 from White Sands, scheduled for 16 April 1946. Therefore his first payload package was kept simple: a single Geiger counter shielded by an inch-thick cylinder of lead, with power supplied by hearing aid batteries and on-board recording by APL's "woodpecker," which punched holes in a moving strip of brass. The package was placed in a pressurized steel cylinder, which also contained a roll of shielded alpha particle photographic film.[1]

When the 16 April missile lost its control systems, dropped a fin, and spun out of control at an altitude of 5 kilometers, Van Allen saw his first package die in a fiery crash as the missile "dug a hole about 25 feet deep and 35 feet in diameter."[2] General Electric's Hermes engineers were doing everything for the first time. At least, someone mused, they got the missile off the pad.

The second flight, on 10 May, carried a duplicate APL package and reached 112 kilometers, crashing 61 kilometers downrange. When the field crews got to it, they found a large crater rapidly filling with water. Army Ordnance Maj. Herb Karsch and J. F. McAllister from G. E. gloomily reported: "There was apparently complete disintegration of the missile because a search at the impact point resulted in no recovery other than bits and pieces totaling 100 lbs in gross weight."[3] The missile, moving at about 1 kilometer per second, had disintegrated at impact. APL's recorders and film were never recovered.[4] Nevertheless, the 10 May flight was a technical success, and hopes were high, even though Van Allen still received no useful data. For the third flight, on 29 May, Van Allen tried to record cosmic-ray data using several dedicated channels from NRL's on-board te-

Impact crater from early V-2 flight, circa June 1946, in the flat sandy areas of the Tularosa Basin. Krause collection. NASM SI 79–13164.

Recovered equipment container shell from the 16 April V-2 flight, which never achieved supersonic velocities. Applied Physics Laboratory, JHU/APL archives.

lemetry service. But again, no useful data were returned by telemetry, and the missile suffered an air burst before it crashed into the mountains.

Krause learned from Van Allen's experience, as well as from his own, that retrieving physical data from the V-2 was not going to be easy. He had placed heavy steel and lead cylinders and boxes on GE's first three flights to study how packages of photographic film would survive impact. Nothing useful was recovered.[5] At its 3 June meeting, the V-2 Panel was not ready to accept the full implication of this problem: "it is difficult to conceive that some of the heavy steel cylinders sent in the 10 May and 29 May launchings have disintegrated."[6] Van Allen, harboring faith in the survival of his cosmic-ray recorder, enlisted the aid of local contract excavators to scour the 10 May impact site. Karsch felt that the absence of fused sand in the area and of any melted shards from the missile gave hope that the woodpecker still lived. Nothing useful was ever found, however, save for a few twisted shards from the cosmic-ray chamber container.

After the first three flights, the problem with physical retrieval turned into a nightmare.

Anyone who had to build devices that would survive impact in the desert must have wondered if they were ever going to succeed. V-2 missiles were not designed to land softly; rather, they were designed to crash and create as much destruction as possible to themselves and to the landscape. And now, as the Germans had warned, many V-2s were likely to suffer air bursts.

The ionospheric and cosmic-ray researchers were not too concerned, as their data could be telemetered. But the solar spectroscopy groups at NRL and APL, as well as those at CFS and Michigan who were planning air-sampling systems, aerial photography cameras, and biological packages, could not proceed without physical recovery. Meeting at APL in early June, V-2 Panel members talked wistfully about parachutes and ejection mechanisms but agreed that nothing like Regener's complex system could be ready for the first round of firings. The film was going to come down with the rocket, had to be protected somehow, and then had to be located.

One possible solution suggested by the Germans at Fort Bliss was to separate the warhead from the rest of the rocket near the end of the flight, a technique Army Ordnance was interested in developing for ballistic warhead reentry. But even if the warhead could be separated from the rest of the missile, it would still hit the ground at a very high speed, and so the same problems remained.

Accordingly, Richard Tousey decided to put a film cassette made of armor-piercing steel into his solar spectrograph, and hope for the best. As Baum recalls, "It was terribly simple conceptually, but it wasn't the easiest thing to make"[7] (figure, p. 138). Since Tousey's device was probably going to be the first photographic instrument to fly, it represented the proof-test for the physical retrieval of data.

While the first spectrograph was being readied at NRL in early June, procurement officers were less than sanguine over its future. In a series of hasty and conflicting telegrams to Tousey's contractor, who was preparing several spectrographs, the NRL supply officer first canceled and then reinstated the contract in a matter of days, implying that even though his people knew that drastic changes to the design were probably needed, they would proceed with the original design.[8]

The supply officer caught the spirit of the situation. There was nothing to do but push ahead and hope for the best. While Tousey's group continued to run integration and calibration tests, Krause's people feverishly prepared the NRL warhead, which did not arrive until 15 May, one month late from the gun factory.[9] They had less than one month before the instrumented warhead had to be air transported from the Anacostia Naval Air Station by the Army Air Forces. Within just a few weeks, they had to install and test Tousey's spectrograph; as well as their meteorological, ionospheric, and cosmic-ray experiments; the NRL telemetry system; and the radio beacons. (See Table 7.1)

Krause's days were filled with coordinating V-2 Panel activities and handling NRL press releases, White Sands liaison problems, Neptune rocket development, and NRL participation in a planned B-29 program at NOTS. His large staff still needed attention, even though he had split them up into six units, each with a hierarchy of managers. Krause no doubt saw little of his home several kilometers to the east in Cheverly, Maryland, and worried constantly over the "bits and pieces" left from the first flights.

Tousey spectrograph showing the polished cylindrical cassette designed to protect the spectroscopic film from impact damage. U.S. Navy photograph, Richard Tousey collection. NASM SI 91–6312.

## The First NRL Launch: In Search of Tousey's Spectrograph

Many surprises awaited the NRL team after their warhead arrived at White Sands on 12 June. The warhead weighed considerably more than the Army limit of 1,000 kilograms, yet balance tests of the rocket with the warhead revealed that another 230 kilograms of steel had to be added to the base of the warhead.[10] Then, as they were integrating their payload onto the rocket, William Baum and DeWitt Purcell had a chance to talk directly with the Germans and realized that the warhead was the wrong place to put the spectrograph. The tail would be a better place; its poor aerodynamics would cause that section to tumble and have a lower terminal velocity before crashing.[11]

After two weeks of checking everything out in the hot White Sands hanger—the mechanisms for emergency cutoff, warhead blow-off, and then rollout; rechecking all the systems on the launch pad; continual radio checks; Doppler beacon checks; and the procedures for evacuating the launch area—the NRL package was ready to fly on the 27th, although the flight was delayed to the 28th. Countdown began at 9:00 a.m. for a launch half past noon. Launch came a few minutes late, delayed by telemetry interference problems.

Baum and the other NRL people, being excluded from the final preparations, could now only watch and worry: "once we had delivered the instrument there and fully assembled it, [and after the rocket was fueled] it was our job to chew our nails until we could locate the crater and see if we could find anything."[12] They could watch in the blockhouse or behind a boondock, uncomfortable in the knowledge that their spectrograph was not in the most likely part of the missile to be retrieved.

Baum recalls that once the "throbbing sound" he felt in his chest died away, either from his heart or from the acoustic pressure of the successful rocket blast, the moment of truth came when the camp sprang into action to find the remains. "We all got in a jeep [and]

a number of us went bouncing out across the desert in the jeep to try to locate the crater."[13]

Reconnaissance aircraft directed them to the crater (figure, p. 136). Standing at its rim, Baum remembers thinking "If the front of the armor-piercing head, the cassette, and all the rest was down there somewhere, it was under tons of rock and rubble."[14] Baum gave up on the spot. But several weeks later, the NRL team felt there might be "a remote possibility that the film container [was lying] intact in the debris and that the film is undamaged."[15]

NRL's V-2 reached just over 100 kilometers during 354 seconds of flight; telemetry was gathered from the NRL cosmic-ray counters and the housekeeping functions for the spectrograph, but the NRL ionospheric telemetry had failed. Telemetry also revealed that the film was wound securely into Tousey's armor-piercing steel cassette, and that explosive bolts mounted at the base of the warhead had detonated. But, as in an earlier firing, the warhead did not separate.[16] The telemetry record convinced Tousey to push for continued excavations at the site throughout the summer of 1946, but to no avail.

Like Van Allen after the 10 May flight, Tousey could not bring himself to believe that his spectrograph would not be found. While the excavation crew continued to dig, work resumed much as before on the retrieval problem. Through the summer, Hermes engineers refined their explosive bolt separation system and tested preliminary parachute ejection devices, but without success. Meanwhile, Tousey decided to place another spectrograph in the tail fin of NRL's next missile.

## A Summer of Hope, Skepticism, and a Search for Alternatives

Although Tousey never broke his stride, others watching the progress of the V-2 spectrographic work were more pessimistic. During the summer, Members of Krause's staff designed a photoelectric spectrometer to telemeter spectroscopic information to the ground,

Richard Tousey inspecting debris from the 28 June 1946 crash site. Richard Tousey collection. NASM SI 91–6305.

and some astronomers and physicists began to think that ultraviolet solar research would never be possible with rockets if they had to depend upon recovery of photographic film.[17] Tousey's experience also gave James Van Allen's APL group second thoughts. Van Allen reported to Jesse Greenstein in July: "We are holding up the firing of our solar spectrograph until recovery of film is adequately in hand. . . . Next Tuesday's V-2 Warhead, therefore, comprises only a series of cosmic-ray experiments with telemetering and steel drum recording in parallel."[18]

Van Allen did not want to sacrifice their spectrograph and cameras to sure oblivion. It did not help that the next missile, fired on 9 July under General Electric auspices, disintegrated in flight. It had contained four packages of cosmic-ray emulsions, an NRL ionospheric experiment, and seeds for Harvard. Nothing useful was retrieved out of the 20 meter diameter crater.

## APL's 30 July Cosmic-Ray Telescope Flight

For the 30 July APL flight, White Sands technical staff, Fort Bliss Germans, and pyrotechnic experts from the New Mexico School of Mines decided to boost the warhead separation system by placing an additional charge of nitro-starch at the base of the warhead. They had little to lose. If the warhead did not separate, it was doomed anyway.

V-2 no. 9 carried APL's elaborate cosmic-ray coincidence telescope. The highly complex system created special technical problems for Van Allen's team. Its 24 Geiger counters all required electronic circuitry that not only directed power but had to coordinate, record, and telemeter information on which pairs tripped as a function of time. The delicate electronics all had to survive the launch and then had to work reliably under difficult environmental conditions.[19]

The 30 July flight was the best yet for the propulsion group. The missile was launched almost on time and reached 160 kilometers. The warhead detonator also worked for the first time, blowing the sections apart, and the tail section fell to earth intact (figure, p. 141). Unfortunately, NRL's 23-channel telemeter system worked for only 160 seconds of the flight, going dead about 70 seconds before the missile reached maximum altitude. All three NRL ionospheric propagation trans-

Recovery of the 30 July missile showing the result of a successful warhead separation. Large portions of the mid and tail sections survived. U. S. Navy photograph, Ernst Krause collection.

mitters failed as well. The separated warhead was never recovered, but the main body of the missile was found in fair condition.[20]

At the sixth meeting of the V-2 Panel in September 1946, Van Allen discussed the spotty cosmic-ray telemetry data his group had received, which he believed showed that the counting rate increased by a factor of 300 at 70 kilometers' altitude when compared to sea level rates.[21] Beyond this, Van Allen said nothing about scientific results, but it was clear that few of his program goals had been met.[22] But his crew did rejoice that at least their telescope system worked; like the General Electric team, they saw any data stream in the first few months of launches as a victory.

The good news from the 30 July APL flight was that the aft and tail sections of the rocket had been recovered. Van Allen's group now decided that it could send up its own spectrograph on APL's next flight, scheduled for 24 October. Van Allen, clearly enthusiastic, reported his successes to Greenstein: "I think it is fair to say that the V-2 is a demonstrated vehicle for scientific investigations."[23]

Others were still not so sure. In early September, Marcel Schein of the University of Chicago, a respected cosmic-ray physicist who had trained two of Krause's cosmic-ray team members, told a Harvard astronomer that the rocket spectroscopy efforts were "not going well."[24] Even Krause remained unsure of the prospects for physical retrieval and continued to have his staff prepare the photoelectric spectrograph, believing, along with several astronomers, that photoelectric spectrographs made more sense.

## Repositioning the Spectrograph Amidst Launch Failures

After the successful recovery of the tail section of the 30 July V-2, there was little question about where to put the spectrograph on the next flight, which would be NRL's and was scheduled for 3 October. With the help of Ernst Stuhlinger (one of von Braun's men) in numerous conferences with Project Hermes staff, Krause finally won them over to the idea of placing the spectrograph in a tail fin.[25] NRL had barely two weeks to build a new metal shroud for the spectrograph, and have it ready for installation at White Sands for V-2 no. 12.[26] The new shroud was fabricated at NRL and rushed to White Sands, where GE was completing a week of work on tail fin modifications. Meanwhile, a second spectrograph had been delivered by the contractor to NRL in September, and Tousey's staff quickly installed all the optics and began the testing cycle, pushing to meet the launch deadline.[27]

When the NRL field crew reached White Sands, they found new rules in force. Three of the last four missiles had been failures, and with a review meeting of the Army Ordnance Advisory Committee on V-2 Firings coming up later in October, Toftoy could not afford another bust. Turner decided that the GE launch crews needed more preparation time, and so he lengthened the embargo period to two weeks before launch. Thus the NRL field crew found that they had arrived too late to have adequate time to properly install and test everything. They objected vigorously to Harold Turner's new schedule, on the basis that they were not warned, and took their case to Toftoy, who agreed to slip the launch date to allow both groups the time they needed.[28] The slip certainly aided the spectroscopic group at NRL. They now had time to make sure that their instrument was happy in the fin of V-2 no. 12.

## Success in October

NRL's second exclusive flight was the most ambitious to date. The arsenal included cosmic-ray detector arrays; the solar spectrograph; ionospheric radio propagation and composition instruments; pressure and temperature sensors all over the rocket, inside the warhead, and inside many of the experiments themselves; numerous cameras to photograph the behavior of the radio antennae, of the rocket trajectory, and the earth's surface; seed packages to study effects of the space environment; photocell aspect indicators; and parachute devices for recovery experiments. All were integrated into the rocket during late September and early October.[29]

On 10 October, work began well before sunup. By the late morning, the spectrograph was loaded with film, and the doors on the rocket were shut. Thirty minutes before launch, the tail fin trailing wires for the ionospheric experiments were drawn out. A final check of all telemetry channels was made 3 minutes to launch, and at the launch at 11:02 a.m. the spectrograph and ionospheric experiments were turned on.[30]

V-2 no. 12 rose slowly into the clear White Sands midday sky and reached an altitude of 173 kilometers in 227 seconds. The falling missile was successfully blown apart at about 35 kilometers altitude. Telemetry from the aspect sensors indicated that the rocket rolled slowly, with a period of about 1 minute, and that the spectrograph's electrical functions worked properly. After 9 to 11 minutes, V-2 fragments fell to the desert some 20 to 30 kilometers from the launch pad. Evidently, the system had worked perfectly. All that remained was to find the exposed film.[31]

The search was agonizingly slow; even though the entire missile with all of its experiments was retrieved, the Army did not find the spectrograph until 16 October.[32] It was quickly removed from what was left of the

tail-fin housing, and flown back to NRL, arriving in Washington on 18 October. The armored film cassette was ceremonially removed for reporters (figure, p. 143) and then taken directly to the Navy's Photographic Services Laboratory for development. There, Tousey, F. S. Johnson, Strain, and Oberly reviewed development procedures with the director of the photographic laboratory and his staff, and, although anxious to proceed, they found that the director would not let his staff work over the weekend. Therefore they had to wait until the following Monday to initiate the planned series of test development runs, film-strip calibrations, and finally the development of the rocket spectra themselves.

After a full day of tests on Monday, the flight spectra could finally be developed on Tuesday afternoon. After what must have seemed like an eternity (but was probably about 5 to 7 minutes) the lights were turned on in the darkroom, and everyone saw for the first time the far ultraviolet spectrum of the sun. Later that day, Charles Strain happily recorded in his laboratory notebook: "spectrograph film came out very well—clearer than the cal[ibration] roll!"[34]

Tousey and Strain worked into the evening, pouring over the dozens of tiny spectra, evaluating them first for missile and instrument performance (roll rate, film movement mechanism, and so on), and then for ozone absorption and solar atmospheric absorption lines. They established that the best spectra reached 80 nanometers beyond the terrestrial atmospheric cutoff, to 210 nanometers, and that indeed, spectra taken at different altitudes revealed varying amounts of telluric absorption by ozone. Strain excitedly recorded that rough data on the solar continuum had been gathered and that many Fraunhofer lines were visible although he added that they really needed "more detail."[35]

After some internal haggling between Tousey and Krause's group, mainly over who would say what, to whom, and when, Strain worked with Newell and NRL public affairs personnel on preparing a press release during

NRL spectrograph and its reconfigured shroud awaiting installation in a V-2 fin (background) for a November 1946 flight. This installation was identical to Tousey's 10 October flight hardware. U.S. Navy photograph, Ernst Krause collection. NASM SI 83–13881.

the rest of the week, while Tousey and his people concentrated on analyzing the spectra, making microdensitometer tracings and line identifications, and planning for their own processing of future film, since the Photographic Services Laboratory machines had broken the film in places.[36]

The NRL press release was met by an enthusiastic media. Although earlier flights had returned scientific data revealing cosmic-ray counts and pressure and temperature information, the successful retrieval of an ultraviolet spectrum of the sun captured the attention of both the scientific and popular press. *The Washington Post* heralded the discovery of the "new ultraviolet" (figure, p. 144) and even saw fit to reproduce samples of two spectra on page 1.[37] The *New York Times*,

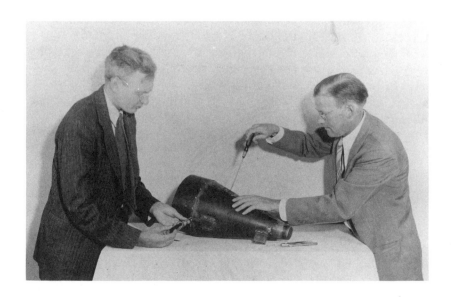

Richard Tousey and C. V. Strain (l to r) ceremoniously remove the 10 October 1946 spectrograph tail fin shroud at NRL. U.S. Navy photograph. NASM 90–6702–0.

## Made From Rocket

# Spectrograms Record Rays Of Sun 65 Miles Above Earth

TITANIUM          IRON

◄──────── OBSCURED BY OZONE ────────►

SPECTRUM AT ONE MILE

◄─────── NEW ULTRAVIOLET ───────►

SPECTRUM AT 35 MILES

PAST THE OZONE—Until a V-2 rocket was shot through the ozone on October 10, man had been unable to graph much of the spectrum of the sun. The spectrum is the image formed by light rays according to their wave-lengths, so that rays of the same wave lengths fall together, while those of different wave lengths are separated from each other, forming a regular progressive series. Ozone, a bluish gas in the air, absorbs the ultra violet rays, and heretofore has limited their recording from the earth. The rocket, equipped with spectographs, recorded the spectrum up to 65 miles. The result was the lengthening of the spectrum, as shown at the 1-mile and 35-mile levels

By Marshall Andrews
*Post Reporter*

Front page story from The Washington Post, 30 October 1946, announcing the first spectroscopic penetration into the ultraviolet. NASM technical files. Copyright 1946, The Washington Post. Reprinted with permission.

*Times Herald*, and *Washington Star* all followed suit. In a few months, specialist magazines such as *Sky and Telescope* proclaimed that the 10 October success was "an event of far-reaching astrophysical consequences."[38] Later articles treated the event in greater detail and with much fanfare. Astronomers reacted by asking for the data. In fact, when Tousey presented his findings at an American Astronomical Society meeting in December, Harvard's Harlow Shapley walked shamelessly up and asked for his slides.[39]

## Looking at Earth

The success of the 10 October flight was really the first clear demonstration that physical data could be returned from a rocket flight. *The Washington Post* and other newspapers had indeed reported launches at White Sands and had paid attention to earlier results, announcing in September, "Cosmic Rays Mystery Pierced with Aid of Nazi Rocket," after APL's 30 July launch. But there was something new and visceral about the tiny photographic spectra. If delicate spectra could survive, maybe people could fly safely, too.

For the same reason, far surpassing the popular excitement of the first solar spectra were the first photographs of earth from space. Easily comprehended by the public, visible proof that the rocket had touched space, earth photographs made for spectacular press.

Unlike prior aerial reconnaissance efforts by balloonists, photography from V-2s was pursued primarily to monitor the behavior of the missile, and only later to study the earth and cloud patterns from space.[40] Clyde T. Holliday at APL and NRL's Thor A. Bergstralh were the two most responsible for direct photography with the first V-2s. Bergstralh, trained in physics, had worked on the use of radar for guided missile countermeasures at NRL during the war, and as part of Krause's section, was in charge of developing NRL's procedures for integrating instruments into V-2 warheads, and techniques for physically retrieving instruments and data. Photography fit perfectly into his retrieval agenda, but Bergstralh recalls that he was directed to aerial photography to solve some frustrating and peculiar problems with the behavior of the V-2s after the first flights. Bergstralh placed standard automatic large-format Navy K-25 cameras in the rocket, fitted with small severely vignetting right-angle prisms to photograph the rocket fins superimposed on the earth during flight to determine directly how the rocket spun and tumbled.[41]

Clyde T. Holliday in contrast, was a photography specialist. During the war, he had worked within APL's proximity fuze project to develop photographic techniques for tracking missiles and recording fuze bursts near aircraft.[42] When he became part of Van Allen's group, he and J. Allen Hynek turned to aerial photography specifically to assess its use for meteorology and long-range reconnaissance. Van Allen recalls the latter application as particularly appealing, although he realized, with his counterparts at NRL, that its immediate application was as a diagnostic tool for recording the performance of the missile.[43]

Holliday used commercial DeVry 35-millimeter motion picture cameras in the early APL V-2s. He chose these cameras because they could be automated using parts taken from a gun director system on a B-29 and had been proven airworthy under rigorous conditions by Army Signal Corps combat newsreel teams. Holliday placed one camera in the midsection of APL's 24 October V-2, nestled between the fuel tanks, and set it to expose three frames per second for a time-lapse record of the motion of the missile. The camera was contained in a steel box, and its armored film cassette was similar to those APL was then developing for its grating spectrographs.[44]

APL's V-2 carried Holliday's camera, as well as Hynek and Hopfield's Bausch and Lomb prism spectrograph and cosmic-ray telescopes to 100 kilometers. A few hours after crashing, the camera was found at the impact site "in almost perfect condition" although the lens had been lost; the other instrument payloads, including the spectrograph cassette, took almost two weeks to find.[45]

Holliday's first pictures were very successful. In November, APL issued a press release showing a series of frames taken from altitudes of 48, 72, and 104 kilometers. Within days and weeks, hundreds of newspapers responded with broad and enthusiastic coverage. The *Herald American* announced "Earth from 65 Miles Up!" and the Trans Lux movie newsreel service proclaimed them to be the "most sensational newsreel pictures of all time." The *Washington Post* caught the significance of the event, showing a single shot covering some 100,000 square kilometers. The *Los Angeles Examiner* let its readers know that "You're on a V-2 Rocket 65 Miles Up!" *Life*, *U.S. Camera*, and many other national and international magazines kept interest up for weeks.[46]

Although the little DeVry cameras worked beautifully, and APL was delighted with the public response, the 35-millimeter format was too small to provide high resolution. Holliday therefore turned to K-25 cameras, placing their optics, shutters, and film transport mechanisms into armored boxes, and by the end of 1948 produced new sets of earth photographs that again gave APL enormous public attention.[47] Holliday continued to refine his technique, presented his best photographs to popular magazines like *National Geographic*, but apparently never subjected his photographs to detailed analysis[48] (figure, p. 146).

The NRL group did not attempt to gain, nor did they receive, the public attention APL did. Their first attempt on 10 October 1946 returned blurred images, but their second flight attempt, on 7 March 1947, was successful.[49] Although intended to check on rocket aspect and instrument performance, Bergstralh also realized that they had done something interesting for "the weather people," as he recalled, "because in that series of photographs, it turned out we had a tropical storm . . . in the Gulf of Mexico."[50] Indeed, both the APL and NRL groups advertised the potential value of their technique for large-scale weather pattern recognition. In the following years, Harry Wexler, the Weather Bureau's observer on the V-2 Panel, made sure that his superior's offices were well adorned with photographs of storms and cloud systems taken from space.[51]

Although they were indeed exciting, the photographic experiments by APL and NRL failed to stimulate more than casual interest in the use of rockets for aerial photography of meteorological conditions or for visual reconnaissance, even though they did demonstrate that physical retrieval problems were being solved. After reading an APL release on the upcoming November 1946 APL flight, *Science Service* director Watson Davis advised his staff that the "really important photos will be a sun—not earth—did they get some of those?"[52] In 1947, F. W. Reichelderfer, chief of the Weather Bureau, chair of the NACA's Subcommittee on Meteorological Problems, and Wexler's boss, concluded at their April meeting that the photography

APL's Clyde T. Holliday displays examples of his converted K-25 aerial cameras circa October 1948. From an Applied Physics Laboratory compilation of press coverage, "So Columbus was Right!" [APL] (1948). Applied Physics Laboratory, JHU/APL archives.

of cloud patterns from V-2 rockets was promising but still "considered to be questionable."[53] Indeed, although APL made much of the potential of aerial reconnaissance from rockets, and the RAND Corporation was at one point interested in securing whatever the groups had collected, Bergstralh felt that the technique was not practical, since rockets could take photographs for only a few moments, and then it would be weeks before another could be fired. "But it certainly made it clear" he added in hindsight, "that if we had satellites, it would be fantastic."[54] Van Allen, on the other hand, predicted at the time that APL's smaller and more cost-efficient Aerobees could ultimately perform weather and military reconnaissance missions. "Under production conditions," his reports attested, Aerobees would become less expensive and easier to handle, "so that routine use in wartime would not be a matter of serious concern."[55]

## The First Season Closes

At the time of the flight of V-2 no. 12 on 10 October 1946, eight more launchings were set through February 1947, and Army Ord-

nance was considering an additional season of 25 firings through early 1948. The 50 percent failure rate of the V-2s to that date helped to slow the Army's timetable, to allow for more time for failure analysis, and GE's launch crews retained complete control of the missiles up to two weeks before launch, to be sure that everything was ready.[56] The atmosphere had changed within the first season of launchings from high-pitched optimism and haste to deliberate planning and caution.

During the first season, the scientific groups gained much practical experience in making devices work on rockets, retrieving data from those devices, and, above all, working together collectively, if not harmoniously, to achieve the maximum diagnostic benefit from each firing. The technical problems everyone faced—ranging from procurement, design, testing, integration, and retrieval of data, to the reliability of the missile itself—all emerged in the first season.

In retrospect, the scientific return from the first season was meager. Tousey had extended the known spectrum of the sun, producing spectra that also revealed the degree of ozone absorption in the ultraviolet, but he knew that these first glimpses were far too crude to celebrate further. And although the first photographs of the earth from space promised future meteorological and geophysical studies, NRL and APL knew that rocket reconnaissance was not around the corner. The ionospheric groups required better radio-frequency isolation and means of improving electronic reliability. Both Krause and Van Allen had achieved some success at making their cosmic-ray telescopes work, but the steel in the missile warhead compromised their results.[57] Van Allen's success, as he later saw it, was in making his electronic system work at all and in isolating his counters from the spurious radiation of the rocket.

The first season demonstrated that there would be no quick scientific return. Research and development teams had to be supported for the long haul. But this would require endorsements and patronage since funding was often not as readily available as one might wish after the leftover wartime accounts dried up. Chapter 9 turns to some of the early mechanisms created to evaluate the conduct of science in the military that had a direct bearing on patronage for upper atmospheric research.

## Notes to Chapter 8

1. See Fraser and Siegler (1948), chap. 3, p. 45.
2. Code 1320 to Code 1300, "First V-2 Firing at White Sands," 17 April 1946. NRL/NARA.
3. "V-2 Report no. 4," 6 June 1946, p. 5. V2/NASM.
4. "V-2 Panel Report no. 4," p. 7.
5. J. J. Bartko to J. G. Bain, 24 April 1946, "Inclusion of Film in Special Containers in V-2 flight of 10 May 1946"; "Rocket Sonde Research Section White Sands Bulletin no. 2," 30 April 1946. NRL/NARA. See also "V-2 Panel Report no. 3," 24 April 1946.
6. "V-2 Report no. 4," 6 June 1946, p. 7. V2/NASM
7. Baum OHI, p. 15rd.
8. On 7 June an urgent telegram was sent to the Baird Instrument Company of Cambridge asking for an "immediate cessation of activity on the High Altitude Spectrograph Model I." Five days later, a second telegram told Baird to ignore the first telegram. T. E. Wright, NRL Supply Officer, to the Baird Associates, 18 June 1946. Box 98, folder 2, NRL/NARA. On 18 June, the NRL supply officer tried to explain to Baird that the first telegram was sent because they thought the spectrograph required "drastic modifications in design to allow for other recovery techniques," and the second telegram reflected further thinking that a redesign would probably not help, "which leaves the problem of recovery about where it was previously." Wright letter, 18 June, ibid.
9. Francis Johnson continued vacuum tests, reaching Lyman alpha in early June, and Tousey designed a special brush photographic processing system. See F. S. Johnson laboratory notebook entries for June 1946. Johnson file, SAOHP/NASM; and Baum OHI, pp. 17–18rd. On the status of the replacement warheads, which were late for all the V-2 Panel members, see F. W. MacDonald to Superintendent, Naval Gun Factory, 17 May 1946. Box 98, folder 2, NRL/NARA. Krause, through Commander Schade, prodded the gun factory to deliver all the warheads on time, even if they went on triple shifts.
10. "V-2 Report no. 5," 15 July 1946. V2/NASM; Baum OHI, pp. 21–22rd.
11. F. S. Johnson, laboratory notebook, 12 June entry, p. 33. Johnson file, SAOHP/NASM. See also Baum OHI, pp. 24–25rd.
12. Baum OHI, p. 29rd.
13. Baum OHI, pp. 26–27rd.
14. Baum OHI, p. 31rd.
15. Garstens, Newell, and Siry (1946), chap. 3, p. 76. F. S. Johnson notes that an identifiable piece of the spectrograph was found, but that it was badly damaged. F. S. Johnson to the author, 30 August 1990, with enclosures.
16. A burst was seen by ground observers, confirming that the bolts exploded. Garstens, Newell, and Siry (1946), chap. 3, p. 34; see also Baum et al. (1946), p. 76.
17. See Chapter 11, and Greenough, Oberly, and Rockwood (1947).
18. Van Allen to Greenstein, 26 July 1946. JGP/CIT.
19. The electronics included coincidence circuitry, cathode followers, telemetering premodulators, recorder drivers, calibrators, and timers, as well as a photoelectric orientation meter to record missile aspect for the directional information Van Allen sought. "V-2 Panel Report no. 5," 15 July 1946, p. 6. V2/NASM.

20. "V-2 no. 9," in Smith (1954).

21. Based on an examination of count-rate curves presented by Van Allen in "V-2 Report no. 6," 10 September 1946, pp. 5–7. V2/NASM. Subsequently, Van Allen found that this result was artificial and due to spurious bursts from the interaction of cosmic rays with the material in the missile warhead surrounding the telescopes. See Fraser (1951), p. 14.

22. He had wanted aspect information to determine if there were any assymetries in the cosmic-ray flux in the east-west direction and also hoped that his telescope results would provide clues to the ionizing nature of cosmic-ray primaries. In a 1948 APL technical review, this flight was regarded as "definitely exploratory" and it was noted that "good exploratory data were obtained" on the behavior of the cosmic-ray telescope, as well as from the photoelectric orientation meter, which indicated that missile roll was an unexpectedly slow 160 seconds. Fraser and Siegler (1948), p. 58, p. 62; "V-2 Panel Report 6," p. 8. V2/NASM.

23. Van Allen to Greenstein, 12 September 1946. JGP/CIT.

24. Walter Orr Roberts to Donald Menzel, 13 September 1946. WOR/UC. Reacting to Schein's opinion, Roberts knew that "photoelectric methods with carefully outgassed and carefully sealed photo surfaces" were more likely to achieve success and advised Donald Menzel of this. Schein preferred aircraft and balloons to rockets for cosmic-ray studies; longer look-times more than compensated for increased altitude. When William Stroud wanted to get involved with rocketry under Schein, he was sent to Princeton. See Stroud OHI.

25. Although everyone agreed that the repositioning was an obvious step to take, Project Hermes staff were not fully ready to agree to the skin modification. "V-2 Panel Report no. 6," p. 12; F. S. Johnson OHI, p. 39.

26. On 26 August, C. V. Strain, with members of the NRL Field Group stationed at White Sands (H. C. Hanks and N. J. Pozinsky), met with White Sands personnel to go over details. The spectrograph had to be placed in fin II or IV to best capture sunlight for firings near noontime. The grating axis had to be parallel to the rocket axis with the entrance aperture axis perpendicular to the fin, so that each bead would look out from opposite sides of the fin.

A light-tight conical steel enclosure that could be fit into the fin to receive the spectrograph had to be easily dismountable and also had to have sufficient light-baffled air vents to allow for quick evacuation of the spectrograph during rocket ascent. [C. V. Strain], "Record of Consultative Services," 12 September 1946, meeting date: 26 August 1946. Box 98, folder 3, NRL/NARA.

27. The second spectrograph arrived at NRL in the second week of September, and Johnson ran through what now was becoming a routine series of focus, exposure calibration, and alignment tests. He with Purcell and Baum, worked out several remaining alignment bugs in the optical system in the third week of September. By the fall, Baird supplied NRL with three spectrographs and was working on a second order for three more. As of 26 October, Baird had delivered four units to NRL. See Lyman Spitzer, memorandum of meeting with Dr. H. M. O'Bryan (of Baird), 26 October 1946. LSP/P; see also Tousey OHI.

28. In this interim adjustment, all experimenter preparation had to be completed before White Sands Proving Ground personnel began the actual physical installation, testing, and rollout. The V-2 Panel agreed to this interim policy change, hoping it would clarify when times of access would occur. See "V-2 Report no. 6," 10 September 1946, pp. 2–3. V2/NASM.

29. At the last moment, Harold Turner granted NRL access to the V-2 on the pad for more tests right up until fueling. Clearly, the rules were in place, but they were to be enforced at the Army's discretion. Leonard E. Zongker to Director, NRL, 5 September 1946. Box 98, folder 3, NRL/NARA.

30. J. B. J. Glanzman to Commanding Officer WSPG, 3 September 1946. Box 98, folder 3, NRL/NARA.

31. Newell and Siry (1946), p. 7, table II.

32. C. V. Strain registered laboratory notebook no. 6353, entry for 17 October 1946, p. 52, reproduces a telex from the NRL field group, through Colonel Turner, that reached Krause on the evening of the 16th: "Spectrograph located. Excellent shape. Film completely wound up." HONRL.

33. Identified as "PSL" in ibid. See also Newell and Siry (1946), chap. 4, p. 55.

34. C. V. Strain registered laboratory notebook no. 6353, entry for 22 October 1946, p. 54. HONRL.

35. Ibid., p. 55. He felt that these data "should be able to repeat [Brian] O'Brien's work [on ozone] except in coarser steps." On O'Brien's ozone studies in the 1930s from balloons, see DeVorkin (1989a).

36. F. S. Johnson to the author, 30 August 1990, with attachments.

37. Marshall Andrews, "Made from Rocket: Spectrograms Record Rays of Sun 65 Miles above Earth," *The Washington Post*, 30 October 1946, pp. 1, 4. This and other excerpts found in HONRL and NASM Technical Files.

38. Dorrit Hoffleit, "Solar Spectrum in Far Ultraviolet Secured," in "News Notes," *Sky & Telescope* 6 no. 3 (January 1947), p. 6.

39. Tousey OHI, p. 71. That astronomers were indeed anxious to obtain high-quality copies of Tousey's spectra has been confirmed by both recollections and contemporary documents. Tousey's invited paper before the American Astronomical Society was nominated by Donald Menzel as the AAS's entry for that year's AAAS prize paper, and this was ratified by the Society Council. See "Council Minutes Volume IV 1945–1947," p. 463. AAS/AIP. See Chap. 11.

40. In 1935, Air Corps Capt. Albert W. Stevens, an established aerial reconnaissance specialist, was the first to photograph the curvature of the earth from *Explorer II*. None on the NRL or APL teams who performed earth photography with rockets had prior interests in photogrammetry, unlike Stevens. See DeVorkin (1989a).

41. Newell and Siry (1947b), pp. 119ff.; Thor A. Bergstralh OHI, pp. 39–40rd.

42. Van Allen OHI, pp. 121–22.

43. Ibid., p. 130–31; and James Van Allen, "Sounding Rockets as Vehicles for Long Range Aerial Reconnaissance," 15 November 1946. APL internal memorandum cited in Fraser and Siegler (1948), pp. 75, 78 n. 8.

44. Ibid., and APL News Release, 21 November 1946. JVA/APL.

45. Fraser and Siegler (1948) and ibid. See Chap. 11.

46. This coverage, stimulated by its 21 November news release, was collected by APL in a pamphlet, "The Earth from 65 Miles Up (A Sampling of the Nation's Press and Radio, Based upon Applied Physics Laboratory Release of November 21, 1946)." See also "APL News Release," 21 November 1946. JVA/APL.

47. The APL director collected dozens of news clippings from a flight in October 1948 and had them bound and sent to APL's friends and patrons. See [APL] (1948).

48. See Holliday (1950a, 1950b).

49. The unsuccessful 10 October attempt was noted in a wire service announcement, 19 October 1946. Science Service files, NASM.

50. Bergstralh OHI, p. 42rd. See also Newell and Siry (1947b), pp. 119ff.

51. Wexler at one point thanked Newell for NRL's pamphlet of cloud photographs, hoping someday to be able to see photographs of entire cyclonic wind patterns. See, for instance, Wexler to Newell, 28 June 1948; and Wexler to Gordon Dunn (WB), 24 April 1949, 15 March 1949. Box 4, folder 1, HW/LC.

52. Watson Davis to RR, 19 October 1946. Science Service files, NASM.

53. Wexler remained sympathetic, however. See "Minutes of the Meeting of the Subcommittee on Meteorological Problems, Committee on Operating Problems, NACA," 30 April 1947, p. 6. NACA folder no. 3, Box 32, HW/LC. R. J. Havens recorded Wexler's sympathetic remarks after attending a 17 September 1948 meeting of NACA's Special Subcommittee on the Upper Atmosphere, chaired by Wexler. See R. J. Havens, "Consultative Services Record" 21 October 1948. Box 100, folder 16, NRL/NARA.

54. Bergstralh OHI, p. 43rd. RAND's interest in securing photographs for "weather forecasting, visual reconnaissance, and communications relay under an Air Force contract" is noted in "Panel Report no. 23," 14 February 1950, p. 20. V2/NASM. On RAND's early entry into aerial reconnaissance from space, see Augenstein (1982); Burrows (1986), chap. 4; and Davies and Harris (1988).

55. See Fraser and Siegler (1948), p. 78.

56. "V-2 Report no. 6," 10 September 1946. V2/NASM.

57. Fraser (1951), p. 12.

# 9

# Advocating Upper Atmosphere Research

*[The V-2] firings were 'dirt cheap'.*

—E. Krause, October 1946.[1]

*Basically, the advent of the guided missile both created a real need for undertaking the investigation of the upper atmosphere, and provided a vehicle with which this investigation could be carried out. Only on the basis of guided missile development could such an undertaking be justified and supported by the military services.*

—C. S. Piggot, March 1947.[2]

*All basic upper atmospheric research is fundamentally and inherently of value to the general problems of national defense.*

—F. L. Whipple, August 1947.[3]

*Space exploration was made possible by no conscious desire to finance the enterprise, as one might, for example, decide to finance the construction of a new telescope. In all its facets, it has emerged as an agreeable offshoot of the persuasive arguments of . . . military strategists. . . . Nor, having benefited from the military investment, is space exploration now freed of its military association and backing.*

—Sir Bernard Lovell, 1973.[4]

On 25 October 1946, Ernst Krause crossed the Potomac to join Holger Toftoy at the Pentagon for a meeting of Army Ordnance's Advisory Committee for V-2 Firings. This interservice committee, created by Ordnance to oversee V-2 activities, was headed by Gladeon Barnes's successor Brig. Gen. H. B. Saylor. Saylor called the meeting to decide if Ordnance wished to support a second season of V-2 launches.[5] The issues at hand were: how many more V-2s could be reconstructed, and would it be worth the effort? This was the first of many meetings where Krause and his successors would argue their case (figure, p. 152).

## Justifying a Second Season

Even though general sympathies were with a second season, Toftoy and Krause lobbied hard for more firings. Toftoy reviewed the 10 October flight, pointing out that the "results to date, including data on the missile and its operation, telemetering, upper atmosphere data, etc. had been better then expected. . . . [I]n the October 10th firing 80% of the total telemetered record was useful."[6]

Krause, speaking for the world of science, added that "in at least three fields more information had been collected from the program than in all the previous ten or twenty

151

**Ernst Krause (r) briefing Naval officers visiting White Sands circa late 1946. U.S. Navy photograph, Ernst Krause collection. NASM SI 83–13890.**

years of work. . . [on] cosmic radiation, solar spectroscopy and temperature and pressure measurements of the upper atmosphere."[7]

No one challenged Krause's contentions, for they offered scientific justification for continuing the program. He argued that the firings were "dirt cheap" and that there was "nothing better than the V-2 for securing the high altitude data," and that more data would be forthcoming, especially if communications improved between White Sands, Aberdeen, and the members of the panel. He went so far as to state that even NRL's planned Viking would not be "quite as satisfactory as the V-2 mainly because it is smaller."[8]

After Krause's testimony, Toftoy called for additional V-2 firings by the members of the V-2 Panel, which would cost Ordnance about

$10,000 each and would increase to $20,000 in the next six months, because replacement parts had to be manufactured. His evaluation was based on knowledge that the existing inventory of parts would yield only 35 V-2 missiles, even after a search by his staff in Europe.[9] By October, Toftoy's staff and Hermes personnel had assessed the costs of replacing missing V-2 parts with American-made reproductions and advised Toftoy that they had enough critical components on hand to construct about 100 missiles without an inordinate amount of effort.[10]

Balancing these costs against the continued expected returns, which were largely those GE and its subcontractors would gain by constructing and refining German parts within systems that had been tested by precursors,

Toftoy called for a continued program. The consensus around the Pentagon table was that sufficient interservice funding was probably available, especially since NRL had already been providing warheads. In the opinion of the committee, such cost sharing would be expected from the other agencies.

The committee adjourned confident in Toftoy, Krause, and their work. But Saylor wanted an explicit endorsement from the world of science, from the body of the V-2 Panel. Krause had no trouble obtaining it at the next panel meeting in November, whereupon Saylor approved a second season of 25 more V-2 missile firings, to extend through 1948.[11]

By now the V-2 Panel was not only a coordinating agent keeping the science under control, but an advocacy group that strengthened Army Ordnance's case for a continued developmental effort within Project Hermes. Even though Krause reported later that "as far as Army Ordnance was concerned the military aspects of the [V-2] weapon were now taking second place to the research aspects,"[12] Charles Green, General Electric's observer, found that the service representatives on the committee had specific military goals that required at least 70 more missiles.

Army Ordnance wanted to test two-stage rockets in at least 10 firings, and the AAF representative wanted more than 50 berths, including six for a "special project the purpose of which he was not free to disclose."[13] By the end of the first season, both the Army and General Electric knew generally what the V-2 could do and how to fire it. But they still needed to test out their own designs and modifications, and the V-2 made an excellent test bed for their engineering research program.

Toftoy wanted the panel to legitimate the Ordnance program.[14] He maintained that Army Ordnance should restrict all effort to building only vehicles for research and development purposes and not rush any missile to operational status: "The field is so new it is essential to place emphasis on fundamental and basic research."[15] The work of the V-2 Panel represented basic research, not only in the corridors of the Pentagon, but to the world at large.

## War Heads and Peace Heads

The Army and Navy portrayed the V-2 firings as both basic research serving civilian interests and ordnance development serving the

Some key members of the V-2 Panel at their 25th meeting, Boulder, Colorado, 13—14 June 1950. (l to r): Homer E. Newell, Jr., Michael Ference, George K. Megerian, Marcus O'Day, James A. Van Allen, and Charles F. Green. Nelson Spencer collection. NASM.

needs of national security. The overall rationale repeated time and again in the popular press was that the V-2, a weapon for war, had taken on a dual role in America: harnessed for scientific pursuit, it would also prepare us for the next war. A June 1946 Navy press release wanted the world to know that the "V-2 Works for Science." Similarly, the June 1946 issue of *Science Illustrated* announced, "Big Ben Works at Peace." Readers of both stories learned how the V-2 had been turned into a flying laboratory, bristling with scientific experiments prepared at White Sands.[16]

Shortly after the first successful launching of a V-2 from White Sands on 10 May 1946, *Life* magazine remarked that the flight: "reminded the U.S. how science multiplies the hazards of future atomic wars. . . . In another war similar rockets might carry atomic bombs anywhere in the world. . . . The main purpose of the Army ordnance tests was to prepare the U.S. for the possibility of such a war." But *Life* then added that: "the V-2 tests were to be more than a development of new weapons and defenses. . . . Later rockets [will] carry aloft instruments which [will] send back information about temperature, gases and cosmic rays in the earth's little-known upper atmosphere."[17] This duality was also reflected by an editorial in *Army Ordnance Magazine*:

> *To accomplish research objectives, the 'war head' of the V-2, with its explosive filling, becomes a 'peace head' filled with scientific paraphernalia for exploring the upper atmosphere and evaluating the performance of the . . . rocket.*[18]

Army Ordnance, according to the editorial, left it to "civilian scientists [to] build up the fundamental research" without any overt policy statement or major internal commitment on its own part.[19] In creating the V-2 Panel, however, Army Ordnance established a relationship that sensitized civilian scientists to their value as entrepreneurs and advocates for the support of basic research that aligned with national security interests.[20] Toftoy's blueprint would soon be put to the

test as members of the V-2 Panel defended their newfound interests in the maze of joint civilian-military consulting panels and boards set up in the wake of World War II to create and adjudicate postwar scientific research and development policy.[21]

## Choosing Sides: The V-2 Panel and the Joint Research and Development Board

In the bewilderment of postwar reorganization, Army Ordnance programs ultimately required the endorsement and authorization of the Joints Chiefs of Staff (JCS). Saylor and Toftoy therefore had to be sure that the JCS's new Joint Research and Development Board (JRDB) knew and approved of its guided missile program. Since the JRDB included elite civilian scientists and engineers, acting in concert with War and Navy Department liaison officers, Army Ordnance made sure that the JRDB also knew about the V-2 Panel.

Created by the Joint Chiefs of Staff in June 1946 out of the administrative remnants of the wartime OSRD and the Joint New Weapons and Equipment Board, the JRDB was to "coordinate all research and development activities of joint interest to the War and Navy Departments."[22] As a "high level policy Board more interested in new fields of endeavor, new weapons and techniques, than in the normal operating level," the JRDB created standing committees, committees to deal with specific subjects and problems, intercommittee panels, and other coordinating mechanisms to oversee developments in all areas of concern to the Atomic Energy Act and to military research and development in general.[23]

Headed at first by Vannevar Bush, the JRDB was not intended to be a statutory body, but one emphasizing "conciliation and voluntary agreement."[24] Bush felt that as long as viable working-level bodies existed, the JRDB would only act as their representative to the War Department and Navy Secretary James V. Forrestal and War Department Secretary Robert P. Patterson. If, however, these working-level bodies found that they could not interact well with one another, Vannevar Bush expected the JRDB to act in a judicial capacity, not so much as in a world of law, but as in a senate of scientists and engineers.[25] In the spirit of the permanent mobilization of science, he worked to enlist scientists and their institutions "as 'full and responsible partners' in the work at hand."[26]

The JRDB absorbed numerous committees and panels in the spring and summer of 1946, but others were hastily formed to wrest control of guided missile research and development and prevent the JRDB from taking over completely.[27] One in particular, promoted by administrators from various Navy and AAF laboratories and their military liaisons, was the "Joint Group to Coordinate Upper Atmosphere Research within the Services."[28] In June 1946 Krause and Van Allen were called to its first meeting and managed to convince those assembled that the V-2 Panel was performing all the coordination needed for the short-term V-2 program. But by August, the ad hoc group concluded that although basic research on the atmosphere was being well-served by the V-2 Panel, a joint Army-Navy committee was still required to coordinate a long-range program in guided missile research; since the procurement of launch vehicles and missile development was "a most expensive undertaking, there should be careful coordination in order to avoid duplication of effort."[29]

Guided missiles were expensive to build and test; basic research performed with them as an allied activity was not. Each of the services was preoccupied with the former activity and were happy to leave the latter at the V-2 Panel's door. Even so, well after the JRDB had formally taken over the Joint New Weapons and Equipment Committee and was becoming familiar with the procurement problems facing the V-2 program, older coordinating bodies still made a bid for control of upper atmospheric research.[30]

In late August, the venerable Aeronautical Board, created to coordinate Navy and Army research, set up the Special (ad hoc) Sub-Committee on Upper Atmosphere Research to determine how the AAF and the Navy

might cooperate under the new JRDB guidelines.[31] And in October 1946, the ONR established the Guided Missile Panel in its Planning Division's Armament section. Although separate from the ONR division that oversaw NRL, it claimed to fill a perceived need to coordinate all Navy missile development.[32]

Faced with the existence of four competing bodies interested in controlling either guided missile development or procurement, as well as upper atmospheric research, Krause and the V-2 Panel chose to align itself with the nascent JRDB. Homer Newell, who attended several Aeronautical Board Subcommittee meetings in the summer, found little evidence to support its claims of competency and impartiality. It could not be allowed to coordinate, he concluded, "the overall upper atmosphere programs when not all interested agencies were represented." The V-2 Panel and ONR recommended that the subcommittee be abolished and that coordination and cooperation be referred to the JRDB, which by September 1946 was well established.[33]

The proposed ONR Guided Missile Panel hit a bit closer to home for Krause. Even so, other members of the V-2 Panel objected to its parochial interests, for it sought to control all forms of missile development within the Navy, including NRL's Viking and APL's Aerobee, and by default to control what was done with these vehicles.[34] After a year of internal wrangling, the ONR Directorate, sensing unending conflict, asked the V-2 Panel to provide a defense for basic research and an endorsement to have the JRDB settle the issue of oversight. Krause had little problem finding the proper words and the proper distance (for this purpose) from guided missile development:

*Research in the upper atmosphere . . . is considered basic and is not carried on primarily for its application to guided missiles. As a matter of fact, in the light of our present knowledge only a few of these fields are foreseeably applicable to guided missiles. . . . In view of the fact that this research is basic I feel that the primary cognizance of the coordination of upper atmosphere research vehicles should lie with the . . . JRDB. The vehicles are as much a part of the research as is the instrumentation.*[35]

Krause wanted upper atmospheric research out of the hands of ONR's Planning Division. The Aerobee and Viking were arguably created as launch vehicles for scientific use, although the technical experience thereby gained was certain to be valuable to guided missile developers. Any ordnance application was better coordinated, therefore, by the JRDB, and not by a special interest group within ONR: "Such a group is not primarily interested in the research aspect of rockets and could very conceivably force the research interests into untenable compromises. . . . It is pure research which deserves and requires special protection and specially assured support."[36]

Krause defended the JRDB in September 1947, one year after it had created a new body that met the panel's needs. Quite aware of the many competing interests trying to gain control over guided missile and upper atmospheric research, Lloyd V. Berkner, as Bush's executive secretary to the JRDB, in September 1946 created the JRDB Panel on the Upper Atmosphere to short-circuit parochial military efforts. He wanted to find scientists of "adequate stature and experience" to fill key positions and was thankful that he had found Thomas Johnson of the Ballistic Research Laboratory to take the position of chair.[37]

## The JRDB Panel on the Upper Atmosphere

Sensing growing competition and confusion and hoping to stabilize all activities related to guided missile development, Bush and Berkner created the JRDB Panel on the Upper Atmosphere before its parent body, the Committee on Geophysics, came into existence. Thomas H. Johnson, who was a well-known cosmic-ray physicist at the Bartol Research Foundation of the Franklin Institute before the war, had tasted large-scale research at Aberdeen and chose to stay after the war. He remained a good friend of upper atmospheric research, however, and was an early and productive participant in the formative months of the V-2 Panel in 1946. Krause was comfortable with Johnson in his new position and approved of his interpretation of JRDB philosophy. He advised his Navy superiors and his own staff that the JRDB panel would "carry weight" because it was composed of, or enjoyed direct links to, top Army and Navy policy makers who advised the Bureau of the Budget.[38]

The new JRDB Panel on the Upper Atmosphere met in early November 1946 in Washington, and Johnson told his membership that there was an "urgent need for immediate coordination of research."[39] Members came from all the services, as well as from the Carnegie Institution of Washington's Department of Terrestrial Magnetism (DTM), Bush and Berkner's home. The NACA had a member, and Johnson represented the BRL. Holger Toftoy was an associate member, Harvey Hall came from BuAer, and Krause was one of two consultants, invited by Johnson as the V-2 Panel's representative.[40]

The JRDB Panel's first meeting got off to an uneasy start when members wrangled over the military-civilian balance and over policy. Johnson, however, managed to identify the panel's primary duties.[41] As an agent of the JRDB, the panel was expected to produce manpower surveys and use them to look for unnecessary duplication, gaps in research, problems with research facilities, and the extent of coordination of research both under way and contemplated by the services. It also had to provide information on planning, review the economics of large-scale weapons systems as they related to upper air research, and in general search for a logistical approach to cooperative science management. But above all, it was responsible for collecting information on the status of upper atmospheric research programs that dealt with the medium above 12 kilometers and bordered "on such aspects of overall planning for national defense as may be needed in formulating an adequate program."[42]

Johnson's panel had little trouble identifying the many areas of scientific research that fell within its jurisdiction. Its first priority was to assess the physical state and composition of the atmosphere, its state of ionization, and the effect of ionization on all forms of radiation propagation. It was to cover work in solar and cosmic radiation, including mechanisms for their absorption by, and influence on, the upper atmosphere. Upper air currents, aurorae, air glow, global ion currents, the earth's magnetic field, meteor research, sound propagation, and light scattering completed the long list the JRDB Panel considered relevant to its scope. Finally, the panel agreed that it should also be interested in the development of any techniques that would assist in the above types of investigations. Simply put, it wanted a hand in launch vehicle procurement.[43] Johnson closed the meeting with a call for an "immediate study of existing programs" and delegated that responsibility to the military representatives, aided by the NACA and DTM members. Reports were due in January 1947.

## The JRDB Panel's Venue: Avoiding Parochial Interests

By its second meeting in March 1947, the JRDB Panel had significantly expanded its sphere of interest and responsibility. Berkner, Johnson, and C. S. Piggot (who was executive director for both the committees on geophysics and guided missiles) had now examined the initial status reports of existing programs and realized that the real challenge for the panel was to justify the technical infrastructure required for upper atmospheric research and to reconcile duplicate functions and operations among the services. The key issue was the proposed development of launch vehicles and the comparative value of each for conducting upper atmospheric research.[44]

The JRDB Panel found that each agency had its preferred vehicle or preferred program. At the second meeting, Marcus O'Day pushed

Project Blossom which would not only perform reentry ballistic studies, but also provide reusable systems for upper atmospheric research—arguing that "considerable savings in instrumentation for additional flights could be made by rebuilding the V-2 and extending its nose to accommodate equipment for parachuting the instrument portions of the rocket to the earth."[45] O'Day and AAF Col. Marcellus Duffy wanted to use the JRDB Panel as a forum to pressure Toftoy for yet another "complete study on the potential supply of V-2 rockets."[46] Toftoy was not interested in an extensive modification of the present stock of V-2s at White Sands for an AAF program. He felt that "improved vehicles for upper atmosphere research will become available from time to time as guided missiles development progresses."[47]

Piggot supported Toftoy's resistance, since the JRDB wished to prevent parochial interests from gaining an upper hand. Piggot also agreed with Toftoy that upper atmospheric research should be defined in terms of the guided missile: "Only on the basis of guided missile development could such an undertaking be justified and supported by the military services."[48] Eventually, O'Day and Duffy were to secure about a half-dozen V-2 missiles for Project Blossom. But this was not to happen for several years. In the meantime, the JRDB Panel faced more serious parochial interests when Toftoy's Corporal missile was pitted against the Navy's Viking.

Here is where the V-2 Panel made a difference. The JRDB Panel on the Upper Atmosphere typically endorsed V-2 Panel positions, which continued in this case. Indeed, the relationship established between the V-2 Panel and Army Ordnance in October 1946, through its Advisory Committee for V-2 Firings, continued between the V-2 Panel and the JRDB. Whereas the JRDB Panel became a forum for service agency representatives to identify upper air data that had bearing on national security, it also acted as a stage on which the V-2 Panel could orchestrate its long-term needs for launch vehicles.[49] At the third JRDB Panel meeting in April 1947, Krause stated that the V-2 Panel needed as many

V-2s as were "practical or economically feasible."[50] Also needed were APL's Aerobees and NRL's Neptune (Viking), or their equivalents, to ensure that an "uninterrupted supply of rockets for the next five years" would be available "to carry out adequately the upper atmosphere research program."[51] Although equivalents included the Corporal E, the Army's surface-to-surface ballistic missile, Krause left no doubt that the V-2 Panel preferred the Neptune because it was designed for research.[52] Even though Toftoy tried to increase the V-2 Panel's interest in Corporal by adding JPL's William Pickering to the panel when Krause left, members consistently endorsed Viking during the next three years when queried by the vacillating JRDB Panel. And although the JRDB consistently followed the V-2 Panel's endorsements, it never made a clear enough statement to settle the issue, which was finally done by competing forces within the Navy itself.[53]

## Assessing Military Needs: The First JRDB Panel Survey

The first survey by the JRDB Panel on the Upper Atmosphere identified the goals each service agency had in mind for upper atmospheric research. All wanted to improve aerodynamic design, propulsion, and the guidance of long-range missiles. They also hoped to identify aerological factors affecting weather; the ionospheric conditions and associated magnetic disturbances affecting long-range radio and wire communication; and to better understand high-energy nuclear processes. In sum, the JDRB Panel had found the right people to show the relevance of virtually any type of research to national security interests.[54]

The JRDB Panel's dual task was (a) to determine how these goals could be met by the various observational programs under way under the aegis of the V-2 Panel, and (b) to assess "the technical requirements for rocket-type vehicles" and take steps to "develop other types of vehicles, the techniques, and the instruments needed for routine measurements of the quantities of interest in upper atmospheric research."[55] The panel solicited the opinions of elite scientists to confirm its findings. For instance, G. B. Dobson of the Gassiot Committee of the Royal Society, London, confirmed the value of the ozone studies being conducted with rockets.[56] Meteor studies, solar studies, and cosmic-ray research were all endorsed. NRL's E. O. Hulburt, for example, testified that the new view of the ultraviolet solar continuum derived from Tousey's spectra improved scientists' understanding of all atmospheric processes depending on solar radiation.[57]

The JRDB Panel was especially interested in cosmic-ray observations from rockets. Arguing that "cosmic radiation furnishes for nuclear investigations a beam of high energy particles which will remain unrivaled for many years by laboratory accelerators,"[58] it listed more than a dozen separate scientific questions in high-energy and particle physics that could be addressed. It treated ionospheric propagation studies in a similar manner, linking them to long-range communications research and evoking the oft-repeated image of the ionosphere as "a gigantic laboratory on the fringes of the atmosphere, wherein solar and cosmic radiation may be studied without hindrance from atmospheric absorption."[59]

The first JRDB Panel survey succeeded in making the military and scientific agendas for upper air research identical. Ultimately, both looked at the same medium and asked the same questions. Thus the panel was able to justify that

> considerable emphasis is currently given to the development of the rocket as a vehicle for upper atmospheric studies. In addition to the upper atmospheric data obtained from their use as a research tool, the experience gained from the construction and launching of rockets will contribute materially to the guided missile program.[60]

Military needs and scientific goals were synergistic as Piggot pointed out, and the existence of the missile not only demanded that upper atmospheric research be pursued, but made it possible to pursue.[61]

## V-2 Panel Members Advise the JRDB: The Second Survey

The history of the JRDB Panel throughout the 1940s was one of repeated surveys and confirmation of the value of its goals. But its surveys were also preemptive strikes against a continual onslaught of proposals by teams of missile system promoters consisting of technical officers and industry contractors. In its first year, the JRDB Panel found its most articulate spokesmen among the members of the V-2 Panel.

In July 1947 the JRDB Panel asked Fred Whipple to chair a working group to identify "reasonable objectives in upper atmosphere research . . . considering military requirements and available resources."[62] With Michael Ference, Homer Newell, L. M. Slack of ONR, and C. G. Montgomery from Princeton, Whipple confirmed the basic vision of the JRDB Panel: "The Working Group is of the unanimous opinion that all basic upper atmospheric research is fundamentally and inherently of value to the general problems of national defense."[63]

The Working Group then filled some eight pages of text with projects that should be conducted with rockets, balloons, and aircraft and by indirect ground-based means. It argued for multiyear funding and a five-year plan, both for instrument development and for long-range statistical studies of slowly varying natural phenomena. All aspects of the research program, it argued, should be "planned with military requirements conceived in the broadest possible construction of this term."[64]

Following the V-2 Panel's philosophy, the JRDB Working Group argued that no single type of rocket or upper air vehicle would satisfy all problems and needs, and identified graduate education and training in upper atmospheric research as "an integral part of the national defense program in this field." University training had to be supported through

research grants, the Working Group argued, because "national security requires a greater number of people trained in these fields."[65]

Whipple's report did not satisfy the Air Forces representative on the JRDB Panel, who expected to see more nuts and bolts recommendations, and not a vague "philosophy of approach."[66] Nevertheless, the JRDB Panel approved the report in September "as formulating a generalized program and objective for upper atmosphere research by the Armed Forces." It also recognized the grumblings of the military officers who wanted more explicit evaluations of how well long-range military interests were being served by upper atmospheric research projects. Air Forces Col. Marcellus Duffy, Capt. H. B. Hutchinson of ONR, and Col. J. S. Willis of the Signal Corps formed a second study group to examine, on a project-by-project basis, all present activities in the services.[67] Their exhaustive survey took a full month, and when completed, it identified almost 100 separate AMC, ONR, and Signal Corps projects and tasks that would cost more than $9 million in projected fiscal 1949 dollars.[68] Money and duplication aside, the important conclusion reached by Duffy and his group was that the military value of all the upper atmospheric research programs was either "direct and self-evident" or, at the very least, they were "essential supporting projects."[69] In other words, they fully supported Whipple's assessment.

Johnson and Piggot combined the Whipple and Duffy reports into one statement, creating the semblance of cohesion demanded by Bush's vision of the JRDB as a collegial body.[70] Their statement, ratified by the JRDB Panel in November 1947, included a resolution that satisfied all parties:

> *The Panel: Resolved that it recognizes the importance of all phases of the well-coordinated V-2 Rocket firing program and the grave consequences of any failures to give adequate financial support to all agencies involved in this program, since the lack of support of the program in any one agency would jeopardize the program as a whole.*[71]

What the statement failed to mention, which neither Duffy nor Whipple wished to address, was the obvious duplication and redundancy in the identified programs. There were six ventures in ionospheric research, nine in solar research, and another six projects for atmospheric composition. Not all were identical, but some (such as ultraviolet spectroscopy), were being pursued by three separate agencies using similar means. This situation did nothing to promote Bush's stated goal for the JRDB, which was to achieve a "rational and disinterested compromise from within the individual services."[72]

On the other hand, it is possible that basic research fell below the JRDB's threshold of concern. We have seen that all the ad hoc groups that were hastily formed to preempt the JRDB in the spring and summer of 1946 ignored duplicate scientific agendas and concentrated on matters dealing with the procurement of vehicles and systems.[73] The upper atmosphere research agendas came under review not out of concern for duplication, but because they provided an additional rationale for assessing guided missile systems. Therefore, in effect, the JDRB left it up to the V-2 Panel to coordinate scientific instrumentation for upper atmospheric research with rockets.

## The JRDB Endorses V-2 Panel Operations

As if to thank Krause for his bold defense of the JRDB over his own patron, in November 1947 C. S. Piggot encouraged Krause to make use of the JRDB Upper Atmosphere Panel's sanction during upcoming congressional budget hearings.[74] Piggot wanted the V-2 Panel to use the JRDB's November resolution in any way it saw fit; in effect the carte blanche blessing came about because the JRDB liked the way the V-2 Panel did business.

Through late 1947 and 1948, members of the V-2 Panel were faced with a tightening budget future, not only for operations but for vehicles to replace the V-2. On several occasions they formed small study groups to

draft a set of statements and proposals to their own agencies, as well as to the JRDB. In every case, when the JRDB was petitioned for support, support was forthcoming. It never failed to respect the V-2 Panel's mission, as articulated by Army Ordnance, to provide "coordinated advice to the Office of the Chief of Ordnance in the conduct of those V-2 firings which are being provided for scientific studies." The JRDB was confident that the V-2 Panel was a truly representative body, because it "comprise[d] representatives of all major research groups in the United States, with the notable exception of the California Institute of Technology group," although the latter had sent a representative to most of the recent panel meetings.[75] And it endorsed the V-2 Panel's original cost-sharing scheme, designed by Toftoy, as "the most efficient way to handle operations; that agencies supplying units and services for the entire program be allowed the necessary expenditures . . . because the budgets of the principal agencies now bearing the financial load have been subjected to severe scrutiny and reduction."[76]

The JDRB's endorsements convinced the members of the V-2 Panel that their future was in good hands. At least that was the consensus in December 1947, when V-2 Panel members decided that their individual budgets were adequate to maintain operations for another year. The V-2 Panel had chosen its allies well; it had secured a second season of V-2 flights and had every reason to expect that as many V-2s as could be assembled would be available for upper atmospheric research. It knew, however, that the V-2s were not going to sustain their needs indefinitely. Domestic replacements were essential for a permanent program.

## Notes to Chapter 9

1. E. H. Krause, "Record of Consultative Services," 29 October 1946. Box 98, folder 4, NRL/NARA.
2. C. S. Piggot (Exec Dir, GMC), "Research and Development Work under Cognizance of the Committee on Guided Missiles," Annex F (UAT 17/1) to [T. H. Johnson], "Minutes,"

of the second meeting on 6 March 1947, 12 March 1947. UAT 3/2 Box 239, folder 3, JRDB/NARA.

3. *Panel on the Upper Atmosphere* 19 August 1947, p. 1. UAT 30/2, attached to Whipple, ibid.

4. Lovell (1973), p. 20.

5. Charles F. Green, "Meeting of Advisory Committee for V-2 Firings," 30 October 1946. Meeting date, 25 October 1946. G E Aeronautics and Marine Engineering Division Report no. 145. Army/G.E. Project Hermes (V-2). USMR(1946–1951) File/NHO.

6. Green, ibid., p. 2.

7. Ibid.

8. Krause reported that each V-2 firing was also costing NRL some $10,000 (including $2,500 for each warhead and $3,500 for the telemetry). He felt that "the V-2 should be used until the cost of the firing was in excess of $75,000." Even their own Viking, under development by the Glenn L. Martin Company, was going to cost at least $100,000 per flight for the first 10. The Martin rocket was initially called the Neptune, and later the Viking. E. H. Krause, "Record of Consultative Services," 29 October 1946, record of meeting of 25 October at the Pentagon. Box 98, folder 4, NRL/NARA. The "dirt cheap" quote also was reported by Green, "Meeting of Advisory Committee for V-2 Firings," 30 October 1946.

9. Maj. James Hamill conducted the search. Charles F. Green, "Meeting of Advisory Committee for V-2 Firings," 30 October 1946. See also Ordway and Sharpe (1979), p. 353.

10. Green, "Meeting of Advisory Committee for V-2 Firings," 30 October 1946, p. 4.

11. "V-2 Panel Report no. 7," 12 November 1946, pp. 2–3. V2/NASM.

12. E. H. Krause, "Record of Consultative Services," 29 October 1946, record of meeting of October 25 at the Pentagon. Box 98, folder 4, NRL/NARA.

13. Charles F. Green, "Meeting of Advisory Committee for V-2 Firings," 30 October 1946. The AAF representative was talking about its planned Blossom series.

14. Holger Toftoy Diary, entry for 5 November 1946. HT/ASRC. On Toftoy's advocacy for guided missile R&D, see Chap. 4.

15. C. M. Hudson, "Minutes of the Meeting of the Ordnance Department Advisory Committee on Guided Missiles—White Sands Proving Ground, Las Cruces, N.M. 13 June 1946," 24 June 1946. 25.32 RAM/CIT. See also Chapter 4.

16. [NRL], "V-2 Works for Science: Captured War Weapon Now Probes Universe," n.d., June 1946; also in enclosure A to G. L. Webb to ORI, 31 May 1946, pp. 1–2. Box 98, folder 2, NRL/NARA; copy in V-2 folder, HONRL; "Big Ben Works at Peace," *Science Illustrated* (June 1946), pp. 41–43. Big Ben was the British code name for the V-2.

17. "U.S. Tests Rockets in New Mexico," *Life Magazine* 20 no. 21 (27 May 1946), pp. 31–35; quotation from p. 31.

18. E. J. Tangerman, "Nazi Vengeance," *Army Ordnance* (September/October, 1946), pp. 153–155; quote on p. 153.

19. Ibid. For a discussion of this pragmatic Army approach during these divisive and indecisive times, see Kevles (1975), p. 44. See also the discussion in Chap. 4.

20. The sensitizing process worked both ways. On such interrelationships, see Forman (1987), Needell (1987), and Leslie (1987).

21. On plans to maintain wartime scientific research in the postwar world and how it was to be managed, see, among many others, Hall (1963) for the control over launch vehicles and satellite studies; and on planning in general, see England (1982); Kevles (1975); Rearden (1984); Bush (1970); Komons (1966); Forman (1987); Sapolsky (1979, 1990); and Stares (1985).

22. "Charter of the Joint Research and Development Board, Secretaries of War and the Navy, June 6, 1946, amended July 3, 1946," excerpted from [Science Policy Research Division, Congressional Research Service, Library of Congress] *United States Civilian Space Programs 1958–1978* (GPO: January 1981), p. 36. Committee print of the report prepared for the Subcommittee on Space Science and Applications of the Committee on Science and Technology, U.S. House of Representatives, 97th Cong. On the origins of the JRDB, see also Rearden (1984), pp. 96ff.; Kevles (1975); and Needell manuscript biography of Lloyd Berkner, chap. 3, "Science and Defense: Institutionalizing the Partnership 1945–1949." Needell also provides an excellent introduction to Vannevar Bush's rationale for the postwar control of science and the resulting structure of the JRDB. A profile of Bush reflected in the nature of the JRDB is given in Reingold (1987).

23. Statement of Col. Wesley T. Guest of the Army Signal Corps, at a meeting of the Radio Propagation Executive Council at the National Bureau of Standards, 13 August 1946. National Bureau of Standards Library, NSBB Central Radio Propagation Laboratory Files, kindly provided by Paul Forman. Reingold (1987), pp. 309–10 points out that the committee structure of the JRDB followed the model Bush created for the OSRD, which itself had been modeled on the committee structure of the National Academy of Science's National Research Council and the NACA. On the NACA structure, see Roland (1985a). See also Needell, "Science and Defense . . ."; Rearden (1984), pp. 96–100, chart 6; Stares (1985), p. 24ff.; and Hall (1963), p. 36ff.

24. Steelman (1947), p. 15.

25. Reingold (1987), p. 309.

26. Bush hoped to extend the collegial contractor-patron character of the OSRD/NDRC. Bush quoted in Rearden (1984), p. 97.

27. The JRDB absorbed the Joint New Weapons Committee and the remnants of the OSRD offices, but also accreted the National Academy of Science's elite Army Ordnance Department Advisory Committee on Guided Missiles. Created by the National Academy for Army Ordnance during the war, it met several times after the war and soon deferred oversight for long-range guided missile development to the JRDB—and put itself out of business. See H. B. Richmond, "Report of National Academy of Sciences Advisory Committee to the Army Ordnance Department on Long Range Guided Missile Weapons," 13 June 1946; Richmond to the "Members of National Academy of Sciences Advisory Committee to the Army Ordnance Department on LRGMW," 23 October 1946; and Toftoy to Richmond, 9 June 1948. 25.32 RAM/CIT.

28. E. H. Krause, "Record of Consultative Services," 11 June 1946. Box 98, folder 2, NRL/NARA.

29. C. H. Smith, "Record of Consultative Services," 1 August 1946, pp. 1–2. Meeting held 19 July 1946. Box 98, folder 3, NRL/NARA.

30. See "Joint Committee on New Weapons and Equipment" folder, 27 September 1946 entry. The transfer began on 30 July 1946 and its Subcommittee no. 7 (German Information) issued a final report, "A Survey of the Dissemination and Use of Enemy Guided Missiles." See also Subcommittee of the Guided Missiles Committee of the JCNW&E "Present Plans at the White Sands Proving Grounds," which indicated that present plans called for assembly of "only 25 V-2's at the average cost of $11,000 each, after which further study will be made as to the practicability of assembling any more." p. 7. Box 59, "Joint Committee on New Weapons . . . ," 27 September 1946 (date uncertain). VB/LC.

31. G. B. H. Hall (USN) and D. L. Hardy (AAF), to "Members, Special (ad hoc) Sub-Committee on Upper Atmosphere Research," 14 October 1946. Box 98, folder 4, NRL/NARA. The Aeronautical Board's chief activity had been to reconcile aircraft procurement between the Navy, Army, and the Army Air Corps. It interpreted this charter broadly and approached guided missile research in steps. In May 1946, it established its Charter of the Launching Panel of the Guided Missile Subcommittee of the Research and Development Committee of the Aeronautical Board, wherein it defined its purview as "Launching, for the purposes of the Panel, is defined to include all devices and operations primarily concerned with starting the motion of a guided missile" other than propulsion. It saw itself as being responsible to the Guided Missile Subcommittee of the Aeronautical Board for "promoting coordination of technical programs in launching guided missiles in which BuAer, AAF, BuOrd and Army Ordnance are involved." "Launching" soon accreted guidance and control, and finally the coordination of all upper air research. See Aeronautical Board Launching Panel folder, charter entry, 15 May 1946. JDRB/NARA.

32. L. M. Slack, "Guided Missile Panel, Establishment of," 30 October 1946. Box 32, NRL/NARA.

33. There were no official representatives from Army Ordnance, the new Office of Naval Research (reorganized out of ORI), and the V-2 Panel. Newell attended two meetings on 3 September and prepared two separate reports. The Navy soon abandoned the Aeronautical Board's Subcommittee. It never came up for discussion by the V-2 Panel again. Homer Newell (for Krause), "Record of Consultative Services," 2; 4 September 1947 [1946]. Box 99, folder 8, NRL/NARA. On

the fate of the subcommittee, see "Joint AAF-Navy Upper Atmosphere Research Projects (ad hoc) Subcommittee" folder entries for 26 February 1947 and for 15 April 1947. UAT 22/1, JRDB/NARA.

34. The Aerobee was initially funded by BuOrd for the Applied Physics Laboratory, with additional funding from ONR to supply NRL with Aerobees. See "High Altitude Research at the Applied Physics Laboratory of the Johns Hopkins University—-A Brief Summary," *The Navy Upper Atmosphere Research Program with Rocket Vehicles* pt. 3, 10 April 1946, pp. 2–3. JVA/APL. The Neptune (Viking) program was funded by the Bureau of Aeronautics. See "The Present and Proposed Program of Rocket-Sonde Research at the Naval Research Laboratory," in ibid., pt. 1, p. 10.

35. E. H. Krause to Capt. H. B. Hutchinson, 3 September 1947. Box 99, folder 8, NRL/NARA.

36. Krause, speaking for the V-2 Panel, concluded his response to ONR with a recommendation that the JRDB reject the Guided Missiles Panel request: "I do object vigorously to their request for *primary* cognizance over coordination of work on vehicles proposed exclusively for high altitude investigations as well as work on guided missile test vehicles." Ibid., p. 2. Since this was primarily an internal NRL/ONR problem, as well as being classified as "Restricted" at the time, Krause's statement does not appear in the minutes of the V-2 Panel. Commentary on Krause's defense can also be found in Bruce Hevly, "Base to Campus: The Influence of Academic Science on Military Research at NRL," paper presented at a "Workshop on the Military and Post-War Academic Science," 17–18 April 1986, The Johns Hopkins University. See also Hevly (1987).

37. L. V. Berkner to V. Bush, "Matters to be Discussed with Dr. Bush," 12 September 1946. Box 59, VB/LC.

38. E. H. Krause, "Record of Consultative Services," 14 November 1946. Box 98, folder 4, NRL/NARA. On Johnson's prewar cosmic-ray research, see DeVorkin (1989a).

39. T. H. Johnson, "Minutes of First Meeting," 26 November 1946, p. 2. UAT 3/1, Box 299, folder 3, JRDB/NARA.

40. Ibid., p. 1. T. H. Johnson to E. H. Krause, 31 October 1946; E. H. Krause, "Record of Consultative Services," 14 November 1946.

Box 98, folder 4, NRL/NARA. The new panel consisted of military agency representatives from the AAF, Signal Corps, DCNO (for air), CNR, and ONR. H. W. Wells was from DTM, and T. L. K. Smull from NACA. Harry Wexler from the Weather Bureau was soon added as another civilian member, as was John A. Wheeler from Princeton. The membership changed frequently.

41. Right off the bat, the ranking officer present, AAF Brig. Gen. A. R. Crawford, wondered why more technical people from the military were not on the panel. JRDB executive staff replied that the panel's "executive or judicial" role "in formulating policies" did not require such membership. Johnson then added that although he felt technical people could indeed "perform the functions of the Committee in a superior manner," their time was better spent doing the actual research, and in any event, the choice of military membership was up to each agency, not the JRDB. T. H. Johnson, "Minutes of First Meeting," 26 November 1946, p. 1. UAT 3/1, Box 299, folder 3, JRDB/NARA.

42. [Lloyd Berkner], "Directive," Formation of the Panel on the Upper Atmosphere, 25 October 1946, pp. 2–3. UAT 1/2, Box 239, folder 2, JRDB/NARA. Later version of a draft dated 15 August 1946.

43. T. H. Johnson, "Minutes of First Meeting," 26 November 1946, p. 1. UAT 3/1, Box 299, folder 3, JRDB/NARA.

44. "Launch vehicles" included rockets and balloons, as well as acoustic soundings and radio reflection. [T. H. Johnson], "Minutes," of the second meeting on 6 March 1947, 12 March 1947. UAT 3/2, Box 239, folder 3, JRDB/NARA. The panel also decided that it should meet on a more frequent basis.

45. Ibid.

46. Ibid.

47. Holger Toftoy memorandum, 19 February 1947. Annex D in ibid. Toftoy did, however, respond to the panel's request. James Bain replied on 21 March that in the time available, no comprehensive report was possible. But they now knew that 100 V-2 missiles could be assembled with General Electric adding the missing parts. Twenty-five of these were for "high priority guided missiles projects" and 75 would ultimately be available for upper atmospheric research. Bain also speculated that an additional 20 to 25 V-2 missiles could be

assembled from available parts, but that critical components such as servos, gyros, and other parts of the control systems were lacking, along with critical valves and piping. These were very expensive to reconstruct, and efforts in Europe to locate parts had revealed that few were available. The British still had 25 missiles but were not ready to give them up. J. G. Bain to OCO, 21 March 1947. Annex B to "Agenda," item 4.1 for the third meeting of the Panel on the Upper Atmosphere. Box 239, folder 3, JRDB/NARA.

48. C. S. Piggot (Exec. Dir. GMC), "Research and Development Work under Cognizance of the Committee on Guided Missiles," Annex F (UAT 17/1) to ibid.

49. The Air Weather Service, for example, reported in March 1947 that it required synoptic studies of the high atmosphere and wanted to secure the means of applying all forms of upper atmospheric data to developing long-range forecasting models, which, among many other things, would aid in determining "meteorological ballistic effects on guided missiles" by various civilian institutions. Annex C (5 March 1947) to ibid; and C. S. Piggot (Exec. Dir. GMC), "Research and Development Work under Cognizance of the Committee on Guided Missiles," Annex F (UAT 17/1) to ibid. Likewise, Newburn Smith, representing the Central Radio Propagation Laboratory of the National Bureau of Standards, reported that their needs paralleled those of the other services as well as those specified by the JRDB Committee on Electronics and its Basic Research Panel. These included data on sky-wave propagation, technical information on antenna loading from ions in the upper atmosphere, problems of radio interference from the missile exhaust and from the ionosphere itself, and, overall, a greatly improved understanding of the way the ionosphere is produced and modified. Newburn Smith, CRPL statement, Annex G (20 March 1947) to ibid.

50. E. H. Krause to T. H. Johnson, 4 April 1947. Box 99, folder 6, NRL/NARA.

51. The V-2 Panel now predicted that remaining German V-2s would last until August 1949 and that WAC Corporals, Corporal E types, Aerobees, Neptune HASR-2 types, and Hermes experimental types could all be replacements. Krause, ibid.; and G. K. Megerian, "V-2 Report no. 9," 31 March 1947, pp. 13–14. V2/NASM.

52. Homer Newell to E. H. Krause, 15 July 1947. EHK/NASM. On the Corporal, see Koppes (1982). The panel's preference for the Neptune over the Corporal is discussed in Chap. 10.

53. On the demise of Viking, see Chap. 10.

54. [T. H. Johnson], "Minutes" of the second meeting on 6 March 1947, 12 March 1947, pp. 1–2. UAT 3/2, Box 239, folder 3, JRDB/NARA.

55. Ibid.

56. Ibid., p. 3, with attached letter, G. B. Dobson (chair of the Gassiot Committee) to the JRDB, 12 March 1947.

57. Ibid., E. O. Hulburt to the JRDB, 25 March 1947. See also Baum, et al. (1946), and Garstens, Newell, and Siry (1946).

58. "Preliminary Statement," 10 April 1947, p. 4. UAT 19/2, Box 239, folder 3.

59. They found 7 specific problems and identified 12 observations to solve them. Ibid., pp. 5–6.

60. Ibid.

61. C. S. Piggot, "Research and Development Work . . . ," Annex F (UAT 17/1) to [T. H. Johnson], "Minutes," of the second meeting, 6 March 1947. UAT 3/2, Box 239, folder 3, JRDB/NARA.

62. Fred L. Whipple to Panel on the Upper Atmosphere, 19 August 1947. UAT 30/2.1, Box 240, JRDB/NARA. This report went through several drafts after a 22 July 1947 meeting.

63. Report of the "Working Group on the Upper Atmosphere Research Program for FY 1949," *Panel on the Upper Atmosphere* 19 August 1947, p. 1. UAT 30/2, attached to Whipple, ibid. E. O. Hulburt, T. H. Johnson, and H. E. Landsberg (Piggot's deputy executive director from the parent Committee on Geophysical Research) advised the Working Group.

64. Ibid, p. 3.

65. Ibid., p. 8.

66. Marcellus Duffy wanted to know how effectively present equipment, manpower, and facilities were being employed. Marcellus Duffy to Panel on the Upper Atmosphere, 29 August 1947. UAT 30/2.4, Box 239, folder 3; also in Box 240, JRDB/NARA.

67. T. H. Johnson, "Panel on the Upper Atmosphere Minutes, Fifth Meeting," 5 September 1947. Box 239, folder 3, JRDB/NARA.

68. Although reports from the three services broke projects and tasks into different levels of res-

olution, the Army Air Force identified about 30 projects totaling $4 million; the Signal Corps a similar number of "tasks" totaling $1.6 million; and the Navy some 40 projects (as determined by other means) in eight separate project groupings, for $3.6 million. Whereas the Signal Corps estimate included procurement costs for Aerobees, the AAF and Navy estimates were for research activities only. "Relations of FY 1949 Proposals to Long-Range Program in Upper Atmosphere," 30 September 1947. UAT 30/3.2 with appendices 30/3.2.1, 30/3.2.2, and 30/3.2.3, Box 240, JRDB/NARA.

69. Ibid., p. 1.

70. On Bush's vision, see Reingold (1987), p. 310.

71. Minutes of sixth meeting, 7 November 1947. Box 239, folder 3, JRDB/NARA.

72. Quotation from Needell manuscript biography of Berkner, chap. 3, p. 35. This at first confirms what a number of historians have recently suggested: that Bush never met his goal, because each service wished to retain its own pet programs, even if they duplicated those in the rival services. See also Reingold (1987); Price (1954), pp. 146–48; Dupre and Lakoff (1962), pp. 36–37, 65–66; and Komons (1966), pp. 7–8. Daniel Kevles identified the ineffectiveness of Bush's JRDB as a regulatory agency in "The Politics of Science," talk given at the National Air and Space Museum, 21 November 1989.

73. Recall that the Army Ordnance Department Advisory Committee on Guided Missiles, the still-born "Joint Group to Coordinate Upper Atmosphere Research within the Services," and the ONR Planning Division's Guided Missile Panel thought only in terms of guided missile development. H. B. Richmond, "Report of National Academy of Sciences Advisory Committee to the Army Ordnance Department on Long Range Guided Missile Weapons," 13 June 1946. 25.32, RAM/CIT. See also E. H. Krause, "Record of Consultative Services," 11 June 1946. Box 98, folder 2, NRL/NARA; and "V-2 Report no. 7," 12 November 1946, p. 4. V2/NASM.

74. C. S. Piggott to E. H. Krause, 20 November 1947, Enclosure D in G. K. Megerian to James G. Bain, "V-2 Report no. 13," 9 January 1948. V2/NASM.

75. [V-2 Upper Atmosphere Rocket Research Panel] to OCO, War Dept. (Chief, R&D Div), "Role of Upgraded Rockets in Research in the Upper Atmosphere," 8 September 1947. UAT 34/1 (2 October 1947), Box 240, JRDB/NARA. The "notable exception" of the Caltech group was soon to be rectified by Toftoy. See Chap. 10.

76. G. K. Megerian to James Bain, "V-2 Report no. 12," 10 October 1947, p. 11; "V-2 Report no. 13," 9 January 1948, p. 11 and Enclosure D. V2/NASM. This was followed in Krause to RDB (attn: UAT), 13 October 1947. UAT 2/6, Box 239, folder 2, JRDB/NARA. The JRDB had just reformed in September as the Research and Development Board in the wake of the National Security Act of July 1947 which created the DoD. Komons (1966), p. 7.

# 10

# New Vehicles for Upper Atmosphere Research

*It seems that the JRDB boys, upon further consideration, are beginning to think that the [Viking] does duplicate the Corporal E after all.*

—Homer Newell to Ernst Krause.[1]

When the V-2 Panel aligned itself with the Joint Research and Development Board (JDRB) and received the Board's endorsements, it hoped that the JDRB would ensure an "uninterrupted supply" of rockets.[2] But important changes were taking place in both bodies and in the over all complexion of postwar rocket development in the wake of program and budget cuts in late 1947. The first operational Aerobee was flown at White Sands in November, 1947, but the Viking was still far from operational. Competing alternative vehicles sponsored by the Army and Air Force threatened to replace it, just as competing oversight agencies threatened to reduce the autonomy of the V-2 Panel itself.

Army Ordnance had its own Hermes missile program, as well as JPL's Corporal series, to offer, and the newly independent Air Force was vigorously promoting its MX series. In addition, military contractors for all three branches of the services were making steady progress on a wide range of missiles; by 1947 there were some 28 contracted programs for tactical ballistic missiles and boosters, and each was struggling to survive in an increasingly hostile climate.[3]

This chapter is concerned with the methods the V-2 Panel used to link its autonomy to its choice of vehicles, and the severe test its initially pluralistic and collegial solidarity was subjected to at the outbreak of war in Korea.

## Major Changes

The passage of the National Defense Reorganization Act in July 1947 created an independent Air Force and placed all three services under the National Military Establishment, reporting to a civilian secretary of defense. The JDRB was reformed in September and renamed the Research and Development Board (RDB). Thereafter it was to be directly responsible to the office of the new secretary, a move that was intended to make it a stronger policy-making and regulatory body. Instead, unification created an even more confused and divisive atmosphere, unsuited for cooperation or coordination. Not only did the change throw missile development policy further into disarray, but it failed to determine how missile development would fit into the emerging American Cold War defense policy.[4]

Changes on a local level were taking place, too. T. H. Johnson, who had moved from the Ballistic Research Laboratory to Brookhaven National Laboratories, announced in September 1947 that he would retire in November as chairman of the JDRB Panel on the Upper Atmosphere, and in late November, Ernst Krause retired as chairman and member of the V-2 Panel, a move reflecting his departure from the Rocket-Sonde Research Section at the Naval Research Laboratory to head up NRL's "nuclear weapons crash program."[5] Johnson eventually was replaced by physicist Charles R. Burrows, new to the JRDB Panel in October, who after 20 years at Bell Laboratories, was now professor of electrical engineering and director of the Engineering School at Cornell.[6] James Van Allen was elected the new chairman of the V-2 Panel and occupied the post for the next ten years.[7] Homer Newell, who became the head of the NRL Rocket-Sonde Section, replaced Krause on the Panel as a regular member, after an unsuccessful bid for the chairmanship.

What began in 1946 as the V-2 Panel became in several months the V-2 Upper Atmosphere Panel, to better reflect its interests. By late 1947, Army Ordnance asked the panel to consider another name change, this time to reflect the fact that (a) the V-2 was in finite supply; and (b) the panel had convinced its patrons that no one vehicle would satisfy its future needs. It would also be expanded to include a new member nominated by Toftoy: the California Institute of Technology, represented by William H. Pickering. There were new players on the field, and the playing field itself was changing.[8]

The panel tabled its deliberations on another name change until February 1948, to give its members time to prepare briefs on the various launch vehicles and missiles available to them.[9] After discussing the matter, panel members decided that a name change was warranted to reflect its broader scope; henceforth it would be called the Upper Atmosphere Rocket Research Panel, or the UARRP. Van Allen was cautious about this change. In no way did he wish to imply that the UARRP was now a regulatory body or that it could allocate missiles. He demanded that it remain "a self-constituted body" devoid of "any *official* duties or responsibilities."[10] All official authority and control would remain the province of Army Ordnance, the participating agencies, and the RDB.[11] Van Allen felt that the panel needed protection.

The principal issue facing the UARRP and the bodies overseeing it was launch vehicle priority. The UARRP often debated the types of vehicles it needed, how many it needed, how it should promote them to the RDB, and who would get how many of them. Of all the possible missiles available for use, the UARRP most frequently mentioned the Neptune and the Aerobee, and it favored both over other missiles. But the Neptune was similar in capability to the Army's Corporal, then being developed at Caltech's Jet Propulsion Laboratory. The RDB was frequently reminded of their similarity and brought this point to the UARRP's attention each time.

In an era when missile manufacturers and the various branches of the armed services were pushing their favorite systems and no clear authority over these activities yet existed in the national military establishment, the RDB had the unenviable role of trying to identify needs on the basis of vague JCS policy, hearing testimony from service officers and contractors and trying to make resolutions concerning vehicle priority.[12] It is easy to see why the UARRP became another political forum for deciding which of the competing missile systems would be used for upper atmospheric research—it had developed strong personal and institutional contacts with the RDB.

Toftoy certainly took this relationship into account when he made Caltech's William Pickering a new member of the panel after Krause's departure. Beyond representation, intelligence gathering was another motive for membership. Toftoy and the academic representatives all wanted to be kept informed about new directions.[13] They were living in a constantly changing R&D world marked by uncertain budgets and periodic questions about missile redundancy. As a result, the members of the UARRP were often at odds over the different missile systems they advocated as launch vehicles.

## The Corporal, its WAC, and JPL's Direction

Army Ordnance had been building missile systems since 1943 (before it established GE's Project Hermes) under a contract with the Guggenheim Aeronautical Laboratory of the California Institute of Technology (GALCIT).[14] On 1 July 1944, in response to

**The WAC Corporal and JPL's Frank J. Malina, at White Sands, late 1945. NASM SI A5048-B.**

the Army's interest, GALCIT reorganized as the Jet Propulsion Laboratory, an engineering contract laboratory tied to Caltech. One of its first Ordnance contracts, ORDCIT (for Ordnance-California Institute of Technology), called for a missile with capabilities similar to those of the V-2.[15] By the summer of 1945, the Army had already invested $3 million but it had not come up with anything even remotely comparable to the V-2.[16]

The original ORDCIT plan was to develop increasingly sophisticated missiles, named in order of military rank. The 2.4-meter rail-launched solid propellant Private missile came first, and then in early 1945, after considerable lobbying from GALCIT, Army Ordnance approved funding for a small-scale liquid-fueled test vehicle called the WAC Corporal (figure, p. 169), which was to lead to the full-sized Corporal. By now the Army recognized that JPL would not be able to produce anything close to an operational sys-

tem for years, but it maintained its support knowing full well that missiles were in its future.

Although the WAC's creators, especially Frank J. Malina, first director of JPL, saw the 4.8 meter, 300 kilogram missile as the fulfillment of their dream of building a sounding rocket, Army Ordnance believed it "was designed to check aerodynamics and propulsion connected with the design of a larger missile." Toftoy said that it could possibly be later used as a meteorological radio sounding rocket for the Army Signal Corps, but this was clearly a secondary interest.[17]

The rail-launched WAC used a modified JATO motor, created at JPL and built by Aerojet Engineering Corporation (an industrial offshoot of GALCIT). It used storable red fuming nitric acid and aniline, and contained no inherent stabilization systems; hence its originators later claimed that the name WAC derived from "without attitude control," although the more popular sexist interpretation stood for Corporal's little sister, the "Women's Auxiliary Corps." After its first successful flight on 11 October 1945, when it reached an altitude of 75 kilometers over White Sands, the WAC went through 11 test rounds at White Sands before JPL decided to make a number of improvements: the most significant change was to reduce the weight of the motor from 23 to 5.5 kilograms.[18]

Douglas Aircraft Company built 14 improved WAC Corporal B missiles for Army Ordnance, four of which were to go to JPL for upper atmospheric research. William Pickering reported to the fourteenth meeting of the UARRP in February 1948 that he was planning to fly two of them for cosmic-ray studies and two for solar spectroscopy, but, for the moment, the whole program was "marking time" because JPL policy regarding upper atmospheric research was still not firm.[19]

Of all the groups, JPL had the greatest momentum in missile development at the close of the war. Partly through Pickering's urging, with Theodore von Kármán's sympathies and Frank Malina's wholehearted support, JPL made tentative plans in early 1946 to integrate upper atmospheric research into its continuing program of rocket development. The March 1946 JPL symposium on rocket technology (see Chap. 4) was the first to have sessions on upper atmospheric research. Pickering hoped that the symposium would be a springboard for his own research plans, centered around the WAC.[20] But JPL moved in quite another direction.

As historian Clayton Koppes has shown, under the direction of Clark B. Millikan, acting chairman of the JPL Executive Board, it was determined that JPL would not support "completely unfettered research," as Frank Malina had urged.[21] Pickering thus had to decide whether to return full-time to Caltech as a faculty member and compete with others on the V-2 Panel for instrument berths or whether to stay at JPL and try to turn the tide for the WAC.[22] He chose to stay at JPL, and throughout the rest of the 1940s, he concentrated on the development of the Corporal system. He continued as a section chief responsible for guidance and control at JPL while retaining his professorship of electrical engineering at Caltech.[23] By 1950, Pickering had transferred to JPL full time, where his attention and that of his JPL colleagues was directed solely at the continued development of guided missile systems.[24]

Pickering, in fact, became one of Toftoy's most vocal spokesmen for the full-scale Corporal E. The first Corporal E flew in May 1947, and in December Toftoy claimed that it was the only large American rocket that could be turned into a tactical ballistic missile system quickly, even though the first successful flight had been followed by two dismal failures.[25] The 11-meter Corporal went through many design modifications in 1948, mainly to the motor's cooling system, to the fuel tanks, and the plumbing. By June 1949 the new motor and fuel system were declared reliable, but control and guidance still were wanting. Even so, William Pickering and Louis Dunn, JPL director, testified for Toftoy at the Pentagon in September 1949 that the Corporal could be controlled reliably with existing technology.[26] It was thus approved for continued development into a tactical weapon system.

The UARRP, dominated by NRL and APL, never warmed to the Corporal or its WAC. At its first official meeting in March 1946, the tiny and inexpensive WAC, which in 1947 cost about $8,000 each in small lots, was simply "not in the same class with a V-2" and could not provide the payload performance characteristics the panel wanted. When the JDRB pitted the Corporal (estimated in 1947 at $45,000 each in lots of 50) against the Viking (which cost at least $200,000 each for the first 10), the UARRP consistently chose the Viking.[27] By February 1949, Newell observed in a rather biased comparative study of the performance-per-dollar characteristics of all available and potentially available missiles (V-2, Viking, Aerobee, Corporal, and the MX-774) that the Corporal lacked stability, payload capacity, and antenna capacity[28] (table 10.1).

Years later, Pickering admitted that the WAC was unsuitable for upper atmospheric research, especially for meteorology.[29] It did, however, provide the technical basis for the successful Aerobee sounding rocket and became famous as the second stage to the Army Ordnance Bumper-WAC series of high-altitude flights.[30] Six Bumper-WACs were fired from White Sands in 1948 and 1949, and two more were flown from the Air Force's Eastern Test Range near Cape Canaveral, Florida, in 1950 (figure, p. 172). These classified flights were not coordinated by the UARRP, although some panel members provided small ion probes and cosmic-ray detectors, and the altitudes reached were widely publicized.[31]

## The Aerobee

The least controversial and most coveted vehicle for upper atmospheric research was the Aerobee, a scaled-up version of JPL's WAC Corporal B. Designed by James Van Allen and his APL staff in the spring of 1946 and contracted to the Aerojet Engineering Corporation and Douglas Aircraft, the Aerobee was, like its predecessor, launched from a tower with a booster and used the same mixture of red fuming nitric acid as oxidizer and aniline

TABLE 10.1

| Missile | Cost (thousands of dollars) | Alt Km | Payload Kg |
|---|---|---|---|
| WAC Corporal | 8 in lot of 15 | 50 | 11 |
|   Improved | n.d. | 100 | 11 |
| Aerobee | 18.5 in lot of 20 | 130 | 70 |
| Neptune | 200 in lot of 10 | 200 | 600 |
| " | | 400 | 70 |
| Corporal E | 45 in lot of 50 | 160 | 230 |
| Bumper-WAC | 500 in lot of 6 | 480 | 11 |
| V2 | | | |
|   Reassembly | 30 at present | 160 | 900 |
|   American made | 175 in lot of 100 | 160 | 900 |
| MX-774 | n.d. | 160 | 230 |
| " | | 250 | 50 |
| MX-770 | n.d. | 290 | 1360 |

*Note*: One of the means the RDB Guided Missiles Committee employed to compare missile systems was to equate their expected or claimed payload/altitude/cost benefits. In October 1947, the rough estimates shown here were used to argue that no unwarranted duplication existed but that very close continued scrutiny was required to be sure that duplication did not emerge later. The Consolidated Vultee MX-774 was an 8,000 kilometer guided missile system being developed for the AAF on a 1.9 million dollar contract, and the North American Aviation's MX-770 was a more powerful winged version of the V-2. These estimates and figures changed continuously, as new missiles appeared and old ones disappeared, and real costs and performance characteristics replaced estimates.

*Source*: "Vehicles for High Altitude Research," 3 October 1947. Agenda item for 8th meeting of JRDB Guided Missile Committee. Box 198, Folder 25, JRDB/NARA. The MX-774 is discussed in Beard (1976); and in Neufeld (1990).

as its fuel.[32] It was, therefore, very much a WAC Corporal at first, scaled up just enough to provide adequate payload for the types of experiments Van Allen envisioned.

"Aerobee," a combination of Aerojet, its prime contractor, and Bumblebee (the APL missile program that gave it life), was a Navy Bureau of Ordnance project that had, as usual, a dual purpose.[33] In April 1946, Ralph E. Gibson, later APL's director, identified the Aerobee as a means and rationale for developing liquid-fuel propulsion techniques for anti-aircraft and surface-to-surface missile systems.[34] ONR's H. B. Hutchinson argued before the JRDB that Aerobee met the Navy's "initial objectives in basic research in areas like aerodynamic design, radio guidance . . . weather forecasting, reconnaissance, and cartography."[35] And the Signal Corps justified its interest in the Aerobee partly to test "locator equipment to be used against guided missiles."[36]

On 17 May 1946, APL signed with Aerojet for 20 Aerobees and an equal number of boosters, to cover its needs, as well as those of NRL.[37] The Aerobee was a bit thicker than the WAC, and, unlike the V-2, its forward sections were made of an aluminum alloy, not only to reduce weight, but to provide a non-magnetic shroud that would not interfere with cosmic-ray or radio frequency experiments. At first, Van Allen hoped the Aerobee could carry a nominal load of 70 kilograms to 110 kilometers altitude using a small solid booster in addition to its main liquid-fueled motor.[38] But his Aerojet advisers showed how this could easily be improved.[39]

Although the Aerobee was built from proven WAC B components, the aerodynamic design had to be revised for the scale-up. Even so, it took only 18 months and three successful dummy firings before the first instrument-carrying rocket was fired on 24 November, 1947.[40] Everything proceeded smoothly through the summer of 1947: new Navy firing facilities—including launch towers, blockhouse, and assembly area for the Aerobee and Viking—were built at White

A Bumper WAC on the launch pad. Note the WAC's enlarged fins, nested into slots in the V-2's modified nose-cone. NASM SI 77–6025.

Sands, funded by BuOrd and BuAer.[41] The only hitch in the process came in July when a wayward V-2 landed in Juarez, Mexico, prompting Army Ordnance to halt all firings of unguided missiles until appropriate safety measures were taken.[42]

The first 20 Aerobees together with the rocket's new launching facilities at White Sands, ultimately cost just over $500,000. Its designers and backers argued that this was far more cost-effective than the multimillion dollar Viking program.[43] By 1950, all agencies engaged in sounding-rocket research were using Aerobees. In addition to APL, which contracted for approximately 20 vehicles in what it hoped would be its first round, NRL in 1946 had 20 Aerobees on order and was considering 20 more, along with an administrative mechanism for using them at White Sands (figure, p. 173). In the next three years,

the Signal Corps also ordered and received 20 Aerobees and was negotiating for more, and the Air Materiel Command acquired 25 Aerobees and was planning for an additional 25.[44]

The Aerobee gained favor with all the scientific rocketry groups in the late 1940s and 50s, for research and development funding in

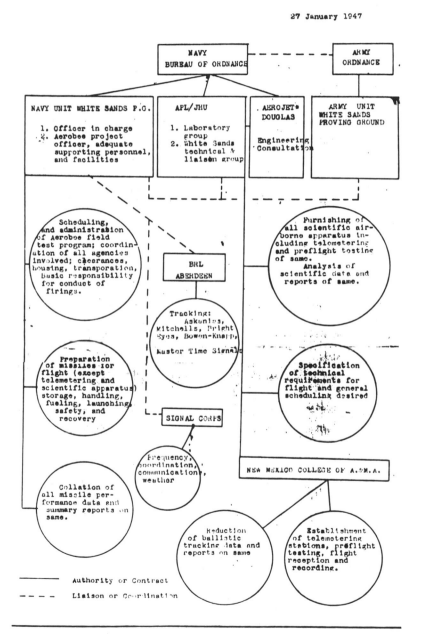

27 January 1947

In January 1947, James Van Allen and his staff provided members of NRL's Rocket-Sonde Research Section with this contract and administrative flow chart identifying how APL would soon be managing its Aerobee flights. NRL had to determine through ONR how it would fit in. "Record of Consultative Services," February 1947. Box 99, folder 5, NRL/NARA.

Note: For five of the presently planned twenty launchings NRL (under ONR)

the national military establishment was shrinking. Even at NRL, three of the four research groups (cosmic-ray, atmospheric physics, and spectroscopy) preferred Aerobees over the V-2.[45] NRL scientists liked the Aerobee because it cost less (in 1949 it cost $14,000) and had a more appropriate design (there was less structural mass and hence less spurious cosmic-ray secondaries). But they still had no trouble criticizing it (it had less stability, a small payload capacity, and less antenna capacity) when arguing for support for their own Viking.[46] Even so, the Aerobee was the sounding rocket of choice for everyone else who could still afford it, with or without the promised Viking.

After 18 months of using the V-2, many found it offered only the illusion of luxury. The economies imposed by the Aerobee often revealed better operational designs: both the APL and NRL cosmic-ray groups learned that the enormously heavy lead-shielded cosmic-ray telescopes they had built for the V-2s yielded less reliable data than the simpler coincidence/anticoincidence counter telescopes flown on Aerobees.[47] The Aerobee could not carry as many experiments on each flight as could the planned Viking or the dwindling V-2, but it was still more cost-effective carrying one or two instruments per flight.

Ironically, the groups most responsible for making the Aerobee possible, JPL and APL, did not get much of a chance to use it. For four years, APL was a good place to do upper atmospheric research. Its legacy in the history of the space sciences is enormous; Van Allen's High Altitude Group participated in 10 V-2 launches and instrumented 18 of the 21 Aerobees it fired as its initial allotment. But, in May 1949, Van Allen found that his own APL directorate was not wholeheartedly behind the effort. Asking for funds for eight additional Aerobees, Ralph E. Gibson, APL director, stated that the APL did not want "high altitude research [to] become a permanent activity of this Laboratory."[48] Gibson added that the Aerobee had certainly proven its scientific worth, as well as its military application "to the field of liquid-fueled guided missiles . . . because of its general application

to service experience in the practical handling of missiles."[49] But APL had no continuing interest in Aerobee itself; as a commentator later observed, the Aerobee was only one system among "the hundreds of test vehicles under the Bumblebee program," and its value to APL's directorate, as the visible and unclassified tip of its largely classified missile effort, was largely spent.[50] To a military contract laboratory dedicated solely to the design and testing of missiles and related ordnance, the Aerobee had taken on purely production status as an Aerojet product.[51]

Times were changing. Expecting a significant decline in military support, BuOrd did not want to spend any more money on basic research, arguing that this was ONR's responsibility.[52] But ONR was already funding the NRL group and had no reason to support the same activity at APL. As a result, Van Allen found himself without Aerobees, and soon without institutional support, when Gibson asked him to assume his wartime managerial duties producing proximity fuzes. Van Allen recalled that "there was nothing in the world I wished to do less."[53] He persevered under the burden for about a year, but the handwriting was on the wall: "If I were to stay at APL, I would probably be more and more absorbed in military type work, which at this time in history I didn't wish to do." Van Allen thus closed down his group and left APL for the University of Iowa. On a university budget, Van Allen could no longer afford to buy Aerobees and so developed his innovative Rockoon program of balloon-launched Deacon rockets in order to continue his studies of cosmic radiation and electrical phenomena in the earth's upper atmosphere.[54]

The Aerobee became the sounding rocket of choice for those who could afford it in the 1950s: NRL, the Signal Corps, and the Air Force at both the AFCRC and the Air Research and Development Command and their university contractors used the Aerobee to continue programs started with the V-2. One measure of the Aerobee's success was that by the early 1950s both the Navy and Air Force had projects afoot to improve its capabilities, while keeping its basic design intact. The first real upgrade came after more than 50 Aerobees had been fired. In 1949, with Air Force support, the motor was improved and became the basic power plant of the Aerobee-Hi, which was capable of sending heavier payloads higher into the atmosphere. All versions of the Aerobee, however, were used by the Air Force, Navy, and Signal Corps for a total of 165 firings by October 1957.[55] Scientific research with Aerobees ended on 17 January 1985 with the 1,037th Aerobee firing, when NASA redefined its suborbital program as a shuttle-based activity.[56]

## The Viking

The Viking, in contrast to the Aerobee, created a great deal of controversy and had to be protected by its backers. A far more expensive and sophisticated program, it therefore competed with similar vehicles being promoted by the other services. Viking, originally referred to as Neptune, was the conceptualization of Carl Harrison Smith and Milton Rosen based upon specifications set by Krause and his NRL superiors. Rosen ultimately became chief scientist for the project after he learned what it took to build a missile in an eight-month apprenticeship at JPL, arranged by Krause. Viking was not only a replacement for the V-2, but a highly refined version, with an aluminum airframe and a servo-controlled gimbaled motor for attitude control during flight. It would retain the V-2's turbine-fed fuel system and alcohol and liquid oxygen mixture.[57]

On 18 February 1946, three ORI officers met with Krause at NRL to discuss rocket procurement. Since Aerojet and Douglas were already providing the Aerobee for NRL and APL, they decided to "try and develop some other source" for their own test vehicle.[58] Reaction Motors, Inc. was a known quantity for propulsion units, and, Krause told his patrons, he recently was told by James Bain that "Martin Aircraft" was interested in getting into the game. After several months of clearing procurement details through channels at the Bureau of Aeronautics, in April 1946, the

Chief of Naval Research approved the $2 million project as a combined NRL, ORI, and BuAer initiative to build 10 high-altitude sounding rockets. The Glenn L. Martin Company of Baltimore won the cost plus fixed fee contract, in concert with Reaction Motors, Inc., of New Jersey, which was to build the propulsion units.[59] NRL was more than a project monitor; it was to share in many of the early design studies its contractors would perform in their search for a suitably powerful motor and stabilized vehicle.

Throughout the summer of 1946, the problem of stabilization took center stage for Rosen and an adviser, MIT stabilization expert Albert C. Hall. Rosen knew that V-2 jet vane breakdowns were a major cause of V-2 launch failures, so they and Martin engineers conducted conceptual design studies of possible gimbal systems, even though they also continued to consider high-density graphite jet vanes, hoping that one of the designs would work.[60]

Rosen (figure, p. 180) studied at JPL from August 1946 though April 1947, learning the basics of missile design and testing, and adopting JPL's philosophy of test firing the entire completed missile, rather than testing each part separately.[61] Upon his return to NRL, equipped not only with his JPL training but the accumulated experience of the first year of V-2 firings at White Sands, Rosen, J. M. Bridger, and others in the NRL rocket group now had at hand detailed analyses of vehicle vibration, stability, and the frictional heating of critical components.

In April 1947, Rosen announced that the Neptune systems and components were at the detailed design and prototype engineering stage. Although Rosen haggled with Martin about testing philosophy at first, eventually they agreed that there would be some testing of its major components, but that the completed Neptune prototype would also be carried through a complete firing cycle to rate thrust and duration.[62] This soon became an important issue as the original delivery schedule had slipped considerably; the first prototype motor was already overdue in April and did not appear until October.[63]

Production version of the improved Aerobee-Hi used by the Navy, Air Force, and Signal Corps. NRL's 42d Aerobee flew on 8 May 1956 to 188 kilometers' altitude as a test of rocket performance. In the 1950s, the Aerobee in its various modified forms became the primary large launch vehicle for upper atmospheric research in the United States. U.S. Navy photograph. NASM SI 90–5329.

Meanwhile, Martin engineers proceeded with component construction and systems design while ONR and NRL representatives touted Neptune before the JRDB (figure, p. 177). By the spring of 1948 Martin was performing dynamic tests of the complex gimbaled steering system. Finally, in December 1948, 11 months behind their original schedule, the first fully assembled Neptune, now designated Viking, was ready for full systems testing.[64]

The schedule slipped another three months at White Sands, where, starting in late January 1949, a combined NRL and Martin field crew met many frustrating delays readying the first Viking for its static tests and launch. Suffering telemetry failures, and most seriously, a series of peroxide leaks in the turbine during static testing, Viking 1 left White Sands for the ionosphere on 3 May and reached 80 kilometers. After 290 seconds of flight, the missile was torn apart by the aerodynamic pressures of reentry, and pieces of Viking 1 landed over a 10-square-kilometer area.[65]

Homer Newell, reporting to the UARRP in August 1949, made little of the three-month delay and frustrations and concentrated on the success of just getting Viking off the ground. He admitted it fell far short of its theoretical height capability of 190 kilometers, but it had performed better than expected, or feared.[66] He also promised that a new and improved Viking motor would boost later Vikings to 300 kilometers altitude carrying 230 kilogram payloads.

Like Viking 1, the second and third Vikings, launched in September 1949 and February 1950, did not reach predicted altitudes, but they, too, demonstrated that the entire system worked; in all flights, the gimbal system operated correctly, providing good stabilization and control. The fourth flight in May 1950 not only was a complete success, but demonstrated the Navy's ability to develop a shipboard-based tactical ballistic missile system. Viking 4, launched from the U.S.S. Norton Sound close to the equator, rose to almost 170 kilometers in a well-stabilized flight carrying a host of cosmic-ray devices and pressure/temperature sensors[67] (figure, p. 179).

Newell emphasized the capabilities of Viking at UARRP meetings and before the RDB. He predicted that it would ultimately reach the $F_2$ layer of the ionosphere (about 500 ki-

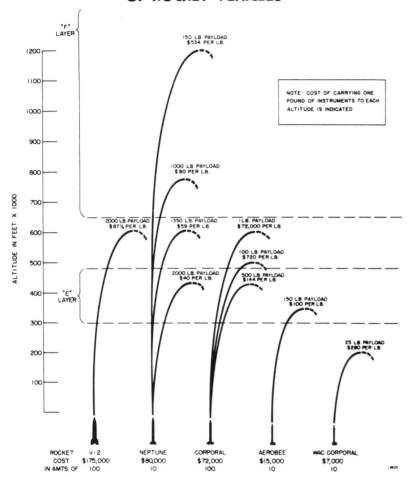

Well before one was operational, ONR compared the payload performance and cost efficiencies predicted for Neptune to those of other missiles considered for use as launch vehicles. Note that ONR claims that Neptune was the only vehicle capable of reaching the ionospheric F layer. From L. M. Slack, "Facts Supporting the Justification for the Development of the Neptune Rocket" (GM 25/1) attached to S. D. Cornell (GMC/JRDB), "Navy Neptune Rocket" (25/1/1), 12 June 1947. JRDB/NARA.

lometers), something neither the Corporal nor V-2 could do. After its first four flights, the Viking demonstrated that it was a viable launch vehicle, and had more flexibility than the old V-2; one could, for instance, vary payload capacity by a great amount without restructuring the air frame or adding lead weights to the warhead. Most of the technical details of Viking, including its expected performance characteristics, remained classified in the 1940s.[68] This was not only due to Viking's potential as a military weapon, but to the efforts of its promoters to extend its life.

## Viking's Dual Role

During the development of the Viking, ONR made sure that the Guided Missile Committee of the RDB knew about Viking's many design innovations. In June 1947, ONR boasted that "the major portion of the information obtained from this program will contribute directly to the design and control of long-range guided missiles."[69] L. M. Slack of ONR took considerable pride in these "all-American" innovations, and with Homer Newell, offered Viking in the dual role as the best replacement for the V-2 as a vehicle for

**Launch of Viking 3 at White Sands, showing its thrust vector control mechanism in action. U.S. Navy photograph, Milton Rosen collection. NASM SI 91–291.**

upper atmospheric research and as a test vehicle for ballistic missile development, claims contested by both the Air Force and Army in hearings before the RDB in the next several years.

Many times these hearings became heated debates, and more than once the RDB would reverse its opinion. In July 1947, for instance, a frustrated Homer Newell reported to Krause: "It seems that the JRDB boys, upon further consideration, are beginning to think that the [Viking] does duplicate the Corporal E after all."[70] In addition to the Corporal, advocated at times by Pickering and Toftoy, the Air Force MX-770 series, specifically its MX-774, built by Consolidated Vultee (later Convair), was promoted by Marcus O'Day.[71] A growing

part of Newell's testimony before the RDB, therefore, beyond its scientific promise, became Viking's "direct military application."[72]

Claims and counterclaims for the superior capabilities of the MX and Viking volleyed across the RDB field throughout the late 1940s. Both missiles used powerplants supplied by Reaction Motors, Inc., and both had gimbaled motors using the same combination of V-2 pressurizing agents and propellants. The MX, however, employed a quartet of smaller motors developed by RMI for the Bell X-1 rocket plane and had a separable warhead.[73] Even though these differences were mitigated by similar claims for performance potential, at the end of the first round of debate, in December 1947, the RDB Guided Missiles Committee adopted the compromise position of the RDB Panel on the Upper Atmosphere, that "there is no unwarranted duplication at present in the development of upper air research vehicles."[74] The RDB would vacillate time and again as the services restated their claims.[75]

In addition to interservice competition, two other factors made Homer Newell pay more attention to the Viking as a military weapon in 1948. There was a strong trend within the RDB to advocate the use of military missiles for upper atmospheric research, rather than vehicles built expressly for science, such as the Viking and Aerobee. At the same time, there was growing resistance within NRL to pay for vehicles built for upper atmospheric research. In May 1948, when Newell asked for additional Aerobees and Vikings, his division superintendent, John M. Miller, shot back, "No funds are being specifically budgeted by ONR or NRL, for 1950, for U[pper] A[tmosphere] rockets."[76] The NRL had a budget for 10 Vikings, and would probably buy more Aerobees, but it would not consider an extended Viking program on scientific merit alone.

In 1948, Newell found that he could still combat any counterclaims for the MX or Corporal, because none had been successful as yet. He marshaled the sympathies of the UARRP in May 1948, whose members were happy to remind the RDB that no "unwar-

**Launch of Viking 4 from the U.S.S. Norton Sound on 11 May 1950. U.S. Navy photograph, Milton Rosen collection. NASM SI 91–294.**

ranted duplication" existed between the Viking, Corporal, or the MX series and that its collective needs were met: "only because the Aerobee and the [Viking] were developed specifically for upper air research. This would not be the case if guided missile development had been relied upon to meet these requirements."[77]

The UARRP's solidarity and the sympathies of the RDB preserved the status quo through 1948, even though NRL continued to refuse support for additional Vikings. Another flap arose in January 1949, when the Air Force once again offered its MX-774 as an upper air research vehicle, encouraged in part by Marcus O'Day, who now was a member of the RDB Panel on the Upper Atmosphere.[78] To settle the matter, the RDB Committee on Guided Missiles asked Holger Toftoy, Rear Adm. D. V. Gallery (BuAer), and Brig. Gen. William Richardson (USAF) to study the matter. In March, they ruled against the MX-774, stating that "The Viking promises to be the most satisfactory vehicle in this class."[79]

O'Day was not satisfied with the proceedings; years later he charged that the ad hoc group was biased against the MX, and that its fate was a good example of how interservice rivalry impeded the progress of ballistic missile development.[80] More to the point, however, was the fact that only the Viking had the endorsement of the UARRP. None of the three missile systems was as yet proven.

JTION
OF 5 MEN
D ON EACH
OF GANTRY
ME TIME

Milton Rosen at White Sands for a Viking launch. U.S. Navy photograph, Milton Rosen collection. NASM SI 91–273.

Corporal's first flight on 22 May 1947 exceeded expectations, but it was a fluke, and the next two total failures drove JPL engineers back to their drawing boards for the next 18 months to redesign the motor.[81] None of the three flights of the MX-774 in 1948 achieved anything close to nominal performance.[82] And at the time of the RDB deliberations, the Martin field crew was tearing down the first Viking motor to fix leaky turbines. Thus, the RDB endorsed a still untested missile over two that had been flown, but had not performed well. Beyond the united front of its BuAer, ONR, and NRL defenders, what evidently sold Viking to the RDB,

therefore, was the interservice support by the UARRP.

Through the rest of 1949 and well into 1950, the Viking enjoyed endorsements from the RDB as its primary large rocket for upper air research.[83] But as war clouds gathered in the spring of 1950 and the National Security Council called for rearmament in anticipation of war, the Bureau of Aeronautics canceled all outstanding contracts for improved Vikings "for lack of funds."[84] Rosen, in a last-ditch effort to save the program, urged John M. Miller, Newell's boss, to take explicit steps to convert Viking into a guided missile.

In December 1950, Miller created the Rocket Research Branch, headed by Rosen, to assess what was required to provide the Viking with an active guidance system.[85] By the spring of 1951, Milton Rosen and J. Carl Seddon had their report ready, which asserted that very little had to be added to the $3.5 million already spent on Viking to equip it with an on-board radio guidance system that could deliver a 230 kilogram warhead to a target 240 kilometers away, with an accuracy of $+/-$ 450 meters. They also claimed that a Viking could be made ready for launch within two hours[86] (figure, p. 182).

Rosen and Seddon convinced F. R. Furth, Director of NRL, to ask the Chief of Naval Research for $2.5 million to produce six modified Vikings within 18 months, using the same contractors and NRL expertise for the guidance system. Their argument was that NRL's six years of experience with the V-2, Aerobee, and Viking programs gave it the "knowledge needed for field testing of missiles, and have proven the value of the Naval Research Laboratory telemetering system as a means of collecting performance data from rockets in flight."[87]

The NRL group had little problem showing how upper atmospheric research with rockets prepared teams for ballistic missile development. This scenario was repeated in Glenn Martin rhetoric, but BuAer did not buy any of it. The Viking did undergo a partial upgrade in 1950, mainly in the amount of propellant it could hold, which increased its overall performance in later rounds. But its publicized potential as a tactical weapon never brought the extension NRL desired. Viking had to compete with BuAer's planned submarine-launched Regulus and BuOrd's later entry, called Triton, which came from the drawing boards at APL. Thus it became only one of several systems within the Navy's three factions interested in building the Navy's ballistic missile capability.[88] Even if the Viking may have been technically one of the more advanced devices in the United States at its inception, as Von Braun later stated, the political uncertainties of the era, when the modes of decision making and program cutting were far from straightforward, did not bode well for Viking's survival.[89] It was a time when many other programs, including Hermes, were being canceled.[90]

Viking's dual role did not escape public notice. The Navy marketed it aggressively in the open literature, to a greater degree than the others. In July 1947, for instance, *Popular Science* announced that Viking would "Double V-2's Record." Weighing only two-fifths of "one of those Nazi dinosaurs," the Viking promised to carry "up to a ton of pay load—either delicate, space-exploring instruments or explosives" to at least 400 kilometers. Glenn Martin made no bones about the meaning of this number: "If it can go that high, figure out what it would do on an incline."[91]

Journalists were invited to view the Navy's first shipboard test when the fourth Viking was launched from the U.S.S. *Norton Sound* in May 1950. *Popular Science* declared that the "Mid-Pacific experiment foreshadows the day when our Navy may add huge guided missiles to its seagoing striking power." The *Norton Sound* constituted a mobile launch platform that could extend the range of guided missiles "far beyond limits imposed by land-based launchings."[92] *Time* pointed out that Viking could carry one of the newly announced "baby atomic bombs."[93] And *Life* reported that Navy officials were delighted with Viking's performance: "they had proved for the first time that big rockets, capable of carrying A-bombs several hundred miles, could be launched from the deck of a ship."[94]

Politics within the Navy and interservice competition worked against the Viking as a military device, and cost stood in its way as a scientific vehicle, especially when there was a more economical alternative.[95] The Viking program was not extended beyond the original 10 rockets, 2 more for guided missile development and another 2 to test components of Project Vanguard.[96] By the time the Navy gained approval to build Vanguard in 1955 as the U.S. launch vehicle for small satellites during the International Geophysical Year,

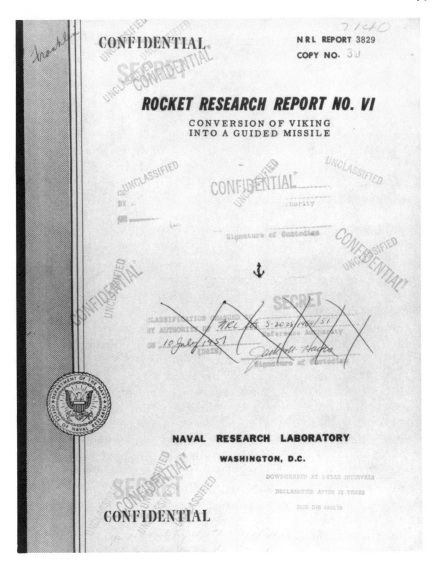

The confidential rating at publication on 1 April 1951 was upgraded to secret in July 1951, and then gradually downgraded. NASM Technical files.

the Martin Company had become the prime contractor for the Air Force's new Titan missile program and so transferred its small Viking project staff into this much larger and more lucrative arena.[97] And just as Glenn Martin staff were turned to Titan, Milton Rosen and his NRL branch moved into Project Vanguard.

## 'Riding on the Coattails of Guided Missiles'

In April 1950, trying to avoid BuAer's imminent cancellation of future Viking improvements, Newell tried to persuade the

UARRP once again to endorse Viking. Van Allen soon found, however, as war in Korea loomed, that traditional UARRP solidarity was being eroded; five members now preferred the RDB policy to let the military provide new vehicles. William Pickering argued strongly that a new improved vehicle for scientific research was not justified. It was more prudent and cost-effective, he added, to wait for what the military would develop in a few years and then use its missiles.[98] Van Allen agreed, adding that the "normal course of military missile developments" would likely make available the means at comparatively moderate cost.[99] William Dow took a harder line, demanding that any improvements had to be justified on military grounds.[100] Even Fred Whipple, who had recently applied to be Hulburt's successor at NRL, was only able to muster a weak endorsement, feeling that the panel should not place any limits on its needs, nor on the "quantity of basic knowledge that can be obtained by the high-altitude studies."[101]

Newell's rebuff sparked what threatened to develop into a serious disruption. Van Allen worked to keep the panel together as three of its most vocal members championed competing systems—Newell for Viking; O'Day for the all-but-dead MX; and Pickering for Corporal. Van Allen knew that the military backers of each system had been stepping up their efforts to apply pressure on the UARRP (through their contacts on the RDB) to force the panel into making a new judgment call. At the time of Newell's request, both the RDB Committee on Geophysics and Geography and the RDB Guided Missiles Committee were calling for UARRP statements. Knowing that he could not produce a consensus, Van Allen did the next best thing: he told the three advocates to hammer out a statement on the scientific objectives that would be served by improved vehicles.[102] With Charles Green of GE acting as coordinator and mediator, the committee gathered statements from panel members and condensed their findings into a draft report to the RDB in September 1950.

Instead of reporting on the scientific value of improved vehicles, O'Day and Green created a wartime survival document. They argued that the collective experiences of the panel members during the previous four years had yielded a wealth of techniques and experiences that would be of use to the military in times of national emergency. Scientific rocketry helped, they added, to refine advanced multichannel matrix systems for telemetry; to gain operational experience in fueling, training, and tactical methods for launching missiles of all types; and to advance the general understanding of rocket flight. O'Day emphasized that his office was contracting with 15 universities, and "these

alone represent a sizable group of people who have gained experience with instrumentation which can be invaluable in time of national emergency."[103]

By their September 1950 meeting, the UARRP found it could no longer justify new vehicles for basic research. It now harbored serious doubts and deep dissension about its immediate course of action. Some panel members warned that "war mobilization could halt all such activity [as upper atmosphere research]" while others felt that "in a national emergency, the Panel [should] not be interested in continuing upper air research unless there is definite value to the military establishment."[104] None, however, felt that such work really had immediate military value, but this caused, in the words of a critical observer from NACA Langley, "several hours of intensive debate," at the end of which, the UARRP "agreed that its funds were being gotten by 'riding on the coattails of guided missiles.'"[105]

At the time of these gloomy deliberations, members of the panel were far from sanguine about their future. Van Allen was in transition to Iowa, had not established a new program, and was about to offer his resignation from the chairmanship of the panel;[106] O'Day was losing support in an internal struggle at AFCRL;[107] and Louis Delsasso and others feared that the firing facilities at White Sands and Holloman were going to be less accessible because of the increased demand for classified firings. The Signal Corps also warned that its meteorological and research groups could well disappear for want of direct service relevance, which was "exactly the type of diversion that the panel wish[ed] to avoid."[108] The UARRP also realized that many of the younger members within its rocket research groups were of draft age, and unless the panel "found some good arguments for staying in existence, the whole staff would be inducted."[109]

Even though members knew that UARRP support had "been based entirely upon the usefulness of the work to the development of guided missiles," and that this usefulness was difficult to pin down, the UARRP felt its best ploy was to emphasize relevance.[110] It simply had no other choice. On the second day of their September meeting, the UARRP members turned the O'Day-Green draft into a formal report to the RDB emphasizing that the UARRP provided "the necessary 'know-how' that would form an excellent nucleus for the training of military personnel in the use of long range guided missiles."[111] Accordingly, at its meetings in January and April 1951, Marcus O'Day suggested, and the panel enthusiastically agreed, that "a handbook on techniques of rocket instrumentation be prepared within one year for the benefit of the Armed Services."[112] In this way the panel might demonstrate its relevance as a resource of technical, as well as scientific information.

UARRP rhetoric thus fell back on the basic motives that had brought it into existence in 1946. Instead of demanding an "uninterrupted supply" of vehicles for upper atmospheric research as it had in 1947, and perceiving that the new war threatened its existence, the panel's only course was to argue that it was directly relevant to military needs and could make do with the vehicles at hand. NRL and AFCRL still provided sufficient support to buy Aerobees, Van Allen found new support to create his Rockoons (with considerable logistical support from the Navy), and both White Sands and Holloman Air Force Base remained open. Because of the UARRP's unofficial status, it was also able to outlive the closest thing it ever knew as a patron: the RDB failed to survive yet another Department of Defense reorganization (in 1953) when Dwight D. Eisenhower's new administration reduced government and military spending, and abolished all statutory DoD staff agencies.[113]

As Van Allen entered academic life at Iowa, and continued as chairman of the UARRP, he pushed to improve panel visibility and its role in upper atmospheric research. Panel members had spent almost five years and many millions of dollars making their instruments work on rockets and were now gathering data on the high atmosphere, yet by 1951 they had failed to make a measurable impact on the scientific or aeronautical com-

munity at large. This failure underscored the gloom of their September 1950 meeting, when members of the panel were doggedly organizing a series of coordinated flights to reconcile their atmospheric data that conflicted with the authoritative NACA tables (see Chapter 15). In the wake of that successful effort in 1951, it was time to make a statement, if only to show that riding on the coattails of guided missiles was indeed proving fruitful for science. Encouraging the panel was the highly catalytic Oxford geophysicist Sydney Chapman, who was spending the 1950 academic year at Caltech and had been attending panel meetings. He would help to guide panel members into the world of geophysics.

Retrenchment at the outset of the Korean War brought panel members back not only to the first principles of a military world, but to a renewed appreciation of the need to move beyond technical achievement to establish authority for their hard-won results. But just as panel members competed for vehicles, their atmospheric data conflicted with those obtained by traditional means. The remainder of this book explores these and other issues surrounding the use of rockets for science.

## Notes to Chapter 10

1. Newell to Krause, 15 July 1947. EHK/NASM.

2. E. H. Krause to T. H. Johnson, 4 April 1947. Box 99, folder 6, NRL/NARA.

3. Stares (1985), p. 27 discusses these programs in the context of early military satellite proposals and notes that only eight survived during fiscal 1947.

4. Roland (1985a), p. 217. On the act of unification (variously called the National Security Act) that led to the creation of the Department of Defense in 1949, see Bush (1970), p. 303; Rearden (1984), p. 98; and Borklund, *The Department of Defense*, noted in Roland (1985a), pp. 204, 374. Stares (1985), p. 27ff., discusses the effect of unification and the creation of the RDB on military space proposals.

5. Krause accepted a new assignment as director of NRL's new nuclear division, and, most

serious for the NRL rocket-sonde program, took "twenty 'first-line people' " with him from his old group. Green and Lomask (1970), p. 9. Both before and after Krause's departure, members of his staff such as Eric Durand developed spectroscopic and electronic detectors for the Pacific tests. See also Krause OHI, pp. 100–04. On T. Johnson, see C. S. Piggot to Whipple, 10 September 1947. Box 10, NRC folder, FLW/HUA. Piggot and Johnson tried to convince Whipple to take over, but Whipple turned it down, citing his responsibilities at Harvard. See Johnson to Whipple, 19 September 1947, ibid. Whipple stalled his decision about the panel chairmanship until mid-October and finally decided that he could not do it. He was fully booked. Whipple to Piggot, 14 October 1947, ibid. Whipple however, did agree to remain active, and in July 1950 became a full member of the RDB Panel on the Upper Atmosphere.

6. *American Men and Women of Science*, 11th ed., p. 691. His activities during the war included radio propagation studies.

7. Van Allen and Newell were the candidates. William J. O'Sullivan to Chief of Research, NACA Langley, "Report on V-2 Upper Atmosphere Research Panel meeting," 21 January 1948. NACA/NHO; Harry Wexler to Hunsaker, 31 October 1947. Box 3, HW/LC.

8. "V-2 Report no. 13," 9 January 1948. V2/NASM.

9. At the January meeting held at NRL, members gave six detailed presentations: Pickering on Jet Propulsion Laboratory's (JPL) WAC Corporal; Newell on plans for the Neptune (or Viking); and Van Allen, Ference, Newell, and O'Day on plans for, and experiences with, the Aerobee. "V-2 Report no. 14," 6 February 1948, pp. 2–9. V2/NASM.

10. James Van Allen to V-2 Upper Atmosphere Research Panel, 17 March 1948, p. 2. Enclosure B to "V-2 Report no. 15," 1 April 1948. V2/NASM.

11. Homer Newell reminisced that the V-2 Panel "was one of the few really working Panels that I've been involved in. It started with an advisory function, but very quickly became the operating Panel for the activity. And yet it never had a charter . . . It worked because the people on it had a job to do in their shops back home, and when decisions were made they could go back and see that the decisions were followed through." Newell OHI, p. 18. AIP/SHMA. See also Newell (1980), p. 35.

12. Membership on an RDB panel or committee was politically useful not only for contractors, but for universities. This was true especially at the University of Michigan, where the administration encouraged its best faculty to push for representation: "It will be a good thing for the University of Michigan to be represented on some of these panels." See J. Ormondroyd to Technical Advisory Board, 24 March 1948, with attachments, including A. P. Fontaine (director, Aeronautical Research Center UM) to Frederick L. Hovde, 3 February 1948. Box 4, folder 3, WGD/BHLUM. Fred Whipple's astronomical colleagues acted in much the same manner when they first heard that he had been asked to become chairman of the RDB Panel on the Upper Atmosphere in late 1947 and then when he joined the panel in 1950. See Kuiper to Whipple, 27 October 1947. Box 3, 1940–50, K-New Mexico Meteor folder, HUG4876.810, FLW/HUA.

13. Large institutions like Michigan, with much invested in new research initiatives, needed to gather intelligence on military interests in order to plot future needs for manpower, equipment, and operations budgets. They had to play this game carefully, reconciling faculty interests as best they could with what was being bought by the military, and they did so by staying in touch with deliberations at meetings of the RDB and UARRP. See, for instance, the deliberations of Michigan's Aeronautical Research Committee when it learned in 1949 that the military services were now looking for a "balance in emphasis" in guided missile research between hardware and pure research, and began to plan accordingly. "Meeting of the Aeronautical Research Committee," 14 March 1949. Box 5, 1949–50 folder, WGD/BHLUM.

14. Koppes (1982), chap. 2, provides a full description of the background to GALCIT and Corporal. The intention to build a working missile system for use during the war was noted by Holger Toftoy sometime after the first successful firing of a WAC in October 1945. See "Section I: Long-Range Rocket-Powered Guided Missiles," pp. 10–11. Box

A763, "History, Planning for Demobilization Period," OCO/NARA.

15. It had to be capable of delivering a 460 kilogram warhead to within 5 kilometers of a target some 240 kilometers distant. Koppes (1982), pp. 21–22. See also Malina (1971), p. 370ff.

16. Koppes (1982), pp. 20–22. On the $3 million value, see Toftoy, "Section I: Long-Range Rocket-Powered Guided Missiles," p. 10. Box A763, "History, Planning for Demobilization Period," OCO/NARA.

17. The prototype WAC missile could carry an 11 kilogram payload to 30 kilometers. Thus, in October 1945, Toftoy claimed: "If the [first] test is satisfactory, this particular model may be adapted to provide a meteorological sounding rocket." Holger Toftoy, remarks at "Joint Army-Navy Meeting on Army Ordnance Research and Development, the Pentagon, 1 October 1945," p. 38. Transcript of meeting between Army Service Forces, R&D Services, and BuOrd, USN. Box 767A, OCO/NARA. See also "Section I: Long-Range Rocket-Powered Guided Missiles," 30 October 1945, pp. 10–16. Box A763, "History, Planning for Demobilization Period," OCO/NARA. Quotation from "Joint Army-Navy Meeting on Army Ordnance Research and Development, the Pentagon, 26 June 1946," p. 39. Box 767A, OCO/NARA. See also Malina (1971), pp. 353, 364. Although Army Ordnance planned to offer the WAC to the Signal Corps, Signal Corps groups ultimately preferred to use cheaper rockets, such as the Loki, and, when the situation warranted, the more capable Aerobee. Bates and Fuller (1986), p. 171.

18. Koppes (1982), pp. 23–24.

19. "V-2 Report no. 14," 6 February 1948, pp. 3–4. V2/NASM.

20. Pickering (1947); Pickering and Wilson (1972), p. 410.

21. Koppes (1982), pp. 26–27.

22. This created a period of "conflicting commitments" for Pickering, who in early 1946 was torn between a career at JPL and his interests in upper atmospheric research. He probably could have obtained berths on V-2s for scientific instruments, but he did not pursue this course. Pickering OHI no. 2, pp. 19–20rd; no. 1, pp. 4–11rd. Harvey Hall to J. W. Crowley, 5 February 1946. File 5–377, JPL/L.

23. Koppes (1982), pp. 32–49, 65; Pickering and Wilson (1972), pp. 391. Pickering was one of the few Caltech professors to remain as a section head at JPL in the late 1940s.

24. Pickering OHI no. 2, pp. 19–20rd; no. 1, pp. 4–11rd. There is some evidence, however, that one spectrograph, built for the Aerobee, may also have been considered for flight on a WAC Corporal by Pickering. In late 1946, William A. Baum left Tousey's group at NRL to pursue a doctorate in physics at Caltech. For a thesis project, he decided to design a compact lithium fluoride solar spectrograph for smaller rockets, advised jointly by Pickering and by Ira S. Bowen and Harold D. Babcock of Mt. Wilson. Two of these instruments were built and fit exactly Pickering's January 1948 description of those planned for WAC Corporal firings. But the spectrograph was flown on NRL's third Aerobee on 1 February 1949, when Pickering was no longer Baum's adviser. Unfortunately, the Aerobee's Tiny Tim booster exploded when it was halfway through the launch tower, and a disappointed Baum left upper air research for good. See "V-2 Report no. 14," 6 February 1948. V2/NASM, pp. 2–9; "Panel Report no. 20," 2 May 1949, p. 5; E. H. Krause to Code 180, "Reply to letter from W. A. Baum . . ." 9 January 1947. Box 99, folder 5, NRL/NARA; and William A. Baum OHI, pp. 20–26rd. Baum has confirmed that his instrument was built for an Aerobee. See Baum to the author, 22 May 1990.

25. Koppes (1982), pp. 38–39.

26. Koppes (1982), pp. 41–42.

27. "V-2 Report no. 1," 7 March 1946, p. 6. V2/NASM. Newell (1953), pp. 13, 24, has noted that the WAC was considered, but its small payload dampened the panel's enthusiasm for it. On the comparative costs of these missiles, see "Vehicles for High Altitude Research," 3 October 1947. Agenda item for 8th meeting of JRDB Guided Missile Committee. Box 198, folder 25; also in Box 240 as UAT 32/2.1 and UAT 32/2, JRDB/NARA. The Neptune was originally estimated to cost $100,000 each in 1946. See E. H. Krause, "Record of Consultative Services," 29 October 1946, p. 2. Box 98, folder 4, NRL/NARA.

28. Homer E. Newell, "The Place of the Navy Viking Rocket in Current Upper Atmo-

sphere Research . . ." 16 February 1949. Box 100, folder 19, NRL/NARA.

29. "The WAC, for example, flashed through the strata of interest too rapidly for sensors to respond; for upper-atmosphere physics and astrophysics, the duration at trajectory's peak, and our ability then to control the vehicle's attitude for instrument pointing, was marginal." Pickering and Wilson, (1972), p. 411. On this occasion and several in 1946 and 1947, Pickering argued that balloons were superior to rockets for meteorological research in all respects save for altitude.

30. As early as August 1946, the JPL ORDCIT program identified a proposed "V-2 WAC Missile" as a V-2 containing a WAC Corporal in its nose. The Bumper Project, proposed by Martin Summerfield of JPL, was authorized by Army Ordnance in October 1946, and in October 1947 Army Ordnance identified it to the JRDB as a program designed to provide information about two-stage rocket systems and associated technical and logistical problems facing long-range ballistic missile behavior. Ordnance testified that it had "no high altitude sounding rocket as such. The vehicles were developed for the purpose of checking theoretical calculations in the development of tactical missiles." See "Vehicles for High Altitude Research," 3 October 1947, p. 5. Agenda item for 8th meeting of JRDB Guided Missile Committee. Box 198, folder 25; also in Box 240 as UAT 32/2.1 and UAT 32/2, JRDB/NARA. See also H. J. Stewart, "Preliminary Considerations Regarding the Proposed V-2 WAC Missile," ORDCIT Memorandum 4–16, 16 August 1946. JPL/L; and Malina (1971), p. 375. Bumper was expected to achieve altitudes in excess of 450 kilometers at a cost of $500,000 per flight, much of the cost going to refurbishing the aging V-2s and to detailed engineering designs of the two-stage system by Douglas Aircraft. " 'Bumper' Missile in Public Showing," *Aviation Week* (24 July 1950), p. 17.

31. "V-2 Panel Report no. 20," 2 May 1949, p. 6. V2/NASM. Notices of Bumper appeared, for instance, in *Science Digest* (June 1949), pp. 76–77: "What Happened to WAC Corporal?" The first flight at the Air Force's long-range test facility in Florida was of a Bumper. It was a public showing, noted in *Aviation Week* (24 July 1950, p. 17). It was also

identified in Siry (1950), p. 420, and noted in Green (1954), p. 41. In May 1949 Bumper no. 5 achieved a world record altitude of 390 kilometers. The scientific results were reviewed in Newell (1953), p. 220.

32. Also as in the WAC, the fuel and oxidizer were pressure-fed into the combustion chamber, thus obviating the need for complex and heavy turbopumps as used on the V-2. Fraser (1948), p. 5. The fuel commonly consisted of 65 percent aniline and 35 percent furfuryl alcohol, to allow for low temperature operations since aniline freezes at 264K.

33. On its origins, see Van Allen, Townsend, and Pressly (1959), p. 55.

34. See R. E. Gibson to Files, 17 and 18 April 1946. "R. E. Gibson Chron Files, 1946." APL Archives, noted in Dennis (1991), n. 31.

35. H. B. Hutchinson, "Extension of the Navy Upper Atmosphere Research Program beyond the Period Covered by Presently Available Rocket Vehicles," 7 May 1947. UAT 24/1 [annex C to UAT 3/3], Box 239, folder 3, JRDB/NARA.

36. "Signal Corps Engineering Laboratories Upper Atmosphere Program FY-48," n.d., 30 September 1947. UAT 30/3.2.2, Box 240, JRDB/NARA.

37. The total cost of $370,000 included $100,000 from ORI for NRL's Aerobees. Van Allen, Fraser, and Floyd (1948), p. 57; Van Allen, Townsend, and Pressly (1959), p. 57.

38. Van Allen, Fraser, and Floyd (1948).

39. See J. M. Bridger and M. Pozinsky, ". . . Joint APL/NRL Problems in Utilizing XASR-1 . . ." (Record of Consultative Services), 22 January 1947. Box 99, folder 5, NRL/NARA.

40. Beyond aerodynamic changes, such as replacing the right circular cone of the WAC with a continuously curved cone, the boosters still had to be designed and more modifications made to the basic APL specifications for NRL's share of the vehicles. See, for instance, M. W. Rosen (Record of Consultative Services), 25 April 1947. Box 99, folder 6, NRL/NARA.

41. Van Allen, Townsend, and Pressly (1959), pp. 58–59.

42. In July, Ordnance warned that "if WSPG should be shut down as a result of some serious mishap with an uncontrolled missile, the entire missile program would be set back

as much as 2 years." "V-2 Panel Report no. 11," 15 July 1947, p. 12. V2/NASM. On the Juarez incident, see Rosen (1955), pp. 38–39; and Ordway and Sharpe (1979). On the safety measures taken, see Chap. 7.

43. See R. E. Gibson to Chief, Bureau of Ordnance, "The Aerobee Program of APL/JHU," 11 May 1949. Box 100, folder 21, NRL/NARA.

44. Ibid., R. E. Gibson to Chief, Bureau of Ordnance "The Aerobee Program . . ."

45. Starting in 1948, ONR funding, and therefore explicit NRL line-item support for the NRL upper atmospheric program (exclusive of Viking) beyond fiscal 1949, was dropping. See Homer E. Newell to Code N1121, "Past and Future Vehicle Requirement for the NRL Upper Atmosphere Research Program," 4 May 1948, attached routing slip commentary. Box 100, folder 14, NRL/NARA. On NRL's overall problems with stable funding, see Green and Lomask (1970), pp. 11–12. See also England (1982), and Sapolsky (1990).

46. H. Newell, "The Place of the Navy Viking Rocket in Current Upper Atmosphere Research," 16 February 1949. Box 100, folder 19, NRL/NARA.

47. Conclusions based on statements in oral histories of Krause, Van Allen, Tousey, and others, SAOHP/NASM. See also H. E. Newell to Code N1121, "Past and Future Vehicle Requirement . . ." 4 May 1948. Box 100, folder 4, NRL/NARA.

48. R. E. Gibson to Chief, Bureau of Ordnance, "The Aerobee Program of APL/JHU," 11 May 1949. Box 100, folder 21, NRL/NARA.

49. Ibid.

50. David A. Anderson, "Missile Program Depends upon What U.S. Can Afford," *Aviation Week* 60 (15 March 1954), p. 86.

51. Ibid. On how this decision reflected APL policy, see Dennis (1986; 1991).

52. On the expected decrease in military funding, see Forman (1987). A. G. Noble (Office of the Chief of the Bureau of Ordnance) to Chief of Naval Research, 15 June 1949. Box 100, folder 24, NRL/NARA.

53. Van Allen OHI, p. 276rd.

54. On Van Allen's fate, see Chap. 14 and Van Allen (1990). See as well Lorence Fraser OHI, pp. 57–58; and Van Allen (1983), p. 21ff., where he states that he left APL in December 1950. On changes at APL, see Dennis (1990, 1991).

55. See "Panel Report no. 27," 31 January 1951, p. 14. Improved Aerobees are discussed in "Panel Report no. 34," 29–30 January 1953, pp. 10–12. V2/NASM. Van Allen, Townsend, and Pressly (1959), p. 60.

56. James Van Allen to the author, 30 May 1990.

57. Rosen (1955), pp. 55–56. NRL followed a Navy code that identified its HASR-2 (High Altitude Sounding Rocket 2) vehicle as Neptune within ONR's upper atmosphere program, called Cosmos, and its overall rocket program, designated Zenith. The Navy also designated Aerobee as Venus, its XASR-1 rocket. See J. M. Bridger to Code 1320, 13 December 1946. Box 98, folder 4, NRL/NARA. This Navy code sheet was limited to ONR programs, and so did not include BuAer's High-Altitude Test Vehicle program for a satellite launch system, which was contracted to the same manufacturers for Neptune. The relationship of Neptune to the short-lived HATV program requires further investigation. Green and Lomask (1970), chap. 1, identify the program, and Hall (1963), p. 411, provides valuable contextual material.

58. E. H. Krause, "Record of Consultative Services," 19 February 1946. Meeting with L. M. Slack, J. J. Baranowski, and O. P. Swecker, 18 February 1946. Box 32, NRL/NARA.

59. Martin was an established airframe manufacturer, and Reaction Motors was an outgrowth of the pioneering Experimental Committee of the American Rocket Society. It had been the first American firm to build and refine a regeneratively cooled liquid-fueled rocket motor and had produced successful liquid JATO motors during wartime. The origins of Reaction Motors are noted in Winter (1983), pp. 14, 85. On the contracting, see L. M. Slack, "Facts Supporting the Justification for the Development of the Neptune Rocket" (GM 25/1) attached to S. D. Cornell (GMC/JRDB), "Navy Neptune Rocket" (25/1/1) 12 June 1947. JRDB/NARA; and F. W. MacDonald (NRL) to Chief, ORI, "Awarding of Contract with Glenn L. Martin Company for High-Altitude Sounding Rockets—recommendations for," 26 July 1946. Box 98, folder 3, NRL/NARA.

60. M. W. Rosen, "Consultation with Dr. Hall . . ." (Record of Consultative Services), 23

August 1946, pp. 1–4. Box 98, folder 3, NRL/NARA.

61. Rosen's relationship to JPL and Caltech is discussed in Clark B. Millikan to F. W. MacDonald, 10 June 1946; Millikan (acting director, Guggenheim Aeronautical Laboratory, Caltech) to J. M. Bartko, 3 September 1946; and F. W. MacDonald to J. G. Bain, 14 August 1946. Box 98, folder 3, NRL/NARA. See also Rosen (1955), p. 28.

62. J. M. Bridger, Record of Consultative Services, 30 April 1947. Box 99, folder 6, NRL/NARA. See also Rosen (1955), pp. 28, 72.

63. See, for instance, the original schedule in C. H. Smith, "Contract Negotiations with Glenn L. Martin, Co." (Record of Consultative Services), 1 August 1946. Box 98, folder 3, NRL/NARA. The first Reaction Motors prototype survived five test firings before it blew up. Rosen (1955), pp. 56–62, provides a detailed accounting of the evolution of the Viking motor.

64. Rosen (1955), pp. 71–72.

65. The first Viking firing is described in Rosen (1955), pp. 73–93. See also Milton Rosen and James Bridger, "Rocket Research Report no. 1: The Viking no. 1 Firings," NRL Report 3583, 19 December 1949, p. 3, Table 1; C. P. Smith (1954), Viking 1 entry.

66. "Panel Report no. 21," 1 September 1949, p. 3. V2/NASM.

67. Following a 1947 launch of a V-2 from the deck of the *Midway*, the Navy refitted the *Norton Sound*, a seaplane tender, to be an experimental "guided missile ship" to demonstrate the feasibility of firing tactical ballistic missiles from shipboard. See Rosen (1955), pp. 96–99. On the Navy's interests in shipboard ballistic missile firings, see Sapolsky (1972), and on their lack of commitment, see MacKenzie (1990), pp. 134–35. The public rationale given for the use of the *Norton Sound* was that it enabled researchers to take cosmic-ray measurements from another geomagnetic latitude even though it was also public knowledge that the test demonstrated the Navy's interest in firing missiles from shipboard. C. P. Smith (1954) and Rosen (1955), chap. 8, recount experiences aboard the *Norton Sound* regarding this test.

68. See, for instance "V-2 Report no. 14," 6 February 1948. V2/NASM; and Code 3420 to All Section Members, 2 September 1948. Box 100, folder 16, NRL/NARA.

69. Viking's servo-controlled gimbal and turbine-fed roll stabilization jet systems were of enough value to guided missile research, claimed ONR, that BuAer incorporated them into its "power plants development program." S. D. Cornell (GMC/JRDB) "Navy Neptune Rocket," 12 June 1947. GM 25/1/1, with attached justification of Neptune by L. M. Slack (GM 25/1), p. 2.

70. Newell to Krause, 15 July 1947. EHK/NASM.

71. Col. Millard C. Young, AAF, deputy member of Upper Air Panel to Secretariat, JRDB, 14 July 1947. UAT 31/1 (dated 29 July 1947), Box 240, JRDB/NARA. The attached proposal, "Recommendations of the AAF on Upper Atmosphere Research Vehicles," argued that more than one vehicle had to be developed and that the MX series was best. On the cancellation of the MX series, see Beard (1976), pp. 55–7; and J. Neufeld (1990), p. 48.

72. Capt. H. B. Hutchinson, USN (UAT/Committee on the Geophysical Sciences, JRDB) to C. S. Piggot (Exec. Dir., C. on Geophysical Sciences), "Vehicles for Higher Altitude Research," 29 August 1947. UAT 32/1.2, Box 239, folder 3, JRDB/NARA. On Panel membership see JRDB Minutes file, entry for the fifth meeting, 5 September 1947. UAT 3.5, Box 239, folder 5, JRDB/NARA.

73. J. Neufeld (1990), p. 47.

74. Committee on Guided Missiles to Committee on the Geophysical Sciences, RDB, 23 December 1947; attached to Karl F. Kellerman (GMC) to CGS: "Vehicles for Upper Atmosphere Research," 17 December 1947. Box 198, folder 25, JRDB/NARA.

75. See, for instance, L. M. Slack to Committee on Guided Missiles, 6 August 1947: "Neptune Research Vehicle—Status Development of," UAT 33/1, Box 240, JRDB/NARA.

76. Newell asked for $300,000 to buy 20 additional Aerobees starting in April 1949, and $2 million to buy a second round of 15 Vikings, starting in January 1951. CAM (Code 1120), noted on routing slip as discussed with JMM (J. M. Miller) and N3420 (Newell), attached to Homer E. Newell, "Past and Future Vehicle Requirements . . . ," 4 May 1948. Box 100, folder 14, NRL/NARA.

77. "Extracts from V-2 Report no. 15," UAT 53/1, item 9, "Missiles for Upper Atmo-

sphere Research," 6 May 1948. See also GM 52/1 and 52/2, JRDB/NARA. See also "V-2 Panel Report no. 15," 1 April 1948, p. 10. V2/NASM.

78. He was impressed with the control characteristics of the MX-774, as well as with its separable warhead, which would be useful in AMC's Blossom program. See Marcus O'Day to Edwin L. Weisl, Special Counsel, Senate Preparedness Investigating Committee, Committee on Armed Services, 31 December 1957, pp. 2–3. MOD/AFGL.

79. Gallery, Richardson, and Toftoy to Frederick L. Hovde, "Report of Ad Hoc Working Group on MX-774 as an Upper Air Research Vehicle," 23 March 1949. GM 25/12, item 10 on the Agenda of the 16th meeting of the Committee on Guided Missiles. RDB/NARA.

80. O'Day was responding here to Edwin L. Weisl's questions on behalf of Lyndon Johnson's Senate committee inquiry into problems in the American ballistic missile program in the highly charged wake of Sputnik. He felt that the MX-774 had been vindicated in Convair's Atlas. Marcus O'Day to Edwin L. Weisl, Special Counsel, Senate Preparedness Investigating Committee, Committee on Armed Services, 31 December 1957, pp. 2–3. MOD/AFGL. Rosen (1955), p. 102, notes that Gallery, a Viking enthusiast, was intimately involved with preparations for the first Viking shipboard test. Edmund Beard (1976), pp. 84–85, identifies Richardson, chairman of a guided missiles panel within the Air Force Office of the Deputy Chief of Staff for Operations, as hardly a champion of long-range surface-to-surface weaponry, preferring air-to-ground missiles. On the debate in 1949 between the MX-774 and the Viking, see also Beard (1976), pp. 63–67; and on the 1957 hearings, see Green and Lomask (1970), pp. 196–97. See also J. Neufeld (1990), pp. 44–50.

81. Koppes (1982), p. 39.

82. J. Neufeld (1990), p. 49.

83. "Panel Report no. 28," 25 April 1951, p. 4, V2/NASM; and Homer E. Newell to Code 1100, 5 December 1949. Box 100, folder 25, NRL/NARA.

84. Rosen and Seddon (1951), p. 8.

85. J. C. Seddon registered laboratory notebook no. 304, 16 February 1951 entry, p. 86, with clipped report, J. C. Seddon "Proposal for Automatic Guidance of a Missile," 29 December 1950. HONRL.

86. Rosen and Seddon (1951), pp. iii, 5. Proposers of any ballistic missile system at the time were keenly aware that the greatest concern was accuracy, rather than range. But the accuracy claims were often wild. As Rosen and his staff pushed hard to demonstrate that Viking's conversion to military use was feasible, he tightened its accuracy tolerances, which strained the faith of his own staff. Nolan R. Best felt that Rosen's push to achieve a $+/-$ 150 meter accuracy at ranges up to 200 kilometers was highly unrealistic. See Nolan R. Best registered laboratory notebook no. 898, pp. 15–17. HONRL. Rosen's entrepreneurship is best appreciated in the political context described in MacKenzie (1990), chap. 3, "Engineering a Revolution." Koppes (1982) has shown that launch setup times for the Corporal were many times those Rosen claimed for Viking.

87. F. R. Furth to Chief of Naval Research, n.d. (circa 1952), found loose in J. C. Seddon, registered laboratory notebook no. 304. HONRL.

88. These included BuAer, BuOrd, and NRL, according to Sapolsky (1972), pp. 5, 16. On the Navy's refusal to fund NRL's request to convert the Viking, see Davis (1967).

89. Von Braun and Ordway (1975), p. 151.

90. On Viking's demise, see Green and Lomask (1970), pp. 11–12; Ben S. Lee, "Missile Super-Agency Fast Taking Shape," *Aviation Week*, 53 (30 October 1950), pp. 12–14; "Report Spells out Guided Missile Plan," ibid., vol. 52 (6 March 1950), pp. 12–13. On how Toftoy temporarily saved the Corporal from burial, see Koppes (1982), chap. 4. Hermes was canceled during this period. Noted by Green (1954), p. 45, and White (1952).

91. "1948 Rocket Will Double V-2's Record," *Popular Science*, (July 1947), pp. 75–77.

92. "Ship-Launched Rocket Zooms 106 Miles Up," *Popular Science* (August 1950), pp. 85–87.

93. "Rocket Away" and "Baby Bombs," in *Time* (22 May 1950). These notes also discussed the possibility of submarine-launched rocket bombs, something the Navy Bureau of Ordnance had been thinking about for some time,

in the knowledge that the Germans had experimented with undersea launches of solid-fueled rockets. There was even speculation about firing V-2s from U-boats. See Sapolsky (1972), pp. 3–4; Klee and Merk (1965), pp. 92, 105–07. In 1947 APL, for instance, won a contract to develop a submarine-launched ballistic missile, code-named Triton. Noted in passing in [APL] (1983), p. 47; and in Sapolsky (1972), pp. 16, 25 and, upon its demise in favor of Polaris, p. 35.

94. "Seagoing Rocket," *Life* (26 June 1950).

95. Indeed, UARRP members such as Van Allen, then at Iowa, could not even afford the Aerobee, let alone the Viking, and newer members of the NRL upper atmospheric effort, such as Herbert Friedman and his group, recall that the Viking's flight instabilities and noisy environment made it undesirable when the Aerobee was available. Frustrated with Viking 9's performance, Friedman recalled thinking that ". . . we were ready for Aerobees. We had had enough of big rockets by then." Friedman, et al., VHI no. 1, p. 37 (02:26:35:000).

96. In June 1952, BuAer gave NRL four additional Vikings for ballistic missile research, but withdrew support within one year and stopped planning for an upgraded, more powerful engine. Rosen (1955), p. 236; Green and Lomask (1970). Leak (1954), pp. 4–6, identifies how Viking 10 and 11 were useful in ballistic missile development.

97. Hagen (1963), p. 440. The politics behind the choice of Vanguard, which incorporated many aspects of Viking, are best examined in McDougall (1985), p. 121ff. Rosen has noted that much of Martin's staff working on Viking were part-timers from other Martin projects because Viking was never large enough to sustain a large and diverse full-time technical staff. See speech, American Institute of Aeronautics and Astronautics meeting, 28 October 1971, panel on Rocketry in the 1950s. Cited in Beard (1976), p. 66, n. 38.

98. W. H. Pickering to J. Van Allen, 18 April 1950. JVA/APL.

99. Van Allen to "All Panel Members," 2 May 1950. JVA/APL.

100. Dow to Van Allen, 13 April 1950. JVA/APL.

101. Whipple to Newell, 11 April 1950. JVA/APL; copy also in FLW/HUA. Whipple had recently applied to be Hulburt's successor as superintendent of NRL's optics division, upon Hulburt's request, and Hulburt knew he probably would not accept the position even if offered, which it wasn't. Nevertheless, Whipple maintained warm relations with NRL staff during this time. See Hulburt to Whipple, 10 January 1950; Whipple to Hulburt, 16 January 1950. FLW/HUA.

102. On Van Allen's inability to rally a consensus, see Van Allen to Landsberg, 26 April 1950. JVA/APL.

103. "Panel Report no. 26," 7/8 September 1950, p. 15. V2/NASM.

104. Comments of O'Day and others, ibid., p. 16.

105. O'Sullivan was an aeronautical research scientist acting as NACA observer. William J. O'Sullivan, Jr., "Report on Upper Atmosphere Research Panel Meeting of September 7 and 8, 1950," 19 September 1950, pp. 4–5. NACA/NHO.

106. See Van Allen OHI, and Van Allen (1990), p. 15. His stated reason for making repeated attempts to resign in late 1950 was because of his move to Iowa. The panel urged him to remain chairman, which he accepted. See "Panel Report no. 27," 31 January 1951, p. 2. V2/NASM.

107. Marcus O'Day's laboratory was eventually dissolved in April 1953 in order to merge the rocket program, which had been under his own electronics umbrella since 1946, with the geophysics program, which had come to AFCRL in the spring of 1948. The purpose of the merge was to create a unified operation. O'Day, however, lost out to Milton Greenberg and others in the reorganization. By 1950, O'Day was finding it harder to gain sympathy for a larger and extended program surrounding Blossom in particular. Marcus O'Day to Commanding General AMC, "Plans, Upper Air Research," 17 June 1949. MOD/AFGL.

108. "Panel Report no. 26," pp. 15–16; O'Sullivan, "Report on Upper Atmosphere Research Panel Meeting . . . ," 19 September 1950, pp. 5–6.

109. O'Sullivan, ibid., p. 6.

110. Ibid., p. 4.

111. Ibid., p. 6.

112. "Panel Report no. 27," 31 January 1951, p. 5. V2/NASM. The handbook did not appear.

113. See "Miscellany," *Physics Today* 6, no. 5 (July 1953), p. 19. The functions of the RDB

were absorbed into the office of the Assistant Secretary of Defense for Research and Development. Daniel Kevles discussed the ineffectiveness of the RDB as a regulatory agency in a paper, "The Politics of Science," delivered at the Contemporary History Seminar, 21 November 1989, National Air and Space Museum. The later years of the UARRP, which lasted until the early 1960s, are briefly outlined in Newell (1980), chap. 4. Green and Lomask (1970), p. 14 n. 25, confuse the UARRP with a late 1950s committee in the National Academy of Sciences's Space Science Board.

# Part 2. Science with a Vengeance

*Almost all of the funds for the various investigations have been provided by the United States Department of Defense.*

—The Rocket Panel (1952).[1]

When the V-2 rockets finally ran out in 1952, after 67 flights, the UARRP continued coordinating the use of the remaining few Vikings and the more numerous Aerobees, along with a growing arsenal of new smaller sounding rockets and a host of tiny solid-fueled rockets carried aloft by balloons. Although UARRP members feared for their professional survival at the outbreak of the Korean War, when all claims of relevance centered around the practical problems of making ballistic missile systems work, the rate of White Sands firings during the early to mid-1950s was affected more by vehicle performance than by war mobilization, although budgets were far from stable both prior to and during this period.[2] (figure, p. 195)

Deeply embedded in the world of guided missile research and development and working in groups led by the various institutional representatives to the UARRP were the scientists, engineers, and technicians building devices that could explore the earth's outer atmosphere, the sun beyond it, and the interaction between the two. We have already identified the questions these groups attacked and have established that although the problems themselves existed before the V-2 era, those who pursued them with sounding rockets were not a part of the disciplines traditionally concerned with the upper atmosphere.[3]

We now examine more closely UARRP research on solar spectroscopy, cosmic rays, and atmospheric and ionospheric physics, keeping in mind the questions raised in the first part of this work. In particular, we wish to determine whether this new group of workers represented a fundamental reorientation of the rationale for performing scientific research on the upper atmosphere, a reorientation stimulated by the emergence of the long-range ballistic missile.[4]

Our approach will be thematic, and thus we will focus only on the details that best typified life in each area. For solar research, it will be the technical choices and the attempts of astronomers to become involved, because this was the area that first attracted their attention and one in which NRL, APL, and Colorado applied diverse instrumental styles to detect the ultraviolet solar spectrum and to bring it back to earth safely.

In the area of cosmic-ray physics, APL and NRL became the exclusive practitioners after Princeton dropped out in the first year, and few on the outside expressed interest in joining in. Despite the enormous talent and technology brought to bear on cosmic-ray problems, the obstacles were daunting and the effort soon metamorphosed into geophysical studies. Thus the story here is mainly one of limited application and the choices practitioners made as they migrated into more lucrative applications of rocketry or into more efficient means of pursuing cosmic-ray physics.

All the services and most of their contractors continually performed measurements from rockets of pressure, temperature, and composition, and developed techniques for studying radio propagation in the ionosphere

**193**

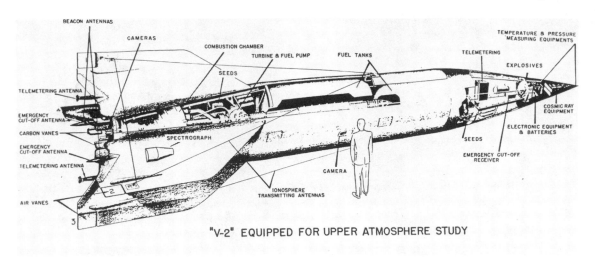

"V-2" EQUIPPED FOR UPPER ATMOSPHERE STUDY

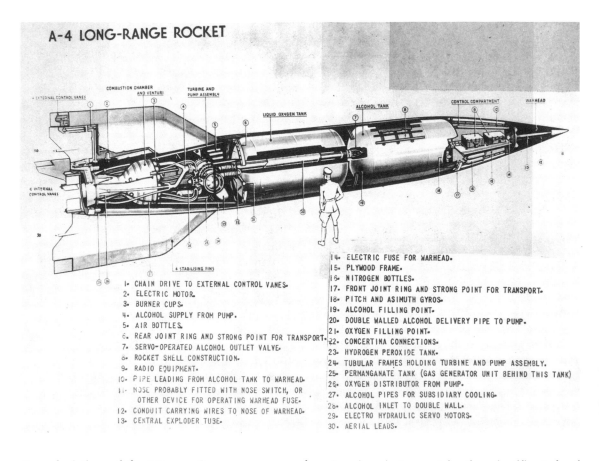

## A-4 LONG-RANGE ROCKET

1. CHAIN DRIVE TO EXTERNAL CONTROL VANES.
2. ELECTRIC MOTOR.
3. BURNER CUPS.
4. ALCOHOL SUPPLY FROM PUMP.
5. AIR BOTTLES.
6. REAR JOINT RING AND STRONG POINT FOR TRANSPORT.
7. SERVO-OPERATED ALCOHOL OUTLET VALVE.
8. ROCKET SHELL CONSTRUCTION.
9. RADIO EQUIPMENT.
10. PIPE LEADING FROM ALCOHOL TANK TO WARHEAD.
11. NOSE PROBABLY FITTED WITH NOSE SWITCH, OR OTHER DEVICE FOR OPERATING WARHEAD FUSE.
12. CONDUIT CARRYING WIRES TO NOSE OF WARHEAD.
13. CENTRAL EXPLODER TUBE.
14. ELECTRIC FUSE FOR WARHEAD.
15. PLYWOOD FRAME.
16. NITROGEN BOTTLES.
17. FRONT JOINT RING AND STRONG POINT FOR TRANSPORT.
18. PITCH AND ASIMUTH GYROS.
19. ALCOHOL FILLING POINT.
20. DOUBLE WALLED ALCOHOL DELIVERY PIPE TO PUMP.
21. OXYGEN FILLING POINT.
22. CONCERTINA CONNECTIONS.
23. HYDROGEN PEROXIDE TANK.
24. TUBULAR FRAMES HOLDING TURBINE AND PUMP ASSEMBLY.
25. PERMANGANATE TANK (GAS GENERATOR UNIT BEHIND THIS TANK)
26. OXYGEN DISTRIBUTOR FROM PUMP.
27. ALCOHOL PIPES FOR SUBSIDIARY COOLING.
28. ALCOHOL INLET TO DOUBLE WALL.
29. ELECTRO HYDRAULIC SERVO MOTORS.
30. AERIAL LEADS.

Navy depictions of the V-2 as a German weapon and an American instrument for the scientific study of the upper atmosphere. Top: U.S. Navy photograph, release date: 22 October 1948. NASM SI 87–8424. Bottom: NASM A5048-C.

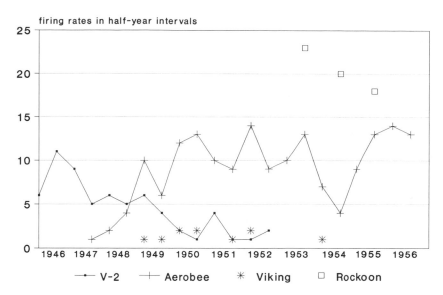

Rocket firing rates in six-month intervals (January to June; July to December), compiled from the UARRP minutes, from Smith (1954), and Smith and Pressly (1959). The decrease in the V-2 firing rate after 1946 reflected the Army's decision to increase preparation times, and the large dip in Aerobee firing rates after 1953 was caused by the Signal Corps' two year hiatus in upper atmosphere firings, as well as growing technical problems with aging first-generation Aerobees. The rapidly increasing rate starting in 1955 included improved Aerobee-Hi firings, supported by pre-IGY activities.

and the nature of the ionosphere itself. But these were also well-established activities in atmospheric and ionospheric physics before the use of rockets, where indirect techniques—mainly the use of radio and radar probes, acoustic probes, and the observation of high-level atmospheric activity such as meteor trails, noctilucent clouds, and the aurorae—had provided much information. Thus those who performed in situ studies of the atmosphere in the context of the rocket had to establish themselves as new authorities on the uppermost regions of the atmosphere. We will therefore concentrate on the attempts of UARRP members to establish authority in these areas.

The workers also had to establish an identity, and the one that emerged can be seen in the audiences the rocket groups wished to address. In subsequent chapters, we will see how different scientific communities responded to the rocketry results and the degree to which those conducting research with rockets ultimately identified themselves with those communities, adjusting their self-image in the process.

Underlying the following discussion will be the constant reminder, as expressed by the UARRP quotation at the outset of this introduction, that upper atmospheric research with rockets existed as a part of a permanent war economy. The first part of this book explored this condition by examining intentions, rhetoric, and proposals. Now we examine actions.

## Notes to Part 2

1. [The Rocket Panel] (1952), p. 1027.
2. On problems with the aging first-generation Aerobees and reduced launch-crew control at the launch pad, see "Panel Report no. 28," 25 April 1951, p. 11; and "Panel Report no. 29," 14–15 August 1951, p. 20. V2/NASM. See also Newell (1959b), p. 242. Research and development budgets within the military fluctuated in the late 1940s, recovering and strengthening

only in the early 1950s. Forman (1987), Fig. 1, p. 153. This period has also been identified by Roland as one where Congress imposed "painfully low budget ceilings" overall. Roland (1985), p. 266. Fragmentary budget figures for NRL during this period show a small drop in the late 1940s in research and development funding estimates and projections, with stability returning in the mid-1950s. Radio Division I upper atmosphere research program projections made in March 1949 for 1949/1950/1951, however, requested an increase in support, although no records have yet been recovered that indicate if their budget was actually increased. See records in HONRL, especially Code 170 to Code 100, 27 May 1947. "Tentative Budget for Direct Costs for Fiscal Years 1948 and 1949;" [Budget and Management Analysis Office], 28 March 1949. "Analysis of 1950 and 1951 Estimates by Source." I am indebted to David van Keuren and Dean Bundy for assembling the materials from which these conclusions are drawn.

3. Needell (1987) has made this point as well for the Sputnik period, just as Edge and Mulkay (1976) have observed this characteristic among the first practitioners of radio astronomy in the 1940s.

4. Paul Forman (1987) has identified such a shift in his important study of the militarization of the American physics community.

# The Sun from Rockets

*The greatest obstacle in the study of the upper atmosphere, is undoubtedly the lack of our direct and precise knowledge of the energy distribution in the near and the extreme ultraviolet radiation of the sun. . . it is highly probable that there are intense line emissions in the extreme ultraviolet . . . The so-called radio fade-outs associated with bright solar eruptions point strongly to such emissions. No direct evidence has, however, yet been obtained. It has been suggested that if solar spectrograms could be taken above the ozone absorption layer, much knowledge regarding ultraviolet radiation from the sun could be gained.*

—S. K. Mitra, 1948.[1]

*I am very sorry indeed that your first instrument didn't work; but as I may have mentioned before it seems that a tough skin and a strong back are the chief requisites of a rocket experimentalist.*

—James Van Allen to Jesse Greenstein, 4 June 1947.

There were many good scientific reasons to send spectrographs above the earth's absorbing atmosphere, yet the first people to conduct solar research with rockets had not formulated any of these scientific goals beforehand. Nonetheless, they were well qualified to build spectroscopic devices for rockets and were, as well, working at institutions ready to support such expensive activities. The techniques they adapted to the spectroscopic study of the sun from rockets included both dispersive and nondispersive optical systems, as well as photographic, electronic, and photochemical detectors.

Although each group had the advice and counsel of a few well-established astronomers who were interested in participating in this early work, few followed the advice of these outsiders. They preferred to be guided by their own sense of what would work best on rockets. This chapter deals with the obstacles each rocket group faced, and the technical choices they made as they designed spectroscopic instruments for rocket flights. The reactions of astronomers to the data gained as a result of these choices and their attempts to become actively involved in rocketry will demonstrate the initial insularity of the activity.[3]

This chapter concentrates on dispersive optical systems and photographic recording. The two chapters that follow treat how technical limitations to this technique led to the development of servo-feedback stabilization systems and to the application of nondispersive electronic detectors to reach the deep ultraviolet, an unknown realm that lay as both a technical challenge and a region where fundamental questions about the ionosphere would be answered.

## Dispersive Optics and Photographic Detectors: Tousey's Design

The first instruments developed by both the NRL and APL groups used dispersive optics and photographic emulsions, along the lines established by Erich Regener. Unlike Regener, American researchers had easy access to high-quality reflection gratings and decided to exploit them as the simplest and most direct means of obtaining a spectrum of the sun. The first successful design emerged at NRL.

On the morning of 26 February 1946, Richard Tousey called his NRL team together to discuss his design decisions. He told F. S. Johnson and William Baum that they would be building an all-reflecting grating spectrograph of the classical Rowland type, that photographic film would be used, and that as an entrance aperture they would employ "beads or cylinders of [lithium fluoride] instead of a slit and diffuser."[4] Tiny optical beads or cylinders would have an extremely wide-angle view, and thus would maximize the chance that the spectrograph, in the nose of the spinning and tumbling V-2 rocket, would catch enough sunlight to expose the film properly. Tousey knew from his wartime work on night vision that "a sphere would have a very wide field of view and still make some kind of point image."[5] Gathering enough sunlight was a problem second only to retrieving the film intact (figures, pp. 199, 226).

With his own staff duly informed, Tousey then led them into a larger meeting with Ernst Krause, Charles Strain, and J. J. Oberly to present their united front. Krause, Tousey knew, had already brought in Harvard astronomer Donald Menzel to advise on spectrograph design. Menzel would not build anything himself, but he suggested that NRL consider developing all-reflecting optical systems using gratings and mirrors; he also said that doubly dispersing devices using two gratings, one feeding the other, would be needed to eliminate scattered visible light.[6] Tousey knew that Menzel's ideas had merit; doubly dispersing systems, although inefficient, still offered the best chance of capturing high-quality spectra free from scattered visible light, but only if the spectrograph could be pointed directly at the sun. These designs, Tousey felt, were "impractical for the earliest flights" because of the space limitations in the warhead and, most important, because of the strict schedule they were working under. The group decided to keep these ideas in mind for later flights. For the moment, simply getting

a spectrograph ready for a June flight, by whatever means available, was top priority.[7]

Krause and Strain doubted that Tousey's radical lithium-fluoride beads would work; conventional slits provided the clearest line profiles.[8] They knew, too, that Menzel was critical of the bead design, fearing that the "result would be a mess." Menzel advised both Krause and APL's E. O. Salant, as well as E. O. Hulburt, Tousey's boss, and his own students (among them Leo Goldberg) that conventional slits and proper light collectors were essential for useful spectra. "Otherwise," Menzel feared, "it is hopeless." Above all, Tousey's design simply wasn't "proper."[9] Goldberg replied to his old teacher: "I agree that the Naval Research Laboratory's proposal is cockeyed."[10]

The APL group was influenced by Menzel's criticism, but Tousey stuck to his guns. Well aware of the pitfalls of the bead design but convinced that it was the best compromise, Tousey pushed forward on the most direct and simple path. The beads were the most likely way to get sunlight into the spectrograph; messy or not, they were going to get a glimpse of the ultraviolet spectrum of the sun (figure, p. 212). By late March, C. V. Strain had compared the flux densities expected for beads and for conventional slits and diffusers and found the former more efficient "by several thousand" if astigmatism were kept low.[11]

F. S. Johnson recalls that Tousey developed a carefully planned and logical experimental process to test 2 and 3 millimeter beads, manufactured by the Naval Gun Factory.[12] Photographic speed was a direct function of bead size, but spectral line definition would decrease with bead size. Johnson's problem was to find the best compromise.[13] While Johnson tested tiny optical beads, Tousey secured Rowland gratings from John Strong of The Johns Hopkins University's physics department in Baltimore. Strong was the leading provider of special Rowland gratings and also was a fount of knowledge on spectrographic design.[14] Although initially reluctant to take on the difficult job, he agreed to provide six 40 centimeter focal-length

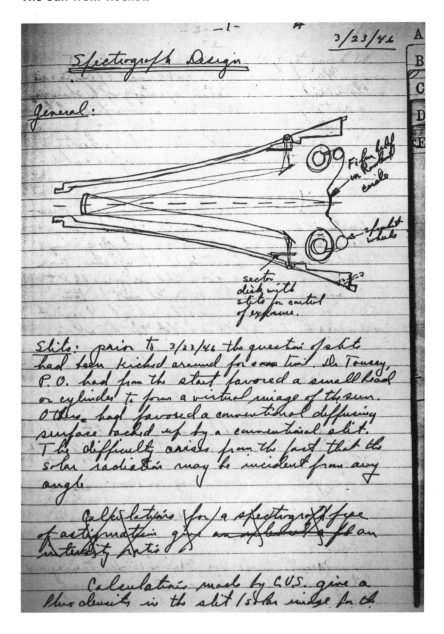

Krause's section distrusted Touseys's beads until Strain determined that the efficiency of a bead would be higher than a slit. In Strain's sketch, entrance beads sit on the skin of the spectrograph, and are indicated by small circles. Sunlight enters through the beads, is deflected by small folding mirrors, and falls onto the Rowland grating at the left. Dispersed sunlight from the Rowland grating then is focused onto film lying on the Rowland Circle. Detail from C.V. Strain registered laboratory notebook no. 6353, entry for 23 March 1946. HONRL.

gratings capable of imaging the far ultraviolet in the Lyman Alpha region, the first line in the resonance series of hydrogen named for Tousey's teacher, Theodore Lyman, and a suspected source of the radiation creating the ionosphere.[15] Tousey and the APL group eventually used some 40 Hopkins gratings over the next 10 years.[16]

The design of the spectrograph took shape in late February. The 40-centimeter grating, which was the longest possible focal length dictated by the geometry of the V-2 nose cone, would be placed in the nose, looking down. Sunlight would be admitted through either one of two beads, sitting on opposite sides of the grating axis. Each beam would bounce off a small mirror up to the grating, and then dispersed light would fall onto properly sensitized 35-millimeter film directly below the grating on the axis of the rocket (figure, p. 199). Tousey wanted to cover the solar spectrum from 100 to 330 nanometers, centered

Typical setup for calibration tests of the NRL spectrograph, circa November 1946. A set of spinning sectors (on table at left) could simulate either different roll rates for the V-2, different exposure times, or a combination when used with the spectrograph's own internal shutters. An arc source sits directly behind the mounted sector. Francis S. Johnson registered laboratory notebook no. 6233, p. 52. F. S. Johnson collection. NASM.

at 215, a region well within the range of standard Eastman Kodak short-wave spectroscopic emulsions.[17] The resulting spectrum would have an average dispersion of 4 nanometers per millimeter, low by astronomical solar spectroscopy standards of the day.

Baum did the detailed design of the mechanical structure of the spectrograph. He and Tousey prepared a contract with Baird Associates, Inc. of Boston, another Theodore Lyman contact, to build the first three samples.[18] Strain and Oberly in Krause's section, working with Baum, were responsible for integrating the spectrograph into the warhead and for determining the electrical and telemetry requirements for its motorized film-advance mechanism. As members of Krause's own section, they were the ones who were eventually to take the spectrograph and warhead to White Sands for the final integration tests and firing. Baird retained the contract through the next three years, eventually providing six spectrographs. After that, a number of spectrographs on the Baird design were built in the NRL shops.[19]

By the end of April, Johnson had Rowland grating samples from John Strong and was busy determining the best combinations of bead and grating for a range of exposure times (figure, p. 200). By early June and the arrival of the NRL warhead in their shops, Tousey's team had succeeded in reaching Lyman Alpha and beyond in an intensive series of laboratory tests of the spectrograph. Tousey declared the spectrograph ready.

## Multiple Agendas at APL

At the 27 March V-2 Panel meeting, Van Allen announced that APL would have a spectrograph ready for its own late June flight; Krause met this challenge saying that NRL would be flying Tousey's spectrograph on its first flight, scheduled for early June.[20] Both groups knew that the first one to be ready would be the first one to fly. So there was a race to get ready, but the real race was simply to meet flight deadlines.

NRL's launch slipped to 28 June, and its fate was not a happy one for Tousey (see Chap. 8). The fact that the entire warhead disintegrated did not bode well for the retrieval of photographic film. This caused Van Allen to cancel the planned first flight of the APL spectrograph on 30 July, dampened the

APL prism spectrograph and armored steel film cassette built by Bausch and Lomb and designed to measure the vertical ozone distribution. The hemispherical entrance aperture sits at the left, feeding light into the narrow spectrograph tube and finally into the large cassette. Applied Physics Laboratory, JHU/APL archives.

enthusiasm of a number of astronomers who were then planning to build instruments to fly on later V-2s, and prompted Krause to have his own staff design a photoelectric spectrometer that could telemeter spectroscopic data back to earth (see Chap. 13).

The spectrograph Van Allen pulled from the 30 July flight was one of three designs his group was then developing. Unlike Tousey, who focused on one design, Van Allen's group sought out different designs for different goals:

to examine the solar ultraviolet spectrum, to determine the vertical distribution of ozone, and to develop a spectroscopic altimeter.[21]

Van Allen wanted to move quickly on the third goal. He asked John J. Hopfield, a senior vacuum spectroscopist, to reexamine research he had conducted at Berkeley in the 1920s on the transmission characteristics of the atmosphere and apply it to the development of an ultraviolet altimeter requested by the Bureau of Ordnance for navigating long-

J. J. Hopfield and Harold Clearman (l to r) and the Bausch and Lomb spectrograph. A gas discharge tube (foreground) is placed in front of the spectrograph entrance aperture to demonstrate their testing procedure. Applied Physics Laboratory, JHU/APL archives.

range cruise-type missiles. Van Allen told his APL supervisors that an ultraviolet altitude meter would be able to measure the height of a missile in flight to an accuracy of +/- 8 kilometers. Either a cosmic-ray counter or a photoelectric UV meter could relay information to a ground station immediately "as a basis for control commands."[22]

The ultraviolet device Van Allen had in mind consisted of a small low-dispersion slit-fed fluorite optical train with a fixed photoelectric detector centered on Lyman Alpha. It would be useful above 100 kilometers, and the varying intensity levels received "could be used to actuate the jet controls so as to maintain this altitude."[23] Van Allen described the dual purpose of his plans to Lawrence Hafstad, Tuve's immediate successor:

> As you know, I have no apologies whatever for working on investigations of no foreseeable application; but such a project will not divert any appreciable effort from our present program, since such evaluations will be essential by-products. And I feel that our thinking and effort will be sharpened up a great deal by having a project of vital military interest "on the ledger."[24]

Van Allen was keenly aware of the need to maintain an entrepreneurial vigilance at APL, a warborn organization devoted to the development of guided missiles.[25]

The first spectrograph developed by APL closely fit Van Allen's description of an ultraviolet meter, because its simple Cornu-type transmission system was designed to work in the region 210–280 nanometers, where ozone absorbed strongly.[26] It differed in the choice of detector only, acting more as a prototype to test the overall concept. J. J. Hopfield would have preferred to use an all-reflecting system like the one Tousey was developing, because he and Hynek wanted to get at the ultraviolet spectrum of the sun to study the ultraviolet energy distribution in the solar chromosphere and corona.[27]

Through the spring of 1946, Hopfield designed the basic ozone spectrograph, assisted by Harold Clearman, a young assistant, and J.

Allen Hynek, an astronomer who had worked at APL during the war and was now a contract consultant to Van Allen's group. The first instrument was constructed by Bausch and Lomb Optical Company of Rochester, New York. The $9,000 spectrograph used a simple lithium-fluoride prism and a conventional slit system, with a light collector made of a lithium-fluoride diffusing hemisphere reminiscent of Erich Regener's design. Like Tousey's design, the APL film cassette was encased in a heavy steel cylinder.[28] The cylinder holding the exposed film was some 25 centimeters in diameter and the entire instrument was 60 centimeters long. The largest single section was the film holder, with the spectrograph a small crooked tube looking like a protruding appendage[29] (figure, p. 201).

The successful recovery of APL's 30 July 1946 flight hardware demonstrated that it was possible to retrieve photographic spectra. Van Allen therefore restored the Bausch and Lomb spectrograph to the roster for the next flight on 24 October.[30] The launch took place on schedule, reaching only 100 kilometers altitude because the rocket motors failed to perform properly. But the flight was successful, and most of the vehicle was recovered within two days of the launch, including a time-lapse motion picture camera. Parts of the fin-mounted spectrograph were also found, but the cassette lay hidden for five more days.[31]

The Bausch and Lomb spectrograph film yielded no data. Hynek, recently returned to the Perkins Observatory in Delaware, Ohio, was not happy with the outcome. Writing to Jesse Greenstein on 1 January 1947, Hynek said that the film was completely fogged and complained bitterly that the APL staff had failed to test the instrument properly before launch. He therefore warned Greenstein, who was then preparing his spectrograph for an APL flight: "Well—see that they don't pull anything like that on you."[32] An April 1947 APL report stated that the spectrograph film cassette completed only a partial rotation so not all the film was exposed. Evidently, the exposure cycle was not completed either, and because a velvet light baffle was not extended across the entrance slot, everything in front

Detail at base of APL grating spectrograph first flown in April 1947. Although similar in overall design to Tousey's NRL instrument, the APL device used slits and external mirrors sitting on blocks to the right and left of the film box at the base. Each mirror sent light through downward facing slits to the Rowland grating at the peak of the spectrograph which then sent dispersed light to film sitting on the focal plane of the grating, the curved platen immediately above the armored film take-up cassette. The external mirror blocks were later replaced by one- and two- axis photoelectric homing mirrors to seek the sun. William Dow papers, University of Michigan. WGD/BHLUM.

of the velvet cover—the exposed film—was ruined.[33] The photographic version of the spectrograph was flown only once; later samples used photoelectric detectors.

Between its flight and the date of Hynek's unhappy letter, two copies of Hopfield's Rowland grating spectrograph were being constructed at APL. By 28 January 1947, one of them had been scheduled for launch aboard V-2 no. 22 on 1 April 1947.[34] Hopfield and Clearman later stated that they built these all-reflecting grating instruments "to avoid some difficulties that were experienced with lithium-fluoride lenses and prisms."[35] Very little is known about the details of the first APL grating spectrograph, however, because it was built completely within the APL laboratory. By early February 1947, Greenstein's own spectrograph was slated for integration in the midbody of a V-2, and the Hopfield-Clearman grating spectrograph was set for the warhead.[36]

APL's all-reflecting grating spectrographs, similar to Tousey's folded Rowland design, were flown on 1 and 8 April and 29 July 1947. Each had slits and progressively more sophisticated means of gathering sunlight; the 29 July sample was the first to have active one-dimensional servo-controlled homing mirrors. Like the NRL design, the APL device used two entrances placed symmetrically on either side of the optical axis of the grating, as well as standard 35-millimeter Kodak spectroscopic film wound into armored steel cassettes (figure, p. 203). APL also turned to John Strong for gratings and found that they could accommodate Strong's preferred models with a focal length of 50 centimeters, providing slightly higher dispersion (3.4 nanometers per millimeter) than Tousey's spectrograph.

As V-2 no. 22 was nearing its 1 April launch, there were last-minute crises at White Sands both in integrating Greenstein's instrument and in correcting problems with the APL grating instrument. The three-man APL field crew—consisting of Howard Tatel, Art Coyne, and S. Fred Singer—had their hands full integrating several cosmic-ray telescopes, spectrographs, nuclear emulsion plate stacks, and aerial reconnaissance cameras into the warhead and mating the warhead to the housekeeping instruments in the missile. They appreciated Greenstein's help during the hectic and frustrating process.[37]

The APL spectrograph was recovered easily after a successful flight to 128 kilometers, and within two weeks Van Allen could report to Greenstein that they had obtained "decent" spectra to about 235 nanometers with good resolution to 250 nanometers.[38] One week later, V-2 no. 23 soared to 102 kilometers carrying another APL grating spectrograph. Although the flight was successful, the spectrograph film drive started while the rocket was still on the launch pad, and all the film was exposed before the rocket was fired.

Jesse Greenstein with all-sky mirror developed in the Yerkes Optical Bureau circa 1946. Greenstein Collection, courtesy of the archives, California Institute of Technology.

## Jesse Greenstein's Yerkes Spectrograph

Hynek's job, in addition to helping APL staff, was to drum up interest among astronomers. By late April 1946, he found that Yerkes Observatory astronomer Jesse Greenstein was willing to design a system for astrophysical studies, on contract to APL.[39] Since Greenstein was the only outside professional astronomer to become directly involved with the design, construction, and flight of a solar spectrograph, his experience bears attention.

During World War II, Greenstein, with Louis Henyey, had formed the Yerkes Optical Bureau to develop optical systems for military use. They continued to develop optical systems after the war for astronomical use, as well as for medical research and diagnosis, while they both returned to pure astronomical pursuits as Yerkes junior staff astronomers.[40] Greenstein's (figure, p. 204) main professional interests were in both theoretical

and observational studies of physical processes in the interstellar medium. But his wartime experience left him with a taste for unusual challenges in optical design. Therefore, when he learned that V-2 flights were being planned at White Sands, and that Hynek was involved, Greenstein wrote to Hynek to find out more, and Hynek responded immediately.[41]

Greenstein considered many designs in his search for sources for ultraviolet photoelectric detectors and film-filter combinations that would eliminate stray visible light. At one point he considered using the exotic Pohl crystals Gerard Kuiper had discussed.[42] But he seemed unable to hit on the ideal combination of reliability and narrow-band resolving power. To make matters worse, the retrieval problems that plagued the program in June 1946 greatly limited his options. Greenstein and Kuiper considered building a photoelectric system, but Yerkes had no "com-

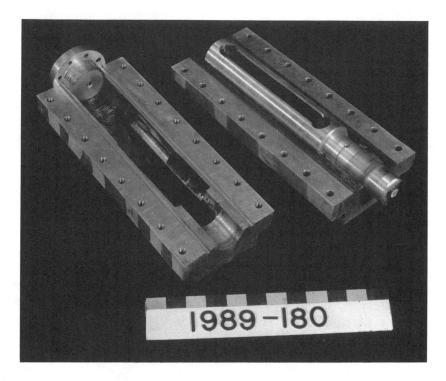

Jesse Greenstein's film holder had 3 centimeter thick aluminum walls. A single photographic plate sat on a 4.5 centimeter steel shaft that was designed to act as a rotating shutter. NASM accession 1989–180, SI 90–8780–0A.

petent electronics expert" who could build an electronic spectrograph, and they feared that it would take at least one year to design and test a reasonable device.[43] To most astronomers, photomultipliers were still exotica. Wartime needs had made photomultipliers available to astronomy for the first time, but their use was at yet experimental.[44] Still, through July, watching the progress of the NRL and APL flights, Greenstein thought only of photoelectric recording and the telemetry of data.[45] Meanwhile, there were budget, personnel, and institutional subcontracting rules to contend with before the $10,000 contract between APL and the University of Chicago was consummated. Greenstein found the procedure a frustrating "hurry up and wait" situation.

Greenstein's spirits lifted after APL's successful 30 July flight. He quickly returned to a simple photographic design, especially since he was unable to secure photoelectric detectors with quartz windows. He ended up choosing a straightforward quartz prism spectrograph design with a single photographic plate capable of detecting radiation longward of 180 nanometers. The spectrograph slit was fed by a diffusing quartz plate set into the skin of the rocket, and a mirror folded the light beam between the slit and collimator to make the instrument smaller. The rest of the instrument resembled a conventional astronomical spectrograph, in the shape of an irregular triangle with outer dimensions of 40 by 91 by 15 centimeters. Including timer and batteries it weighed about 45 kilograms, and so was too heavy to fit into the fin, especially with its heavily armored cassette (figure, p. 205). Greenstein's spectrograph was designed to provide clear line profiles at a dispersion of 3 to 5 nanometers per millimeter, an astrophysical priority not shared by any of the other rocket spectrograph designs.[46]

Although Greenstein was cordially received at APL, and Van Allen offered him every consideration, the Yerkes astronomer found that designing and building his spectrograph was easy compared with the red tape and bureaucracy he experienced in having it tested, integrated, and flown under security restrictions. Greenstein was not heartened by the failure of the Bausch and Lomb spectrograph, which convinced him that his device needed to be shake-tested at APL and all systems thoroughly tested in a vacuum and at low temperature. But he also began to wonder whether the whole effort was worth it.[47]

In late January 1947, as he prepared for a final testing schedule at APL, Greenstein hoped that Van Allen would let him put the instrument through a series of rough tests, not only on the shake table, but under the stress of a one-ton load "to see whether the [shutter] bearings can stand the resulting deformation."[48] On a snowy first week of February in Silver Spring, Greenstein found that his instrument failed even the shake test, which damaged one of the optical components and sent him on a mad dash to the Perkin-Elmer Corporation in Connecticut to obtain a replacement. Constantly pressed for time and hampered at every turn by security clearance problems at APL, Greenstein cried out to his friends back home: "There is no end to military secrecy!!"[49]

By the second week of February, Greenstein was en route from APL to White Sands on the Pennsylvania Railroad with his shaken but intact spectrograph. Finally, he had time to think over the past months, and his memories were not warm ones. He wrote to Otto Struve:

> After innumerable trials, difficulties, emergencies, and disasters, the instrument is now in good condition and adjustment. . . . The last two months have been extremely strenuous and trying for me, as well as for the many at Yerkes who helped me rush the instrument to completion.[50]

Unlike those who were fully within the sphere of rocket research at NRL and APL, Greenstein had to answer to his director, to his colleagues and students, and to his career. He was thankful for the help the Yerkes staff and Perkin-Elmer had provided. But he felt guilty about neglecting his observatory duties: he had not even begun to prepare lectures for a new graduate course or prepare

for an upcoming observing run at McDonald Observatory in Texas. This caused "continuous strain," and relief was not in sight.[51]

## At White Sands

There was certainly no relief at White Sands in February 1947. Greenstein found that mounting the spectrograph required the help of a machinist and draftsman, as well as materials for some unforeseen integration problems. Unlike NRL experimenters, Greenstein lacked the institutional support on site to get these last-minute details done. "Pappy" White, the Project Hermes site foreman, could not help, claiming it was not the responsibility of General Electric until the situation became an emergency. Eventually, APL and NRL personnel came to the rescue, and Greenstein reciprocated by helping the APL team make last-minute adjustments to their own spectrograph.[52]

V-2 no. 22 successfully flew to 130 kilometers on 1 April and, although the Hopfield-Clearman slit spectrograph worked, Greenstein's spectrograph failed. Only the plateholder was brought back to Yerkes. It was opened and the film developed, but the film strip was clear; it was never exposed to solar radiation. Greenstein telegrammed Struve with the unhappy news, speculating that "either the electrical circuit broke down, or the switch initiating the cycle did not operate on take-off."[53]

While Greenstein picked up the pieces of his long-neglected observatory activities, the 1 April telemetry record was analyzed at APL. On 4 June 1947, Van Allen relayed their findings. The timing motor did indeed operate, sending proper signals to the system at the designated times, but the motor driving the film cylinder, although activated, failed to budge the cylinder. Van Allen reported that he regretted not having had the time to give Greenstein's spectrograph a full mechanical test at APL: "I am very sorry indeed that your first instrument didn't work; but as I may have mentioned before it seems that a tough skin and a strong back are the chief requisites of a rocket experimentalist."[54]

But Greenstein had indeed had access to the shake table and had put his instrument through a severe test. He tested the motor, film holder, and batteries, and they all worked on the shake table.[55] He never fully reconciled the failure. "Tough skin" or "strong back" aside, Greenstein's future in rocket research now seemed unlikely.

### Greenstein Pulls Out

When he heard of the failure, Yerkes Observatory Director Otto Struve confided in Gerard Kuiper, "we have [endured] bravely through Greenstein, but the results have not been very encouraging."[56] Struve did not cancel the APL contract but insisted that it become secondary to Greenstein's regular observatory responsibilities. As a result, Greenstein continued to consult for APL and provided a design for a three-channel photoelectric UV spectrometer that was built at APL and flown successfully as an APL experiment.

Greenstein also helped Kuiper organize a planetary atmospheres symposium, held in September 1947, and invited Eric Durand from NRL and Harold Clearman from APL to report on the status of solar spectroscopy with rockets. He also agreed to have Clearman study astrophysics at Yerkes for several months, as a favor to Van Allen, to better analyze the spectroscopic data Clearman and Hopfield had gathered from the first rocket flights.[57] Van Allen felt the visit would be good for both institutions:

> On our side, it would tend to lift us out of the "strong back-weak mind" position in this field; and on yours it would open the way for continuation of this arrangement in connection with any further results that we may get in subsequent work.[58]

There would, however, be no further Yerkes interest: Greenstein had neither personal nor institutional reasons to continue to be an active participant in scientific rocketry. The pressures on him, in fact, dictated just the opposite because astronomical institutions had little sympathy for nontraditional instrumentation projects whose goals lay outside programmatic boundaries.[59] Greenstein's talents and experience in the Yerkes Optical Bureau stimulated him to build a rocket spectrograph, but his interests and traditional responsibilities quickly removed him from further involvement.[60]

## Astronomers on the Periphery

Watching closely both Greenstein and Tousey's progress was a small group of astronomers at Michigan and Yale who were keenly interested in scientific rocketry. Unlike Greenstein, they were not instrument specialists, so their methods of getting involved and their subsequent experiences were quite different. At the outset, they did not lack for enthusiasm.

In September 1945, still deeply involved in finishing war work, Leo Goldberg at Michigan confided his dreams to his former Harvard professor, Donald Menzel:

> If anyone asked you what technological development could, at one stroke, make obsolete almost all of the textbooks written in astronomy, I am sure your answer and mine would be the same, namely, the spectroscopy of the sun outside of the earth's atmosphere. . . . [I]t seems to me that we have reached the stage where at least serious developmental work along these lines is decidedly practical. As I understand it, the V-2 Rocket has attained a height of 60 miles, and with the host of control mechanisms that have come out of the war it should be possible to point the rocket at the sun within the required limits of plus or minus 1/4 of a degree. . . . I am just wondering if anyone is doing anything about it. I would like nothing more than to be involved in such a project, even if it meant shaving my head and working in a cell for the next ten or fifteen years.[61]

Goldberg was a specialist in solar physics at the McMath-Hulbert Solar Observatory in

Michigan. Under Menzel at Harvard, Goldberg had studied physical processes in the solar atmosphere and in gaseous nebulae and realized that Bengt Edlén's analysis of emission lines in the solar corona "provided the major scientific justification for putting solar telescopes above the earth's atmosphere."[62] Menzel would soon help to bring Goldberg and Yale's Lyman Spitzer together in Washington to see what could be done about sending solar spectrographs into space.

Lyman Spitzer, like Menzel and Goldberg adept as a theorist with strong sensitivities for observational problems, was interested in the physics of gases in interstellar space. During the war he had worked for the NDRC in New York and Washington on problems of underwater sound and undersea warfare.[63] As with many bright and ambitious young astronomers, the war had put Spitzer's research on hold, but at the same time gave him a passion for research on a scale far larger than he had dreamed of before. At war's end, Spitzer became a consultant for Project RAND, carrying out a theoretical analysis of conditions in the earth's upper atmosphere at altitudes in excess of 300 kilometers. He also provided RAND with a study of the types of astronomical observations that could be made from an earth-orbiting satellite by telescopes of various apertures.[64]

Spitzer returned to civilian life and his professorship in astronomy at Yale, but Goldberg did not enjoy faculty status at Michigan. He therefore felt that he "had lost five years in the war" and wanted to catch up and achieve some stability by becoming a full-fledged faculty member in "a decent position."[65] Knowing this and aware that the two young theorists were interested in rocketry, Menzel used his excellent contacts in Washington to help them get started, primarily at the Navy's Office of Research and Invention (ORI).[66] Goldberg and Spitzer had their own contacts through the remnants of the OSRD.[67] But among them all, Spitzer's ace was his old Yale professor, Alan T. Waterman, who was soon to be chief scientist at ORI.

Spitzer wanted to transform Yale astronomy into a modern, competitive astrophysics research group, and Waterman was looking for people just like Spitzer who represented the best scientific minds in the nation.[68] In March 1946, after talking about it with Waterman and Goldberg, Spitzer proposed that ORI fund an astrophysical research unit under his direction to be operated jointly by the Physics and Astronomy departments at Yale. Spitzer at first thought $10,000 per year from ORI was sufficient to establish the new astrophysical unit and to engage in high-altitude spectroscopy with rockets. Thinking of Goldberg, he wanted to add to the Yale staff "a first-rate astrophysicist who would be keenly interested in working full time in this important program. . . . and to formulate a detailed scientific program for consideration by ORI."[69]

At Menzel's urging, Goldberg responded warmly to Spitzer's definition of a center that would emphasize theoretical analysis and discussion of results, within the broad context of solar physics.[70] He agreed with Spitzer that what rocketry groups like NRL needed, they could provide: sound astrophysical advice from astronomers. NRL, Spitzer had argued, could not be expected to handle the whole job, including a full astrophysical analysis: "I am sure that there is plenty of work to go around. . . . There is no doubt that research in rockets will be supported indefinitely by the government, unless we all blow ourselves up first."[71]

Above all, Spitzer and Goldberg agreed that they required autonomy in their work. They could not be subject to either V-2 Panel or NRL direction. As Goldberg concluded: "If several groups are involved I would find it infinitely more attractive to be associated with the coordinating institution rather than with one of the 'farms,' since, as you say, there would be a much wider scope afforded for ideas and initiative."[72] Thus Goldberg and Spitzer's early proposals to ORI through the summer of 1946 were to build an astrophysics team at Yale that could make spectrographs, fly them, and analyze the results. Goldberg would be hired as an associate professor, and several technical experts in optics, mechanical control, and electronics would be needed. This combination of talent, Spitzer predicted, could also help NRL design improved spectroscopic instruments.

While Spitzer and Goldberg were working on their plan, the first flights of spectroscopic devices on rockets were dismal failures. ORI advised Spitzer and Goldberg that "it may be necessary to telemeter the spectroscopic information back to the ground." But Spitzer feared that "this would require work with photocells and electronic equipment, concerning which I am almost completely ignorant."[73] Goldberg and Spitzer needed to find people who could build these devices, using the latest technology. This set the two apart from the V-2 Panel members, whose priority was instrument building.

Concerned about these technical issues, Spitzer and Goldberg visited Washington in July 1946 to promote their program, finding enthusiastic support from Waterman, J. C. Boyce, and others at high levels who were attracted to the idea of securing connections with elite science.[74] They found considerable resistance at the working level, however, especially from Richard Tousey at NRL. Their three-year contract proposal to ORI was by now $100,000, and Tousey likely saw it as a threat to the autonomy of the NRL program.[75]

After meeting Tousey, Goldberg grew worried about the whole enterprise. First, given the poor chances for recovery of any device from a V-2, Goldberg couldn't imagine sacrificing a $5,000 spectrograph "every time a rocket went up." He felt that simpler, low-cost spectrometers would be more reasonable than sophisticated spectrographs. He also did not want to have to be "taking orders" from NRL, "unless their function is chiefly one of liaison." NRL, in short, was not an acceptable project monitor for Goldberg.[76]

In the fall of 1946, four factors led Spitzer and Goldberg to reduce the scope of their plans. First, Tousey's October 1946 spectrograph flight captured the first ultraviolet spectrum of the sun, which vindicated his design in the face of Menzel's and Goldberg's earlier criticisms. Second, they were now aware of the considerable technical demands

actually making such devices entailed, partly by watching what Greenstein was then going through. Third, by late October 1946, Goldberg had been selected as the next director of the observatory and chairman of the Department of Astronomy at Michigan, partly because he had been courted by Spitzer. And fourth, Spitzer himself was about to leave Yale, which had chilled to his plans, to become director of the Princeton University Observatory. Both faced the considerable task of rebuilding observatories in the postwar era.[77]

Throughout the remainder of 1946 and well into 1947, Goldberg, Spitzer, and Menzel warmed to the achievements of the rocket groups at NRL and APL.[78] Goldberg was now "tremendously impressed" with APL's and NRL's organizational ability to build instruments quickly and felt that if university astronomers still were to participate, they would act "most effectively as astronomers . . . by providing the skills and training needed for interpretation of the data."[79] Tousey and Strain, however, felt strongly that "the agency which obtained the data should complete whatever analysis they wished to undertake and then publish their results."[80] They did not look kindly upon anyone who felt that the data would fare better in other hands.

ORI, however, now reorganized as ONR, still wanted to support Spitzer and Goldberg: "NRL would proceed to make whatever measurements they feel capable of making, after which the spectra would be made available to us," Goldberg recorded.[81] In November 1946, Goldberg and Spitzer discussed their plans with Charlotte Moore Sitterly at the National Bureau of Standards, reigning world's specialist in spectral-line identification, but were surprised to find that the NBS had not yet been contacted by NRL or APL for its critical knowledge. Goldberg saw this insularity as evidence of incompetence:

*I am sure the NRL people do not realize the tremendous amount of labor that will be involved in simply identifying the lines. The spectrum is a jumble of blends and probably detailed identi-*

*fications will not be feasible for more than a few of the strongest lines. What must be done is to predict the spectrum, making more or less trial and error adjustments until the computed spectrum agrees in appearance with that shown by the microphotometer tracing.*[82]

Mrs. Sitterly would soon contact Tousey and offer her help, which was warmly accepted. But she was also sympathetic to her astronomical colleagues' frustration, especially Menzel's concern that the confused and densely packed spectral region that was now being revealed by the first NRL spectra was going to require far more powerful instrumentation and theory to unravel, for a "complete analysis of the abundance, excitation temperatures, ionization temperatures, electron pressures, and f-values for all of the important lines in the solar spectrum." Menzel soberly concluded that "many difficulties are going to arise. The definition appears to be poor and the problem of blending is going to worry us a great deal."[83]

For her own part, Charlotte Moore Sitterly expressed unabashed glee over the prospect of mining a new region of the solar spectrum. Her first step was to plan an ultraviolet extension of her monumental *Multiplet Table of Astrophysical Interest.*[84] She looked forward to fuller discussions with Goldberg and the others at meetings of the American Astronomical Society (AAS) in Cambridge in December and also took steps to bring Tousey into the astronomical fold. To this end, she arranged a coming out party for Tousey by finding a place for him on the AAS program on December 29.[85]

Menzel, as vice-president of the society, saw to it that the AAS Council nominated Tousey's paper for a $1,000 prize offered by the American Association for the Advancement of Science (AAAS). In a letter to NRL Director H. A. Schade, Menzel expressed his delight in the work of the NRL group, adding that Tousey's paper "was literally the 'highlight' of the meeting—and no pun intended."[86] Soon after the AAS meetings, Joseph C. Boyce found Henry Norris Russell

"studying with great pleasure and excitement a print of this spectrum [October 10] that he just received from us," as Boyce later recounted to Tousey. Russell had also been at the December AAS meetings, and commented to Tousey that "we were now seeing the beginning of a new field of astronomy."[87]

Tousey's appearance at the AAS in December brought him into closer contact with astronomers, and because of their positive reception of his work, he warmed to the idea of sharing data. On 31 January he wrote cordially to Menzel upon learning of Menzel's letter to Schade: "I know that he was very pleased to receive it and I am grateful to you for sending it."[88] Tousey enclosed a copy of the best print he had at the time, and promised to send better ones "if I ever receive them."

Indeed, as Mrs. Sitterly made plans to analyze Tousey's first spectra, she realized that they were seriously wanting in resolution and detail. By the end of January 1947, she told Tousey that "it would be a waste of time to analyze the present spectrum. The higher the resolution, the simpler the analysis."[89] She also cautioned that determining even the relative intensities of the unblended lines was fraught with difficulties, and that trying to do them at NRL would "require the full time of a trained physicist."

The advice Tousey and Durand received from Mrs. Sitterly confirmed that their priorities in solar spectroscopy were correct. In the first years, all effort was focused on getting better and better spectra, through building, testing, and flying continually refined spectrographs. Analysis was secondary: throughout these early years, to 1950 at least, it was limited to determining the vertical terrestrial ozone distribution, the spectral energy distribution of the sun, and the identity of the major lines in the rocket ultraviolet.[90] The first two did not require high-resolution spectra.

University-based astronomers, however, although excited by the spectra Tousey and the APL group were producing, needed something more, so that they could address more central problems in stellar atmospheres. In the late spring of 1947, Goldberg and Spitzer formally inaugurated their ONR-supported Astrophysical Consulting Group, based at Michigan, in the hope that improved spectra would be forthcoming from the rocket groups. Goldberg's first step was to hire Keith Pierce, a young Berkeley astrophysicist, to construct synthetic spectra for the rocket ultraviolet, extrapolating from what was known in the visual region.[91] The goals of the Astrophysical Consulting Group were to improve knowledge of solar element abundances in general, "to re-examine the entire problem of the formation of absorption lines, with special attention to physical blending."[92] Only then, using the 'curve of growth' technique, could they determine in any comprehensive manner the number of elements in various states of excitation in the solar atmosphere.

Goldberg subcontracted with Menzel to support Harvard graduate students and invited Ira S. Bowen and Robert King at Mount Wilson to participate.[93] King's laboratory furnace expertise would greatly aid the Michigan curve of growth analysis of the equivalent widths of lines in the solar ultraviolet. Goldberg, always optimistic, felt in August 1947 that they must "take the most favorable view and assume that the spectra of the future [V-2 flights] will show high resolution and that equivalent widths measurement will be possible," even though "with the spectra presently available, I am sure that the resolution will be insufficient to permit a curve-of-growth type of analysis."[94]

By late 1947, both NRL and APL had secured hundreds of ultraviolet solar spectra on about a half-dozen flights. Their data confirmed the existence of major spectral features in the region 220 to 290 nanometers, revealed that the ultraviolet solar continuum was fainter than expected, and provided a useful first reconnaissance of the distribution of ozone in the highest regions of the atmosphere. In 1947 Tousey modified his Baird spectrograph by replacing its beads with slits, and the April July 1947 APL instruments all utilized slits, as well as crude light collectors. But still, none returned high-quality spectra (figures, pp. 212, 213). Through early 1948,

line definition started to improve, but it still was far below the level astronomers needed.

Bowen and King pulled out in late 1947; after reviewing the rocket data, they decided that their involvement was "premature" and that there were more pressing tasks at Mount Wilson.[95] Their departure highlighted the marginality of what Goldberg, Spitzer, and Menzel were trying to do.[96] Goldberg did not have to be convinced; he and his staff were already getting more excited about exploring the sun's infrared spectrum than its ultraviolet depths.[97]

The situation did not improve in 1948. Although the spectra were marginally better, both resolution and dispersion were insufficient for astronomers. Line identification was possible, which the NRL and APL people took relish in making, aided by Charlotte Moore Sitterly. But blends were still severe and plagued them for several years until highly stable pointing controls became available in the 1950s, finally making feasible high-dispersion solar spectroscopy from rockets.

By October 1948, the time had come to renew the ONR contract, and now Goldberg confronted the futility of the effort. He wrote Spitzer:

> *Frankly, I have been somewhat disappointed in the development of the rocket project since its inception. It had been my hope and I think also yours that the rockets would open a new field of solar research which would justify the existence of an analysis group on a long-range basis. At least thus far, however, the rockets have made an interesting contribution to the field of solar spectroscopy, [but] they have hardly opened up a new field of investigation. That, of course, may come with time but at the moment I should not want to ask for a renewal of the contract solely on the basis of the V-2 investigation.*[98]

Spitzer felt about the same as Goldberg. He had many other demands on his time and wanted to use ONR money for more immediate goals. Spitzer, however, looked to the future: "I still think it likely that in the long

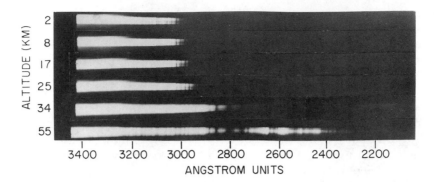

Tousey's first solar spectra revealing the gradual penetration of the rocket through the ozone layer. The best spectrum displayed, taken at an altitude of 55 kilometers, revealed hundreds of spectral lines and features to 230 nanometers, but most were blended in the low dispersion image. U.S. Navy photograph. NASM SI 87–8423.

run astronomical observations from above the atmosphere may revolutionize astronomy. At the moment, I must admit, this development seems some number of years in the future." Spitzer, too, admitted that their rocket-related research was not going anywhere. Al-

though he was reluctant to break ties, he agreed that "it is the sensible step to take."[99]

Thus, after 18 months of ONR support, Spitzer and Goldberg, now deeply involved as directors of major ground-based observatories, decided to fold up their Astrophysical

## 3200-2300 ANGSTROMS

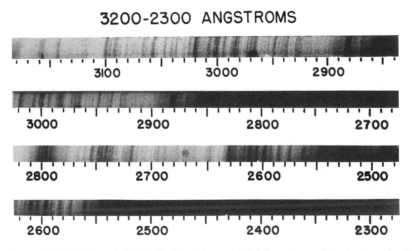

Solar spectrum from NRL's 7 March 1947 flight, taken at 75 kilometers altitude. For the first two years, the best NRL spectra came from this flight of the first NRL spectrograph to use slits. Although these spectra showed hundreds of lines, few were unblended, but the resolution was still between 0.1 and 0.15 nm, and many unusual spectral features were revealed along with about 140 new lines and blends. Under magnification, the magnesium doublet at 280 nanometers showed internal structure, and distinct emission reversals in both line cores. The spectral lines appear curved because the rocket rolled during the exposure. From Durand, Oberly, and Tousey (1947). U.S. Navy photograph.

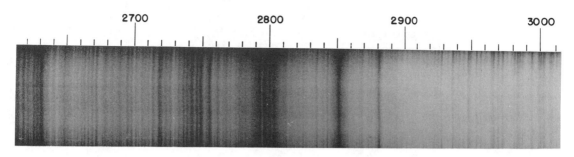

**Small portion of the solar spectrum from the 29 July 1947 flight of APL's grating spectrograph. Taken at an altitude of 135 kilometers aboard V-2 30, this spectrum shows improved detail in the region around the magnesium doublet at 280 nanometers. From Hopfield and Clearman (1948), fig. 2. Applied Physics Laboratory, JHU/APL archives.**

Consulting Group and both applied independently to ONR for stellar and solar research support. Goldberg kept up with the ever-improving solar ultraviolet spectra produced by Tousey, as well as by other groups, and recalls that even though he bowed out, he saw the V-2 period as critical because it "was a way of getting started quickly and involving scientists who otherwise would have turned to other things." It gave the United States "a head start on what I call the space age, which I really date from that time, rather than from October 1957." Goldberg remembers that at the time, his opinion was that "the best scientific work wasn't done at government laboratories," but that the rocket work at NRL,

namely Tousey's long series of successes, "started a kind of renaissance."[100]

## Traditional Astronomical Practice

The experiences of Greenstein, Hynek, Goldberg, and Spitzer offer insight into how and why those who succeeded in using rockets for scientific research formed a group distinct from any in the disciplines they served. Astronomers who were closely tied to main-

**Henry Norris Russell and Lyman Spitzer Jr., circa 1950. Margaret Russell Edmondson Collection, AIP Niels Bohr Library.**

**Donald Menzel, Robert R. McMath, and Leo Goldberg in the mid-1950s. News and Information Services, University of Michigan, Observatory Collection, box 14, folder "R. R. McMath groups," Michigan Historical Collection, Bentley Historical Library, the University of Michigan.**

stream astronomy were reluctant to commit themselves to the long-term effort of developing astronomical instrumentation for sounding rockets, but they were also uncomfortable with the fundamental nature of the activity.

Astronomers of that day were more at ease with proven instruments that would provide new data quickly and efficiently and that would last for years of use and reuse. Traditionally, an astronomical instrument's value increased with age: as it accumulated observations of natural phenomena, its characteristics became better known and could be accounted for. It became a trusted, refined, and reliable tool for research, and not an end unto itself.

In distinct contrast to astronomers, some of the experimental physicists at NRL delighted in the prospect of constantly building and improving on their instrumentation, being driven to do so by the fact that their instruments were likely to be destroyed upon use. Years later, Ernst Krause still recalled with enthusiasm what rocket work meant to the tool builder:

> Now, this is a good way to do some experimentation. We're going to get away from this business of having a complicated, costly set of apparatus in a physics laboratory in a basement in some university, and because it is complicated and costly, it lasts for 50 years and generation after generation grinds out theses on that same equipment because it's expensive and new equipment is more expensive. We've got a setup here which by its very definition is going to get destroyed each time. How good can you have it?[101]

Both NRL and APL were in the business of producing a capability with guided missiles; as we have seen, Krause and the Navy did not distinguish between the research done with rockets and the research that led to better rockets, and APL regarded the Aerobee as an appropriate test bed for improving the breed. Krause always looked for new and fresh ideas and projects. Thus, his remarks apply equally well to the rockets as to the instruments placed on board.

In the 1940s, most astronomers were simply not comfortable with the degree of funding and manpower commitments required to pursue active experimental programs with rockets, nor were they comfortable with the style of research that required a great dedication to building devices that were likely to be destroyed upon use.[102] Certainly, senior observatory directors like Shapley or Struve did not show much patience for the effort, and when Goldberg and Spitzer realized that it would be quite some time before useful data would come from their program, they lost the initial enthusiasm they had for studying the sun from rockets.

## Early Analyses of Photographic Solar Spectra

The "strong back, weak mind" syndrome Van Allen identified was a reflection of the chief task of scientific rocketry, as Goldberg and Spitzer realized, which was to demonstrate that spectra could be secured from rocket flights. But both Van Allen and Tousey appreciated the fact that this demonstration was not going to get them too far. Analysis was essential.

Although J. J. Hopfield's background served him well for the reconnaissance of both experimental and natural spectra, as a classical laboratory spectroscopist he was not too enthusiastic about analysis based upon physical theory.[103] Thus, Van Allen sent Harold Clearman to Yerkes for three months to learn enough astrophysical theory from Greenstein to produce a preliminary analysis of the best APL solar spectra taken on the April and July 1947 flights. Upon Clearman's return to APL, he and Hopfield limited their discussion to line identifications, a listing of probable transitions causing each identified line, and a series of preliminary microphotometer tracings of the density profile of the spectra, to detect blends.[104]

Tousey, with Durand and Oberly, did little more than Hopfield and Clearman at first.

The best of their 10 October flight spectra revealed that the ultraviolet solar continuum was heavily depressed below that of a theoretical 6,000K perfect radiator, and that many prominent spectral features predicted by astronomers indeed existed. With Charlotte Moore Sitterly's help, spectral analysis at NRL extended to the use of multiplet structure to identify lines.[105] Their most significant conclusions from these first years of data, however, dealt not with the sun, but with the earth's atmosphere. Both NRL and APL were able to derive a rough picture of the vertical ozone distribution, based upon how far their spectra penetrated into the ultraviolet as a function of increasing height.[106]

Even though Tousey had fended off Goldberg, Spitzer, and Menzel in 1946 and 1947, by 1948 he was starting to have serious concerns about the inadequacies of his staff. Durand and C. V. Strain had other duties under Krause, as did Charlotte Moore Sitterly at the NBS.[107] After Baum's departure in late 1946 and after both Durand and Strain left in the fall of 1947, Tousey was "without Ph.D. assistance." On occasion, Tousey asked astronomical contacts for "any suggestion[s] concerning a capable and congenial physicist who might be persuaded to join this project."[108] Tousey's reluctance to allow outsiders to take over reduction and analysis and the difficulty of securing physicists or astrophysicists experienced in the analysis of complex spectra all contributed to delays in completing data analysis, and kept his work insulated.

Nevertheless, Tousey's clear priority was to advance experimental techniques in rocket spectroscopy, first with V-2s and then with Aerobees and the NRL Viking—a priority characteristic of all groups of rocket experimenters at NRL, APL, and elsewhere. In summing up his V-2 era efforts, therefore, it is not surprising to hear from Tousey that: "We [eventually] got a reasonably good [solar] spectrum, much improved over the very first. Our analysis work, however, was always behind, in fact, it always has been."[109]

## Notes to Chapter 11

1. Mitra (1948), pp. 518–19. Mitra notes in his preface that the book went to press before he knew about the V-2 upper atmospheric program in the United States.

2. Van Allen to Greenstein, 4 June 1947, p. 2. JG/CIT.

3. Portions of this chapter are adapted from DeVorkin (1989b).

4. F. S. Johnson, "V-2 Rocket Spectrograph," 28 February 1946 entry, p. 7, in his registered laboratory notebook no. 6233 (Francis S. Johnson, February 6, 1946 to October 6, 1947). FSJ/NASM.

5. Tousey OHI, p. 119rd.

6. C. H. Smith, "Record of Consultative Services," 19 February 1946, meeting held 14 February 1946. Box 32, NRL/NARA.

7. C. V. Strain registered laboratory notebook no. 6353, p. 1, entry dated 23 March 1946. HONRL; F. S. Johnson OHI, p. 28.

8. F. S. Johnson OHI, p. 30.

9. Menzel to E. O. Salant, 20 February 1946. LG/HUA.

10. Goldberg to Menzel, 28 February 1946. LG/HUA.

11. C. V. Strain registered laboratory notebook no. 6353, p. 1, entry dated 23 March 1946. HONRL.

12. On the testing process, see F. S. Johnson, "V-2 Rocket Spectrograph," registered laboratory notebook no. 6233, entry for 1 April 1946, p. 16.

13. Johnson OHI, p. 33.

14. On John Strong and the Hopkins tradition in optics, see Henry, Beer, and DeVorkin (1986).

15. Strong recalls that, because he was a good friend of Hulburt, he foolishly agreed to provide the gratings, for they were extremely difficult to make properly. Some gratings were poor, but at least two of them were fair enough to work. According to Strong, "It was with these fair ones that [Tousey] drew his first blood." Strong OHI, pp. 63–64rd. See also Tousey, "Record of Consultative Services," 17 January 1947. Box 99, folder 5, NRL/NARA.

16. Tousey OHI, p. 256rd. The 40-centimeter focal length gratings were another compromise. Tousey had originally wanted to use faster, 25-centimeter gratings, whereas Strong wanted to provide 50-centimeter elements, since they were easier to blaze.

17. F. W. MacDonald to Lynn Brown, Eastman Kodak Co., 6 June 1946. Box 98, folder 2, NRL/NARA. See also C. V. Strain Regis-

tered Notebook no. 6353, pp. 2–4. HONRL. See also Baum OHI, pp. 15–16rd; Johnson OHI, p. 30.

18. Contracting with Baird was a necessary expedient. Tousey, wizened from past problems with NRL's technical services, such as design and drafting, or even the shops, knew that they could not be counted on in a hurry, and so he turned to Walter Baird, an old-time Cambridge friend who was also close to Theodore Lyman. Baird held a doctorate in electrical engineering from Johns Hopkins and went into business in 1936 to provide universities and industry with analytical devices for spectrochemical analysis. Since spectrographs were his firm's principal product and since it had recently moved into infrared gas analyzers (making its first one for Kellex, which was responsible for Baird changing the name of his company to Baird Atomic) and had demonstrated that it could, "under pressure," build "a specialized instrument before [it] had created an all-purpose laboratory model," it was the ideal contractor for Tousey. See "Instrument Makers of Cambridge," (December 1948), pp. 136–40, 139. The initial choice of Baird Associates as sole source did not keep Tousey and Strain from discussing similar contracts with the Farrand Optical and Instrument Company of New York. Charles V. Strain, "Record of Consultative Services," 9 May 1946. NRL held two contracts with Baird: N6ORI-75 and later N1730–11081. See Wright to Baird, 16 June 1946. Box 98, folder 2, NRL/NARA.

19. Purcell OHI, p. 49.

20. See "V-2 Panel Report no. 1," p. 6; "V-2 Panel Report no. 2," pp. 16–17. V2/NASM.

21. In late February, Van Allen identified APL's agenda for solar spectroscopy, in Van Allen to M. H. Nichols, 25 February 1946. JVA/APL. See also C. V. Strain, "Record of Consultative Services," 16 April 1946. Box 98, folder 1, NRL/NARA.

22. Van Allen to L. R. Hafstad, 29 March 1946, p. 1. Box 118, "Bumblebee Series of Missiles" folder, MAT/LC. See also Appendix I, in [Van Allen], *High Altitude Research at the Applied Physics Laboratory of the Johns Hopkins University—A Brief Summary* (10 April 1947). Copy no. 204—Gibson. JVA/APL.

23. J. A. Van Allen, "Rough Quantitative Considerations in the Use of Cosmic-Ray and Ultra-Violet Altitude Meters," 2 April 1946. Box 118, "Bumblebee Series of Missiles" folder, MAT/LC.

24. Van Allen to L. R. Hafstad, 29 March 1946, p. 2. Box 118, "Bumblebee Series of Missiles" folder, MAT/LC.

25. See Dennis (1986, 1990, 1991). Tousey was definitely aware of the needs of the fleet, as were his colleagues at NRL. As William Baum made plans to return to Caltech for graduate training, he discussed the idea of an ultraviolet altimeter with W. Pickering. See Pickering (1947), p. 11. Further, Tousey also identified military justifications for what he did, and occasionally performed duties that were "not our primary interest." Unlike Van Allen, however, he felt no need to search for military application, since his institution did not demand it. See Tousey OHI, p. 74rd.

26. Both Hynek and Hopfield were never fully satisfied with lithium-fluoride optics. As Hynek told Leo Goldberg at a meeting of the American Optical Society in Cleveland in March, its light transmission at Lyman Alpha was only 20 percent. Goldberg to Menzel, 11 March 1946. LG/HUA.

27. See Hopfield (1946).

28. On the design, see Mann (1946), p. 79.

29. See [news note], *Sky & Telescope* "Astrophysics and V-2 Rockets," 5, no. 12 (October 1946), p. 7.

30. Van Allen to Greenstein, 12 September 1946. JG/CIT.

31. Van Allen to Greenstein, 31 October 1946; Clearman to Greenstein, 5 November 1946. JG/CIT.

32. Hynek reported to Greenstein: "Nothing on most [film] strips—fog on the rest. (Hopfield is here for a few days) One thing made me quite mad—I learned from Hopfield that the instrument wasn't tested in a vacuum before sending it to White Sands. And after I cajoled and insisted that it be so tested. I borrowed a vacuum chamber for them from John Strong so they could do just that." Hynek to Greenstein, 1 January 1947. JG/CIT. Hynek's criticism did not match his published statements in the reliable journal *Sky & Telescope*, based upon a presentation he gave at the September meetings of the American Astronomical Society, where he stated that "the entire plateholder is encased in a light tight steel cylinder which preliminary drop tests have shown to be extremely

rugged." See "American Astronomers Report," *Sky & Telescope 5*, no. 12 (October 1946), p. 7.

33. The strips that had been exposed "were open to ground light" and "strips of film which had not reached the exposure position in flight showed the pre-flight spectra in good condition, upon development." Quotations from C. P. Smith, *Upper Air Rocket Research Report XXI,* V-2 no. 13, p. 2; Hopfield and Clearman (1948), p. 887; "Listing of Experimental Equipment Flown in V-2's during 15 April 1946 to 15 April 1947, Appendix A," p. 4, in [Van Allen], *High Altitude Research at the Applied Physics Laboratory of the Johns Hopkins University—A Brief Summary* (10 April 1947). Copy no. 204—Gibson. JVA/APL. An incomplete exposure cycle would indeed cause fogging of those strips already exposed. This is supported by the description of the film advance mechanism in Mann (1946), p. 80.

34. See "V-2 Panel Report no. 8," 4 February 1947. V2/NASM.

35. Hopfield and Clearman (1948), p. 877.

36. Megerian to Bain, "V-2 Panel Report no. 8," 4 February 1947, p. 12. V2/NASM.

37. After the flight, Van Allen thanked Greenstein for his "excellent assistance . . . in preflight overhaul of the APL spectrograph," adding that "I understand this was probably essential to its success." Van Allen to Greenstein, 18 April 1947. JG/CIT.

38. Ibid.

39. Hynek to Greenstein, 23 April 1946. JG/CIT. Hynek also plied Greenstein with questions about optical design, especially how to design an efficient light collector like Regener's, or sunfollowers so that they could use doubly dispersing systems to reduce scattered visible light, as Menzel had suggested. Hynek to Greenstein, 24 April 1946. JG/CIT.

40. Formed as a contract shop for the Office of Scientific Research and Development, the Yerkes Optical Bureau of the University of Chicago designed rugged optical periscopes, sights, and wide-angle projection systems for use on torpedo dive bombers and tanks, as well as sophisticated panoramic optical systems for flight training. On the Optical Bureau, see DeVorkin (1980).

41. J. Allen Hynek to Jesse Greenstein, 24 April 1946. JG/CIT.

42. On Greenstein's search for Pohl crystals, which could be designed to sense ultraviolet radiation while remaining insensitive to visible light, see I. Estermann to J. Greenstein, 24 June 1946. JG/CIT. By July 1947, Kuiper had cut through red tape and classification and received six halogenid potassium bromide Pohl-type crystals and gave them to Greenstein to test. Little came from this work, because the crystals proved to be difficult to calibrate and stabilize and were sensitive only to the wavelength range 170 to 180 nanometers. Therefore, they were useless for the detection of Lyman Alpha at 121.6 nanometers. See Greenstein to Van Allen, 14 July 1947. JG/CIT; Goldberg to Spitzer, 23 June 1947. LG/HUA; see also Daniel B. Clapp to Field Service, Chief of Naval Research, "Pohl Crystals Requested by Dr. Leo Goldberg," 2 May 1947. OANAR:NA-4S2 Office of the Assistant Naval Attache for Research, American Embassy, London. LG/HUA.

43. [Kuiper] "The Rocket Project," "High Altitude Spectroscopy," 18 June 1946, p. 1. JG/CIT. Kuiper was not exaggerating. He was then collaborating in a project to build a photoelectric infrared spectrometer, but the electronics expertise came from R. J. Cashman and W. Wilson of Northwestern University. See Kuiper (1949), p. 351, n. 65.

44. On the early use of photoelectric detectors in astronomy, see DeVorkin (1985).

45. Greenstein to Van Allen; and to Hynek, 16 July 1946. JG/CIT.

46. Clearman to Greenstein, 18 November 1946. JG/CIT; J. Greenstein, "Progress Report: J.H.U. Purchase Order no. 13728, Yerkes Observatory Spectrograph," n.d., folder R. JG/CIT.

47. Greenstein to Van Allen, 8 January 1947; Greenstein to Hynek, 8 January 1947. JG/CIT.

48. Greenstein to Van Allen, 24 January 1946. JG/CIT.

49. Greenstein to Miss Ness, 9 February 1947. OS/AIP.

50. Greenstein to Struve, 13 February 1947. OS/AIP.

51. Ibid

52. Greenstein to Van Allen, 24 February 1947; Van Allen to Greenstein, 18 April 1947. JG/CIT.

53. Greenstein to Struve, copy of telegram text, n.d. OS/AIP. Filed after a 22 March letter to Struve.

54. Van Allen to Greenstein, 4 June 1947, p. 2. JG/CIT.

55. Greenstein to Van Allen, 12 June 1947. JG/CIT.

56. Struve to Kuiper, 13 June 1947, written on Greenstein to Struve, 11 June 1947. OS/AIP.

57. See, for instance, Greenstein to Van Allen, 2 June 1947. JG/CIT; and Kuiper (1949).

58. Van Allen to Greenstein, 18 April 1947. JG/CIT.

59. On the tradition-bound views of observatory directors after World War II, see DeVorkin (1991).

60. Greenstein was also interested in the galactic radio noise observations by Grote Reber, and helped him prepare a proposal to ONR. Greenstein also worked on preliminary theoretical analyses of the character of the radio radiation Reber detected. See Jesse Greenstein, "Report, Academic Year, 1946–47, Jesse L. Greenstein." Folder R, 1946–47, JG/CIT; and correspondence in the Struve collection with Shapley, T. J. Killian, Grote Reber, H. G. Bowen, and Frank B. Jewett, between 1941 and 1947. OS/AIP. See also Greenstein (1984).

61. Goldberg to Menzel, 28 September 1945. DHM/HUA; copy in Goldberg file, SAOHP/NASM.

62. Leo Goldberg (1981), p. 16.

63. Lyman Spitzer OHI, p. 2. See Spitzer (1946).

64. Lyman Spitzer OHI, pp. 4–17; esp. pp. 5–6; see also L. Spitzer, "Astronomical Advantages of an Extraterrestrial Observatory," 5, pp. 71–75 in *Project RAND Report*, Douglas Aircraft Company, 1 September 1946. SAOHP working files, Spitzer collection. Rough draft dated 30 July 1946, copies in LGP/HUA and in LSP/P. Project RAND's interest in the nature of the atmosphere above 300 kilometers was to determine satellite lifetimes. On early U.S. satellite proposals, see Hall (1963), and on Spitzer's involvement, see Smith (1989).

65. Goldberg OHI no. 2, p. 16rd.

66. Donald Menzel was well connected in Washington. As part of his ionospheric research at the Interservice Radio Propagation Laboratory, under the Joint Chiefs of Staff, and in his role as chairman of the Wave Propagation Committee, Menzel had made many close contacts within the Navy. In August 1945, Menzel linked solar studies and ionospheric research in his lobbying effort at the nascent ORI, pointing out to the Navy that Harvard's High Altitude Observatory, a coronographic observatory near Climax, Colorado, could predict the quality of long-range radio communication and would act as an alert to launch sounding rockets that would study the chromospheric and coronal phenomena causing radio disturbances. Menzel to Op-20-G, 7 August 1945. DHM/HUA. Menzel also linked the High Altitude Observatory to ionospheric research and to scientific rocketry in discussions with Admiral Furer and E. U. Condon in late 1945 and suggested that a "three-way program involving the Navy, the National Bureau of Standards, and the high-altitude station of the Harvard College Observatory at Climax, might be very useful." Menzel to Gordon Dyke, 19 November 1945, HUG 4567.5.2, DHM/HUA. On Menzel's interest in continuing ionospheric work after the war, see Menzel to Struve, 31 August 1944; Struve to Menzel, 15 October 1945. OS/AIP; and "U.S. Navy-Correspondence and other Papers, 1942–1955" folder, especially Menzel to Capt. H. T. Engstrom, n.d. DHM/HUA; see also Alfred C. Lane to Walter Orr Roberts, 18 December 1945. WOR/UC.

67. On Goldberg's contacts, see Goldberg to Menzel, 12 October 1945. LG/HUA. See also Goldberg to Gordon Dyke, 5 December 1945. HUG 4567.5.2 "General Correspondence," DHM/HUA. During the war, working at Michigan, Goldberg helped to provide solar data to the astronomical network that fed into the Interservice Radio Propagation Laboratory. Through Menzel, Goldberg was well known to the Wave Propagation Committee, and on his own to the Office of the Chief of Naval Operations for his observatory's OSRD contribution to the application of "various gyroscopic techniques to fire control problems and the development of equipment for testing fire control devices." See Vice Adm. George F. Hussey, CNO, presentation of the Naval Ordnance Development Award to McMath-Hulbert Observatory, 23 February 1946. Box 6, RRM/BHL. There were also awards from the Navy Bureau of Ordnance for work on bombsight calibration devices. On the low-altitude bombsight itself, see Boyce (1947), p. 49.

68. On ONR's interest in astronomers, see DeVorkin (1991).

69. Spitzer to "Planning Division," 1 April 1946. Spitzer also asked for support from other related military organizations. See Spitzer to J. C. Boyce, 10 May 1946. LSP/P.

70. Goldberg did not need much urging. He felt pinned under Robert R. McMath's foot at Lake Angelus and longed for a real astronomical position. McMath was an adept and innovative amateur astronomer who was chairman of the Board of the Motors Metal Manufacturing Company and through his wartime work entertained many close Washington contacts. Goldberg was a junior member of the staff and resented the way McMath ran his observatory. It was distant from the campus of the University of Michigan and run "like a factory" more than a university observatory. Goldberg OHI no. 2, p. 10rd.

71. Spitzer to Goldberg, 28 May 1946. LSP/P.

72. Goldberg to Spitzer, 17 May 1946, LSP/P.

73. Spitzer to Goldberg, 1 July 1946. LSP/P. This feeling of inadequacy was shared by Spitzer's colleagues in the Physics Department, who shied away from photon counter development. See Roland E. Meyerott to Marcus O'Day, 13 January 1947. LSP/P.

74. Goldberg to Menzel, 18 July 1946. HUG 4567.5.2, DHM/HUA.

75. See Spitzer OHI, p. 13. Confirming contemporary evidence is from Goldberg to Menzel, 18 July 1946. HUG 4567.5.2, DHM/HUA. Tousey recalls as well that "we have always felt that ONR was trying to run our things. And this [bringing Spitzer and Goldberg in] was a good case." Tousey OHI, pp. 75–8; p. 146.

76. Leo Goldberg to Donald Menzel, 18 July 1946, p. 2. DHM/HUA.

77. On Goldberg's challenge at Michigan, see Goldberg OHI no. 1, pp. 47–8. SHMA/AIP; and Rudi Paul Lindner "From Hussey to Goldberg," manuscript draft of talk given at the 14 June 1989 meetings of the Historical Astronomy Division of the American Astronomical Society at the University of Michigan. Spitzer also found that his Yale director Dirk Brouwer was less than sympathetic to his plans. See Spitzer to Goldberg, 12 October 1946. LSP/P; LG/HUA. On Brouwer's reaction, see Dirk Brouwer to Spitzer, 12 August 1946. LSP/P. In late October 1946, Spitzer was approached about taking the chairmanship at Princeton and by 27

January 1947 knew he had the job. See Shapley to Spitzer, 2 November 1946; Spitzer to Menzel, 27 January 1947. LSP/P; LG/HUA.

78. Leo Goldberg, "Memorandum of Visit to Cambridge and Washington, November 14 to 20," n.d., p. 3. LG/HUA.

79. Goldberg, ibid., pp. 3–4.

80. Tousey and Strain, "Record of Consultative Services, 20 November," 3 December 1946, p. 2. Box 98, folder 4, NRL/NARA.

81. Leo Goldberg, "Memorandum of Visit to Cambridge and Washington, November 14 to 20," n.d., p. 4. LG/HUA.

82. Goldberg to Menzel, 21 November 1946. LG/HUA.

83. Menzel to Goldberg, 6 December 1946. LG/HUA; LSP/P.

84. C. M. Sitterly to Leo Goldberg, 25 November 1946. LG/HUA. See Moore (1945).

85. Phone conversation, Sitterly to the author, 8 March 1982.

86. The AAS was then meeting with the AAAS, and member societies were encouraged to nominate papers. Tousey's nomination did not win; the AAAS chose instead a biological contribution. Tousey OHI, pp. 71–72rd.; See also Tousey to Menzel, 31 January 1947; Menzel to Schade, 9 January 1947. DHM/HUA. See also Council Minutes of the American Astronomical Society IV 1945–47, p. 463. AAS/AIP.

87. Tousey (1967), p. 240.

88. Tousey to Menzel, 31 January 1947. DHM/HUA.

89. Tousey and Durand, "Record of Consultative Services," 30 January 1947. Box 99, folder 5, NRL/NARA.

90. Among the many sources cited here, see Tousey OHI, pp. 74rd.; 89rd.; Krause to Code 180, 9 January 1947. Box 99, folder 5, NRL/NARA; Durand and Tousey (1947), p. 7.

91. See Goldberg OHI no. 2, pp. 29–30rd; Goldberg to Menzel, 27 May 1947. LG/HUA.

92. Goldberg to Struve, 23 April 1947. OS/AIP.

93. Leo Goldberg, "Report on Work During Month," 1 July 1947. LSP/P; Spitzer to Goldberg, 19 July 1947. LG/HUA.

94. Goldberg to Spitzer, 11 August 1947. LSP/P.

95. King's furnace was in constant demand, and Bowen was totally preoccupied getting the

new 200-inch telescope working properly. See Goldberg to Urner Liddel, 6 October 1947. DHM/HUA, copy in LSP/P; Robert King to Lyman Spitzer, 9 October 1947. LSP/P.

96. Menzel to Goldberg, 6 February 1948. LG/HUA.

97. Their intent was to observe the sun at wavelengths of up to two microns using a modified Littrow spectrograph at the McMath/Hulbert Observatory. See A. Keith Pierce "The Infrared Spectrum of the Sun," no. 428 (February 1965), p. 5; see also Goldberg to Spitzer, 5 February 1948. LSP/P; and Leo Goldberg to the author, 4 December 1985.

98. Goldberg to Spitzer, 5 October 1948. LG/HUA.

99. Spitzer to Goldberg, 14 October 1948. LG/HUA.

100. Leo Goldberg OHI no. 2, pp. 26–33rd; quotations from pp. 55rd. and 43rd.

101. Krause OHI, pp. 55–56; see also D. Purcell OHI, pp. 43–44.

102. In 1947, the annual budget of NRL's Rocket-Sonde Research Section for fiscal 1949 was estimated at $2,340,000 within which $500,000 was earmarked for upper atmosphere research alone. In contrast, an American Astronomical Society elite committee recommended in 1950 that the needs of astronomy could be met if the proposed National Science Foundation provided $540,000 annually, in addition to what their home universities traditionally provided. See E. H. Krause to Code 100, 21 February 1947. Box 99, folder 5, NRL/NARA; and "To the Members of the American Astronomical Society," 16 February 1948. JGP/CIT. See also C. D. Shane, "Preliminary Proposals for Support of Astronomy by the National Science Foundation," *The Astronomical Journal*, 55 no. 1183 (1950), pp. 84–86. These matters are examined in detail in DeVorkin (1991).

103. On this shared trait, see Kenat and DeVorkin (1990).

104. Eventually, Clearman was accepted as a thesis candidate by Allan Shenstone at Princeton, and in 1951–1952 completed his dissertation, which was a reevaluation of the best V-2 spectra from both APL and NRL. Clearman to Greenstein, 3 August 1948. JG/CIT. See also Clearman (1953); Hopfield and Clearman (1948).

105. Durand, Oberly, and Tousey (1947, 1949). On the theory of multiplets, see Kenat and DeVorkin (1990).

106. Baum et al. (1946); Durand and Tousey (1947); Krause (1948); Durand, Oberly, and Tousey (1949).

107. See "Structural Organization Rocket Sonde Research Section Radio Division I, NRL," 10 February 1947. EHK/NASM.

108. Tousey to Menzel, 15 October 1947. DHM/HUA; see also March 1947 Krause memorandum to NRL Director on plans for establishment of high-energy physics section. EHK/NASM.

109. Tousey OHI, p. 169rd.

# 12

# Pointing Controls and the Race to Lyman Alpha

Although Richard Tousey had not had a successful spectrograph launch from 7 March to July 1947, he was more concerned about not being able to break the 200 nanometer barrier, reach Lyman Alpha, and say something useful about the solar influence on the ionosphere. APL was already experimenting with heliostat homing mirror systems, and both Tousey and Eric Durand realized that NRL should be doing the same.

Indeed, there was considerable room for improvement, as discussions during the first two years of V-2 Panel meetings made clear. The leading Air Force representative, Marcus O'Day, realized that to make a mark in this area, he had to find a group able to develop a means of pointing instruments directly at the sun during the rocket flight: such control would not only increase available exposure times, but would provide the means to stabilize other devices pointing back at earth.

## Attempts by NRL and APL

APL concentrated on tiny servo-driven photoelectric homing mirrors to direct light into their spectrographs, whereas NRL created a large servocontroled cradle that could direct Tousey's entire Baird spectrograph toward the sun. Neither group achieved a full measure of success; even though the miniature APL heliostats worked on several flights, they did not produce any breakthrough spectra.[1] The bulky NRL device based on a tracking unit from an SCR-584 radar antenna was never given a full test: after several years of expensive and frustrating experimentation, the two

prototypes that were built were lost in a series of V-2 launch failures (figure, p. 222).[2]

Although the NRL sunfollower was never given a proper test, some doubted that it would ever work. Tousey laments that "we never had the money or the courage to try it again." It was attempted "so to speak, . . . to get ahead of Van Allen."[3] The NRL group responsible for the device was, in F. S. Johnson's opinion, not "attentive to miniaturization." As a result, the original pointing control they designed "was something of a monstrosity."[4] At NRL, heavy radar servosystems constituted available and familiar technology, and NRL's future, they hoped, lay with the similarly sized Viking. APL's solar homing instrument was born of a tradition of building little devices for proximity fuzes, miniaturizing everything to keep inertial mass as small as possible (figure, p. 223). Also, the future they hoped to have in space lay not with the Viking, but with the far smaller Aerobee.

## The University of Colorado

By the fall of 1947, Marcus O'Day, anxious to succeed where the Navy had failed, searched for a group interested in building a device that could stabilize a solar coronograph on a rocket.[5] He found no interest among established research groups, but through Donald Menzel and his Harvard staff, learned that physicists at the University of Colorado were ready to give it a try.[6]

O'Day wanted to be able to image the corona for several reasons. He was influenced

Door ejection test for NRL sunfollower, White Sands Proving Ground, circa 1950—51. A modified Tousey spectrograph can be seen cradled inside the massive sunfollower, whose azimuthal motion was accomplished by rotating the entire cone. F. S. Johnson collection. NASM SI 90—13029.

The fate of one of NRL's sunfollowers, 14 June 1951. This was V-2 55 at the instant its nosecone ejection system detonated. The ejection system was designed to allow the NRL sunfollower to be recovered. Kennedy (1983), p. 61. The scene does not match exactly an account by Herb Karsch, "Panel Report no. 29," 14—15 August 1951, p. 15, but it is the only record of a launch pad explosion in Smith (1954).

APL grating spectrograph flown on V-2 30, 29 July 1947, with miniature two-axis sunseeker homing mirror visible on the underside, just above the instrument base. This system provided resolution to 0.07 nanometers between 210 and 320 nanometers, according to "V-2 Panel Report no. 12," 10 October 1947, p. 7. V2/NASM. Applied Physics Laboratory, JHU/APL archives.

first by Menzel, who wanted the work to complement the agenda of his High Altitude Observatory in Climax Colorado, which was devoted to coronal studies that could aid ionospheric radio research. Even though coronal studies did not require access to the far ultraviolet, a coronograph in space would not suffer from the severe light scattering of the atmosphere. At the same time, O'Day knew that the Air Force was interested in building stabilized platforms for aerial reconnaissance studies from balloons, and eventually from spacecraft. Here was a convenient way to get started, because the pointing requirements of a coronograph were similar to those required to image the earth from space.[7]

In fact, the coronograph had to be kept so stable that Tousey thought the whole idea absurd at the time. A coronograph is a high-precision optical telescope with a small occulting disk on its optical axis to block out the intense light of the solar surface, so that the far fainter outer atmosphere can be examined. Ideally, the angular diameter of the occulting disk should be just a bit larger than the solar angular diameter, 1/2 degree, and the pointing accuracy a small fraction of that.[8]

Unrealistic as it may have seemed, O'Day funded the project, and between April and May 1948, University of Colorado Physics Department chairman William B. Pietenpol set up his staff and began to look at various possible designs for the coronograph and pointing control servomechanisms.[9] Pietenpol brought together physics faculty who specialized in electrical engineering; some had worked at the Armour Research Foundation and at the OSRD Underwater Explosives Laboratory during the war. Pietenpol also had access to mechanical and aeronautical engineers, good instrument shops and a staff of general technicians with backgrounds in mechanical engineering, plastics technology, radar electronics, and engineering physics. Only

Pietenpol held a physics doctorate, but Walter Orr Roberts. director of Harvard's High Altitude Observatory and members of his staff (for example, J. W. Evans), had extensive experience in optical design and were happy to help out. In total, there were 20 people in the group, and, by July 1948, they had set up shop and were making preliminary design studies.[10]

## Initial Designs for the Coronograph and Pointing Control

The Colorado group, assisted by O'Day and the astronomers, had little trouble finding potential applications for their anticipated pointing control.[11] Beyond the coronograph, they first planned to build a solar disk imaging monochromater, as well as spectrographs. Their basic pointing control design, then, had to be versatile, and so they decided to follow NRL's lead and build a two-axis cradle that could point whole instruments. But wary of the problem of inertial mass, they also followed APL's style and made the cradle as light as possible for instruments that could be flown on Aerobees.

O'Day brought Colorado team members into contact with other rocket groups sponsored by the Air Force. Pietenpol painfully realized that this was no normal engineering venture: it was one "aggravated by the fact that the device has to work in a rocket."[12] Pietenpol's staff knew that they would have difficulty achieving the arc minute accuracy required for coronal studies because the stabilization device also had to have a coarse homing mode capable of searching for the sun rapidly over wide arcs of the sky as the rocket tumbled. They knew that these two requirements conflicted. Thus as they started to learn about servomechanisms, they decided that their pointing controls would require coarse and fine guidance modes, controlled by two separate photoelectric systems.

Early ideas for hydraulic controls soon gave way to electrical servomechanisms that could be modeled theoretically. By the fall of 1948, the laboratory was testing breadboard models, some with three sets of searching, homing, and guiding systems they hoped would bring

the optical axis of the coronograph to within 1 arc minute of the center of the 30 arc minute solar disk.[13] In December, the laboratory announced that it had built one breadboard model whose coarse azimuth servosystem could correct a 90 degree error to a plus or minus 1 degree error in less than one-half second. But after six more months of refinements, their laboratory mock-ups could do no better.[14] To overcome this next level of response time and accuracy, the team decided that they needed improved theoretical analyses, even if these slowed the pace of experimentation. These deliberate efforts at theoretical modeling contrasted with the developmental styles in the early APL and NRL rocketry groups.[15]

Managing the program at Colorado became a complex affair itself. In 1949, as the AMC Upper Air Program at the University of Colorado grew, it developed into a combined program effort of the Physics Department and the university's Engineering Experiment Station. The new entity was called the Upper Air Laboratory, with sections specializing in electronics, servosystems design, a rocket field testing group, and a solar group that examined problems in solar radiation, disk imaging, solar prominence observations, and coronal observations. A separate Physics Instrument and Research Laboratory came directly under the Physics Department and included engineering design, drafting, and the instrument shop. It was also connected to the Upper Air Laboratory. Before it had any operational successes, the UAL accepted additional AMC subcontracts for sunfollowers to support other AMC rocketry research groups, including a one-axis azimuthal pointing control for a Denver University skylight spectrograph and a two-axis system to hold a monochromater that could return images of the sun in the light of the Lyman Alpha emission line. Yngve Öhman, a research associate of the Harvard College Observatory working at Climax with Roberts, helped to design the monochromatic camera, which soon became an internal UAL project.

This more complex administrative structure drew Pietenpol away from the daily operation of the laboratory, and others were brought in at all levels. Russell Nidey was added to the engineering and technical staff to develop the pointing controls, and William A. Rense was brought into the Physics Department from Louisiana State University in June 1949 to specialize in far ultraviolet spectroscopy. Rense had taught in the University of Colorado's summer school for several years before his full-time appointment, and Pietenpol knew that he was adept at designing and using vacuum ultraviolet spectroscopic equipment. By the time Rense was on the staff in the summer of 1949, his impression was that Pietenpol's highest priority was "a vacuum spectrograph to photograph the hydrogen Lyman Alpha line."[16]

## William A. Rense and Lyman Alpha

Rense had some prior interest and student experience in astronomy. As an undergraduate physics major at Case Institute of Technology, Rense took astronomy courses from Jason Nassau and conducted some astrometric research. In 1935, Rense entered graduate studies in physics at Louisiana State University but then switched to Ohio State University, which had both physics and astronomy and where he studied terrestrial molecular absorption in the solar spectrum under J. Allen Hynek.[17] His doctoral thesis, however, was in experimental physics, concentrating on vacuum ultraviolet spectroscopy.[18] After obtaining his Ph.D. in 1939, Rense held several teaching positions, and during the war he studied ultraviolet reflectance analysis techniques before coming to Colorado. Rense's background therefore paralleled closely that of Hopfield and Tousey in the field of the vacuum ultraviolet.

Rense was hired at Colorado to develop laboratory vacuum spectrographs in order to test the monochromatic camera, but he quickly became a leading contributor to the design and construction of all the Colorado payload instruments.[19] He first helped with the coronograph. Pietenpol's team always knew that the coronograph was "inherently a difficult problem."[20] But by late 1949, they all realized just how difficult it was to make a co-

ronograph work on a rocket; at least by then they knew that they were not likely to add anything to what Roberts' and Menzel's High Altitude Observatory could already do from earth.[21] This, together with the fact that the first-generation Aerobees were not too stable, and their first-generation pointing controls were not much better, made the coronograph seem out of the question for the near future.[22]

To ease themselves out of the coronal observation business, Rense, with J. M. Jackson, S. C. Miller, and W. E. Behring, concentrated on the monochromater. They wanted to use it in conjunction with a grazing incidence spectrograph to examine the solar disk in the Lyman Alpha emission line. They chose a modified Wadsworth design because it was optically the most efficient and hence produced the brightest image. Just as important, it was the simplest to produce and was also the most interesting to build for general laboratory use.[23]

Three separate teams built the monochromatic camera, the grazing incidence spectrograph, and the prototype two-axis pointing control in the summer of 1950. Rense, responsible for the first two, recalls that as he planned his grazing incidence device he assumed that the Lyman Alpha line was going to be narrow. But no one knew the intensity or even the half-width of Lyman Alpha with any precision; they therefore had to plan for a wide range of exposure times.

The grazing incidence spectrograph was straightforward using a Rowland grating, slit, and scatter-reducing baffles. Rense was familiar with grazing incidence spectrographs since his days at Ohio State and naturally decided that this was the most efficient design to employ if Lyman Alpha and its surroundings were the primary target. The dispersion of Rense's instrument was 1.3 nanometers per millimeter and resolution was 0.01 nanometers at Lyman Alpha—moderately high dispersion compared with that of contemporary instruments from NRL and APL. While Rense attended to the basic optical design and layout, others worried about its mechanical features, specifically, how the two instruments would fit back-to-back into the biaxial pointing control, with a combined total weight of less than 12 kilograms[24] (figure, p. 229).

The first Colorado biaxial pointing control to be flown carried the combined solar monochromater and grazing incidence spectrograph. It was launched on 12 April 1951 from Holloman Air Force Base aboard an Aerobee.[25] Thirty seconds after launch, a fuel line broke, and the rocket reached a height somewhere between 30 and 37 kilometers. The nosecone was ejected, and an Aerojet-designed parachute system brought the 115 kilogram nosecone and payload to earth some

10 kilometers downrange, with but slight damage. An initial analysis of telemetry led O'Day to report at the next UARRP meeting that the biaxial sunseeker worked properly during the flight. But later reports indicated that no reliable record of performance had been returned because the telemetry commutator failed during the flight.[26] Another University of Colorado pointing control was aboard an Air Force Aerobee launched a few months later, but again rocket failure compromised its performance. The pointing control did seem to work "during the one second after it was swung out and before [rocket] failure."[27]

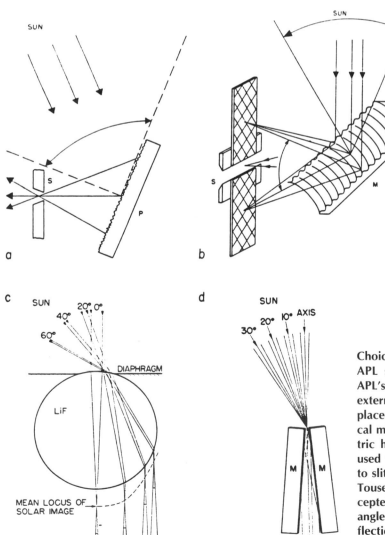

Choices for entrance slits for NRL and APL spectrographs through 1949: a) APL's stationary diffuse reflector sitting external to the spectrograph; soon replaced by b) APL's corrugated cylindrical mirror, part of an active photoelectric homing system; c) Tousey's bead, used throughout the 1940s in parallel to slit spectrographs; and d) Baum and Tousey's mirror-jawed slit which accepted light from a comparatively wide angle and fed it by multiple internal reflections to the grating. From Tousey (1953b), fig. 24. U.S. Navy photograph.

## INSTRUMENT IN UPPER ATMOSPHERE

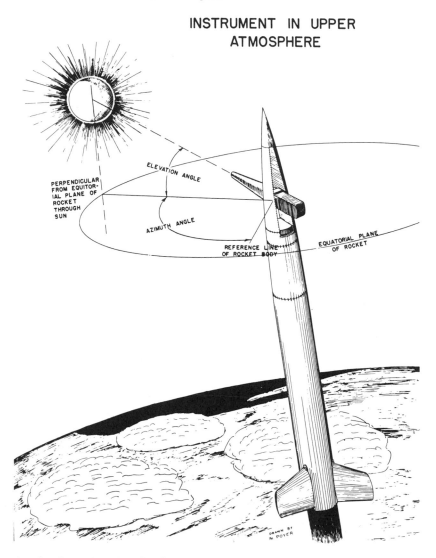

Basic configuration for the University of Colorado pointing control. Sets of photoelectric eyes would search out and lock onto the sun, sending signals to servomotors that would keep the spectrograph oriented toward the sun while the rocket tossed and spun. In spite of the original design specifications of one minute of arc pointing accuracy, operational field testing in 1949 and 1950 rarely exceeded accuracies of +/- one degree. As of April 1953, Colorado noted that "a delivered accuracy of +/- 30 minutes is a conservative figure with which to work." Their fourth flight of a biaxial pointing control did hold to within +/- 15 minutes of arc, the tolerance Leo Goldberg believed could be achieved in his musing in September, 1945. See Goldberg to Menzel, 28 September 1945. DHM/HUA, and Stacey (1953), p. 2.07.03. From University of Colorado Upper Air Laboratory Progress Report no. 8. 29 April 1950, fig. 2.

The Aerobees were consistently failing just when the Colorado biaxial system was being tested. Since there had been several recent Aerobee failures at Holloman and White Sands, the Air Force and the UARRP asked Aerojet to evaluate the problem; eventually, older Aerobees were returned to Aerojet for refurbishment, and a newer Aerobee-Hi was developed to replace them.[28] Nevertheless,

firings continued. In October 1951, O'Day outlined for the UARRP his planned Aerobee firings for 1952, which included sunfollowers in January, May, and June, among 10 scheduled flights.[29]

The first Aerobee to fly successfully for quite some time (after five failures in a row) was an NRL solar ultraviolet experiment in mid-February 1952, equipped with a heliostat sunfollower. Although the NRL sunfollower worked, it was misaligned and kept the instrument pointed consistently 5 degrees away from the sun.[30] AFCRC Aerobees were successful in May and June when payloads from Denver and from the University of Colorado both flew, but sadly this time, both payloads failed.

O'Day announced that the next biaxial system would fly in December 1952, and even though the device had malfunctioned in June, he endorsed it "as a foundation for an extensive Air Force radiation program of measurement of the solar constant, the spectroscopic study of the solar ultra-violet, and the University of Denver spectroscopic studies of the airglow."[31] But to date, not one sunfollower, from any source, had worked completely to specification, after over five years of effort.

## Lyman Alpha Attained

By the fall of 1952, Rense canceled the monochromater, feeling that the grazing incidence spectrograph had a better chance of capturing the Lyman Alpha line itself, even if the pointing control provided only minimal performance. On 11 December, after a week of final tests and simulations at Holloman, the Colorado group fitted the spectrograph into the pointing control one last time and inserted two protective doors over the system in the warhead. They then turned over the payload to the waiting Holloman launch crew.

This time, everything worked perfectly as the Aerobee climbed to 85 kilometers.[32] One 28 second exposure revealed the solar continuum from 200 to well past 500 nanometers, with well-defined absorption line structure. Blueward of 210 nanometers, continuum brightness fell off rapidly, and the next object to be seen clearly was Lyman Alpha in emission, standing quite alone at 121.5 nanometers[33] (figure, p. 230). The total intensity of the line was estimated at about 0.05 microwatts per square centimeter above the earth's atmosphere, which Rense found agreed with Tousey and Kenichi Watanabe's recent

measurements using thermoluminescent $CaSO_4$ phosphors activated by manganese.[34]

Before Rense's successful flight, the structure of solar Lyman Alpha was unknown. Tousey's and Herbert Friedman's rougher nondispersive observations with phosphors and filters indicated that emission had to exist, but the character of the emission remained unknown.[35] Tousey supported the phosphor work by Watanabe as an expedient, but he always found photographic "spectra . . . more convincing,"[36] and it was Rense's spectrum, produced by the first use of a grazing incidence spectrograph in a working biaxial pointing control, that proved that Lyman Alpha was indeed an emission feature.[37]

Both Rense and Tousey were aware that knowing the intensity and character of Lyman Alpha would help solve many problems both in solar and ionospheric physics. It turned out to be the strongest single emission feature in the extreme ultraviolet: "Its intensity," Tousey later noted, "is at least as great as the entire integrated solar flux from Lyman [alpha] to zero angstroms."[38] Therefore its reconnaissance was likely to contribute to a better understanding of the structure of the chromosphere and corona.[39] Its strength and variability certainly played a part in controlling terrestrial ionospheric phenomena, although Herbert Friedman's electronic observations of the extreme ultraviolet and X-ray region at about the same time showed that the more significant solar driver of the ionosphere existed in the higher-energy coronal realm, which both Rense and Tousey then set out to conquer with dispersed spectroscopic photographic techniques.

# Continued Refinements in the 1950s

With pointing controls an operational reality in the 1950s, spectroscopic groups at NRL, the University of Colorado, and the Air Force Cambridge Research Laboratory could now redesign their instruments for longer expo-

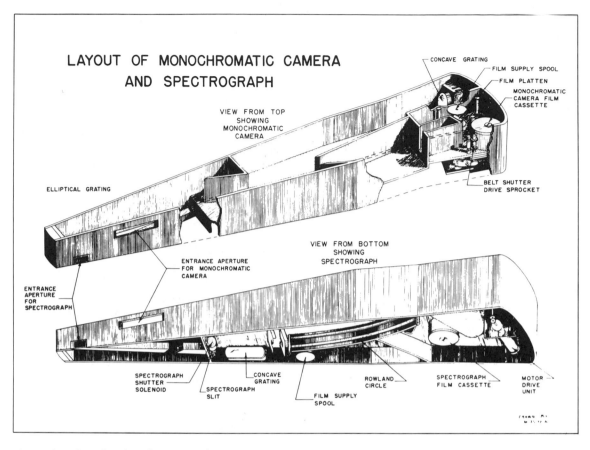

LAYOUT OF MONOCHROMATIC CAMERA
AND SPECTROGRAPH

The University of Colorado's monochromatic camera and grazing incidence spectrograph were mounted back-to-back in the same box and flown on 12 April 1951 cradled in a pointing control. Sunlight entering from the left passes through the spectrograph slit and is dispersed and focused onto a long strip of film drawn along the grating's Rowland circle. From University of Colorado Progress Report no. 8. 29 April 1950, fig. 18.

sure times and higher dispersion with finer resolution. After a visit to Rense's laboratory to examine the Colorado pointing controls, Tousey recalls that "once we found that they really worked, we were very anxious to get them."[40]

Tousey had already had some success with an NRL one-axis pointing control. In September 1952, he was able to record the solar continuum to 190 nanometers with good line resolution. A new lightweight Rowland spectrograph recorded second-order spectra at the comparatively high dispersion of 2.2 nanometers per millimeter, and was able to produce an image of the magnesium II doublet at 280 nanometers that showed the in-

tense emission cores better than any had before.[41] Tousey looked forward to even better performance with the Colorado pointing system.

After his successful capture of Lyman Alpha in December 1952, Rense wanted to improve his grazing incidence spectrograph using new collecting mirrors that cut down the inherent astigmatism of the grating. Through the 1950s, he produced a series of refinements as he worked with T. Violett and numerous students to elaborate on the Colorado grazing incidence system.[42] Tousey also began to consider more sophisticated devices, notably the doubly dispersing optical designs that Menzel suggested in 1946. The possibility of longer exposure times also made stray visible light problems more immediate: it set, Tousey recalled, "a detection limit on the weak emission lines and continuum in the extreme ultraviolet."[43]

Both Tousey and Rense adhered with almost religious zeal to photographic recording, which was especially susceptible to scattered visible light. Nevertheless, they felt that the information density of a photograph far exceeded any other available means of recording and was the only practical means of obtaining information on spectroscopic line structure. Their parallel effort through the 1950s therefore can best be depicted as an ever-deeper probing into the ultraviolet with finer resolution and higher definition. With pointing controls that allowed them to increase exposure times from seconds to minutes and with improved short-wave photographic emulsions, Tousey and Rense added doubly dispersing spectrographs, imaging spectroheliographs, and, in the early 1960s, high-dispersion echelle spectrographs and coronographs. They both found ways to overcome the stray light problems and eventually developed standardized light-weight spectroscopic systems for Aerobee flights. Tousey found the greatest degree of instrument flexibility in a dual-spectrograph design employing two normal incidence grating instruments

sensitive to the range between 50 and 300 nanometers, whereas Rense preferred to fly single instruments.[44] And finally, by the mid-1950s, Aerobee parachute systems had been perfected to the point where Krause's one-shot philosophy, that instruments would be destroyed upon use, was replaced by a sense that all effort should be made to reuse these expensive devices, gathering invaluable diagnostic as well as scientific data from them as they became mature tools for research (figure, p. 231).

## The Fate of the Colorado Group

In January 1953, Homer Newell and the UARRP congratulated O'Day, and especially his Colorado contractors, for the first photographic image of Lyman Alpha. They all now knew that reliable sunfollowers had arrived and would be available to anyone capable of buying them. To aid in their marketing, the University of Colorado Upper Air Laboratory produced a training manual for the use of the pointing control.[45]

With an operational pointing control and Pietenpol's retirement at age 70 in 1954, a number of changes took place in the Upper Air Laboratory. Without the UAL's chief protector, Rense found that his "number-one

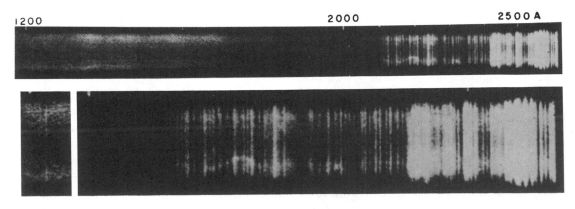

The image of the Lyman Alpha emission line, at 121.5 nanometers, captured during a 28 second exposure by William Rense's grazing incidence spectrograph, flown on 11 December 1952. Rense (1953), 300—01. See also Pietenpol, Rense, Walz, Stacey, and Jackson (1953), 156; and Tousey (1953b), 674. Examination of the Lyman Alpha line by microphotometer revealed both a central emission core and two broader emission wings. Richard Tousey collection. NASM.

Recovered payload from a Laboratory for Atmospheric and Space Physics flight at White Sands, circa early 1960s. LASP was an outgrowth of Rense's group at the University of Colorado. White Sands Missile Range photograph. Fred Wilshusen collection.

task was to keep the physics department interested in the program so that we could stay in the physics department, and I had a tough time there."[46] Some of the physicists, he felt, "thought that any effort that was centered primarily on instrument development, of an engineering nature, should not be in the physics department."[47] Rense did not have Pietenpol's seniority and could not prevent the removal of the pointing control group to Colorado's Research Services Laboratory.[48]

It apparently made little difference to the physicists that the UAL brought in income from outside groups buying pointing controls. The university felt otherwise, and sought mechanisms to preserve this lucrative activity. In the transition, however, a number of members of the Upper Air Laboratory and the Research Services Laboratory decided to leave the university and develop a commercial venture, primarily to exploit their collective talents and experience, but also to get away from the stigma of second-class status on campus. Not everyone left; in fact, for about 10 years, pointing controls were produced both by remaining Research Services Laboratory personnel and by the splinter group, which reorganized as a subsidiary of Ball Brothers Corporation of Muncie, Indiana.[49]

Solar research was not a mainstream astronomical activity in the United States, even after World War II. Its links to ionospheric research gave it, in Walter Orr Roberts' words, "a demand [and] a constituency."[50] But this did not create many converts in established centers of astronomy or physics. Beyond the University of Michigan, the High Altitude Observatory, the rapidly weakening solar division at Mount Wilson, and the rocket groups at NRL and Colorado, little observational solar physics was being done, in spite of the strong support of the Navy and the Air Force for both ground-based and rocket-based solar research.[51] In addition, beyond Rense's feeling that faculty elitism worked against the rocket work at the University of Colorado and that few were sympathetic to solar research, according to Roberts, the faculty was also generally adverse to a military presence on campus, both from an administrative stand-point and from a philosophical aversion to what military funding implied for the future of academics in America. Roberts agreed with this view, aligning with Harlow Shapley rather than Donald Menzel. Shapley constantly worried that if directed by military patronage, his staff would become preoccupied with instrumentation rather than with research and publishing results. Indeed, although Air Force funding transformed a portion of a quiet Colorado physics department into a major research facility, it was many years before solid contributions were made to solar physics. Shapley's fears were not without substance.[52]

## Notes to Chapter 12

1. On APL's designs, see James A. Van Allen, "High Altitude Research at the Applied Physics Laboratory of the Johns Hopkins University—A Brief Summary," 10 April 1947, Appendix D, p. 2. See also Hopfield and Clearman (1948), p. 879; Fraser and Siegler (1948).

2. Durand and Tousey turned to Harry L. Clark to build the servosystem. Clark was then the head of an infrared project in Hulbert's Optics Division and had electronics expertise and the resources to coordinate the effort. While Clark designed and built the sunfollower, Durand and Tousey contacted the Air Materiel Command at Wright Field for parachutes, and the Bureau of Ordnance Proving Ground at Dahlgren for an ejection device. Such a costly system had to be recoverable, even in the Navy. See Clark (1950), p. 71.

3. Tousey OHI, pp. 156–58rd.; quotation from p. 158rd.

4. F. S. Johnson OHI, pp. 44–55; quotation from p. 55.

5. K. W. Patrick to Com. Gen., Air Materiel Command, 29 September 1947, "Request for Construction of Parachute for Naval Research Laboratory Sun-follower Mechanism," p. 2; and Eric Durand, "Record of Consultative Services," 30 September 1947, p. 2. Box 99, folder 9, NRL/NARA.

6. Vacuum ultraviolet physicist William Rense, who later became part the group, recalls hearing that O'Day had been visiting departments of physics around the country looking for people who were interested in applying their talents to scientific rocketry but was not meeting with much success. Rense OHI, pp. 16–17rd.; and Walter Orr Roberts OHI, pp. 99–100rd. Roberts declined to get involved with rocketry directly. He had his hands full maintaining Harvard's solar observatory at Climax. Roberts recently recalled, "It seemed to me a waste of talent to learn to do things in rocket flight, because we conceived of the corona research [at the High Altitude Observatory] as integral and very essential to [measurements] from outside the atmosphere with rockets." Roberts OHI, p. 99rd.

7. The connection is at present only circumstantial. See Davies and Harris (1988).

8. Tousey OHI, p. 155rd. Coronographs were first successfully flown on rockets in the early 1960s. On the development and use of coronographs in solar physics, see Hufbauer (1991), pp. 91–96; and Evans (1953).

9. Pietenpol received his doctorate in physics from Wisconsin in 1916 and had been teaching physics at the University of Colorado since 1920. He formed his team at about the same time Clark was designing his first two-axis sunfollower.

10. See W. B. Pietenpol, "Airborne Coronograph for Rocket Installation Progress Report no. 1, April 1, 1948 to July 1, 1948," University of Colorado, 30 June 1948, pp. 8–9. CF/AFGL.

11. Pietenpol (30 June 1948), ibid, p. 15.

12. Pietenpol, ibid., p. 12. O'Day knew that his real political strength lay in his contractors and that it was in everyone's best interest to keep them in close contact and informed about AAF upper air research activities. Accordingly, in May 1948, O'Day sponsored an AMC Upper Air Research Symposium at the Watertown Arsenal, inviting all contractors to attend, to participate, and to learn. The four-day symposium included tutorial sessions on the AMC and CFS upper air programs, White Sands operations, military security, travel and transportation, procurement, and other administrative matters, as well as technical sessions on the theory and design of ionospheric beacons and beacon triangulation techniques, on V-2 common power supply systems, and on data transmission services. The Watertown arsenal symposium became more than a meeting of contractors and patrons. It was a mechanism for university-based scientists to become familiar with

the infrastructure of missile research and development. It was also an introduction to a new world for laboratory experimenters used to buying or building devices for the workbench, a world of new tool-building possibilities and experimental capabilities made possible by Air Force interests and needs. W. B. Pietenpol and his staff came away from this symposium with a heightened sense of what lay before them second only in value to direct site visits to White Sands. See W. B. Pietenpol, "Airborne Coronograph . . . Report no. 1," pp. 8–9. CF/AFGL.

13. W. B. Pietenpol, "Airborne Coronograph for Rocket Installation Progress Report no. 2, July 1, 1948 to October 1, 1948," p. 10. CF/AFGL. The Colorado logic diagram was identical for each axis; both the fine and coarse photoelectric cells worked simultaneously, one overriding the other through a differential. For the technical details, see [Pietenpol], "Historical Report," 1 April 1948–31 December 1948, p. 5. CF/AFGL.

14. [Pietenpol], "Airborne Coronograph for Rocket Installation Progress Report no. 4, January 1, 1949 to April 30, 1949." CF/AFGL.

15. It is quite possible that similar theoretical studies took place within the latter two shops, but they were not compelled to report them.

16. William Rense OHI, pp. 16–19rd.; quotation from p. 18rd.

17. Ibid., pp. 4–7rd. He took his physics from J. Miller and R. S. Shankland at Case.

18. Rense at first hoped to pursue experimentally an interpretation of the coronal spectrum, but ended up studying the laboratory spectra of yttrium and zirconium in the extreme ultraviolet using both Schumann and phosphorescent oil emulsions. Ibid., pp. 8–10rd.

19. See "Progress Report no. 5," University of Colorado Upper Air Laboratory, 1 January 1949 through 1 July 1949, p. 46. WAR/SAOHP.

20. See *University of Colorado Upper Air Laboratory Historical Report* no. 2, 1949, p. 11. WAR/SAOHP. In all, 15 optical transmission surfaces and two reflection surfaces were involved in the coronograph's design, and every one of them had to be flawless, to reduce inherent instrumental light scattering.

21. *University of Colorado Upper Air Laboratory Progress Report* no. 6, 31 October 1949, pp. 22–23. WAR/SAOHP.

22. See W. B. Pietenpol, "Airborne Coronograph for Rocket Installation Progress Report no. 2 July 1, 1948 to October 1, 1948," pp. 6–8. CF/AFGL; and O'Day's complaints in "Panel Report no. 28," 25 April 1951, p. 11. V2/NASM.

23. *University of Colorado Upper Air Laboratory Progress Report* No. CL-1, 30 June 1950, p. 7. CF/AFGL; see also: *University of Colorado Upper Air Laboratory Special Report* no. 2 "An Ultraviolet Monochromatic Camera," vol. 1. 15 December 1950, pp. 81–84; and *University of Colorado Upper Air Laboratory Progress Report* no. 8, Contract No. W19–122ac-9, 29 April 1950, pp. 27–28, WAR/SAOHP; and Behring, Jackson, Miller, and Rense (1954), pp. 229–31.

24. *Upper Air Laboratory Progress Report* no. 8, 29 April 1950, p. 33. WAR/SAOHP.

25. Holloman became the Air Force's launch site for Aerobees. See Meeter (1967), and Benton (1959).

26. Smith (1954); Report EHO-24, Holloman Air Force Base, 31 May 1951; Report MHTH-133, Holloman Air Force Base, December 1951; *University of Colorado Upper Air Laboratory Progress Report* no. 11; and no. 12, Contract W19–122ac-9, 31 January 1951 to 30 April 1951. WAR/SAOHP; "V-2 Panel Report no. 28," 25 April 1951, pp. 5–6. V2/NASM.

27. "V-2 Panel Report no. 29," 14–15 August 1951, p. 18. V2/NASM. O'Day's description of the fate of Aerobee no. 14 in the minutes of the 29th meeting of the V-2 Panel does not agree with C. P. Smith's later chronology. Smith (1954).

28. In addition to the limitations of the first-generation Aerobee, panel members found many operational problems, each unique to each flight failure. Incorrect fuel mixtures, improper fueling procedures, contamination, leakages, and other problems associated more with launch procedures and lax launch crews than with design, created a most frustrating profile of relaxed standards and little attention to detail at the launch pad. See "V-2 Panel Report no. 29," 14–15 August 1951, p. 20. V2/NASM.

29. "V-2 Panel Report no. 30," 21 October 1951, p. 12. V2/NASM.

30. "V-2 Panel Report no. 32," p. 5. V2/NASM.

31. Ibid.

32. Ninety seconds into the flight, the payload doors blew off and the pointing control ser-

vomechanism came to life. The pointing control jumped into its search mode and sought out the sun in a matter of seconds. Once the sun was found, the fine error sensors kicked in and the servomechanism started counteracting the motions of the missile. Scenario based upon commentary in "A Typical Flight" in David Stacey, (Colorado "Special Report," April 1953). See Stacey (1953), p. 1.02.01.

33. Rense (1953), pp. 300–01. See also Pietenpol, et al. (1953), p. 156.

34. Ibid., Rense, p. 301; Tousey, Watanabe, and Purcell (1951).

35. Tousey (1953a), p. 250.

36. Tousey (1967), p. 242.

37. Tousey (1963), p. 3.

38. Tousey (1967), p. 242. An Ångstrom is 1/10 nanometer.

39. Rense OHI, p. 26rd.

40. Tousey OHI, p. 221rd; see also pp. 119–121rd.; Rense OHI, p. 29rd.

41. Tousey (1953b), p. 670; Johnson, Purcell, Tousey, and Wilson (1953); "V-2 Panel Report no. 33," 7 October 1952, p. 10. V2/NASM.

42. Rense and Violett (1959).

43. Tousey (1963), p. 7.

44. Ibid., p. 6, see Fig. 2.

45. Stacey (1953). Copy courtesy Fred Wilshusen, SAOHP Files.

46. Rense OHI, p. 50rd. See also Bartoe commentary in Ball Brothers OHI no. 2, p. 26rd.

47. Rense OHI, p. 51rd.

48. Bartoe commentary in Ball Brothers OHI no. 2, p. 26rd; Rense OHI pp 20; 26–29.

49. See oral histories of Rense, Roberts, Dolder, J. Simpson, and the Ball Brothers Research Corporation group. SAOHP/NASM. Ball Brothers Research Corporation became a major supplier of pointing controls and stabilization systems for the Air Force as well as for NASA rocket payloads, for the first half-dozen satellites in the OSO series, and for many more recent satellite systems.

50. Roberts OHI, p. 86rd.

51. In a general review of observatories engaged in some form of optical solar research, no matter how small, 51 observatories were identified worldwide, but only nine of them were in the United States. R. Coutrez, "Table 1. Solar Optical Instrumentation," in Kuiper (1953), pp. 728–33. Also, of the 19 observatories engaged in solar radio observations throughout the world, only three were American: at the National Bureau of Standards' Central Radio Propagation Laboratory in Boulder, at Cornell University, and at the Naval Research Laboratory. J. P. Wild, "Table 2 Solar Radio Instruments," in ibid., pp. 734–37. On solar physics immediately after World War II, see Hufbauer (1991), chap. 4.

52. Menzel was singularly anxious to exploit military support, but Shapley was against it, which created considerable rancor and confusion within the Harvard department and at Climax. Shapley to Roberts, 25 July and 8 August 1947; Roberts to Shapley, 13 August 1947. WOR/UC. See also Roberts OHI, p. 159rd. The situation among the faculty at Colorado was similarly described by Rense OHI, pp. 20; 51, and requires further attention. See DeVorkin (1991).

# 13

# Beyond Lyman Alpha and Ozone

In Chapters 11 and 12, we saw how photography was exploited to study the sun from rockets. Alternatives to photography—both electronic and photochemical—were available, but the groups we have been examining at NRL, APL, and Colorado were led by people trained in vacuum ultraviolet spectroscopy, which was traditionally wedded to photographic recording. At two points in the first decade, however, there were compelling reasons to turn to these alternatives. The first came in the spring and summer of 1946, when physical retrieval problems created a summer of doubt. And the second came in the late 1940s, when the 200 nanometer barrier had still not been breached.

## Alternative Detectors and the Push to the Extreme Ultraviolet

In 1946, worried about data retrieval and the problem of severe visible light scattering, both APL and NRL, as well as Spitzer, Goldberg, and Greenstein, toyed with the photoelectric recording of spectra and the use of thermoluminescent crystals that could record high-energy radiation while remaining insensitive to visible light. The astronomers knew from Gerard Kuiper's intelligence reports that halogenid crystals developed by R. W. Pohl would turn color or change in internal electrical resistance when exposed to far ultraviolet solar radiation. After much haggling and negotiations through various military channels, Whipple and Menzel managed to secure examples of these crystals for Jesse Greenstein and Leo Goldberg in July 1947. Greenstein received six halogenid potassium bromide crystals activated by potassium hydride. They were sensitive to radiation of wavelengths less than 228 nanometers, and so were potentially useful for James Van Allen's ultraviolet altimeters for V-2 rockets.[1]

The Pohl crystals were difficult to fabricate and to use. Pohl had advised Kuiper that while halogenid crystal chemistry was theoretically straightforward, "there was a great deal of 'know-how' in the successful preparation of these crystals."[2] This did not diminish astronomers' enthusiasm for them; and in January 1948, Van Allen announced plans to fly a set of crystals.[3] Van Allen's group indeed flew Pohl crystals on a V-2 on 27 May 1948. The flight itself was a success, but after surviving impact, the crystals met an ignominious fate when they were dissolved by water that had collected in the impact crater before the recovery team could get at them.[4] Van Allen thus returned to more certain technology for his hoped-for altimeter, and Pohl-type crystals ceased to be considered by any of the rocketry groups.

Krause was not impressed with Pohl crystals. In the wake of the June 1946 flight, he directed J. J. Oberly, who had been somewhat involved with Tousey's spectrograph work, and two electronics specialists, M. L. Greenough and C. C. Rockwood, to design a photoelectric spectrometer to telemeter spectroscopic information to the ground.[5] They developed a simple fixed multidetector spectrometer to sample two regions of the solar spectrum, one centered at 290 nanometers, near the edge of the Hartley ozone band, and the other at 121.6 nanometers, centered on Lyman Alpha. Tousey's lithium fluoride beads acted as substitute slits, and the grating was also Tousey's. Thus in every respect save for the photoelectric detectors—standard RCA tubes—the device was intended as a brute-force tool to maximize the chances that something would be seen. The feat was more important than the fidelity of the result.[6] One instrument was built during the summer of 1946, but when Tousey's 10 October flight returned successful spectra, the photoelectric project was abandoned, even though several junior staff wanted to see it through.[7] The rapid retreat from this line of instrumentation speaks to the technical priorities of the group, and most of all to Tousey's tenacity and penchant for photography.

Tousey's focus on photography was tested again in the late 1940s when the 200 nanometer barrier refused to yield.[8] Unlike the near ultraviolet, the region beyond the barrier was unknown territory. Theorists were unsure whether continuum intensity dropped naturally, because one was looking at a realm dominated by chromospheric emission, or whether it remained a black-body spectrum, but was rendered dim by the parlayed effect of terrestrial and solar atmospheric absorption.[9] Tousey's group had found that the black-body continuum of the sun was depressed in the 200 to 300 nanometer region, but neither they nor anyone else could extrapolate beyond that point.

Although initially promising, by the late 1940s, dispersive techniques employing photography had failed dismally to answer these questions, which underscored the compelling problem of the solar source of the ionosphere. Thus, as a side effort starting in 1948, Tousey asked Kenichi Watanabe to prepare nondispersive phosphorescent and thermoluminescent detectors that might reach Lyman Alpha.[10] Watanabe had joined Tousey's staff as a physicist in 1948 after eight years of college teaching and a 1940 doctorate in physics from Caltech under John Strong, where he came under the influence of Strong's creative laboratory style and interest in detectors for infrared atmospheric ozone research. Watanabe was therefore well suited for adapting thermoluminescence techniques to ultraviolet spectroscopy for similar ends, and soon found that calcium sulfate phosphors doped with

manganese ($CaSO_4$:Mn) were sensitive only to radiation short of 140 nanometers. This made them nicely immune to scattered visible light. After exposure to the hard ultraviolet or X rays, these phosphors would, when heated in an oven, release visible radiation in an amount proportional to the amount of hard radiation absorbed.[11]

Martin J. Koomen helped Watanabe find suitable filters to place in front of the phosphors to further isolate the Lyman Alpha region, as well as other bandpasses.[12] Watanabe, Koomen, and DeWitt Purcell prepared sets of phosphors that were flown on four V-2 flights between 1948 and 1950. Those that were successful (that is, those recovered intact) demonstrated that Lyman Alpha dominated the far ultraviolet solar spectrum.[13]

Although Watanabe's phosphors apparently detected soft X-ray radiation in one instance, Tousey contends that another NRL experiment had the honor of the first "quick and dirty" confirmation that the sun emitted X rays.[14] In August 1948, T. R. Burnight, in Krause's section, placed standard photographic Schumann plates behind beryllium filters and flew a set of them on an NRL V-2. The processed plates showed darkening that could have meant exposure to radiation shorter than 0.8 nanometers.[15] Burnight repeated his experiment on 9 December 1948, using both beryllium and aluminum filters, but was not able to reproduce the results of the 5 August flight. Tousey later argued that Burnight's 5 August results were real, because on that day "there was evidence for unusual solar and ionospheric activity."[16] Whatever the result of Burnight's August 1948 experiment, Tousey agrees that his techniques were crude and preliminary. They indeed were "quick and dirty" compared with Watanabe's more refined phosphors, or Friedman's electronic photon counters, which, Tousey later admitted, "offer[ed] the best means for studying soft X-rays from a rocket."[17]

The NRL Rocket-Sonde Research Section did not turn to photoelectric designs largely because of Tousey's preference for photography. James Van Allen's interest in an ultraviolet altimeter, however, directed him to photoelectric techniques. A photoelectric spectrometer could telemeter intensity information instantly and continuously and could also make possible "firing from shipboard at other latitudes where physical recovery [would] not be possible."[18] Since ozone absorption was the primary indicator of height, studying its distribution would also suggest how it could be used to guide a missile. Accordingly, in 1949, Harold Clearman and Van Allen built three- and four-channel quartz prism spectrometers based on Greenstein's 1947 designs, using optics he supplied from Yerkes. The design was straightforward. A spinning mirror sent flashes of sunlight 20 times per second into the spectrometer, which dispersed the light onto three slits set in the Rowland grating focal plane at points where ozone absorption dominated, in the Hartley ozone band at 260 nanometers, and at two points where the solar continuum was relatively free of absorption. Behind each slit sat a standard RCA photomultiplier.[19]

APL's first ozone spectrometer flew in June 1949, and provided a continuous profile of ozone absorption, which showed a double maximum not confirmed on later flights of their four-channel model, which had wider bandpasses and a refined rotating sunlight collector.[20] APL's penultimate Aerobee flight, on 25 January 1951, carried the spectrometer successfully to 88 kilometers. Hopfield and Van Allen obtained a smoothly varying ozone profile that compared with NRL's later measurements showing a real maximum at about 25 kilometers and a nearly continuous profile of ozone concentration from 25 to 60 kilometers.[21] Even though they had demonstrated the great value of this simple means of determining ozone profiles, this was APL's last spectroscopic flight; APL by this time was going out of the high-altitude research business.

## Herbert Friedman

The failure of photography to breach the 200 nanometer barrier and to reveal the solar source of the ionosphere captured the attention of Herbert Friedman, a solid-state physicist who headed the Electron Optics Branch within E. O. Hulburt's Physical Optics Division at NRL. In contrast to those whose prism and grating spectrometers relied on commercial photoelectric detectors or photographic plates, Friedman and his group built their own detectors, as their identity at NRL was as a detector development team.[22]

Herbert Friedman (figure, p. 240) was trained in experimental solid-state physics at The Johns Hopkins University under James Franck, R. W. Wood, and Joyce A. Bearden. Friedman had already obtained a taste for experimental physics from his undergraduate courses at Brooklyn College, and since Bearden was a specialist in X-ray spectroscopy, Friedman moved in that direction. Friedman quickly found that available detectors for X-ray spectroscopy were far from adequate. Laboratory electrometers used with ionization chambers were fine for high-energy cosmic-ray and alpha particle analysis, but were too insensitive for lower-energy X-ray event analyzers. At Bearden's suggestion, Friedman modified Geiger counters for the lower-energy X-ray region by building them with "a long absorbing column so that the X-ray quanta could enter the counter and have a good probability of being absorbed before they got to the end of the counter."[23]

With the help of a good glassblower in the Physics Department, Friedman managed to produce highly sensitive thin-window counters that could hold a small amount of quenching vapor made of argon gas and alcohol within a moderate vacuum. The counters worked well, and within a few weeks he was detecting soft X rays and, with a spectrometer, was outlining structures of absorption edges for the metals. "By the end of the month," Friedman recalls, "I knew I had a thesis."[24]

After graduation in 1940, Friedman continued with detector development. He recalls having no luck landing a university position in an anti-Semitic academic world, but through the good services and contacts of the Hopkins Physics Department chairman, by 1941 Friedman was working in NRL's Metallurgy Division on proximity fuzes.[25] Fried-

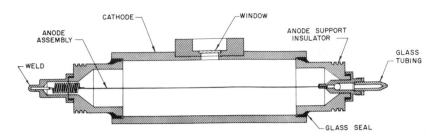

Friedman's counters acted as simple photoelectric amplifier circuits. Their outer shells, typically 2 cm diameter chrome-iron cylinders, functioned as electron-donating cathodes for the circuit with a thin isolated wire drawn down the 5 cm cylinder axis as the anode. In the prototypes built at NRL, glass caps sealed at either end of the cylinder supported the wire; in manufactured versions, ceramic caps replaced the glass. The entrance aperture for the illustrated side-fire version, preferred because the anode was supported at both ends, was covered by various filter materials depending upon the specific spectral region to be isolated. The filters acted in concert with various quenching agents (electro-negative gas fillings) to isolate desired spectral regions. Nitric oxide, chlorine, neon, bromine, or argon provided the different ionization thresholds, and, for the ultraviolet and soft X-ray regions, windows could be made of sapphire, beryllium, calcium fluoride, lithium fluoride, quartz, or barium fluoride. The substance and thickness of the window and the ionization potential of the gas filling the chamber determined the spectral response of the counter. U.S. Navy photograph. From Chubb and Friedman (1955), fig. 1.

man soon moved into radiography to use his efficient X-ray detectors for the stress analysis of metals and also developed detectors for quality testing and control devices. He constantly looked for quicker, more accurate, and more efficient techniques and soon caught Hulburt's attention, who then invited him into the Physical Optics Division to head up a new branch for electron optics.[26]

During the war, Friedman found many uses for his detector expertise, ranging from electron microscopy for materials testing and quality control to goniometry with proportional counters for the quality testing of crystal oscillator plates. For this Friedman received the Navy's Distinguished Civilian Service Medal in 1946.[27] He also continued to develop detectors, especially ruggedized ultraviolet halogen counters suitable for operation in harsh environments, and, immediately after the war, photoelectric dosimeters for nuclear tests in the Pacific. Friedman's development of a reliable counter for the Bureau of Ships, designated the BS-1, which was reproduced in the millions for use in the Navy's standard Radiac dosimeters, garnered enormous equity for his programs in the postwar era; he felt he could write his own ticket.[28]

In August 1945 many at NRL were excited by Krause's war stories of V-2 missiles, and Friedman recalls feeling that his highly efficient halogen counters were well suited for solar work. But he had more pressing interests and duties as he was deeply involved in radiation dosimeter development and in evaluating atmospheric atomic bomb tests using his halogen counters.[29] But with each new vehicle failure or spectrograph lost in the desert throughout 1946 and 1947, Friedman recalls becoming "more and more itchy to try my way of doing it. . . . But I really could not begin to concentrate on it until I could begin to see [my] way out of the bomb program."[30] He points out that his involvement in what later was called Project Rainbarrel, an interservice system of collecting and detecting radioactive rainwater residue around the world as a brute-force method of nuclear test detection, kept him from trying out his counters on a V-2 until 1949.[31]

Friedman cannot recall exactly when he decided to prepare halogen counters for rocket flights; he remembers being anxious to do so after he found that all of the detectors in his secret Rainbarrel network responded to a solar flare event.[32] But it is also likely that his entry was a response to knowledge that the solar corona was extremely hot and was a likely source of bountiful X-ray emission. His early associate, E. T. Byram, believed that Friedman's interest had been stimulated by an article or news account in the late 1940s that discussed the new image of a hot corona. After reading the article, Byram realized that people would be looking for X rays, and discussed this with Friedman.[33] Just as likely a stimulus were the hints of solar X rays provided by Watanabe's phosphors and Burnight's film-filter combinations.

Friedman's upper atmospheric research interests evolved gradually and were organically tied to the counter development activities of his staff. After Byram arrived at NRL in December 1947 from an engineering job at Glenn L. Martin Company in Baltimore, one of the tasks Friedman gave him was to develop an electronic discriminator circuit to eliminate cosmic-ray background, using the "coincidences between the Geiger tubes that were bundled together for [a nuclear radiation] monitor that had been already designed and built."[34] Within a month, Byram recorded in his laboratory notebook that Friedman's counters were "evidently photo-sensitive" as he sought out all sources of background noise.[35] By February, using small copper-bodied prototypes, Byram was detecting daily cosmic-ray background variations as he continued to develop suitable discriminators.[36] But he was also learning how Friedman's counters could be applied as photosensitive devices.

Friedman's laboratory notebooks for the same period were brimming with ideas for applying his counters. During the war, possible applications included fuel gauges, flow meters, paint chemistry analyzers, aircraft icing detectors, fungicide inhibiters in optical instruments, and metal surface-quality testers.[37] Through 1945 and 1946, Friedman explored ways to improve and market his Geiger tubes, using, for example, X-ray fluorescence as a broad diagnostic tool. And there certainly was a market: In early 1947, as they contemplated spectroscopic instruments for rockets, academic scientists like Lyman Spitzer and his Yale colleagues realized that they

might need photon counters, "a technique which has not yet been developed" (and which they had little taste for). This was just when Friedman was working hard to extend his photon counter detectors into new realms.[38]

By April 1948, Friedman and his staff were developing collimators to make their X-ray counters directional and were starting to develop filter-gas combinations to isolate specific bands in the ultraviolet and infrared.[39] In May 1949, they submitted "preliminary records of invention" for counter tube designs capable of becoming "Monochromatic Photon Counters" to detect specific ultraviolet spectral bands between 10 and 20 nanometers wide.[40] Thus, by May, Friedman was taking definite steps toward extreme ultraviolet and X-ray spectroscopy, but his many other lines of activity under the umbrella of detector development continued unabated.[41]

## Friedman Enters Rocketry

As Friedman's staff continued radio-dosimeter development, a few started to prepare rugged counters for solar observations (figure, p. 238). Friedman's three-man rocket group included Byram, from the bomb detection group; Samuel W. Lichtman, from one of the Radio Divisions in which he had contributed to NRL's pulse-code modulation telemetry systems; and a mechanical engineer, Joseph J. Nemecek, who was responsible for building the experimental package and making it rugged enough to withstand the vibrations expected during launch.

Friedman recalls that he had to push his way onto the firing schedule through the Rocket-Sonde Research Section. By August 1949 section head Homer Newell announced at a UARRP meeting that NRL's first application of Friedman's counters for "investigating the solar spectrum from 50Å into the ultraviolet" would be on V-2 no. 49, scheduled for 29 September 1949.[42] Their first payload contained six photon counter tubes in two pressurized boxes, as well as cosmic-ray counters and photoelectric aspect indicators.[43] The tubes were sensitive to four

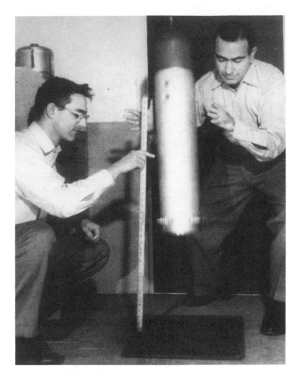

**Herbert Friedman (r) and Talbot Chubb (l) testing a balloon payload as part of Operation San Diego Hi, 1956. U. S. Navy photograph, courtesy Herbert Friedman.**

well-defined X-ray and ultraviolet spectral regions. Friedman's bank of counters therefore became a low-dispersion, low-resolution spectrometer able to discriminate intensity levels of solar soft X rays, Lyman Alpha, and the intensity of the Schumann region beyond the limit of Tousey's or APL's conventional spectroscopic systems.

Friedman's two counter boxes flew to 150 kilometers on 29 September and returned data during most of the flight. The ultraviolet counters reported increased flux starting at only 7 kilometers and quickly became saturated, but a rough reduction of the flux recorded in the Lyman Alpha band, 115—135 nanometers, agreed to within an order of magnitude with the phosphor flights of Tousey, Watanabe, and Purcell.[44] The soft X-ray counters responded above 87 kilometers and reported increased flux as the rocket penetrated the E layer of the earth's ionosphere, increasing by a factor of two between 100

and 150 kilometers. Using atmospheric pressure data obtained by Ralph Havens' NRL gauges during the flight and assuming that absorption was from oxygen and nitrogen, Friedman later concluded that the observed X-ray spectrum peaked in the range 0.5 to 0.7 nanometers and that its flux was between 1/100th and 1/1,000th of that required to sustain the E layer.[45] This still constituted strong evidence for the X-ray origin of the terrestrial ionosphere, since his filter-gas combination had evidently detected only a "small portion of the entire solar x-ray spectrum."[46]

Friedman's first flight was exploratory, to test out his system; and even though he concluded that useful data were obtained, he knew that he had to extend his observations deeper to measure a more representative portion of the soft X-ray spectrum. He was impressed by the recent theoretical successes of Fred Hoyle and D. R. Bates, and those of Hannes Alfvén, who had been able to show that Bengt Edlén's solar coronal temperatures produced sufficient X-ray flux to account for the level of ionization of the E layer. Friedman's first results left little doubt that X rays were a significant source of ionization in the earth's atmosphere, but they were not as yet comprehensive enough to be declared definitive.[47]

The first successful flight was followed by several launch and vehicle failures of V-2s and Aerobees, but finally, an NRL Viking flight on 15 December 1952 provided the confirmation Friedman wanted. His staff had prepared an elaborate set of counters that detected the wavelength range between 1 and 6 nanometers. Although the rocket had an excessive roll rate and there were problems with telemetry, Friedman's detectors measured a solar flux 10 to 100 times what was needed to produce the E-layer ionization.[48]

At about the same time that Friedman measured the X-ray flux in the solar spectrum, showing it to be a significant contributor to the maintenance of the ionospheric E layer, both Friedman's and Tousey's independent measurements of the flux strength of Lyman Alpha showed that it alone was sufficient to gradually dissociate both molecular oxygen and nitric oxide in the lower D layer, and so the ultraviolet remained a source of ionization in the earth's atmosphere as well.[49] The complexity of the solar source of the observed structure of the earth's ionosphere was finally identified through these observations.

While Friedman in 1951 credited the earliest detection of solar X rays to Burnight and both X-ray and Lyman Alpha detection to Purcell, Tousey, and Watanabe, he later concluded that Burnight's observations were spurious, something Tousey contests.[50] But the important, and indeed, critical point to be made here is that Friedman's photon counters provided far more reliable and retrievable information about the extreme ultraviolet and X-ray regions of the spectrum than did Burnight's photographic emulsions or even Watanabe's phosphors. Friedman's observations were as much a confirmation that his detectors were a new, versatile, sensitive, and reliable technology as they were a confirmation of the theory of the ionosphere elucidated by Hoyle and Bates.

Friedman's success has been attributed to the simplicity of his detectors.[51] Photon counters provided spectral information directly through telemetry, in contrast to photographic, thermoluminescent, or electrochemical devices. Counters were also far more rugged and reliable than photoelectric designs. The economy and simplicity of Friedman's solid-state detectors and his ability to build a wide array of prototypes, and then have a great many samples on hand for testing and evaluation, provided just the right combination of factors for a extended two-pronged program of continuous development and flight testing. Through the early 1950s, his core rocketry staff—consisting of Byram, Lichtman, Talbot Chubb, Robert Kreplin, and others—devoted their primary energies to laboratory and field testing, on which success depended. As Kreplin noted recently, "The philosophy was just to test it and test it and test it until you were convinced that it was going to work." Byram recalled that the testing process was an exciting challenge; in some ways it was a test of his own abilities of concentration: "I enjoyed fighting them. It wasn't a frustration, it was a challenge. It was mind over Geiger tube."[52]

## Watanabe Moves to the Air Force

Kenichi Watanabe did not enjoy the independence at NRL that Friedman did. Accordingly, he left in 1951 for the Air Force Cambridge Research Center to work on the development of photoelectric techniques for ultraviolet detection of solar radiation and to apply this equally new technology to studies in upper air composition and ionospheric physics. Watanabe, long an advocate of photoelectric or electronic recording, even while working for Tousey at NRL, dedicated his own AFCRC laboratory to developing techniques for the electronic detection of dispersed sunlight, something he was not able to do at NRL. With Hans Hinteregger, Watanabe argued in 1953 that "the use of photoelectric techniques in place of photographic detection is very often not just a matter of convenience, but almost a necessity in absorption spectroscopic or photochemical studies."[53]

Watanabe and Hinteregger knew that conventional phototubes and multipliers were, like photographic emulsions, sensitive to visible radiation. And since they were encased behind either glass or quartz windows, they were also unable to sense far ultraviolet radiation. Watanabe and Hinteregger therefore directed their effort toward developing windowless photoelectric cells and special high work-function photocathodes that would be sensitive to the extreme ultraviolet and insensitive to scattered visible radiation. By the mid-1950s, they had developed high work-function tungsten cathodes that could detect radiation below 100 nanometers, and were designing photoelectric scanners using grazing incidence optics and fixed detectors set behind slits that moved along the Rowland grating's focal plane. This combination became the basic AFCRC rocket spectrometer that was flown in scores of flights from the late 1950s through the 1970s.[54]

## Research Styles in Ultraviolet Rocket Spectroscopy

Since none of these experimentalists had been active in solar, atmospheric, or ionospheric research earlier, their drive into the ultraviolet was more an exercise in spectroscopic technique than an exploration of the ultraviolet solar spectrum and the structure of the upper atmosphere. Historian Peter Galison has noted that it is typical for experimental physicists to concentrate on one familiar technology even when facing different observational problems.[55] Certainly, Friedman's and Watanabe's techniques reflected training and a background quite different from that of Tousey, Rense, and Hopfield. And each brought a particular instrumental research style to the solution of problems in astrophysics, atmospheric physics, and ionospheric physics. Thus, Tousey, along with Rense and Hopfield, applied techniques from vacuum ultraviolet spectroscopy, whereas Friedman employed those from solid-state physics. Only when pushed to the limit did any of them devote more than cursory attention to the techniques they were not familiar with by training or experience. F. S. Johnson saw his work for Tousey as an obvious application of his training and talent, although these did not derive from any previous professional interest in the subject. He added that the same held true for the others:

> Maybe Friedman's work typifies that best of all. He was working with various types of counters, and he had the ability and the people who could assemble such things and put them in rockets. So the interest developed there because of the opportunity. It wasn't that he had been waiting a long time for an opportunity to make these measurements on the sun, or on astronomical objects.[56]

Friedman became interested in solving the mysteries of the ionosphere because he knew that his detectors stood a better chance of doing the job. As he recalls, his nondispersive filter-detector combinations were an ideal means of providing information on "com-posite flux: what the broad band energy input was at each height level in the atmosphere."[57] Since Tousey's strengths led him naturally to dispersive techniques, his original goal was to design a brute-force device, maximizing the chance that a spectrum would be obtained, with secondary priority given to the quality of the spectrum.[58] His first year studying the sun from space, therefore, was devoted to establishing the feasibility of a technique: that an instrument could be built that could gather spectroscopic information during a rocket flight.[59]

Even though technical choice preceded feasibility in most cases presented here, nontechnical factors played an important shaping role as well. Van Allen's interest in an altimeter may be the best example, but Tousey also remained sensitive to the need to adapt when absolutely necessary. Thus, in late 1952, just before Rense's success in photographing Lyman Alpha, Tousey felt that he had "reached a point of diminishing returns on the ozone problem."[60] Since identifying the solar source of the ionosphere was the most important objective of his group at NRL, he was especially concerned that, thus far, photography had failed "because of the low intensity of the radiation and the difficulty of obtaining a long enough exposure from an unstable rocket."[61] He indicated then that "less selective" methods were more promising, such as thermoluminescent phosphors, or Herbert Friedman's electronic counters, and for a moment, Tousey seemed willing to move in that direction. Even though the shift evaporated when word of Rense's success appeared, it demonstrates that even the most tenacious adherents to a technical style could be influenced by institutional priorities.

Tousey, of course, was in no way ignorant of photoelectric techniques. During the war, his branch had tested both domestic and captured German infrared devices, and Tousey used photoelectric laboratory equipment to calibrate his rocket spectrographs.[62] But as he and others have consistently argued, photography could capture a far greater density of spectroscopic information than could Friedman's photoelectric techniques. Friedman never tried to identify the ultraviolet line spectrum of the sun in the V-2 era. And save for his attention to continuum spectra for ozone absorption analysis, Tousey similarly limited his attention to problems that could be resolved by obtaining line spectra. The research interests of each group leader were, therefore, guided by their personal technical preferences and strengths; when this was not the case, conflicts arose, as they did when Tousey supported and Friedman refuted Burnight's alleged detection of solar X rays. The desire to avoid competition therefore caused each group to limit the problems they attacked to those best examined with the technologies they favored—the ones in which they had invested the most effort and equity.

One limitation existed for all the major groups we have thus far examined: however they chose to conduct solar, atmospheric, or ionospheric research, the effort depended on what observations could be performed from a rocket. The Navy and Air Force supported the research choices of each group, and did support solar research beyond rocketry. But the rocket groups conducted solar research exclusively in terms of rocket observations, and no ground-based solar observatory was supported in the first decade to conduct solar observations from rockets. There is no evidence that the service agencies would have inhibited such a mix; they were as interested in technical capability as they were in the solution to specific questions about the nature of the solar atmosphere, or its effect on the earth's ionosphere. We explore this situation further in the next three chapters.

# Notes to Chapter 13

1. Greenstein to Van Allen, 14 July 1947. JG/CIT. Leo Goldberg's samples were sensitive only to the wavelength range 170 to 180 nanometers, and therefore were useless for the detection of Lyman Alpha at 121.6 nanometers. Goldberg to Spitzer, 23 June 1947. LG/HUA; Daniel B. Clapp to Field Service, Chief of Naval Research, "Pohl Crystals Requested by Dr. Leo Goldberg," 2 May 1947. OANAR:NA-4S2, Office of the Assistant Naval Attache for Research, American Embassy, London. LG/HUA. See also Kuiper to Spitzer, 17 June 1947. LSP/P.

2. Ibid., Clapp memorandum, 2 May 1947, p. 1.

3. Megerian to Bain, "V-2 Panel Report no. 13," 9 January 1948. V2/NASM.

4. Megerian to Bain, "V-2 Panel Report no. 17," 16 June 1948. V2/NASM.

5. Greenough, Oberly, and Rockwood (1947), pp. 73, 78–84.

6. The spectrometer was supposed to provide a continuous record of intensities in these bands during the rocket flight. A third photoelectric channel, centered on the undispersed zeroth order from the dispersing grating, was added to calibrate the two other channels.

7. Greenough, Oberly, and Rockwood (1947), p. 78. See also C. V. Strain registered laboratory notebook no. 6353, entry dated 21 August 1946, p. 43. HONRL. Strain indicated that the photoelectric spectrograph was put on hold "for a later firing" and that at first, F. S. Johnson did not agree with this decision.

8. Tousey OHI, pp. 224–26rd.

9. Bates and Massey (1946).

10. Tousey had learned about these alternatives from his Harvard teacher, Theodore Lyman, who in the 1930s experimented with ultraviolet and X-ray sensitive phosphors, trying to isolate the extreme solar ultraviolet with potassium phosphors and a small Hilger quartz spectrograph. His experiments were successful only in laboratory tests. Tousey (1967), p. 239.

11. On Watanabe, see John Holmes, "Kenichi Watanabe," *Journal of the Optical Society of America* 59 no. 12 (December 1969), pp. 1686–87; and Tousey OHI, p. 86rd. John Strong had been at Caltech before moving to Johns Hopkins. See John Strong OHI, pp. 20–22; and F. S. Johnson OHI, pp. 47–48. On the thermoluminescent phosphor technique, see Newell (1953), pp. 163–64.

12. To fully evaluate the relative importance of Lyman Alpha and X radiation, Tousey needed to isolate three broad spectral bandpasses: blueward of 0.8 nanometers (soft X rays); between 104 and 134 nanometers (the Lyman Alpha region); and between 123 and 134 nanometers. Koomen had already been looking for such filters, and eventually found that beryllium, lithium fluoride, and calcium flu-

oride produced the desired phosphor-filter combinations. See Martin J. Koomen registered laboratory notebook no. 6987, entries for 23 June 1947 through March 1948. HONRL.

13. The first flight, on 18 November 1948 (V-2 44), did not record any soft X-ray emission short of 0.8 nanometers, but some was recorded on the second flight on 17 February 1949, which coincided with a period of strong ionospheric activity. This series of flights confirmed the existence of significant radiation in the Lyman Alpha range and also hinted at the existence of broad absorption adjacent to Lyman Alpha, because the longest wavelength band was always weakest. Tousey, Watanabe, and Purcell (1951), pp. 792–97; Newell (1953), p. 167.

14. Tousey OHI, pp. 159–60rd.

15. Burnight (1949, 1952).

16. Tousey (1953b), pp. 667–68.

17. Tousey (1953b), p. 668.

18. Megerian to Bain, "V-2 Panel Report no. 21," 1 September 1949, p. 14. V2/NASM.

19. James Van Allen, "A Direct Determination of the Altitude Distribution of Atmospheric Ozone," 24 October 1949. JAVA/APL. Prepared as a preliminary report for the UARRP. The bandpasses were at 264.3, 300.1, and 372.4 nanometers, and RCA 1P28 photomultipliers were used. Van Allen planned to build at least five spectrometers. Clearman to Greenstein, 6 January 1948, in reply to Greenstein letter of 12 December; Van Allen to Greenstein, 28 September 1949. JG/CIT.

20. Van Allen and Hopfield (1952), p. 181. The first spectrometer was completed by Clearman in early 1949, as he left APL for the academic world. It flew on Aerobee A-14 on 23 June. James Van Allen, "A Direct Determination of the Altitude Distribution of Atmospheric Ozone," 24 October 1949, p. 3. JAVA/APL. The double maximum did follow roughly a preliminary analysis of the first NRL photographic spectra from 10 October 1946, even though discussion of it was missing from a later analysis by Johnson, Purcell, and Tousey in 1951. In 1951, careful remeasurements of the NRL spectra by Johnson "indicated that the double maximum was produced largely by difficulties in the preliminary densitometry. A single smooth curve . . . is all that the final data support with certainty." Johnson, Purcell, and Tousey (1951), p. 593.

21. Balloon flights had actually detected the maximum, but rocket flights were required to confirm and refine both the nature of the maximum and the overall character of the ozone distribution. In particular, no balloon observations could have hoped to provide, as APL's Aerobee A-20 did, the continuous profile to 60 kilometers. Van Allen and Hopfield (1952); Fraser (1951), p. 22.

22. Hevly (1987), p. 190. The exception was, of course, Watanabe, who prepared his own chemical and crystalline sensors.

23. Friedman OHI, pp. 11–17, quotation on p. 17.

24. Ibid., p. 18.

25. Ibid., pp. 21–23. Friedman already had experience with proximity fuzes at Hopkins on an early NDRC project.

26. Friedman OHI, pp. 36–38. See also Hevly (1987), pp. 188ff.

27. Friedman OHI, p. 32. See Hevly (1987), p. 79.

28. See Hevly (1987), p. 199; and Herbert Friedman, Talbot Chubb, E. T. Byram, and Robert Kreplin, "Early X-Ray Astronomy: Session One," videohistory interview, 12 December 1986, pp. 3, 15. SIVP/SIA.

29. Friedman OHI, pp. 43–44, 51.

30. Ibid., p. 52.

31. Ibid., pp. 54–56. "Rainbarrel" was developed by a Navy nuclear chemist, Peter King, and was tested during a series of nuclear explosions in the Pacific starting with "Operation Sandstone" in April 1948. Lin Root, "The Long Range Detection System," unpublished manuscript pp. 13–16. Charles Ziegler has noted that Rainbarrel, along with AEC and British detection programs, was secondary to the JCS-designated Air Force system, which was adopted after Sandstone. Thus, after Sandstone, Friedman may have had good reason to look for new areas of application for his detector expertise. See Ziegler (1988), pp. 216–17, 227, n. 53. On the Physical Optics Division involvement in nuclear test detection, see Hevly (1987), pp. 196–97.

32. Friedman OHI, p. 59; and Hevly (1987), pp. 197–98. No major solar flare event on record readily coincides with his entry into rocketry. The biggest solar prominence on record occurred on 4 June 1946, noted by Menzel (1950), pp. 167–69, and there was a series of large events throughout the 1940s. On 20 November 1946, Grote Reber detected an

"enormous outburst of solar energy at 480 mc, at which time the solar radiation increased by a factor of several thousand." Greenstein to Struve, 27 November 1946. JGP/CIT. So other events through 1948 and 1949 could probably have been detected by "Rainbarrel," such as a large solar flare on 8 May 1949, or the ground-based worldwide detection of cosmic rays from a solar eruption on 19 November 1949, but both of these events occurred after Friedman initiated his rocket experiments. See Forbush, Stinchcomb, and Schein (1950), p. 501; and *Proceedings of the Institute of Radio Engineers* 37 (1949), p. 1489.

33. Commentary by E[dward] T[aylor] Byram during a videohistory interview with Friedman, Talbot Chubb, and Robert Kreplin present. See "Early X-Ray Astronomy: Session One," December 12, 1986, 02:00:55:00, p. 25. SIVP/SIA. Byram may have seen notices of the events in ibid, or may have read articles like Joseph L. Pawsey's 1946 paper "Observation of Million Degree Thermal Radiation from the Sun at a Wavelength of 1.5 metres," *Nature 158* pp. 633–4.

34. E. T. Byram, "Early X-Ray Astronomy: Session One," 12 December 1986, 01:18:40:00, p. 8. SIVP/SIA.

35. E. T. Byram, registered laboratory notebook, 181–87–36, entry for 19 January 1948, p. 3. HONRL.

36. Ibid., p. 9.

37. Herbert Friedman, registered laboratory notebook no. 4076, 181–67A6325 no. 116, entries between November 1943 and February 1945, especially the section "Geiger Counter Applications to Aircraft Instruments." HONRL.

38. Roland E. Meyerott to Marcus O'Day, 13 January 1947. LSP/P. Meyerott was a Spitzer colleague in Yale's Physics Department who was keenly interested in building spectrographs with Spitzer, until he learned from a theoretical study by David Bates and Harrie Massey (1946) that the spectral features he hoped to detect lay beyond the limit of photographic detectability but could be detected by photon counters. Meyerott, in deciding not to submit a proposal to O'Day, simply pointed out that they lacked the manpower to develop such devices.

39. Friedman, registered laboratory notebook no. 4164. HONRL.

40. Although his notebooks record steps leading to this capability starting in April 1948, his first public discussion of them came in October 1949 at an NRL-sponsored "Symposium on Geiger Counters," which reviewed the status of their work as of May 1949. See Herbert Friedman and L. B. Clark, "Preliminary Record of Invention," 11 January 1950. Contained in Friedman, registered laboratory notebook no. 4164. HONRL.

41. This point is developed in Hevly (1987), pp. 210–11.

42. "V-2 Panel Report no. 21," 1 September 1949, p. 16. V2/NASM. On competition for rocket berths within NRL, see H. Friedman OHI, pp. 37–38. See also F. S. Johnson OHI, p. 22; Richard Tousey OHI, pp. 39rd., 52–55rd.

43. Friedman, Lichtman, and Byram (1951).

44. Ibid. (1951), p. 1029.

45. Ibid.

46. Ibid., pp. 1029–30.

47. Hoyle and Bates (1948). On Edlén's contributions, see Hufbauer (1991). See also Hevly (1987), p. 205.

48. Byram, Chubb, and Friedman (1954a), pp. 274–75.

49. Byram, Chubb, and Friedman (1954b), pp. 276–78; and Friedman, OHI, p. 71.

50. Friedman OHI, p. 66, argues in retrospect that the height of Burnight's 5 August 1948 rocket, when the exposure was made, was too low for measurable X-ray penetration, that his beryllium window was far too thick, and that Schumann plates were very sensitive to contact pressure and would blacken simply upon being touched. Tousey has argued, however, that Burnight's 5 August detection was made possible by an unusually bright solar event and subsequent large X-ray flux. Tousey (1953), pp. 667–68. See "Enclosure A" attached to: Director, NRL to Naval Unit, WSPG "Request for firing of Aerobee Venus II," 20 December 1948, p. 2. Box 100, folder 18; and attachments to: Philip G. Blackmore to OCO, 26 July 1948, "Preliminary Information on V-2 Rocket no. 43 to be fired at 0507 hours, 5 August 1948." Box 100, folder 15, NRL/NARA.

51. See Hevly (1987), pp. 215–16, and the references contained within.

52. Herbert Friedman, Talbot Chubb, E. T. Byram, and Robert Kreplin, "Early X-Ray Astronomy: Session One," videohistory interview, 12 December 1986, SIVP/SIA. Kreplin

comment on p. 80 (03:46:40); Byram comment on p. 33 (02:18:03:00).

53. Hinteregger and Watanabe (1953), p. 604.

54. See AFCRC *Annual Reports* during the decade. The basic AFCRC rocket monochromater is discussed in Hinteregger, Damon,
Heroux, and Hall (1960).

55. Galison (1985), p. 310.

56. F. S. Johnson OHI, pp. 27–28; see also Richard Tousey OHI, p. 70rd.

57. Friedman OHI, pp. 71–72.

58. Tousey OHI, p. 144rd.

59. This characterization reflects similar priorities among those who worked to establish designs for guidance systems for ballistic missiles. See MacKenzie (1990), p. 149.

60. Tousey (1953a), p. 249.

61. Ibid.

62. Tousey OHI, pp. 112–14rd. The highest cited
publication in Tousey's group in its first decade (garnering over 100 citations in the
1950s) dealt with means to improve the ultraviolet efficiency of photoelectric cathodes.
See Johnson, Watanabe, and Tousey (1951).

# 14

# Migrating from Particles to Fields

*[The earth's geomagnetic field acts] as a huge [magnetic] rigidity spectrometer for all charged cosmic-ray primaries.*

—James Van Allen (1952).[1]

Since their first unambiguous detection in 1911–1912, confirmation of their celestial origins in the 1920s, and the identification of their character as a class of high-energy charged particles in the early 1930s, cosmic rays became the philosopher's stone for the particle physicist who wished to probe the mysteries of matter. The highest-energy cosmic rays invariably proved to be the most effective probes.[2]

## The Elusive Cosmic Ray

When a cosmic ray enters our atmosphere, it invariably collides with atmospheric atoms and molecules, creating showers of lower-energy secondary or daughter particles. These remnants sometimes reach the ground, and, although very useful as indicators of the total energy involved in the original interaction, they reveal little of the original particle, or "primary" cosmic ray itself. In the 1930s, groups in Germany, England, and the United States started to use balloons to carry automatic ionization chambers, Geiger counters, and photographic emulsions into the high atmosphere in the hope of capturing glimpses of the original act. The maximum altitudes gained by balloons flown by the German and American groups averaged some 24 to 30 kilometers routinely, which was still within the region in which unadulterated primaries had already reacted with the atmosphere to produce secondaries.[3] Even so, by the late 1930s there was growing evidence that primaries

were highly accelerated protons; the first confirmed observations came just at the outbreak of the war, from balloonsonde observations by Marcel Schein in Chicago.[4]

After the war, even with Schein's success, the problem remained to observe the full range of particles that made up the primary cosmic-ray flux. This goal, in Ernst Krause's memory, made cosmic-ray physics "a very hot subject during the previous decade."[5] It was the one he would choose to pursue as he planned NRL's rocketry program. James Van Allen, who also had training and experience in nuclear physics, was attracted in the same way, as were those at Princeton who headed the V-2 work after Myron Nichols left in July 1946.

The attraction was simple: rockets could fly far higher than balloons or aircraft, and thus enter the region where cosmic-ray primaries dominated. But from the outset, there were serious drawbacks: flight times were very short, and the V-2 was far from a pristine environment for cosmic-ray research. Payloads aboard balloons and aircraft could be flown frequently and for many hours at a time, thus increasing the comparative odds of capturing something interesting. Cosmic-ray physicists understood this: most who were involved with ballooning before the war remained so, rejecting rocketry at the outset.[6] Marcel Schein, Van Allen recalls, was a friendly adviser who influenced the APL group's early designs for cosmic-ray telescopes. But Schein "was not persuaded that rockets had a future in his own career. He

247

[Schein] thought that balloons, because they had longer flights, and maintained known altitudes in space, were more fruitful. . . . So he was supportive but not a participant."[7] Bruno Rossi at MIT was also not interested in using rockets. Van Allen recalls that "he didn't think that [rockets] had much of a future at all. He was really quite cynical and critical about spending money flying cosmic-ray equipment in rockets."[8] Although two of the most established practitioners of cosmic-ray physics in America did not wish to get involved with scientific rocketry, a few of their former students became members of the NRL and Princeton cosmic-ray groups.

This chapter focuses on the factors that worked against cosmic-ray research with rockets and led some group members to abandon rocketry and others to migrate to more fruitful areas of rocket research, such as the earth's magnetic field. Since much of the basic character of cosmic rays was learned from how they interacted with the geomagnetic field, it is not surprising that such migrations took place.[9] James Van Allen at first used the earth's magnetic field to study the intensity and energy spectrum of primary cosmic rays and then turned his attention from the particles that were being affected by the field to the field itself, using the particles as probes. Understanding how and why he made this shift will elucidate the professional choices facing those who defined their research interests in terms of the rocket.

## Group Profiles and Motivations

The first two years of cosmic-ray research with rockets, as Van Allen recalls, were a "very controversial period . . . [when] . . . the haze of battle was covering the whole scene."[10] Each group not only had to battle the instrumentation, but felt considerable pressure to be the first to discover the nature of cosmic-ray primaries. APL, NRL, and Princeton all built their large telescopes with that one problem in mind.

Although Van Allen determined the direction of the APL group, he let nuclear physi-

cist Howard Tatel guide the daily activities of junior physicist R. P. Peterson, engineer Jim Jenkins, and later S. Fred Singer, a physics graduate student from Princeton who had been doing war work at the Naval Ordnance Laboratory. Influenced by Tatel, Van Allen first hoped to study nuclear interactions using cosmic rays, but soon concentrated on seeking out the identity of cosmic-ray primaries.[11]

Similarly, Ernst Krause let Serge E. Golian and Gilbert J. Perlow organize NRL's effort to determine the identity of the primaries. Golian and Perlow had worked in cosmic-ray and nuclear physics at Chicago before the war; Golian worked for Schein, whereas Perlow received his Ph.D. there in 1940 and taught at the University of Minnesota briefly before moving to the Naval Ordnance Laboratory, and then to NRL in 1942.[12] Beyond detecting primaries, they also wanted to discover what types of reactions took place as the primaries encountered the earth's atmosphere, and they argued that rockets were the only way to conduct direct observations.[13] As at APL, the NRL group flew collections of Geiger counters and photographic plate stacks provided by others. But NRL was the only group to prepare and fly a series of small Wilson cloud chambers to examine the particles directly, producing their first chamber photographs from 145 kilometers in 1948–1949.[14]

Although both APL and NRL staff members had some experience in cosmic-ray research, the Princeton Physics Department was a strong center for high-energy and nuclear physics and attempted the most sophisticated observations using detectors and arrays its members had used previously in the laboratory and field.[15] John A. Wheeler, John F. Brinster, Niels Arley, and Donald J. Montgomery, among others, responded to Ladenburg's and Nichols' invitation and prepared several cosmic-ray telescopes for flight in late 1946.

All three groups were aware of the many potential applications of cosmic-ray physics, especially in the new world of the atomic bomb. The Joint Research and Development Board (JRDB) frequently heard claims for improving atomic weapons technology. Lawrence Hafstad, who succeeded Merle Tuve as APL director, rationalized that Van Allen's research was: "frankly intended to be one of the Navy's contributions to pioneering physics, without regard to practical end-results. However, we are keeping our eye on the matter of very energetic particles, carrying energies much greater than those involved in the fission of uranium."[16] Cosmic rays and atomic energy were frequently linked. Van Allen's friends created a mock newspaper article to spoof the connection: In October 1947, he received one titled "V-2 Helps Army Swell Stockpile of Cosmic Rays" which was subtitled: "Rocket Wrests Rays from Sky; U.S. Soon May Possess Enough for Cosmic Bomb."[17] Bombs aside, Van Allen pursued a more prosaic application: just as he explored the use of a photoelectric spectrometer as an ultraviolet intensity meter for missile guidance, he wanted to determine if cosmic-ray intensities could perform the same service.

The NRL group later argued that cosmic-ray research had a "practical application because cosmic radiation represents a potential hazard to operations of high-flying manned vehicles and to instrumentation, particularly transistors, to be used in earth satellites."[18] But most of all, as Ernst Krause recalls, the type of technology required for cosmic-ray research was the same that NRL traditionally was interested in, such as on-board logic circuitry for command, guidance, and control and pulse techniques employed in high-frequency radar. The activity itself therefore would help NRL increase its control over, and understanding of, guided missile electronics systems. Princeton's John Wheeler made the same points to his APL patron in 1946.[19]

## Typical Instrumentation

Many small differences existed between the various types of detectors employed by the cosmic-ray groups, but they all used Geiger counters, ionization chambers, and proportional counters alone, or in combination as coincidence and anticoincidence telescopes. In addition, some prepared nuclear emulsion plate stacks and chemical absorbers, and NRL built Wilson cloud chambers. Virtually anything used on the ground was tried in some form or another aboard a V-2. As S. F. Singer recalls, Van Allen, as a dynamic competitive group leader intent upon establishing a niche for his program, opted for any experiment that could exploit the payload space of a rocket. Indeed, it was Singer's strong impression that the primary reason for the existence of upper atmospheric research at APL was to learn how to instrument rockets: "We really were constrained to only do work that could be done on rockets."[20]

### Ionization Instruments

Cosmic-ray detectors used on the V-2s were typically ionization chambers, Geiger (or Geiger-Mueller) counters, and proportional counters. The principles on which this family of detectors were based had been known for several decades, but their practical application required many years of effort. Even in the 1930s, one of the first physicists to apply them to cosmic-ray research, Bruno Rossi, considered them "a kind of witchcraft."[21]

With the acceleration of wartime technology, however, counter development after the war reached the point where none of the groups had any difficulty finding suitable detectors from commercial sources.[22] APL, NRL, and Princeton all applied the standard technique created by Rossi that placed Geiger counters in coincidence and anticoincidence circuits to provide information on the direction and intensity of incident radiation. The solid angle defined by the extremities of the counter chambers became the effective field of view of the array, commonly called a cosmic-ray telescope.

The availability of counters and the carrying capacity of the V-2 naturally stimulated all cosmic-ray observers to create large complex devices, piling on absorbent lead slabs over long trays of counters to gain as much information as they could about the energy ranges and directions of incident cosmic radiation during the short flight. To compensate for the short observation times of a ballistic flight, the detectors were made as large as possible to increase the chances of collecting something interesting.

### Photographic Emulsions and Cloud Chambers

The simplest detectors were photographic emulsions, which had been in use since the 1930s to record the passage of cosmic rays and their interactions with matter. Regular photographic plates, however, had thick glass backings that interrupted the record of the cosmic ray's passage. After the war, British cosmic-ray physicist Cecil F. Powell, working with the Ilford Company, developed a thick, self-supporting 'nuclear' emulsion. With his Bristol University colleagues, Powell discovered the pi-meson (or pion) in 1947 in the first continuous records of nuclear disintegration processes in the atmosphere.[23]

Few of the rocket groups were familiar with preparing or analyzing nuclear emulsions. Krause's successor, Homer Newell, still felt in the 1950s that "the whole process [of analyzing nuclear emulsion tracks is] as much an art as it is a science."[24] NRL therefore flew photographic plate stacks prepared by physicist Herman Yagoda and others, who also performed their own analyses. APL flew shielded rolls of alpha particle emulsions and plate stacks of nuclear emulsions, but gathered in no useful data.[25]

Although they were not interested in the relatively simple application of plate stacks, the NRL group pioneered in the difficult task of building Wilson cloud chambers to determine the composition of the cosmic-ray flux. Cloud chambers periodically create a supersaturated vapor by expansion cooling. An incident particle in the chamber will dissociate or ionize particles in the vapor, which in turn act as condensation nuclei upon which the supersaturated vapor condenses into a visible trail. Usually, a trail is photographed when the expansion cycle is triggered by the passage of the particle through a Geiger-counter array.[26]

## The Seductive Luxury of the V-2

APL's first large cosmic-ray telescope, destined for the 30 July 1946 flight, was based on a design developed by Marcel Schein and consisted of two independent telescopes set in a V pattern on either side of the base of the V-2 warhead, just above the battery boxes and beneath the original placement of the APL spectrograph (figure, p.251). The two dozen counters in the telescopes were obtained from the Geophysical Instrument Company of Arlington, Virginia. They were arranged under varying thicknesses of lead and were connected to cathode follower circuits, scaling circuits, and coincidence circuits based on designs developed by Bruno Rossi. The telescope communicated through an NRL telemetry system, which had to handle sampling rates that were at the limit of its capability.[27]

Although some telemetry was received from the 30 July payload, not all channels worked, and the warhead was lost at the crash site. Lorence Fraser reported in 1948 that his group obtained good "exploratory" data, but there were still many problems to work out, particularly the interpretation of the counting rates.[28] Both the APL and NRL cosmic-ray instrument arrays demanded better telemetry, but Van Allen's preoccupation in the summer of 1946 centered on building arrays that could survive the launch of the V-2; even though there were no moving parts, 6-g vibration tests on the APL shake table usually loosened electrical connections and produced spurious signals, so "Lorry Fraser and I used to say that we'd just go out and kick the tires and hope that it would work."[29]

APL believed at first that it had achieved some success with coincidence arrays in April 1947, when for the first time all telemetry and counter channels worked. The APL group were confident that they had found an "enormous (and monotonic) increase in counting rate" with altitude in their complex coincidence circuits, from 5 counts per hour on the ground to 4 counts per second at 50 kilometers.[30] During the same period, they flew identical cosmic-ray telescopes on B-29 flights for a total of 85 hours to fill in data for the lower atmosphere (figure, p. 255). With several reasonable flights under their belts, and with confirming and calibrating data from the B-29 flights, the APL group verified the ex-

APL staff member A. E. Coyne readying a Schein-type cosmic-ray payload for a V-2 flight set for 17 December 1946. The counters are hanging in boxes on either side of the electronics rack containing mixers and recorder amplifiers. Applied Physics Laboratory, JHU/APL archives.

istence of the region of maximum secondary radiation between 16 and 20 kilometers that had been detected first by Regener and Pfotzer. But they also obtained data far beyond the maximum region where the counting rate diminished continuously until about 50 kilometers, when it leveled off onto a 'plateau.'

Van Allen knew, however, that they were still receiving confusing and conflicting data from their counters, so the general picture, and in particular the meaning of the plateau, remained unclear.[31] Was it cosmic, instrumental, or environmental? Gradually, the group came to believe that many sources of error, even in simple counting rates, lay in the design of the experiment and the environment in which it flew. By the summer of 1947, Van Allen and his group had confirmed that the complex telescopes, buried within the steel warhead of the V-2, produced spurious data; he noted that in the first four successful cosmic-ray flights APL had conducted, no "care was taken to separate the counter from the mass of material in the rocket itself."[32]

Along with the NRL group, who had experienced the same problems with similarly complex devices, the APL group had to iso-

late their counters from the steel of the V-2 warhead. By late 1947, APL and NRL had flown seven elaborate counter telescopes, and both groups found that they had created systems that were too complex to work reliably; they taxed the telemetry channels unduly. The people charged with interpreting the data stream faced confusing signals from the hundreds of electronic components that made up the telescopes. Lorence Fraser, who managed procurement and integration problems for Van Allen, remembers that "when we got the data back [from the first flights] there was so much cross talk between circuits that we couldn't make head nor tail out of the results."[33] Both groups also found that they could not account for the spurious secondary showers created by the heavy steel V-2 warheads. Krause hoped at first that the warhead's thick skin would act as an additional absorbing filter, but Van Allen was not so optimistic and soon took steps to simplify the experiments and shield them from the steel mass of the V-2.[34]

Thus in preparing for a V-2 flight in July 1947, Van Allen and Howard Tatel abandoned their large telescope design, and with Jim Jenkins' assistance placed a set of four

**Serge Golian and an early NRL cosmic-ray igloo showing counters before they were encased in lead absorbers. U. S. Navy photograph, Ernst Krause collection. NASM SI 83–13933.**

Geiger counters in a narrow cylinder of cloth-filled Bakelite at the tip of the warhead. The cylinder was capped by a wooden cone, and just under the wooden tip lay a single counter; the remaining three were bundled just below the single counter to act as a coincidence shield for spurious particles created by the warhead mass. Van Allen called this his "primary flux experiment."

Launched aboard V-2 no. 30 on 29 July 1947, the array reached 160 kilometers in a smooth well-stabilized flight. Counting rate data were summarized in one figure in a 1948 *Physical Review* article (figure, p. 256). The secondary maximum of just under 50 counts per second was reached at 20 kilometers, and

the counting rate then dropped to about 20 per second and stayed constant in the region between 50 and 160 kilometers. This was five times the rate deduced from earlier coincidence observations.[35] Van Allen and Tatel refined the picture of the high-altitude cosmic-ray plateau and provided an estimate of the primary flux rate at 22.4 +/- 0.2 counts per second in what they thought was the interplanetary medium. In a few months, Van Allen speculated: "It is very *tempting* indeed to conclude that we are dealing only with the primary radiation in this region and to conclude further that this plateau value of counting rate would continue for many hundreds of kilometers higher."[36]

Van Allen was greatly encouraged by his success and in 1948 sang the virtues of scientific rocketry in the popular scientific literature. He identified the rocket's ability to achieve greater heights as its primary asset, and, bolstered by his successful application, argued that what he and his staff were doing was largely exploratory and had to remain relatively unhindered by theory. "Most importantly," he felt that "no experimentalist has such an implicit faith in the completeness of current theories as to fail to take advantage of any opportunity to extend the realm of actual observations, even though only in an exploratory way."[37] The rockets were there to be used and Van Allen looked forward to solving many of the most pressing problems in cosmic-ray physics with them.

## Reception and Rebuttal

When Bruno Rossi reviewed the state of cosmic-ray physics at the spring 1948 Solvay conference, he did not look kindly on the results from the early rocket flights. Rossi flat out did not trust their data or Van Allen's identification of the cosmic-ray plateau.[38] He felt that all rocket data from White Sands were compromised both by the presence of secondaries reflected up from lower portions of the atmosphere, called the cosmic-ray albedo, and by secondary radiation from the rocket itself.[39] Rossi also believed that a geo-

James Van Allen and James Jenkins with a coincidence counter circuit. Applied Physics Laboratory, JHU/ APL archives.

magnetic latitude effect could be contributing to the discordances he had found in Van Allen and Tatel's observations.

In response to Rossi, Van Allen decided to take observations at latitudes far removed from White Sands. Simply put, to better understand the nature of cosmic-ray primaries, he also had to know more about how the earth's geomagnetic field affected those primaries and how secondaries captured in the field might contaminate the count. Van Allen had already been involved in a series of polar expeditions code named Nanook and High jump in 1946 and 1947 sponsored by the Bureau of Ordnance. APL's participation was then justified by APL management and by BuOrd as being useful for obtaining "data of general interest in the fields of radio propagation and cosmic rays which may have ultimate bearing on guided missiles development."[40] Although this effort was minimal, it motivated Van Allen to study geophysical effects later in 1948.

Supported by a growing Bureau of Ordnance interest in shipboard firings of missiles, the APL group launched three Aerobees in March 1949 aboard a converted 10,000 ton seaplane tender, the USS *Norton Sound*. Sitting on the earth's geomagnetic equator at a point about 960 kilometers off the coast of Peru, the APL crew shot small cosmic-ray telescopes and magnetometers into the ionosphere and then repeated similar studies in January 1950 during another voyage to the Gulf of Alaska. Van Allen and Albert Gangnes concentrated on the Geiger counters and S. Fred Singer on the magnetometer observations.

In early 1948, Van Allen, prompted by Harry Vestine, asked Singer to prepare total field magnetometers for Aerobee rockets to determine if postulated current sheets in the E layer generated magnetic fields that were strong enough to alter the earth's general field. Through Vestine's contacts, Singer collaborated with W. A. Bowen and E. Maple at the Naval Ordnance Laboratory to adapt a lightweight NOL underwater mine device for use on Aerobees.[41] The March 1949 flight from the geomagnetic equator ultimately established the existence of a discrete region of concentrated current—an electrojet—residing far lower, and with more intensity, than anyone expected. Singer's task was to explain the result, which he did only after leaving APL in 1950.[42]

Along with Singer's successful magnetometer observations, the need to know more about the earth's magnetic field for cosmic-ray studies drew Van Allen further into geophysical questions, but his goal remained to determine the nature of the primary cosmic-ray energy spectrum. The *Norton Sound* flight data gave Van Allen a three-point spectrum that revealed a fivefold cosmic-ray intensity increase from the geomagnetic equator to the southern coast of Alaska. The consistency of his results helped to establish that what he was observing was a real characteristic of the primary radiation.[43]

In effect, Van Allen was doing physics, using geophysics as a tool. He had been primed for this by senior DTM colleague Harry Vestine, but at APL, the theory came from physicist Ralph Alpher, whom Van Allen had asked to provide a theoretical study of how to relate cosmic-ray observations to contemporary geomagnetic theory. The theory of Carl Störmer showed how the motion of charged particles was affected by the earth's magnetic field. Alpher applied Störmer's theory and made it accessible to APL experimentalists.[44] But this association also brought geophysics to center stage.

## Ultima Thule: Van Allen's "Rockoons"

By the summer of 1950, Van Allen had decided to leave APL to become chairman of the Physics Department at his *alma mater* in Iowa.[45] When Van Allen moved to Iowa, the APL high-altitude research group dissolved, and its remaining members were moved to other parts of the laboratory or left entirely. Van Allen states that "Everyone was sort of scratching around, trying to decide how he was going to continue his research beyond this point."[46]

Regrouping at Iowa, Van Allen secured enough support from the Research Corpo-

ration in New York City to spend part of 1951 flying simple cosmic-ray balloonsondes with a few graduate students as assistants. As he recalls "We didn't have anything very strongly in mind at that point."[47] Indeed, Van Allen used the time to analyze the backlog of APL rocket data he had brought with him. Analysis continued through the summer of 1951 when he took up a Guggenheim Fellowship at Brookhaven National Laboratories.

Van Allen went to Brookhaven to study both experimental and theoretical high-energy physics techniques, but he also used the time to think about what he was doing and where he was going. The text of a paper he prepared that summer for a November aerospace conference reveals his thoughts on geophysics: to Van Allen, the earth's geomagnetic field acted "as a huge [magnetic] rigidity spectrometer for all charged cosmic-ray primaries."[48] Looking on the earth's magnetic field as a physicist looked on an accelerator, Van Allen wanted to complete the geophysical mapping of the intensity of the primary cosmic-ray spectrum that he had started at APL, which had reached only to the upper mid-latitudes. Now he hoped to examine their intensities at high geomagnetic latitudes.[49] The theories of Störmer, Lemaitre, and Vallarta, as interpreted by Alpher, provided a rough model for how charged particles moved in the earth's magnetic field. They showed that at high geomagnetic latitude, particles of lower momentum, or generally less magnetic rigidity, would arrive at the top of the atmosphere without being deflected.[50] So a greater energy range of primaries would be observable as one moved north, and Van Allen planned to go all the way to the geomagnetic pole, close to Thule, Greenland, which was accessible in the summer by ship.

Van Allen's Brookhaven summer was also a time to think about alternatives. That he emerged from the effort still wedded to upper atmospheric research indicates he had not stopped thinking of research in the context of a rocket. But his Research Corporation support could not handle a continuing research agenda like the one he had enjoyed at APL; Aerobees were completely out of his price range. He needed a low-cost substitute.

Starting in the summer of 1946, a "Joint Army Air Forces/Navy Upper Atmosphere Research Program" operated by the Naval Ordnance Test Station and the Muroc Air Base (now Edwards AFB), flew a wide variety of scientific packages aboard B-29 aircraft. In addition to cosmic-ray groups at Bartol, National Geographic and MIT, groups from Harvard, NRL, the Bureau of Standards, the Weather Bureau, Princeton, Johns Hopkins, Rochester, and DTM provided a wide variety of remote sensing, atmospheric, and astronomical packages. Brookhaven National Laboratories photograph P8–8-8, courtesy A. Needell.

After a season flying balloons in Iowa, Van Allen decided to combine the economy of ballooning with the capability of the rocket as a cost-effective means of regaining the altitude he needed. The idea was simple: a balloon would carry a small inexpensive rocket to about 16 kilometers, where a pressure-sensitive switch would trip and cause the rocket to fire directly through the balloon. Although the combination had been suggested before, Van Allen and his Navy friends thought of it anew as a means to send tiny counters to between 80 and 100 kilometers. Van Allen was able to take advantage of ONR's highly successful and accessible Skyhook balloon program, as well as his continuing good relations with Navy personnel.[51] With backing from ONR, the Atomic Energy Commission, the Research Corporation, and with the Navy's logistical support and surplus solid-fueled Deacon and Loki rockets, Van Allen and his graduate students built up an arsenal of what were later called "Rockoons" to study the latitude variations of high-altitude cosmic-ray phenomena. His goals were still seated in cosmic-ray physics, and, as the character of his funding agencies suggests, nuclear physics.[52]

Van Allen sold the Rockoon idea to his UARRP colleagues as the most economical

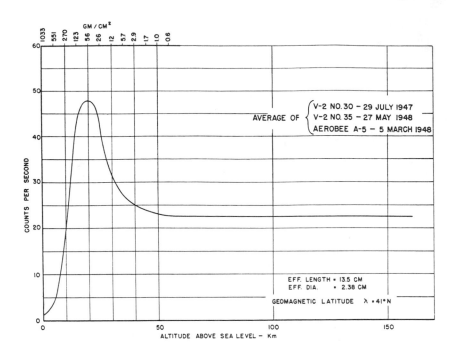

AVERAGE OF { V-2 NO. 30 – 29 JULY 1947
V-2 NO. 35 – 27 MAY 1948
AEROBEE A-5 – 5 MARCH 1948 }

EFF. LENGTH = 13.5 CM
EFF. DIA. = 2.38 CM

GEOMAGNETIC LATITUDE λ = 41°N

COUNTS PER SECOND

ALTITUDE ABOVE SEA LEVEL – Km

APL's composite cosmic-ray plateau based upon single-counter data. From Fraser (1951), p. 13.

means to measure the total cosmic-ray intensity above the atmosphere. In addition, he argued, Rockoons made possible "synoptic observation on pressure, temperature, ozone, etc., due to its low cost. It is also adapted to geographical surveys because of the simple launching arrangements."[53] And so Thule, Greenland, became the goal of his first season of firing Rockoons.

After Van Allen completed tests of the system at White Sands in the summer of 1952, he and two of his Iowa colleagues, Leslie Meredith and Lee Blodgett, traveled with their contingent of Rockoons aboard the Coast Guard cutter *Eastwind* to Thule. In August and September 1952, they fired seven Rockoons with varying degrees of success. Five of the seven systems worked successfully; the highest altitude attained was 100 kilometers. Two Geiger counter flights returned data from 90 kilometers and below, and two others carried ionization chambers that reported cosmic-ray data to 64 kilometers. In follow-on expeditions during subsequent summers, the Iowa team found something more than cosmic rays, however. Between geomagnetic latitudes 64 and 75 degrees, and at elevations above 50 kilometers, they also found a high

intensity of easily absorbable (or 'soft') radiation, which they soon realized was the first direct detection of auroral electrons. In July 1953 they repeated their observations aboard the *U.S.S. Staten Island* (figures, pp. 257, 259). Van Allen had touched the Northern Lights.[54]

Thus, between 1949 and 1957, while searching out the nature of the primary cosmic-ray spectrum, Van Allen and his group made solid contributions to geophysics, ranging from their discovery of the equatorial electrojet to their detection of auroral radiation. Although conquests such as these made geophysics alluring, Van Allen's migration from cosmic-ray physics into geomagnetic studies was neither abrupt nor complete. It took well into the late 1950s to make the transition, as elucidated later in this chapter.[55]

## NRL's Counters and Cloud Chambers

Unlike APL, NRL persevered with highly complex experiments, trying to design circuits to isolate and reveal the separate effects of true secondary showers and those caused

Control panel of APL's cosmic-ray altimeter, circa June 1947, likely created for the Bureau of Ordnance. Applied Physics Laboratory, JHU/APL archives.

by the warhead. NRL preferred heavy "lead igloo" designs, as some at both NRL and APL called them, because they helped the group improve its expertise with electronics on rockets. Electronics development was one of Krause's primary rationales for engaging in such research, which paralleled NRL's goals better than breakthroughs in cosmic-ray physics did.[56] Sharing the APL experience, the NRL group, led by Perlow and Golian after Krause's departure, found that its data provided only a confusing picture. There were simply too many variables to sort through, even if everything seemed to work right, which it almost never did. About the only result they could establish in the first flight season was

Final preparations are made to a Deacon rocket payload by Iowa staff member Robert A. Ellis (l), assisted by a Navy crewman aboard the U.S.S. Staten Island, stationed in the auroral zone, July 1953. U.S. Navy photograph. NASM SI 90–5328.

that the counting rate was greater above 60 kilometers than it was on the ground.[57]

NRL's instrumentation policy, like APL's, was to explore any means available to study cosmic rays from rockets. After the successful retrieval of photographic data from Tousey's spectroscopic flights in late 1946, Krause and Perlow planned to add photographic cloud chamber experiments to their cosmic-ray ensemble.[58] They first thought of using a standard coincidence counter to trigger their cloud chambers. But by 1947, they decided that the vastly greater counting rates at operational heights were sufficient to catch something interesting during random expansion cycles.[59]

The first cloud chamber, a miniaturized version of an NRL laboratory device, flew on a V-2 on 22 January 1948. It was a glass cylinder 15 centimeters in diameter and 8 centimeters high, set within a far larger structure that held the camera, lighting system, power supply, control electronics, and shielding. One complete chamber cycle—including expansion, exposure, clearing, equilibrium, and film advance—took 25 seconds, a good chunk of the flight time.

Charles Johnson was a member of the NRL team for the 22 January flight. Arriving at the crash site, he helped to drag the sealed camera cassette from the shards of the destroyed chamber, under a meter of earth, and ultimately found that one usable stereophotographed particle track was obtained at an altitude of 145 kilometers. The particle penetrated both lead plates and so NRL's initial conclusion was that it was an alpha particle with an energy in excess of 500 MeV: therefore it was assumed to be a cosmic-ray primary[60] (figure, p. 266).

The 22 January cloud chamber was the first of several launched in the next three years, but it was the only one that returned data worth discussing.[61] Johnson recalls that the process of building the cloud chambers was a good example of "trial and error engineering."[62] As with the spectroscopic flights, all options were worth trying, but only the most efficient designs, or those most promising of future improvement, would be continued. The cloud chamber effort lost some support when

Krause left; Newell was willing to continue development, but their continued failure, and Perlow's departure in 1952, ended NRL's efforts in cosmic-ray physics studies with rockets. Remaining NRL staff such as Charles Johnson migrated into other sections under Newell and to areas of greater institutional interest, such as ionospheric and upper atmospheric physics.

## The Princeton Group

Of all the groups, Princeton's had the most frustrating experience and left rocketry the quickest. At the outset, Princeton was given three slots for experiments in cosmic-ray and ionospheric physics, under the general direction of Myron Nichols. Rudolf Ladenburg and Nichols were the ringleaders; both encouraged other members of the Physics Department to participate, such as John A. Wheeler, Donald J. Montgomery, George T. Reynolds, and John F. Brinster.[63] By the summer of 1946, Princeton's APL contract for telemetry equipment for guided missiles had been amended to include cosmic-ray research in support of telemetry studies and involved some 40 people on Princeton's staff, mainly specialists in electronics and radio instrumentation.[64] H. D. Smyth, chairman of the Physics Department, oversaw the contract, although the primary coordination was done by junior faculty such as Nichols in the Palmer Physics Laboratory, and the work, recalls William Stroud, was the responsibility of graduate students.[65]

Princeton's first scheduled V-2 flight was set for 27 June 1946. As of 6 May, Nichols and Brinster had almost completed an ionospheric radio propagation experiment, and pressure and temperature sensors were also being built by Nichols' staff with Montgomery's advice. Originally, Marcel Schein's Chicago group was to build the cosmic-ray detectors for Princeton, but the "difficulty of planning spatial arrangements at a distance" made this impractical. In some haste, both groups decided that one of Schein's graduate students, William Stroud, would come to

An APL Rockoon balloon at launch from the U.S.S. Staten Island, July 1953, near Thule, Greenland. U.S. Navy photograph. NASM SI 89–1154.

Princeton to construct the proper instruments.[66]

Princeton was not ready for the 27 June slot, which went to NRL. The V-2 Panel rescheduled Princeton for launches on 31 October 1946 and 23 January 1947. But when Nichols left for Michigan, taking a good chunk of the Lark telemetry group with him, APL became concerned that Princeton was falling behind in its primary contract responsibilities to develop radio telemetry systems for missile guidance and wanted chairman Smyth to put some of his best remaining men on it.[67] APL's technical supervisor for contracts, Ralph Gibson, also met with Wheeler and Brinster on 26 July to determine how the V-2 cosmic-ray work was progressing. The Princeton group was ready for the first flight, but Wheeler remained cautious about overcommitting themselves to the rocketry, especially because Gibson was not happy with their telemetry progress. Princeton, Wheeler noted,

would continue if they were "getting good results and advancing the subject." They would evaluate their progress in September.[68]

Wheeler was the general supervisor of cosmic-ray research at Princeton. His arsenal included balloonsondes, ground-based observing stations, and laboratory research, as well as the V-2. He never fully warmed to the V-2, especially when faced with delays and criticism from APL. As he noted in a private diary, he wished above all to "protect integrity of commitment" by searching for the "best way" to observe cosmic rays: there was always the question, "if better results some other way[,] do it the other way."[69] Wheeler, unlike his NRL or APL colleagues in 1946, did not think only of rockets as vehicles for studying cosmic rays.

For reasons unknown, Princeton's 31 October slot was advanced to 8 August, but by then they were ready, under the direction now of John Brinster as the technical supervisor of the rocketry project. The warhead was filled with neutron counters for Ladenburg, trays of Geiger counter telescopes, shielded and unshielded pressurized ionization chambers to examine shower activity, ionospheric radio propagation equipment to measure the ion density in the E layer, and a single unshielded Geiger counter set in a brass tube at the tip of the warhead cone to measure the total cosmic-ray intensity profile.[70] To satisfy the terms of their APL contract, Princeton also used its own dedicated four-channel Lark telemetry system for recording cosmic-ray counts.[71]

Fifteen members of the Princeton staff and faculty spent two weeks at White Sands in August, mounting and integrating their payload. Although they were thankful that the launch was slipped one week to the 15th to fix a steering control mechanism, some were far from happy to have to sit in Princeton's electronics trailer an extra week, where "it was hot enough to fry an egg."[72] Also, the extra week did not help. Scant moments after launch the recalcitrant servosystem balked once again, and the V-2 went out of control, settling into a flat spin at an altitude of 6.5 kilometers. The White Sands safety officer hit the fuel cutoff switch, and the flight was a washout, even though the Lark telemetry system reported that the Princeton package was working. Wheeler reported to APL that the flight provided confirmation that the Lark telemetry system worked and tested favorably against the rival NRL system. Aware of APL's priorities, he added, "This point is essential as part of Princeton's obligation to the picture on telemetering development."[73]

This first failure did not deter the Princeton group; indeed, the first priority of the payload effort, as Wheeler claimed, was to test the NRL and Lark telemetry systems for comparative reliability, signal strength, and freedom from interference. The preparation alone for these experiments brought them much needed practical experience, and Princeton was anxious to continue. At the time, more than half of all V-2 launches were failures, and each group was well aware of the hazards. Princeton also had two more large

cosmic-ray payloads planned and under way for a 7 November firing, were collaborating with NRL in interpreting their recently obtained cosmic-ray data, and were planning to cooperate with NRL in a third cosmic-ray experiment on 9 January 1947.[74]

Plans for Princeton's 7 November firing proceeded on schedule. Their two cosmic-ray packages, now shock-mounted on rubber pads, were even more ambitious than before. To minimize spurious secondary particles produced by the steel skin of the warhead, the directional telescopes looked through glass windows set into doors in the V-2 warhead. And as before, the Princeton Lark telemetry system would be tested against NRL's.[75]

If the 15 August failure did little to deter the Princeton group at first, the failure of AMC's 22 August missile gave them pause, and they were happy when Army Ordnance ordered a halt in the firing schedule until the problem was found. Meanwhile, the Princeton group continued cosmic-ray balloon flights, built laboratory cloud chambers, and maintained theoretical analysis groups "to correlate and explain experimental results."[76] Through October they looked forward to their next flight, as White Sands firings on 10 October and 24 October both were launch successes. On 29 October, Brinster, Stroud, and about four others from Princeton shipped their second warhead to White Sands, amidst fanfare created by a Princeton University public affairs proclamation that "next month's launching of the German terror weapon" would help to keep the nation "first in the search for new fundamental knowledge."[77]

Princeton's second V-2, launched on 7 November, was another total failure caused by a faulty servo system. Immediately after launch, the rocket swerved wildly and leveled off in a southerly course at an altitude of 360 meters. Again, the range officer hit the emergency fuel cutoff control, and again, Princeton suffered another total loss.

Now, Princeton began to have serious doubts about how it should continue. One week later, Smyth, Wheeler, and others traveled to Washington for a meeting at ONR with Alan Waterman and his staff. Gibson was there along with Ernst Krause and Gilbert Perlow from NRL, and L. M. Mott-Smith from BuOrd. It was a coming to terms. Princeton wanted out from the V-2 program but wanted to discuss continuing support from BuOrd, NRL, and ONR "for doing basic research work in cosmic rays," as well as developmental research in telemetry.[78] Brinster argued that "the probability of having completely successful firings is progressively decreasing since the best components of the missiles are being used up first."[79] After some discussion, BuOrd, ONR, and NRL all agreed to continue Princeton's support. Krause particularly wanted to continue the association "to assist [NRL] . . . in the theoretical aspects of this work."[80]

Gilbert Perlow stayed in touch with Wheeler and Brinster, trying to encourage them to provide cosmic-ray payloads. But Princeton passed up the opportunity, formally declining any participation in NRL's 10 January launch of V-2 no. 18, which, of course, worked beautifully, reaching 115 kilometers and returning data from NRL's counter telescopes. Brinster attended the next V-2 Panel meeting on 28 January 1947 and the minutes revealed no regret in their decision. For the present, they simply were not "ready to meet schedules for field tests."[81]

Losing Princeton as an active participant removed the only mainstream physics department participating in cosmic-ray research with rockets. Van Allen publicly regretted the "very unfortunate rocket flights" Princeton experienced, which "forced [it] to withdraw from rocket research without having obtained flight results." Marcus O'Day used the Princeton experience to call for a reduction in the frequency of flights: the two-week firing schedule maintained by the Army was too fast to allow outside groups to participate.[82]

Through 1947 and 1948, Princeton continued to fly detectors on Skyhook balloons, aircraft (especially the Navy's B-29 program based at Inyokern), and blimps and developed more sophisticated mountain and sea-level detector arrays.[83] In the summer of 1948, all six of their allotted Skyhook flights were successful, and by the end of July they had collected much useful data on nuclear star production in the atmosphere and were busy analyzing it as a nuclear cascade process.[84] As the Princeton group under Wheeler moved more and more into basic studies that would illuminate nuclear fusion processes, they still remembered that their general approach had been "an outgrowth of Princeton's wartime development work on telemetering devices."[85] But by 1949, they pointed out to ONR's Nuclear Division that continued support for cosmic-ray observations would extend "knowledge of the transformations produced by elementary particles" with the long-range goal of acquiring "understanding and control of fundamental nuclear processes." By April 1949, a full 40 percent of their effort was devoted to balloon and B-29 observations.[86]

The experience of the Princeton group may well have further isolated rocketry from mainstream cosmic-ray physics. But it also provided training for several people who became valuable additions to other balloon and rocket groups. John Winckler, a physics faculty member on the Princeton research steering committee, later joined Minnesota's growing balloon program, while William Stroud went to the Signal Corps to continue rocket work, and Fred Bartman, another graduate student who had been in the ground-based nuclear star observation group, became a welcome addition to Myron Nichols' rocket group at Michigan.[87]

## Migrations

Homer Newell admitted at the end of the V-2 era that cosmic-ray research with rockets was difficult to justify. He recalled in 1953 that "The rocket cosmic-ray program began with considerable enthusiasm and backing, and a number of interesting results were obtained. . . . In spite of the successes, however, the disadvantages of the rocket vehicle have gradually militated against extension of the work."[88]

Newell added that the much longer observing times of well-stabilized Skyhook balloons could provide better directional information than could similar experiments on unstabilized rockets. And given what was known about the flux density of primaries above White Sands and the rate of their absorption with depth in the atmosphere, it was actually more likely to catch a heavy primary at balloon altitudes over a period of hours than from a rocket during only a few minutes of operational look time. Indeed, the first unambiguous detections of heavy cosmic-ray primaries and the identification of a heavy component in the primary spectrum (such as helium nuclei and particles with atomic numbers up to 40) came in 1948 from unmanned balloon flights of cloud chambers and nuclear emulsion plate stacks flown by groups in the Minnesota and Rochester Physics Departments that participated in ONR's Project Skyhook.[89]

Newell's opinion reflected the fact that during the V-2 era, the major discoveries in cosmic-ray physics were made by balloons from Powell's group at Bristol and from the Minnesota and Rochester balloon groups. His NRL associates shared his view. In August 1953, Charles Johnson, L. R. Davis, and J. W. Siry argued on the basis of six years of work that "the advantage of height is greatly compromised by a flight time measured in minutes, varying aspect of the rocket after powered flight, and undesirable secondary radiation produced in the rocket's mass." Johnson bluntly added that their results had been severely compromised by the steel V-2 warheads: "a majority of the particles detected were secondary particles generated in the rocket. These secondaries made it difficult to obtain a value for the primary intensity from the data."[90]

Thus cosmic-ray studies from rockets proved to be a relatively short-lived activity. Princeton was gone by 1947, APL by 1950, and NRL by 1951. Results from the rocketry effort were interesting, but had no real influence on mainstream cosmic-ray research. For the cost and effort, it was, Van Allen later felt, a "good but thin body of results."[91] All the rocket groups became frustrated with the small impact they were having on the field;

their efforts garnered very few citations in the cosmic-ray review literature during this time. For this reason, as well as the technical frustrations already discussed, cosmic-ray research disappeared in the surviving rocket groups during the early 1950s. Many members of Gilbert Perlow's group migrated into other more active fields such as spectroscopy, ionospheric radio propagation, or atmospheric physics, and some, like Perlow himself, simply left rocketry altogether for more traditional physics research at Minnesota. At Iowa it was the same, where Van Allen leaned more toward geophysical research with Rockoons. At one point in 1953, Van Allen considered leaving Iowa and rocketry altogether; he was then on leave at Princeton working with Lyman Spitzer to design an experimental 'stellerator' for controlled thermonuclear fusion. When it failed to show rapid progress, Van Allen decided to return to Iowa and to do "research I did understand and had good prospects for results."[92]

## Van Allen's Migration Toward Geophysics

The difficulty in tracing Van Allen's (or any physicist's) migration from cosmic-ray physics to geophysics is that both employed virtually the same techniques, instrumentation, and procedures: their discoveries could be applied in either area. His Rockoon flights at high latitudes through the mid-1950s continued to bring back information on cosmic-ray intensities, but they also revealed much about the character of the earth's magnetic field and the complex current sheets in the equatorial regions. In particular, with his growing staff and student force at Iowa, Van Allen's first in situ observations of enhanced electron flux in the polar auroral regions became a major focus of his continuing flights during the International Geophysical Year (IGY).[93] Many of Van Allen's reasons for arctic flights in the late 1940s and 1950s reflected the rhetoric he used to justify conducting cosmic-ray observations as part of Operation **Nanook** in 1946.[94] The identification of his migration has to be sought in other social factors.

Van Allen maintained the chairmanship of the UARRP through the 1950s, and incidentally also served on the NACA Subcommittee on the Upper Atmosphere from 1948 through 1952. By 1954 he had, in addition, become deeply involved with planning for the International Geophysical Year as a member of many of its technical panels. His first exposure to the idea of an IGY came from a dinner he hosted in 1950, in honor of Sydney Chapman, and from that point on, the growth of collaborative geophysical efforts by the UARRP foreshadowed the way in which upper atmospheric research would be conducted during the IGY.

Throughout the 1950s, Van Allen continued to publish mainly in physics journals. Even though he became an associate editor of the *Journal of Geophysical Research* in 1959, his first publication there was not until 1956. He continued as an adviser for nuclear physics to ONR and NSF and was also an associate editor for *Physics of Fluids* throughout the decade.[95] Rather than migrating from one problem area to another, then, Van Allen at first accreted new problem areas amenable to the types of instrumentation and experimentation that he knew how to build and to conduct and that showed the greatest promise for intellectual contributions. Only when he began to sense a lack of kinship and sympathy for his work at meetings of the American Physical Society in the late 1950s, and when, coincidentally, a paper submitted to the *Physical Review* was summarily dismissed as unsuitable for publication in that journal, did Van Allen realize that his friends were in astrophysics and geophysics. "That was the scientific community that we were really talking to."[96]

## Reception by the Communities of Science

By the late 1950s, Van Allen's papers, wherever they were published, were gaining more attention in the world of geophysics and very little in the world of cosmic-ray physics. Indeed, in a lengthy retrospective memoir by senior members of the cosmic-ray fraternity, Van Allen's name was the only one to appear from the early rocket work, and this was for his later discoveries from the Explorer satellite series.[97] In Vitali Ginzburg's 1964 review of cosmic-ray astrophysics, none of the 524 references is to work done on rockets in the V-2 era. In late 1947, as L. Jánossy in Dublin was completing his comprehensive review of theory and practice in cosmic-ray physics, he regarded ballooning and conventional aircraft as the proper means of exploring the upper atmosphere and made no mention of rocketry.[98] In 1949, when the International Union of Pure and Applied Physics provided a worldwide listing of cosmic-ray laboratories and physicists, none of the rocket groups was mentioned, save for Princeton.[99] Cosmic-ray physicist H. Victor Neher of Caltech did not even consider the use of rockets in 1949 when he applauded the use of the B-29 aircraft as a means of "performing experiments on cosmic rays at high altitudes." Finally, modern historians rarely mention the role of sounding rockets in cosmic-ray physics.[100] By and large, then, the contributions of rocket-borne efforts to cosmic-ray or particle physics garnered little attention in the review literature, and when it did, as in Rossi's 1948 review, it was not integrated into any analysis or general conclusions. Van Allen, Singer, and Perlow were occasionally invited to give addresses at conferences and symposia in the late 1940s, but it was more in the context of sponsored research at sites such as the Echo Lake Colorado conferences, which highlighted the efforts of ONR-sponsored research.[101]

In contrast, reviewers and coordinators of research on the physics of the upper atmosphere certainly did invite rocket scientists to their meetings and did give them space in their reviews. Not only did J. A. Ratcliffe and others include rocket work in extensive reviews in the 1950s and 1960s, but invited their leaders—such as Van Allen, Friedman, and Newell—to contribute as well. Astrophysicists such as Gerard Kuiper made sure that the upper atmosphere researchers at NRL and APL were represented at his 1947 Yerkes

Observatory symposium on the atmospheres of the earth and planets, and in 1953 the Gassiot Committee of the Royal Society invited the UARRP to co-organize a major international conference on the rocket exploration of the upper atmosphere. Clearly, by the 1950s, geophysical and geomagnetic studies of the upper atmosphere and its environs by rocket had become part of the growing tool kit accepted by the geosciences community.[102]

In the early 1950s, just as accelerators were attaining energies that competed with those available from cosmic rays, cosmic-ray physics was finding a warmer home in astrophysics or geophysics.[103] As some physicists recall, geophysical considerations that had long been considered peripheral became more central issues for discussion at international conferences, whereas cosmic-ray physicists who wished to continue to discuss their contributions to particle physics were encouraged to meet somewhere else.[104] Even so, those working at NRL or elsewhere who were now seeing greater value in geophysical applications still sensed, as Charles Johnson pointed out, a "grass roots" feeling that geophysical

studies were not welcomed by the American Physical Society.[105] People like Johnson and S. Fred Singer had been, along with Van Allen, trained in physics and so focused on the journals of physics as their favored means of communication. To Singer, as well as to Herbert Friedman and Van Allen, the *Physical Review* was the most prestigious outlet available. Singer felt that his efforts would be lost in the *Journal of Geophysical Research* among the long listings of terrestrial and atmospheric data. Professionally, Singer found little incentive to "identify with hydrologists and soil scientists and others who were publishing in the *JGR*."[106]

## The Institutional Context for Cosmic-Ray Research with Rockets

Van Allen recalled that experimentation in the V-2 era was conducted at a frantic pace, and, as a result, there were "a lot of short cuts in the testing and preparation of the equipment." Also contributing to their early problems was their "general conceptual mistake" of building highly complex devices.[107] There

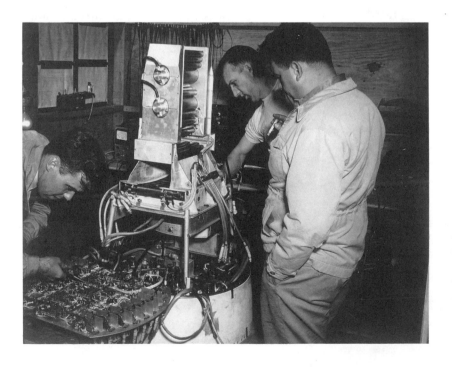

C. A. Schroeder, Charles Y. Johnson, and Gilbert J. Perlow (l to r) make prelaunch adjustments to NRL's cosmic-ray telescope aboard the U.S.S. Norton Sound, for the flight of Viking 4, 25 April 1950. U.S. Navy photograph, Charles Y. Johnson collection. NASM SI 89–19746.

Charles Johnson and Serge Golian retrieving the cloud chamber film record from their 22 January 1948 V-2 flight. Frames from "Upper Atmosphere Research." NRL Motion Picture Service Project No. 1 (1950). U.S. Navy photographs. NASM film archive.

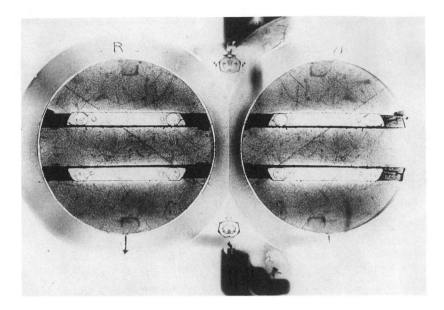

Stereophotograph (negative print) of cosmic-ray tracks from the 22 January 1948 flight of the NRL cloud chamber, revealing one that penetrated both lead plates. U.S. Navy photograph, Charles Y. Johnson collection. NASM SI 89–19737.

were too many variables in the data stream, too many problems with heavily taxed telemetry, the local mass of the rocket, and variations in the counters. It was an unfamiliar new way to do science that removed control from the scientist.

But complex counter telescopes placed on rockets served institutional goals at APL, as well as at NRL and Princeton: they served to test telemetry systems and served as a means for improving them. Counters demanded stable on-board electronics and logic circuits, which would help improve guidance and control systems. More directly, Van Allen's cosmic-ray altimeter prototype, built in 1947, was designed to guide "long range supersonic missiles" at altitudes between 12 and 32 kilometers[108] (figure, p. 257). The problems facing cosmic-ray research on rockets, therefore, were the problems rocket developers faced when trying to improve the breed.

For military patrons, the problem of the cosmic ray became the promise of better missile systems. Van Allen certainly appreciated this fact, since he was working within an institution defined completely by missile research.[109] To his peers at the American Physical Society in 1947, Van Allen left little doubt about the problems of cosmic-ray research with rockets:

*Rocket research imposes peculiar problems of its own. The flight is short. The equipment is inaccessible and no checks are possible after impact. Accelerations, vibrations, and rapid changes in temperature are encountered. Much effort has been expended in making the equipment reliable and independent of external conditions.*[110]

APL was just the place to expend that effort; its tool-building priorities were largely satisfied in Van Allen's quest for cosmic-ray primaries. Van Allen did not stay at APL, of course, and was free to pursue anything he liked at Iowa. But the research momentum he had established at APL continued at Iowa: it was what he knew best.

The activities of both the APL and NRL groups left little doubt that their main concern was making instruments work on rockets. At NRL and APL, at least, that was their job. If equipping rockets for cosmic-ray research prompted them to meet their institutional responsibilities, then it could be pursued with less concern for gathering scientific data of immediate interest than if the group sat within an academic physics department whose criteria for success were far more rigidly confined to physics, as was the case at

Princeton.[111] In like manner, if the cosmic-ray research on rockets bore little fruit, these groups were free to shift their attention to more interesting or lucrative fields, such as geophysics. The overriding concern was not so much that science was being done, but that useful things were being done to inform the needs of guided missile development.

# Notes to Chapter 14

1. Van Allen (1952), p. 246.
2. Useful reviews of cosmic-ray research activities from the period are Auger (1945); Braddick (1939); Jánossy (1948); LePrince-Ringuet (1950); Rossi (1953, 1964). Histories of earlier periods include Kargon (1981); Sekido and Elliot (1985); Ziegler (1986, 1989); De Maria and Russo (1989). DeVorkin (1989a) has examined how cosmic-ray studies were part of both manned and unmanned balloon programs in the 1930s and 1940s.
3. In 1935, Erich Regener and his associate Georg Pfotzer found that the maximum rate of cosmic-ray secondary production occurred around 20 kilometers and dropped off above that point. See also Bowen, Millikan, and Neher (1938); Schein, Jesse, and Wollan (1941), p. 615.
4. See Johnson (1939), p. 208; Schein, Jesse, and Wollan (1940; 1941).
5. Krause OHI, p. 70rd.
6. Regener, for example, never attempted cosmic-ray experiments with V-2s at Peenemünde, although their observation had been a prewar passion.
7. Van Allen OHI, p. 187rd; and Singer OHI, no. 1, p. 9rd. Van Allen's 30 July 1946 telescope was in fact a close copy of a ground-based array created at Chicago. See Fraser and Siegler (1948), pp. 56–58; and Schein, Jesse, and Wolan (1940).
8. Van Allen OHI, pp. 187–88rd.
9. In the late 1920s and early 1930s, physicists first realized that cosmic rays were charged particles by determining that they were deflected by the earth's magnetic field.
10. Van Allen OHI, p. 186rd.
11. Ibid., no. 2, pp. 109rd; 174rd. See also Van Allen (1983), p. 17. These were problems Van Allen recalls hearing about in Iowa and at the Department of Terrestrial Magne-

tism. On Van Allen's shifting interests, as sensed by a junior member, see S. Fred Singer OHI no. 1, pp. 7–11rd.
12. Gilbert Perlow to the author, 23 February 1987.
13. Krause (1949), p. 194.
14. For the cloud chamber work, see Golian, Johnson, Krause, Kuder, Perlow, and Schroeder (1949). The plate stacks were prepared by Herman Yagoda and later by Maurice M. Shapiro.
15. On the development of the group's interests, see Nichols to Tuve, 26 November 1945, p. 1. Box 119 "Unmarked Black Folder," MAT/LC; see also: H. E. Newell, Jr., "Record of Consultative Services," 29 October 1946; E. H. Krause, "Record of Consultative Services," 12 November 1946; G. J. Perlow, "Record of Consultative Services," 27 November 1946. Box 98, folder 4, NRL/NARA.
16. L. R. Hafstad and G. R. Tatum, "Report of the Director of Research to the Subcommittee on Cooperative Research," 3 October 1946. Box 27, folder 6, Office of the President Papers, Johns Hopkins University Archives. Cited in Dennis (1991), pp. 9–10.
17. From JVA/APL, copy in SAOHP/NASM.
18. Townsend, Friedman, and Tousey (1958), p. 15. Even though this statement was made after the fact, it reflects the same rationale (save for the mention of transistors) described in planning documents in the 1940s.
19. Krause OHI, p. 90rd. J. A. Wheeler, "General Survey of the Princeton Project Program—Cosmic Rays and Telemetering," 28 August 1946, p. 2. Appendix IV to H. D. Smyth, "Annual Report NOrd 7920," n.d. A-475 folder (Wheeler), PP/P.
20. Singer OHI no. 2, p. 25rd.
21. Rossi (1985), p. 56. On the reliability of these detectors in the 1930s, see also Galison (1987), p. 83; Rossi (1981), pp. 34–41; and Weisz (1941), p. 18. See also Neher (1985), p. 91; and DeVorkin (1989a), chaps. 3 and 8. All three detectors worked by the same principles. An ionizing particle penetrates a sealed, gas-filled chamber containing a positively charged electrode insulated from the negatively charged chamber walls. Electrons, liberated by the collision of the penetrating particles within the gas then migrate to the electrode, producing a current, which in the case of an ionization chamber will be a measure of the total ionizing power

of the incident flux of particles. If the potential difference between electrode and walls is increased, however, the ion-collecting rate will increase. Within a certain range of potential difference, the counter output will be proportional to the ionization energies of single particles, and so this type of counter is called a proportional counter. If the potential between cathode and anode is increased well beyond this bounded range, each cosmic-ray event will produce a cascade of ions within the gas-filled chamber, enough to make each event measurable, or countable. This is a Geiger or Geiger-Mueller counter. Full descriptions of detectors designed for rocket-borne instruments appear in: Newell (1953), pp. 248ff.; and Ehmert (1953), pp. 711–15. On the earliest cosmic-ray detectors used during balloon ascents, see Ziegler (1986).

22. See, for instance, Krause, Fraser, Singer, and Van Allen oral histories, SAOHP/NASM.

23. Cosmic rays produced tracks of darkened silver grains in Powell's nuclear emulsions; the number of affected grains, or the number of side tracks caused by the ejection of electrons within the emulsion, was a measure of the charge of the particle. If a stack of emulsions stopped a particle, the depth of the track indicated the original energy of the particle, and if the orientation of the plate stack was faithfully recorded at the time the event took place, its incident vector could be retrieved. Furthermore, the shape of the particle tracks after impact within the emulsion revealed the mass (or momentum) and charge of the ejected particles. Sandström (1965), pp. 11–17; Jánossy (1948), pp. 402–04; Simpson (1986), p. 7. See also Sekido and Elliot (1985), pp. 127, 210.

24. Newell (1953), p. 250. Newell was trained as a mathematician, and had experience in field geology when he joined NRL.

25. Fraser and Siegler (1948), p. 45.

26. Once physicists learned to trigger cloud chambers with Geiger counter coincidence circuits, cloud chambers became efficient and effective tools. On early detectors, see Ziegler (1986, 1989). On seminal discoveries using cloud chambers, see Brown and Hoddeson (1983), pp. 8–9.

27. Each counter had to be sampled rapidly to provide sufficient time resolution for coincidence, expected to be on the order of 0.005

seconds. This was only one of the many problems with the complex device; the NRL 23-channel telemetry system could sample at a rate of 200 samples per second where each sample contained 24 pulses, or 23 differences. See Durand (1949), p. 138. For a general description of the APL telescopes, see Fraser and Siegler (1948), pp. 45–63.

28. Fraser and Siegler (1948), p. 62.

29. Van Allen OHI, p. 181rd.

30. Fraser and Siegler (1948), p. 62.

31. Ibid., p. 64

32. Van Allen and Tatel (1948), p. 246.

33. Lorence Fraser OHI, p. 41. See also Fraser (1948), p. 62; Fraser and Siegler (1948); and Fraser (1951).

34. Krause (1949), pp. 194–95.

35. Van Allen and Tatel (1948). See also Van Allen (1948), pp. 171–75.

36. Emphasis in original. Van Allen and Tatel (1948); Van Allen (1948), pp. 172–73.

37. Van Allen (1948), p. 172.

38. Rossi (1948), p. 574.

39. These included data taken under a special aluminum shrouded telescope flown by NRL's Golian and Krause in 1947. Rossi (1948), pp. 559–60, p. 574.

40. These two expeditions carried APL technicians, balloon-borne cosmic-ray counters, and radio propagation experiments. Nanook was conducted in the summer of 1946 in Alaskan waters, and Highjump sailed to Antarctic waters in December 1946. This application and the justification for the costly expeditions were provided by L. R. Hafstad, director of research at APL. See L. R. Hafstad and Woodman Perine to Commander, Task Force SIXTY-EIGHT, 13 December 1946. Polar Programs files, JVA/APL.

41. S. F. Singer, E. Maple, and W. A. Bowen, "Naval Ordnance Laboratory Memorandum 10261," 19 May 1949. Also cited as Section T Internal Memorandum APL/JHU CF-1324, 7 December 1949. APL folder, HT/DTM. See also S. F. Singer Aerobee magnetometer file, 1948; and his "University of Maryland Historical File, 1107." SFS/NASM. Singer recalls that although he was supportive of the work, Vestine believed that it was a "long-shot." See Singer OHI no. 2, pp. 27–29rd.

42. Singer OHI, ibid.; Singer, Maple, and Bowen (1951); Singer (1954).

43. See Van Allen and Gangnes (1950), pp. 50–52; Van Allen (1983), p. 19; and Van Allen OHI, pp. 254–58rd.

44. Alpher (1950), p. 466, thanks Van Allen for suggesting this work. On Alpher's early influence, see Singer OHI no. 1, pp. 7–8rd. It was also Alpher's advice, Singer feels, that directed Van Allen's attention to the fact that a latitude survey of cosmic rays would reveal the energy spectrum of the primary radiation.

45. On Van Allen's move to Iowa, see Van Allen OHI, pp. 272–90rd; and Van Allen (1983), chap. 3.

46. Van Allen OHI, p. 295rd.

47. Van Allen OHI, p. 285rd. He had an initial grant of $5,000 from the Research Corporation.

48. See White and Benson (1952), chap. 21; and Van Allen OHI, pp. 181–82rd; Van Allen (1952), p. 246. On this analogy, and on Alpher's influence, see Singer OHI no. 1, pp. 7–8rd.

49. Van Allen (1952), p. 244.

50. Ibid., pp. 245–47. The magnetic rigidity of a particle, the ratio of the particle's momentum to its electrical charge, is a measure of the difficulty of curving the trajectory of the charged particle by a magnetic field. Particles with high magnetic rigidity would be unaffected by the earth's magnetic field, and the suspected cutoff was in the range of sixty billion volts. But very few particles possessed such high values. Particles with lesser rigidity, which were far more common, would be affected, and so their spectral intensity (the range of energies encountered among them) above the atmosphere would become a function of their initial direction and the geomagnetic latitude of first encounter.

51. No recovery or reuse of these inexpensive ($1,000 to $2,000 in 1950 dollars) rockets was required, since all data were transmitted by radio. The technical aspects of Rockoons are covered in Van Allen and Gottlieb (1954); Van Allen (1983), chap. 3; and Newell (1959a). On Skyhook, see DeVorkin (1989a), chap. 10; and Moore (1952).

52. The term "Rockoon" was coined after the first flight season in August 1952. See "Panel Report no. 33," 7 October 1952, p. 14. V2/NASM. On the necessity of graduate student manpower, see "Panel Report no. 35," 29 April 1953, p. 11. V2/NASM. On ONR/AEC interests, see DeVorkin (1989a), and E. P. Ney OHI, pp. 7, 10–11.

53. See "Panel Report no. 32," 30 April 1952, p. 10. V2/NASM.

54. They interpreted their results as evidence of "auroral soft radiation." See Van Allen (1957) and (1983), pp. 24–25.

55. Van Allen OHI, pp. 109rd, 174rd. See also Van Allen (1983), p. 17.

56. Krause OHI, p. 90rd. An NRL retrospective publication later argued that the NRL cosmic-ray group had gathered useful information on the seasonal variation of cosmic-ray intensity from the White Sands region and the composition of the primary component, especially the intensities of the two major constituents: protons and alpha particles. The NRL group also examined gamma radiation in the range 0.1 to 90 MeV, showing that it constituted less that 0.1 percent of the total cosmic-ray flux. Townsend, Friedman, and Tousey (1958), p. 15. See also "History of the Upper Air Rocket Research Program at the Naval Research Laboratory 1946–1957," NRL Report 5087 draft, p. 14. HONRL. For a detailed early account, see Newell (1953), pp. 260–81. "Lead igloo" was a descriptive term based upon the shape of the detector arrays. James Van Allen to Commanding General, White Sands, 25 June 1948; L. W. Fraser to Commanding General, White Sands, 7 January 1949. JVA/APL.

57. The NRL coincidence telescope had a different geometric configuration than did the APL telescopes and yielded discordant results. Golian, Krause, and Perlow (1946). See also Krause (1949).

58. Krause and Perlow (1947), pp. 14–15.

59. Wilson cloud chambers, however, required a gravitational gradient for operation; the action of gravity on the condensed particles would clear the chamber quickly. In free fall, the chamber would not clear. Thus NRL specified that their cloud chambers had to be mounted far from the axis of the rocket, and the rocket had to spin to produce artificial gravity so that the experiment would work. Golian, Johnson, Krause, and Kuder (1948), p. 58.

60. Golian, Johnson, Krause, Kuder, Perlow, and Schroeder (1949), p. 524; Golian, Johnson, Krause, and Kuder (1948); film footage of the search at the crash site, NRL Public Information Office.

61. Johnson, Davis, and Siry (1954), p. 306.

62. Charles Johnson OHI no. 1, p. 34rd.

63. S. F. Singer recalls Nichols as an especially energetic organizer. See Singer OHI no. 1, pp. 3–4rd.

64. The amendment provided an additional $250,000 from 1 March 1946 through 31 March 1948, to supplement the original $335,000. D. J. Montgomery, "Annual Report on Project Assisted by Outside Funds," 23 July 1946. A-475 folder, PP/P.

65. Stroud to the author, 19 July 1990.

66. "Elementary Particle Projects as of 6 May 1946," n.d. Fragment in A-701 (Smyth) file, PP/P. William Stroud OHI by James Capshew, 11 January 1989, pp. 3–4. GSFC/NASA.

67. R. E. Gibson to H. D. Smyth, 23 July 1946. A-475 folder (Wheeler), PP/P. On Nichols' move to Michigan, see W. Dow to R. F. May, 2 July 1946; and Robert P. Weeks, "The First Fifty Years," October 1964, p. 21. Typescript draft of history of the Department of Aeronautical and Astronautical Engineering. WGD/BHLUM.

68. "Agenda of Meeting 26 July 1946." A-475 folder (Wheeler), PP/P.

69. J. A. Wheeler, fragmentary notes from a 29 July meeting with Gibson, Tatel, and Hafstad. A-475 folder (Wheeler), PP/P.

70. J. F. Brinster to L. R. Hafstad, "Report on First Princeton V-2 Firing," 12 October 1946. Bumblebee Communication CM-357. HT/DTM.

71. See Megerian to Bain, "V-2 Panel Report no. 3," 24 April 1946; and "V-2 Panel Report no. 5," 9 July 1946, pp. 7–8. V2/NASM. See also J. A. Wheeler, "General Survey of the Princeton Project Program—Cosmic Rays and Telemetering," 28 August 1946. Appendix IV to H. D. Smyth, "Annual Report NOrd 7920," n.d. PP/P.

72. They traveled in a convoy of electronics-filled trucks and autos. See J. F. Brinster, "Notes Concerning Trip to White Sands," 1 July 1946; D. J. Montgomery, "Annual Report on Project Assisted by Outside Funds," 23 July 1946, p. 2. A-475 folder, PP/P. Comment by Walter Van Braam Roberts (1990), p. 77.

73. J. A. Wheeler, "General Survey of the Princeton Project Program—Cosmic Rays and Telemetering," 28 August 1946, p. 2. Appendix IV to H. D. Smyth, "Annual Report NOrd 7920," n.d. A-475 folder

(Wheeler), PP/P. Flight record from Smith (1954), and "V-2 Panel Report no. 6," 5 September 1946, pp. 14–15. V2/NASM.

74. "V-2 Panel Report no. 6," ibid; and Homer E. Newell "Record of Consultative Services," 29 October 1946. Meeting of 21 October held at NRL between Krause, Newell, and Perlow with Wheeler, Arley, and Kusaka. Box 98, folder 4, NRL/NARA.

75. "V-2 Panel Report no. 6," ibid.

76. D. J. Montgomery, "Annual Report on Project Assisted by Outside Funds," 23 July 1946, p. 2. A-475 folder, PP/P.

77. [Princeton University Public Affairs Office] 28 October 1946. Science Service Files, NASM.

78. They asked for $200,000 per year for the cosmic-ray studies alone. E. Krause, "Record of Consultative Services," 12 November 1946. Date of conference: 12 November. Box 98, folder 4, NRL/NARA.

79. J. F. Brinster to L. R. Hafstad, "Report on Second Princeton V-2 Firing, November 7, 1946," 12 December 1946, p. 1. HT/DTM.

80. E. Krause, "Record of Consultative Services," 12 November 1946. Box 98, folder 4, NRL/NARA.

81. Megerian to Bain, "V-2 Panel Report no. 8," 28 January 1947, p. 13. Report dated 4 February 1947. V2/NASM. Smith (1954), record of V-2 no. 18. H. D. Smyth also indicated in late 1946 that they wished to remain in contact with the APL and NRL rocket efforts, as the Princeton theoretical group expected that their data might well provide fresh insight into fundamental problems. See H. D. Smyth, "Memorandum to Committee on Project Research and Inventions," n.d., circa late 1946, p. 3. A-701 file, PP/P.

82. Van Allen (1948), p. 172. O'Day pushed for a reduction not for Princeton, but for the growing number of AFCRC/Air Materiel Command contractors that were gearing up to use V-2s, as well as his own growing Blossom Project. Megerian to Bain, "V-2 Panel Report no. 8," 28 January 1947, p. 14. V2/NASM.

83. See G. Reynolds to Urner Liddell, "Monthly Report," 15 June 1948. PP/P. This short-lived but major program warrants further historical attention. Whereas most participants found the program very useful, a few, such as William Swann from the Bartol Research Foundation of the Franklin Institute,

found the B-29 program frustrating because of flight schedule changes caused by constant maintenance problems with the aircraft, so his Bartol group soon returned to ballooning. On the "Joint Army Air Forces/Navy Upper Atmosphere Research Program," see D. F. Rex "Agenda for Conference" with enclosures. B-29 folder, DHM/HUA; see also Wallace Brode to C. V. Smythe, 20 July 1946, with attachments. PP/P; and Melvin Payne to W. F. G. Swann, 9 May 1946; W. F. G. Swann to W. F. Swann, 10 July 1947; W. F. G. Swann to J. Piccard, 14 February 1948; 2 March 1949; "Research & Development Project Card (JRDB)," N6ori-144, 3 February 1947, attached to T. H. Johnson to W. F. G. Swann, 23 January 1947. WFGS/APS.

84. "Monthly Report" June 1948, dated 12 August 1948. PP/P.

85. See "Proposal for Continuation of Research Project in Cosmic Rays," Contract N6-onr-270—Task II, for the Year 1948–49, 6 March 1948. Princeton Committee on Project Research and Inventions, Department of Physics. A701 folder, PP/P.

86. "Proposal for Continuation of Research Project in Cosmic Rays and Elementary Particle Physics," 10 April 1949 draft, pp. 1–2. Princeton Committee on Project Research and Inventions, Department of Physics. A701 folder, PP/P.

87. On staff responsibilities, see H. D. Smyth, "Memorandum to Committee on Project Research and Inventions," n.d., circa late 1946. A-701 file, PP/P. See also Stroud OHI by James Capshew, pp. 3–4. GSFC/NASA.

88. Newell (1953), pp. 260–61.

89. See Freier et al. (1948a, 1948b). On their participation in Skyhook, see DeVorkin (1989a), chap. 10.

90. Johnson, Davis, and Siry (1954), p. 306.

91. Van Allen OHI, p. 193rd.

92. On the fate of the NRL group, see C. Y. Johnson OHI no. 1, p. 35rd. Van Allen spent 15 months at Princeton in 1953 and 1954. He recalls being initially stimulated by Lyman Spitzer's optimism that significant confinement would soon be achieved, but that when it did not, Van Allen became discouraged by the prospect of a long and frustrating process. One of Van Allen's associates, Melvin Gottlieb, replaced him at Princeton to continue in the controlled fusion experi-

ments. See Van Allen OHI, pp. 191–96rd; quotation from p. 195. On Project Sherwood and Princeton's Stellerator, see Bishop (1958) and Bromberg (1982).

93. See Van Allen (1983); and "James Alfred Van Allen, Vita," p. 9. SAOHP/NASM.

94. See L. R. Hafstad and Woodman Perrine to Commander Task Force SIXTY-EIGHT, 13 December 1946. Polar Program folder, JVA/APL.

95. See tabulated data in Chapter 17.

96. It is not known if the paper rejected by the *Physical Review* was on cosmic rays or was geophysical. Van Allen OHI, pp. 297–98rd. Van Allen, OHI no. 6, p. 252, feels that his conversion to geophysics came during the IGY. Herbert Friedman recalls that in the early 1950s, the editor of the *Physical Review*, Samuel Goudsmit, thought his solar X-ray work better suited for a geophysical journal, although he eventually agreed to publish it and later contributions. See Friedman OHI, p. 62.

97. Sekido and Elliot (1985). See their Preface. For citation studies, see Chapters 16 and 17.

98. Ginzburg and Syrovatskii (1964). Similar treatments were made by contemporary reviewers. No citations to rocketry appeared in two review articles on evidence for solar cosmic rays by J. A. Simpson at Chicago and by A. Ehmert, Regener's old associate at the Max Planck Institut für Physik der Stratosphäre, Weissenau, Germany in 1952. See Kuiper (1953), chap. 9, pt. 13, pp. 711–21. Jánossy (1948).

99. [Auger] (1949). This was a compilation based on replies to a questionnaire, however, and must be considered along with the omissions such a process entails.

100. H. V. Neher to W. A. Gustafson, 29 November 1949. HVN/CIT. Even though B-29s could reach only 12 kilometers altitude, without them, Neher considered that "the only other way" to achieve high altitude observations "would have been with balloons and this would mean carrying up heavy loads if the same information were desired." Neher felt that the B-29 flights, offering long look times and heavy payload capacity, as well as their ability to cover a range of latitude in a short time, were invaluable for determining how the cosmic-ray flux varied with time. Brown and Hoddeson (1983) note discoveries made by balloons and accelerators exclusively.

101. See "Proceedings of the Echo Lake Conference, June 23–28, 1949" (ONR: November, 1949). Echo Lake, near Idaho Springs, Colorado, was a cosmic-ray observing station operated year-round by the University of Denver Physics Department, with cooperation from cosmic-ray groups at Chicago, Cornell, MIT, NYU, and Princeton. ONR sponsored this research and the conferences jointly with the AEC and the Research Corporation.

102. See also John A. Simpson OHI no. 1, pp. 78–79rd. DSH/NASM.

103. Not only were the energies becoming comparable, but the flux of particles available on call was far greater than what occurred naturally. See, for instance, Marshak (1983), pp. 398–99; Winckler and Hofmann (1966), p. 1. By the end of the 1950s, comparatively few physicists were still employing cosmic-ray studies for nuclear or particle physics research, and many among them had already turned to astrophysical studies based on cosmic-ray phenomena. See Sekido and Elliot (1985), Preface; and Ginzburg and Syrovatskii (1964), p. 3. At Chicago, for instance, stimulated by his work on the atomic bomb, John A. Simpson began in cosmic-ray physics in 1946 but soon migrated to cosmic-ray astrophysics, searching for particle origins, while Schein's larger group continued to study high-energy nuclear interactions and particle production. See Simpson (1985). Marcel Schein and NRL's Maurice Shapiro continued to use balloons to study heavy primaries and associated astrophysical phenomena. Shapiro argued in 1958 (according to a AAAS abstract) that "despite the keen competition of huge atom smashers, the cosmic radiation still provides particle energies a million times higher than those generated in the laboratory." M. M. Shapiro, Abstract, AAAS, Washington D.C., 1958 (Trends in Cosmic-Ray Research). See also Linsley (1963).

104. Robert Leighton, quoted in Marshak (1983), p. 399. On the increased interest in geo-physical and astrophysical concerns, see Brown and Hoddeson (1983), p. 7, Table 1.1; and John A. Simpson OHI no. 1, pp. 78–79rd. DSH/NASM. The advocacy of influential physicists such as P. M. S. Blackett and of the geophysicist Scott Forbush at the DTM aided in this shift, especially after Forbush, Schein, and others detected the first solar-flare-induced cosmic rays in 1949, a relationship long suspected by Forbush. See Forbush, Stinchcomb, and Schein (1950). H. D. Smyth, in late 1946, sensed the growing connection between cosmic-ray research and the physics of the upper atmosphere: "Such relations may be expected to evolve into quantitative means for measurement of atmospheric conditions as a by-product of the further development of cosmic-ray physics." H. D. Smyth, "Memorandum to Committee on Project Research and Inventions," n.d., circa late 1946. A-701 file, PP/P.

105. Charles Johnson discussion with the author, 16 July 1990.

106. Singer OHI no. 2, pp. 35–37rd., quotation from p. 36rd.

107. Van Allen OHI, p. 186rd.

108. In 1947 they built a 10-counter double-coincidence circuit based on a Rossi design, tested the instrument against barometric readings in high-altitude aircraft, and predicted that they could get instantaneous altitude readings accurate to +/- 150 meters when above 3,600 meters. Fraser and Siegler (1948), p. 68.

109. Singer OHI no. 2, p. 25rd.

110. Fraser, Peterson, Tatel, and Van Allen (1947), p. 173. Abstract for an APS paper.

111. The Princeton Physics Department committed far more effort and manpower than either Yerkes, Harvard, Princeton, or Michigan astronomers did to spectroscopy, but left quickly and for much of the same reasons as their counterparts did. As Wheeler pointed out early on, if there was a better way to gather data, they would do it the other way. J. A. Wheeler, fragmentary notes from a 29 July meeting with Gibson, Tatel, and Hafstad. A-475 folder (Wheeler), PP/P.

# 15
# Gaining Authority for the High Atmosphere

*"It would look as though the fault may lie with the meteors themselves."*

—Jacchia and Whipple, 1956.[1]

Fred Whipple was keenly interested in what the V-2 observations would reveal about the density and temperature profiles of the high atmosphere. Since the 1930s, he and his Harvard College Observatory colleagues had been studying meteor trails in the earth's atmosphere to determine the solar orbits of this class of celestial debris by the parallactic plotting of their fiery remnants. But Whipple knew that as these tiny invaders heated the earth's atmosphere to incandescence, they revealed in their passage the nature of the resisting medium they encountered. During the war, Whipple reviewed all that was then known about the relationship of meteor trail analysis and conditions in the upper atmosphere, and, finding himself on the V-2 Panel after the war, focused intently on comparing his indirect photographic reconnaissance technique with the in situ observations of panel members, specifically those from NRL and Michigan.

In the first years of firings at White Sands, there were promising signs of agreement between Whipple's indirect observations and those taken directly by rockets. But by the end of the 1940s, differences appeared as the data on both sides became more refined. Whipple and his panel colleagues knew, of course, that many unknowns were at play: Whipple had made assumptions about the character (density, composition, tensile

strength) of the meteorites that were disturbing the earth's atmosphere, and William Dow, Nelson Spencer, Ralph Havens, and others were never able to completely decode the conditions their various sensors were subjected to during flight.

Accounting for the differences between the two camps was touchy. First, both sides had programs to maintain and justify to patrons in the Navy and Air Force; and second, both knew that there was a practical consumer of their data and conclusions: the American aviation industry and their military patrons. The chief arbiter of conditions in the high atmosphere was the National Advisory Committee for Aeronautics (NACA). Since the 1920s, its 'Standard Atmosphere' had been the gauge that the government and industry used to assess the variation of temperature and density with altitude, essential for the standardization of aircraft instruments and performance. But not only did the rocket data differ among the various rocketry groups, as well as with Whipple's observations, but they all differed from the NACA model. Who was to gain authority for the high atmosphere?

In this chapter we see how the members of the V-2 Panel learned to work as a team—one that included Whipple, and was inspired by him as well—to reconcile their differences and to present a united front whose primary purpose was to establish a new standard atmosphere.

273

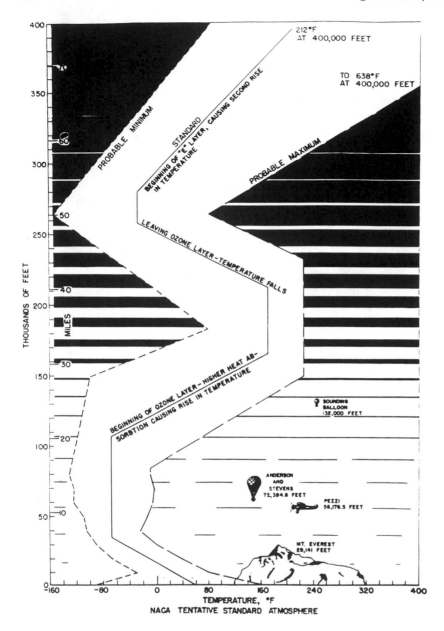

National Advisory Committee on Aeronautics' Tentative Standard Atmosphere, which extended its Standard Atmosphere to operational guided missile heights. In the late 1940s this was the official model for how temperature varied with height. Harry Wexler Papers, Box 23, NACA folder, Library of Congress.

## Meteors and their Medium

'A meteor is a flash of light, made by a falling meteorite,' so the old saying goes. Small chunks of iron and rock called meteorites frequently enter the earth's atmosphere and become frictionally heated and thus heat the medium as well into a combined local plasma. The resulting luminous meteor trail, together with spectral analysis of the plasma, reveals the nature of the meteorite and of the heated medium itself. If that small chunk somehow survives passage through the atmosphere, it could be examined directly, its composition determined, and hence subtracted from its spectroscopic signature during flight. The result would be a measure of the medium through which it traveled.

All parts save one of this scenario had taken place in the 1930s. The one part that eluded all attempts was the recovery of a meteoritic sample that had been detected spectroscopically while in flight. Still, meteorites as probes of the ionosphere became an important tool in the 1930s because astronomers assumed typical compositions, on the basis of samples that had been collected for centuries from unrecorded falls.

Meteors also disturb the ionosphere creating bursts in radio energy.[2] In 1929, H. Nagaoka in Tokyo proposed that meteors were the cause of such curious effects. Two years later, A. M. Skellett, who was then a part-time graduate student in astronomy at Princeton working at the Bell Telephone Laboratories on radio meteor echoes, showed that meteors could increase the level of atmospheric ionization when they passed through the ionosphere.[3] Through the 1930s, both photographic and radio studies confirmed these characteristics and also revealed that the heights of meteor trails varied seasonally and that there were several maxima, which themselves fluctuated. By the end of the 1930s, continued studies of meteor trails, spectra, and echoes provided data for estimating temperatures and densities in the high atmosphere.

During World War II, radar networks looking for aircraft, and later V-1 and V-2 missiles, often encountered meteor echoes. After the war, the existence of radar heightened the need to better understand the high atmosphere, and all aspects of radio propagation through it, which revitalized meteor research and the study of the ionosphere. Increased attention to meteor research brought together astronomers, radio engineers, and ionospheric physicists who found that they shared similar interests. Although they communicated with one another, few interacted closely, although some, like Whipple, established productive networks of patronage that stimulated collaboration in what was considered by most astronomers at the time to be a borderline occupation.[4]

## Fred Whipple's Reconnaissance of the Upper Atmosphere

Fred Whipple became interested in meteor orbits in the early 1930s, partly through his contact with the Estonian astronomer Ernst Öpik. By the mid-1930s, as a Harvard faculty member, Whipple refined the old photographic technique of determining meteor heights and trajectories by photographic triangulation, and, with the aid of theory developed by Öpik, used these astronomical data to determine the nature of the earth's upper atmosphere.

In 1942, working at the Radio Research Laboratory at Harvard, Whipple reviewed what was then known about atmospheric densities and temperatures from the brightness distribution of photographic meteor trails. He devised a quantitative method for converting the observed beginning, middle, and end points of the trails into densities and temperatures of the upper atmosphere up to 110 kilometers. He had only 12 good photographic plate pairs, but they provided what he felt were sufficient data to demonstrate the internal consistency of his method, which he then used to derive a relationship between density and altitude from assumed ambient temperatures based on Weather Bureau statistics. He found good internal agreement, and his trail method thus became the basis for his future work.[5]

At war's end, Whipple obtained support from the Navy Bureau of Ordnance for meteor observations and reductions, in part because he argued that his work was relevant to hypersonic ballistics studies. This support enabled him to set up a large photographic meteor network in New Mexico and buy into an extensive computing program at MIT. It also allowed him to plan for vastly improved photographic camera systems, based on the Schmidt optical system, that later came to be known as Super-Schmidts.[6] Whipple had to make many assumptions about the hypersonic meteorites his teams had photographed, and about the medium through which they traveled. First, the shape, size, mass, and density of the original body were unknown. Were they rotating, and how well did they trap or release the frictional energy impinging on their surface? How efficient was the heat transfer from the heated meteoritic body to the atmosphere? What were the processes through which the meteorite released its energy? And how much of this released energy was transformed into light, how much into heat (the luminous efficiency constant)? Whipple deftly manipulated these unknowns and assumptions in his application of Öpik's theory in 1943, and assumed that all of the meteorites had similar properties.

There were many inherent difficulties with Whipple's technique that could not be overcome at the time. But his method soon won endorsements from the astronomical community and from members of the V-2 Panel, who cited this work frequently.[7] In May 1946, James Van Allen advised a New Mexico colleague that "one needs only to study Fred Whipple's [1943] paper to appreciate that observations of meteor trains provide perhaps the most powerful method we have for investigation of density, temperature, and composition of the upper atmosphere."[8]

Even though Whipple's work represented established authority, he realized that he would eventually have to reduce the number of unknowns embedded in the analysis. Thus, when a proposal appeared from Fritz Zwicky offering such a possibility, Whipple was interested.

## Attempts to Calibrate: Zwicky's Pellets

At the third meeting of the V-2 Panel in April 1946, Caltech physicist Fritz Zwicky proposed hypersonic ballistic studies of artificial meteors during V-2 flights. Zwicky wanted to observe the spectra and luminosities of vaporizing shaped charges shot into space by rocket grenades carried by a V-2. Such observations might be useful for hypersonic missile aerodynamics, but could also yield information on the upper atmosphere.[9] Whipple at first was skeptical of Fritz Zwicky's proposal, but Van Allen was highly enthusiastic and volunteered APL's assistance if Whipple would act as adviser.

Whipple knew, however, that Zwicky's pellets could provide the all-important calibration data he needed, plus information for the experimental ballistic studies of his associate R. N. Thomas, who wished to establish an "astroballistics laboratory" to study problems in heat transfer, mass loss, and lu-

minous efficiency.[10] Whipple therefore called for a special night launching of a V-2, committed Harvard to cooperate with Caltech and Aerojet in the project, and provided manpower for operating ground cameras in Arizona and New Mexico to track the particles.[11] Whipple hoped that a coordinated study of the artificial meteors by optical, spectroscopic, and radio means would provide data on the ionization and luminosity effects of a known mass moving at a known velocity: "The scientific value of this experiment is to calibrate the luminous efficiency constant, now highly theoretical." Whipple felt that this goal would "fully justify the experiment."[12]

By November, plans for a December firing were progressing smoothly. APL had subcontracted with the New Mexico School of Mines to set up the grenade launching system for the V-2, and Whipple and his Harvard meteor program field staff were preparing an impressive arsenal of wide-field reconnaissance cameras for use in the desert. Whipple's enthusiasm remained high; he let the world know that "if the December experiment is successful . . . any element of guesswork concerning the actual size of natural meteors—and, even more important, the reactions of the upper atmosphere to moving bodies passing through it—will be largely eliminated."[13]

The War Department invited a group of news agencies to attend the "spectacular shoot" in December.[14] The news people soon found many astronomers gathered in the desert. Whipple led the Harvard contingent. Harvard staff member Dorrit Hoffleit, still part of Aberdeen's DOVAP team, arrived to perform initial trajectory analyses. APL, Zwicky, and the New Mexico School of Mines set up the camera network, which had swelled to some 30 cameras over hundreds of square kilometers. Zwicky recruited outlying observers and had many meteor enthusiasts primed for the shot standing ready in the White Sands area, as well as in Tucson, Albuquerque, Flagstaff, and at Palomar, where a 46-centimeter Schmidt camera was called to duty.[15]

Standing outside the trailer to view the launch on 17 December, Hoffleit viewed the first night firing of a V-2 with the critical eye of a ballistician: "Before I lost sight of it, it was a decidedly stellar image, though still red. I was under the impression that just before vanishing it had changed its course just perceptively. However, an impulsive jerk of the head would have produced the same impression."[16] About 20 seconds after she lost sight of the missile in the darkness, Hoffleit saw a meteor flash across the sky. It seemed to be far from the missile trajectory, but she wasn't sure; even though she knew that the artificial meteors were going to be far below the visual threshold, the specific velocity vector she recorded for the interloper could not rule out a relation to the V-2. She saw no other aerial reports until warhead blowoff, which was quite visible. Even so, "excitement, conversation, confusion as to the time of observation, and geometric projection effects [made] the clear association of corresponding events extremely doubtful."[17]

The press briefing at the launch pad earlier that day (discussed in Chapter 7) heightened the tensions that evening. Fritz Zwicky chafed at his treatment at the hands of Col. Harold Turner (figure, p. 277), commanding officer at White Sands, and the tension continued that

night as false reports of seeing the artificial meteors stirred the field crew. This was an APL flight of cosmic-ray telescopes, nuclear emulsions, and a National Institute of Health fungus and spore colony, but all speculation centered on Zwicky's pellets, and speculation abounded.

What Hoffleit and other BRL observers thought was warhead blowoff was actually the V-2 exploding at 440 seconds. After that night, she reported: "Sadly, the continuous rush of events of the day finally showed, and my head wouldn't work properly anymore."[18] The spores were lost in space as the missile vaporized. Zwicky's meteor experiment, from which nothing was seen or heard during the flight, also came to naught. The first speculation was that the grenades simply did not fire, but Ordnance specialists and astronomers later concluded that the artificial meteor trails were too faint to observe even if the grenades did fire.[19]

Whipple realized that too many variables still existed to prevent the experiment from being a real test. Watson Davis of Science Service had asked for a long report on the firing, but a disappointed Whipple wired Davis early the next morning that "Long feature

Harold R. Turner demonstrating for the press how to load the artificial meteor grenade launchers in the V-2, set for flight on the evening of 17 December 1946. U.S. Army photograph. NASM SI 80–3835.

article on meteor experiment scarcely justified. I need sleep." In 1952 Whipple told a colleague that "the idea . . . is still not practicable."[20] Zwicky, of course, wanted to try again, as did a few sympathetic colleagues. He did not appreciate Whipple's lack of support and harbored a resentment for years. At a retrospective astronautics symposium in 1971, Zwicky sounded off:

> *Our disappointment was enormous. Indeed, the failure of our experiment turned out to be a disaster, because further launchings of this sort were subsequently blocked for a full eleven years. Some so-called experts on (natural) meteors, among them Professor F. L. Whipple of Harvard, reported to the cognizant agencies of the US government that the experiment which I proposed could not possibly succeed and should not be supported.*[21]

Indeed, Whipple did nothing afterward to promote Zwicky's efforts in the V-2 era. In theory, he appreciated that experimental calibration would be valuable, but it would not happen in the manner Zwicky had proposed. Even without calibration, and even if based on many untested assumptions, Whipple's meteor trail technique was still the accepted standard in 1947.

Whipple continued to build his meteor photography network through the late 1940s, supported by Naval Ordnance and by ONR. Whipple assured his Naval Ordnance contract monitors that the network was central to his "program of research on the physical nature of the upper atmosphere and on the ballistics of objects moving at extremely high speeds through rarefied air."[22] To his old wartime radio colleague F. E. Terman at Stanford, Whipple added in May 1947 that his optical techniques were at present the best for "the study of the ionosphere."[23] But to his astronomical colleagues, Whipple continued to emphasize that his optical network would be used for "the most fundamental problems," such as determining meteor orbits and the atmospheric profile.[24]

In sum, Whipple facilely adapted his meteor studies to a wide range of problems, as he continued to shepherd the development of James G. Baker's Super-Schmidt cameras through the Navy's cumbersome procurement process, which, along with the many intrinsic technical obstacles facing the production of these new telescope cameras, delayed their introduction into his meteor network until the early 1950s.[25] In April 1950, optical fabrication had been taken over by the National Bureau of Standards, and Whipple informed an impatient Clyde Tombaugh that the NBS had at last been successful in molding the steeply curved correctors: "I am certainly looking forward to this with great anticipation in view of the years I have spent holding my breath for it"[26] (figure, p. 95).

Whipple's Harvard Meteor Project did not wait for the Super-Schmidts. He sent graduate students and field hands, including Harlan J. Smith and Richard E. McCroskey, to New Mexico to observe with conventional cameras and in 1947 started to obtain data that were then reduced by his computing team, led by Jacchia and Kopal at MIT. The results Whipple was obtaining by late 1948, however, did not fully support his first optimistic impressions in 1946.[27] He now found that the temperatures and densities derived from his meteor trails agreed with the direct measurements coming from the V-2s only up to about 70 kilometers; by the 90 kilometer level they diverged considerably. The question soon became, of course, who would be the arbitrator for authority on conditions in the upper atmosphere? The differing rocket results came mainly from teams at NRL and Michigan, none of whose members were known or recognized as practitioners in meteorology or atmospheric physics.[28]

## Rocket Observations of Temperature, Pressure, and Composition

### Ionization Techniques at Michigan

William Dow's Michigan group thought of studying the ionosphere as if it were a laboratory exercise:

*Most of our knowledge of the properties of ionized gases has been derived from experiments with electrical discharges through gas-filled tubes. It is natural to try to carry over as much as possible of this knowledge and experience to the study of the ionized layers of the upper atmosphere.[29]*

Dow's upper atmosphere group at Michigan, all trained and experienced in electronics and electrical engineering, were busy building thermionic ion gauges for V-2 flights in the summer of 1946. They also looked for ways to turn measured ion densities within the ion gauges into ambient atmospheric pressure and density values for the medium through which the rocket traveled.

William Dow knew that a subgroup of Krause's NRL section, consisting of Ralph Havens, Herman LaGow, and Nolan R. Best, were going to place Pirani gauges and platinum wire resistance circuits in and around their warhead, in the mid section and in the fin section for temperature measurements on their 28 June 1946 missile. After the flight, they received enough telemetry to know that their simple devices had worked; temperature information was gathered until the telemetry quit. But the pressure readings were compromised because the telemetry failed to function actively during the period when the expected mean free path of the ambient gas was in the right range. Also, unexpected voltage drains in both the high- and low- pressure systems produced nonlinearities rendering the measurements useless save for diagnostics.[30] As the NRL team planned for its second flight, Dow's group geared up for its August launch, using the NRL experience as a guide.

The Michigan group developed thermionic ionization gauges to detect pressure ranges lower than the limit of NRL's standard Pirani gauges.[31] Thermionic ionization gauges were vacuum tubes consisting of a heated tungsten filament and helical grid partly surrounded by a nickel-coated plate. These were familiar triode devices, known to provide an accurate means of making measurements in

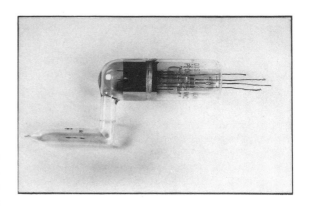

**Michigan ionization gauge for the side wall of their truncated cone probe. Michigan's thermionic ion gauges worked only under low pressure and density conditions where the mean free path of the ambient gas was large. They were therefore useful only in the high atmosphere, above 80 kilometers, and all laboratory calibration of the tubes had to be performed under conditions simulating this region. Nelson Spencer collection. NASM SI 89–1141.**

high vacuums in the laboratory. Dow's specialty in vacuum tube technology played a strong role in the selection of this style of instrument, even though making it work on a rocket required a few extra steps, and some necessary "gadgetry" like a means to expose the inside of the tube to ambient conditions in the high atmosphere.[32]

Dow's plan was defined by his talents and experience, as well as by the conditions of his AMC contract. He could not duplicate the efforts of other groups, so by preference as well as contract, Dow worked from his strengths. Dow's ionization gauge system (figure, p. 280) was fully developed and ready for Michigan's first flight in August, which we have seen (Chapter 7) came to a dismal end in a fiery launch failure. But Dow, Spencer, and the others were already testing their next gauges for flight by the time they had learned of the unfriendly conditions at White Sands. Subsequently, they enjoyed three successful flights, on 21 November 1946, 20 February 1947, and 8 December 1947, using sets of up to five ionization gauges within each warhead. In the fourth flight, the Michigan group added standard Pirani gauges similar to those used at NRL, for measurements in higher-pressure low-altitude regions.

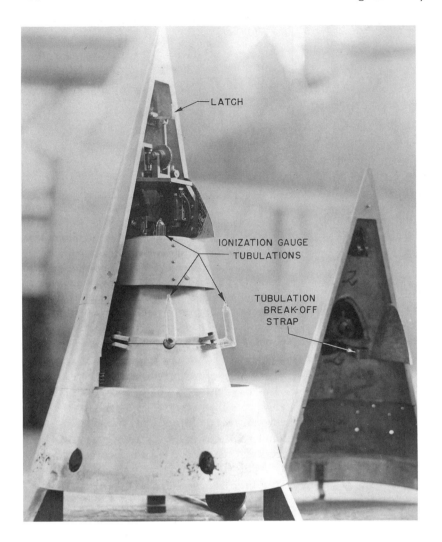

LATCH

IONIZATION GAUGE
TUBULATIONS

TUBULATION
BREAK-OFF
STRAP

A set of Michigan's ionization gauges set within their protective shroud. Dow's group developed a mechanical triggering device to break the tube seals and open the warhead chamber to expose the sampling port when the missile was at operational altitudes. The upper truncated cone acted as well as an isolated collector for Dow's 'Langmuir probe.' Nelson Spencer collection. NASM SI 89–1132.

Michigan's pressure readings from the February 1947 flight closely matched those obtained by the very different NRL instruments from a March 1947 flight in the region of 85 to 105 kilometers. Both deviated, however, from the NACA 1947 standard atmosphere; starting at about 65 kilometers, both gave lower pressures, which increased to a worrisome factor of 10 at 100 kilometers[33] (figure, p. 281). In general, the two 1947 Michigan pressure—altitude curves agreed better with themselves and with the best NRL 1947 curve than any of them did with the NACA standard atmosphere.

By mid-1948, Dow was prompted by Marcus O'Day to have Spencer, aided by Schultz and Reifman, prepare a statement for publication so they could reach a larger audience than that served by their contractor reports to the Air Materiel Command.[34] Dow's staff was familiar with Whipple's 1943 analysis and respected his judgment. They turned to Whipple to comment on their results and to suggest a journal receptive to their work.[35] In May, Whipple told Spencer that their manuscript "Pressures in the Upper Atmosphere" was "exactly what we wanted and will be extremely valuable to us." But he added that "it will be extremely difficult to come to any firm decision as to density, temperature, pressure measures in the upper atmosphere at the present time."[36] Whipple saw so much vari-

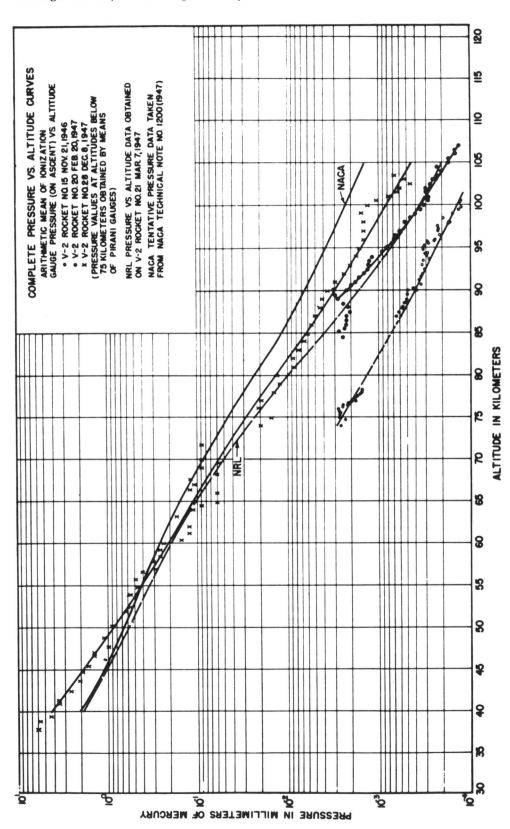

Summary pressure—altitude curves prepared by the University of Michigan that demonstrate greater internal agreement between the NRL and Michigan data, and considerable divergence from the NACA Standard Atmosphere starting at 65 to 70 km. Michigan's February 1947 flight data agreed closely with NRL's March 1947 curve, indicating that comparisons between the techniques of the different groups required closely coordinated flights. From Schultz, Spencer, and Reifman (1948), p. 10.

ation in their own data from flight to flight that he speculated it might be due to seasonal effects. But he was happy to see that at least the slopes of their pressure-altitude curves were similar; only the zero points varied. Whipple promised Spencer that he and his colleagues would be subjecting their data to further analysis and that he would be writing again soon.

Three weeks later, Whipple wrote to Dow with more thoughts. He felt it was not critical where their paper was published; it would probably be better in a geophysical publication, but he added that he would rather see it in the *Physical Review*: "Someone ought to break in on the nuclear physicists and include some other type of physics in this journal."[37] Whipple suggested that they cut their paper down to a note supporting their one summary pressure-altitude chart, which would no doubt be the interest, he advised, of any reader. He also cautioned them that there was little chance that the curve from their first flight of November 1946 could possibly be correct: "To be very frank, we are extremely suspicious of the low measures. . . . To obtain such low pressures . . . would necessitate such an amazing change in the temperature distribution in the lower atmosphere that it could not fit in with any material that we have to date."[38] Whipple agreed that in the upper atmosphere, beyond 70 kilometers, the NACA curve was certainly too high, but even so, the error could not be great enough to justify believing those early flight results.

Although Whipple critiqued the Michigan results in a friendly and collegial tone, it gave Dow and his staff second thoughts. Their second contractor report concluded that the 21 November 1946 results "may not be too reliable" and they canceled all plans for early publication of the results.[39]

## Improved Technique at NRL and Michigan

Even though Whipple's criticisms amplified their concerns, both NRL and Michigan had been constantly looking for ways to improve their instrumentation and to seek alternative means of determining ambient temperature,

pressure, and density profiles. Both Dow's and Myron Nichols' groups at Michigan searched for not only improved thermionic designs, but for ways to determine temperature from the bow shock waves made by the V-2 and means of taking air samples. Nichols' High Altitude Engineering Laboratory, supported by the Signal Corps, confined their interest to the region below 80 kilometers, whereas Dow's was most interested in the region above that point, in accordance with the capabilities of their chosen technologies.

At NRL, Ralph Havens concentrated on improving Pirani gauges, which could measure only small ranges of pressure because of inherent radiative cooling limitations. By 1949, Havens, who had trained in physics and knew how to build electromechanical devices, had created what his colleagues have called the Havens cycle gauge, which was capable of measuring the very large pressure ranges expected during a rocket flight, from $10^3$ atmospheres down to $10^{-5}$ millimeters of mercury.[40] Havens placed the heated filament of a Pirani gauge within a motorized bellows chamber that expanded and contracted in a short cycle and acted as a reasonably controlled environment. The bellows itself had a small hole, which allowed the Pirani gauge to sense ambient pressure. As the bellows cycled in and out, the Pirani element voltage would be modulated by that cycle, which could be amplified and calibrated as a function of the pressure of the bellows chamber itself. The average pressure of the chamber, over one cycle, was in theory equal to the ambient pressure in the rocket, as long as the pumping time allowed by the small hole in the bellows was long compared with the cycle of the bellows.[41] Havens knew that he had not perfected the gauge in the first several years, but his goal was to make simple mechanical devices work on rockets, and he was happy that he had found a mechanical solution to what his NRL colleagues thought was an electronics problem.[42]

The Michigan and NRL groups met mainly at V-2 Panel meetings, but at one point Michigan members visited NRL to compare notes and to talk about new delicate techniques for measuring tiny pressure changes.[43] Havens and LaGow questioned Michigan's dependence on thermionic ionization gauges to make observations in the ionosphere, arguing that ion gauges, like the Philips gauges NRL resorted to for the lowest-pressure regimes, disturbed the local gas, created ions and electrons of their own, and could be compromised by residual gas within the rocket warhead chamber. Michigan's design was, LaGow concluded, more prone to error than the more docile NRL gauges, which were less sensitive to the same environmental problems.[44]

Dow's group was well aware of the environmental problems that could affect their measurements, and through 1948 and 1949 built improved ionization gauges and searched for more realistic calibration techniques. The latter effort led them to a generally improved system called the "alphatron," which used a radioactive source as an ionizing agent, rather than a heated filament.[45] The alphatron vacuum gauge was first used at Michigan as a laboratory calibration standard, but its builders grew to believe that it was a better way to make pressure measurements on rockets; starting in 1949, Dow's group built and flew a series of alphatrons and found that their results had better self-consistency than did their old thermionic ionization gauge measurements. In May 1951, to an audience of electrical engineers, Spencer expressed considerable pride in the Michigan group's ability to amplify the tiny currents the alphatron produced.[46] As with the thermionic ionization gauges, the alphatrons were encased in sealed glass chambers that had small tubulations that were broken when the rocket achieved operational heights. For Michigan's V-2 and Aerobee flights, alphatrons dotted what was by now the standard Michigan mechanical structure that included all housekeeping functions. Typically, two gauges were set to measure cone wall pressure, and one was placed at the tip of the warhead to sense overall chamber pressure.

The Michigan group also used Langmuir probes to collect information on ambient electron temperatures and densities, and for a brief period subcontracted with the Cook Research Laboratories of Chicago to build mass spectrometers for use on the AMC Blossom series to measure gas ratios (argon to helium, and nitrogen to oxygen) that would indicate the degree of mixing in the upper atmosphere.[47] Dow's group initiated these parallel activities mainly to aid their observations of temperature, pressure, and density in the high atmosphere, but these devices also satisfied the Air Force goal of addressing the full problem of radio propagation in the ionosphere. Composition profiles were important for understanding just how ionizing agents acted to create, maintain, or destroy the various radio-reflecting regions.

## Alternative Techniques

Parallel to Dow's activities, Myron Nichols' comparatively larger group in Michigan's aeronautical engineering department developed a radio-mechanical density probe for Aerobees, supported by the Signal Corps. The probe was an inflatable sphere that was ejected from the Aerobee nosecone and then fell to earth carrying the miniaturized components of a DOVAP system.[48] The time delay of a radio-frequency pulse that had a frequency slightly higher than the critical frequency of the E layer would indicate the average density of the medium between the falling sphere and the ground. This method, suggested by Nichols in 1946, required a large ground-based radio system and transmitters in the sphere itself.[49] By the early 1950s, Nichols' group, including Leslie Jones, F. L. Bartman, and about 20 other engineers and physicists had successfully used the device on several Aerobees.[50]

At the same time, the Signal Corps Engineering Laboratory had been conducting sound grenade experiments from rockets, following suggestions by Beno Gutenberg, Fred Whipple, and Harry Wexler of the NACA Special Subcommittee on the Upper Atmosphere.[51] After several attempts with V-2s, where failed telemetry, inaccuracies in both DOVAP and optical tracking systems, and other problems foiled the precise determination of the times of each grenade blast, the

Signal Corps turned to Aerobees.[52] By 1950 the SCEL declared its experimental design to be ready for an intensive series of night flights of Aerobees. Michael Ference and his SCEL colleagues, including William G. Stroud (whom we met first at Princeton in the cosmic-ray group), developed techniques for converting the flash and acoustic report of these grenade explosions into both atmospheric wind data and temperatures. On a typical flight, a number of grenades would be set to explode at different heights above sensitive listening stations. The acoustic stations would record when each explosion was heard, and sets of optical tracking stations would photograph the explosion flashes, determining by triangulation exactly where the rocket was at the instant of each explosion. The delay of the acoustic signal behind the optical flash provided an estimate of the average speed of sound in the air at any one point, and from this Ference and his staff could determine the average density and hence the temperature of the air mass between the rocket and the station. By comparing the different delay times between two successive explosions, they could also derive the average temperature of the horizontal layer of air lying between the positions the rocket had taken at those two times.[53]

By 1950 then, four separate rocket groups were pursuing upper atmosphere temperature, pressure, and density profiles. Although ample data were accumulating after more than 50 flights, there remained many nagging inconsistencies and strong suggestions that both diurnal and seasonal variations were somehow conspiring with instrument error to influence their results. The realization that this was happening emerged in late 1949 and grew serious in early 1950 and through the year as the UARRP meetings became a forum for coordinating a coherent attack on the upper atmosphere.

## A Period of Adjustment: Facing NACA's 'Standard Atmosphere'

Although discrepancies persisted between each group's results, with Whipple's meteor results, and with profiles derived from radar echo data, the most serious differences—and the most contentious—were with the National Advisory Committee for Aeronautics' (NACA) tentative tables for conditions in the upper atmosphere, a 1947 extension of its 1925 Standard Atmosphere.[54] Considered a legal document for the aeronautical industry, the NACA Standard Atmosphere was a carefully guarded commodity, which, because of its function, represented an appropriately conservative consensus of the body whose responsibility it was to see to its well-being: the NACA Special Subcommittee on the Upper Atmosphere, formed in 1946.

Since February 1945, the NACA knew that the existence of the V-2 would create new demands on its aeronautical expertise in postwar America. For example, it realized that it had to extend its profiles of temperature, pressure, and density beyond the traditional 20 kilometer limit, at least to a height of 160 kilometers, "to cover rocket type flight."[55] Through January 1946, NACA could not decide how it should extend its atmospheric tables. Some members argued that the extension could not be done by extrapolating the usual atmospheric formulas, and that the best way was to use the missiles themselves to conduct atmospheric research.[56] Thus, when it learned that Army Ordnance was sponsoring the use of V-2s for this purpose, the NACA received Army Ordnance's blessing to send an observer to the V-2 Panel meetings. Calvin Warfield attended the first meetings along with other NACA specialists, and by May, Harry Wexler, the new chairman of NACA's Special Subcommittee on the Upper Atmosphere, was the official observer at V-2 Panel meetings.[57]

Wexler was chief of the Special Scientific Services Division of the Weather Bureau, a close associate of Francis Reichelderfer, chief of the bureau and chairman of the NACA's regulatory Subcommittee on Meteorological Problems. Reichelderfer helped Warfield set up the special subcommittee as a collegial study group under the NACA Committee on Aerodynamics. They both knew that NACA had to be capable of collecting and appraising all available observational and theoretical information on the upper atmosphere, but it also had to have the authority to establish the new extended Standard Atmosphere. Therefore the special subcommittee had to contain as many elites as possible, as Reichelderfer bluntly told Oliver Wulf of the California Institute of Technology, a recognized specialist on the photochemistry of ozone: "having your name included on its membership list would certainly add greatly to the stature of the Committee."[58] Accordingly, in addition to Wulf, NACA invited Fred Whipple, Beno Gutenberg, professor of geophysics at Caltech, and Lloyd Berkner, ionospheric physicist at the Carnegie Institution's Department of Terrestrial Magnetism (and soon the executive secretary of the JRDB). They would provide all-important authority, while others, such as Ernst Krause of NRL, would provide contact with those groups collecting the data.[59]

The subcommittee's original intention was to gather all available information and add to it new data coming from the missile tests. Through the spring, Wexler and his NACA and military liaison took a first look at what they had available for standardizing the characteristics of the upper atmosphere to 160 kilometers, and were not satisfied with what they found. Data they had at hand from NACA studies (both observational and theoretical) were internally inconsistent and conflicted with intelligence reports on what the Germans had learned about the upper atmosphere during the war.[60] The subcommittee therefore hoped that the American rocket groups would meet the NACA's need for improved temperature, pressure, and composition information.

But the subcommittee also knew that these data would not be forthcoming for some time, whereas in the spring of 1946 the Technical Services Command and the Navy Bureau of Aeronautics were both pressuring the NACA Panel to "take vigorous action in coordinating this work, to establish a set of atmospheric tables that can be adopted as standard by all concerned."[61] NACA's patrons wanted quick action and were willing to fund it out of related NACA, Army, and Navy

contracts. Thus the special subcommittee decided to have Warfield collect all available data for a draft report called "Tentative Tables for the Properties of the Upper Atmosphere."[62]

The subcommittee met in June 1946 to evaluate Warfield's report and found it inadequate for their endorsement. Members also agreed that "one of the most important things which this committee could do would be to recommend experiments which might be done to increase the accuracy of the present data and to extend present data to regions not yet known. . . . [M]uch use could be made of the V-2 in this field." Still hoping the V-2 would provide the data requested, the subcommittee turned to the specific experiments planned by the members of the V-2 Panel. But after a presentation by Newell and comments by Whipple, some, particularly Gutenberg, pointedly criticized their plans.[63]

Wexler felt compelled to move the subcommittee to action, citing the political necessity of preparing a tentative standard atmosphere with the data at hand. They dutifully drew up a rough schematic of what such a standard atmosphere might look like, based on what Warfield presented, and then closed their meeting by outlining more than one dozen types of observations that had to be made for temperature and composition to improve the situation.

Inspecting the draft minutes after the meeting, Whipple found that the subcommittee failed to identify his meteor technique among those desired. Whipple recognized the political value of the widely distributed NACA minutes, so this was no trivial matter; he acted immediately to reinstate meteor studies on the list, arguing to Wexler that, to date, his techniques provided "the most complete source of information on upper air densities."[64] Krause shared Whipple's sensitivity, and was equally vigorous in having the NACA minutes reflect the efforts of NRL and the V-2 Panel. Wexler complied.

Although the NACA Special Subcommittee and members of the V-2 Panel shared the goal of better understanding the upper atmosphere, pressures on them were different and gave them a somewhat different view of life.

The subcommittee was a political forum with intellectual undertones created to evaluate existing data and to promote new ventures for gathering data. And the panel was a mechanism for producing that new data. Panel members knew that their futures were tied to the ultimate success and acceptance of their efforts, whereas none on the subcommittee had livelihoods that depended on its existence.

Directed by the NACA Committee on Aeronautics and its patrons, Warfield, with members of his staff, refined their tentative tables for temperature, pressure, and density profiles to 32 kilometers, and in September 1946 presented *NACA TN 1120* to the subcommittee.[65] These interim tables were based on direct observations from radiosonde balloons, extrapolations from known temperature profiles, astronomical observations of meteor trails, and radio probes of the ionosphere, all reconciled with the best theory available. After the academic members of the subcommittee ratified the effort, Wexler prepared a final version, "Tentative Standard Properties of the Upper Atmosphere," extending to 120 kilometers, which was published in January 1947 as *TN 1200*.[66]

Through the summer and fall of 1946, Wexler and the subcommittee hoped that rocket observations could be incorporated into *TN 1200*. But as Wexler attended meetings of the V-2 Panel, he soon realized that this would not be possible in the near future. In October 1946, Wexler advised Reichelderfer that the rocket work was potentially useful, but progress was going to be very slow, and the situation did not warrant immediate heavy involvement from the Weather Bureau.[67] He looked forward to the day when American-designed rockets built to take meteorological soundings would be available, but knew this was not going to happen soon:

> *The day of a widespread network of rocket sounding stations seems to be quite distant and it is therefore appropriate for the meteorologists, while awaiting development of the rocket-sonde, to consider the possibility of set-ting up their own program to learn more about the conditions in the upper atmosphere.*[68]

Wexler had lost interest in the immediate use of rockets for synoptic meteorology, even though he thought that the rocket scientists had produced some good data on pressures and temperatures in their first flights. In April 1947, at a meeting of Reichelderfer's NACA Subcommittee on Meteorological Problems, he "noted that results of recent pressure measurements using the V-2 have shown remarkable good agreement with the tentative standard characteristics. In view of the dearth of data available at the time the subcommittee made its recommendations, the agreement was surprisingly good."[69]

In his enthusiasm, Wexler had overstated the case for agreement. Although the early rocket data seemed to confirm the NACA tables, both sets were at best tentative, and Reichelderfer's committee "considered [the rocket data] to be questionable."[70] To raise his boss's level of appreciation of what the rocket groups were up against, Wexler invited Krause and O'Day to testify about the many technical problems that prevented the quick return of scientific data. But their efforts before both Reichelderfer's committee and the special subcommittee mainly drew criticism and reinforced the view that little if anything useful would come of the rocketry. In reply to Krause's contention that the discrepancies in their data were caused by varying atmospheric conditions, Oliver Wulf countered that they were more probably due to improper instrument design.[71] The best they could do for the present, the subcommittee concluded, was to endorse with caution *TN 1200* to the NACA and to the American aircraft industry.[72]

Through 1948 and 1949, the NACA Special Subcommittee on the Upper Atmosphere continued to watch the progress of the rocket observations. At meetings in September 1948 and October 1950, Newell, O'Day, Van Allen, and members of their groups and their contractors presented their latest findings. Whipple did the same, promoting his meteor program and ancillary research conducted by his associates Kopal, Jacchia, and Chaim Pekeris.[73] These data allowed for more detailed comparisons with *TN 1200*, but there was still no consensus. Consistent temperature results from three NRL flights agreed closely with the median line of the NACA standard to 40 kilometers, but then from 50 to 70 kilometers the NRL results yielded temperatures outside the NACA "probable minimum" range.[74] Even less consistent results came from the Michigan and meteor observations.

By 1950, the NACA was under considerable pressure, both from its patrons and from the International Civil Aeronautical Organization (ICAO), to codify its tables into a new and acceptable standard. The subcommittee's tables had remained tentative for three years, while a new ICAO standard was rapidly gaining authority, mainly through its use of refined theoretical work in Europe.[75] The question for NACA now was not when they would have better rocket data, but if such data would ever be available (figure, p. 288).

## T-Day at White Sands

In February 1950, NRL's Ralph Havens expressed what had been on the minds of many UARRP members: what was needed was a coordinated series of flights of each group's instruments, along with ground-based sound-propagation studies, and balloon-sonde observations. The UARRP should sponsor a T-Day, or Temperature-Day as he called it, that would provide a comparison of the Michigan, NRL, AFCRL and SCEL atmospheric measurement techniques[76] (table 15.1).

Although the UARRP immediately endorsed Havens' idea, T-Day took one year to get off the ground. Not until June would the UARRP get rolling, doing so only when convinced that its results lacked internal consistency and differed greatly from Whipple's results, as well as from those of Beno Gutenberg and the infrared studies of ozone by Arthur Adel at Michigan. Whipple's newer results now strongly suggested seasonal variations and differed greatly from the NACA curves as well as from the rocket results. In the face

of these many conflicts, the UARRP concluded in April, "In general, the rocket data do not yet appear to be sufficiently accurate or comprehensive to yield any conclusive information on seasonal or diurnal variation."[77]

Sydney Chapman, who was visiting Caltech that year from Oxford, attended the 25th meeting of the UARRP in June 1950, held in Colorado (figure, p. 306). By this time, Chapman, a leading geophysicist, along with Van Allen, Lloyd Berkner, and others had started thinking about a coordinated worldwide assault on geophysical problems, what became the International Geophysical Year, and so cooperation and coordination were exciting themes in the air.[78] The UARRP revived Havens' proposal and appointed a panel to organize T-Day, setting it for the late summer of 1950, the earliest time that a bright meteor shower was expected and that the panel could have ready the wide array of vehicles and instruments its members required.

Michael Ference, Havens, and O'Day caucused in Colorado and concluded that this effort would lead to an effective "interchange of experiments and techniques" and ultimately to a revision of the present NACA tentative curves.[79] Although Whipple was absent from the Colorado meeting, he endorsed "the joint attack on the upper atmosphere."[80] In August, Whipple invited Ference, Chapman, O'Day, Spencer, and Havens to Harvard, along with the venerable S. K. Mitra, visiting from Calcutta, to plan T-Day. The summer deadline had long vanished, and now they set the night of 11–12 December 1950 as their target, a peak meteor shower night. They also hoped to secure the participation of the large Stanford radio group, led by L. A. Manning and O. G. Villard, to make radar-meteor observations in coordination with the rocket flights, as well as extensive SCEL balloon flights throughout the day. By September and the 26th meeting of the UARRP, where its members remained divided over their fates during the Korean War, four Aerobees and one Viking had been committed to T-Day. Several weeks later, UARRP members told the NACA Special

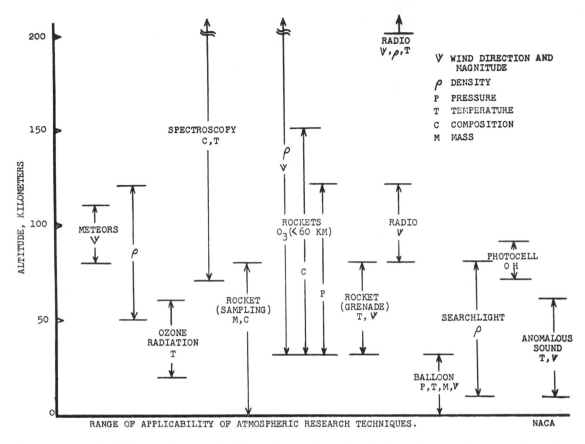

RANGE OF APPLICABILITY OF ATMOSPHERIC RESEARCH TECHNIQUES.                    NACA

The NACA Special Subcommittee on the Upper Atmosphere view of where data on the upper atmosphere must come from, including indirect ground-based studies, balloon and rocket measurements, and meteor observations. This presentation was part of a NACA resolution in support of establishing another rocket launch site in the auroral zone, at Fort Churchill on Hudson Bay in Manitoba. Appendix A to the minutes of the 23 October 1951 meeting of the NACA Special Subcommittee on the Upper Atmosphere. Harry Wexler Papers, Box 23, NACA folder, Library of Congress.

Subcommittee that T-Day would be their "maximum effort" to settle the many problems facing acceptance of their results.[81]

Although Stanford bowed out, T-Day proved to be quite a night. The Signal Corps launched three Aerobees: two with Ference's grenade experiments and the third for Nichols' falling sphere. The AMC fired one Aerobee for Dow's alphatrons and ion gauges, and NRL launched Viking no. 6 with Havens' instruments. All the while, the Signal Corps sent aloft meteorological balloons, and Whipple's meteor camera teams worked every clear night between 9 and 15 December.

A high fraction of the rocket experiments worked as planned, and Whipple's network bagged some 78 Geminid meteor trails. By the end of January, Whipple felt that "in the [preliminary] comparison of data obtained by means of rockets on atmospheric density with data obtained by meteor studies, the latter have resulted in higher values although no serious errors are evident."[82] On 31 January, the UARRP held its 27th meeting at NRL, and invited Sydney Chapman along with others to evaluate the "T-Day Operation." Although some members immediately declared the operation a success and called for T-Day no. 2, the prevailing opinion was that it would take some time to digest the results completely. This feeling continued through the spring, as data reductions continued, but Van

TABLE 15.1   T-Day Experiments

| | |
|---|---|
| a. Rocket-borne pressure guages for ram and ambient pressures | NRL, UM (AMC) |
| b. Bolometric radiation detectors | NRL, URI (AMC) |
| c. Mass spectrographs for composition—atomic and molecular oxygen to nitrogen | NRL, APL, Cook (AMC) |
| d. Sampling for composition and mixing | UM (SCEL) |
| e. Sound grenades | SCEL |
| f. Oscillating probes for Mach angle | UM (SCEL) |
| g. Balloon sondes | SCEL, WSPG |
| h. Ozone | NRL. APL |
| i. Ground experiments on anomalous propagation of sound | AMC |
| j. Meteor luminosity and retardation | Harvard |
| k. Ambient temperature | Cook (AMC) |

*Source*: Listing of experiments and agencies contributing to studies of the temperature, pressure, density, and composition of the high atmosphere, as identified by the UARRP as it began planning for T-Day. Based on listing in "Panel Report no. 23," 14 February 1950, p. 25. See also succeeding panel reports through 1950 for additions. V2/SAOHP. URI (University of Rhode Island); UM (SCEL) (Nichols and Jones, Department of Aeronautics, University of Michigan group); UM (AMC) (Dow and Spencer, Department of Electric Engineering, University of Michigan group); Cook (AMC) (Cook Laboratories).

Allen knew that time was running out for the NACA.

## Digesting the Results: The "Rocket Atmosphere"

In April 1951, Van Allen, now at Iowa and faced with reestablishing his research momentum, felt it was time that the UARRP produce a "Standard Scientific Atmosphere." Even though the T-Day results still differed greatly from the NACA standard, Van Allen believed that the rocket data for temperature now showed "a remarkable measure of agreement" to 100 kilometers. The UARRP, he argued at its 28th meeting, had better make some definite statement on the matter, because there was worldwide interest in it, and, evidently, because patrons were watching. Van Allen reminded his colleagues, including Harry Wexler, that the NACA Special Subcommittee on the Upper Atmosphere had decided at its 1950 October meeting that the new rocket data "were not yet sufficiently comprehensive . . . to justify revision of the NACA tables."[83] The UARRP now had to reverse this impression by publishing its results collectively and quickly. The time had come for the panel to establish once and for all not only its identity, but its authority in the high atmosphere.

Harry Wexler reacted coolly to Van Allen's call to action, explaining that the NACA needed standards with the highest internal consistency, but that everyone recognized that the NACA standard did not represent "true" conditions. "Nor does the aeronautical industry appear to require at present," he added, "a standard of greater validity in the altitude range 30 to 120 km than is represented by the NACA TN 1200."[84] The rocket data, Wexler pointed out, were valid only for White Sands, whereas American industry required consistent tables for a large latitude range. Most serious, the rocket temperatures at 60 kilometers were almost 80K cooler than NACA values, whereas "Whipple's meteor data . . . show substantial agreement with rocket determined densities over New Mexico and with the NACA atmosphere over Massachusetts." This, Wexler contended, was an example of the type of confusion that could be due to seasonal effects and that made the rocket data incomplete. Even so, Wexler encouraged the panel to present a summary tabulation of their findings so that the NACA subcommittee could take them up at its next meeting in the fall of 1951.[85]

The UARRP agreed to prepare a report for the NACA's review, but also prepared the report for publication. The UARRP met in August 1951 in a two-day seminar to coordi-

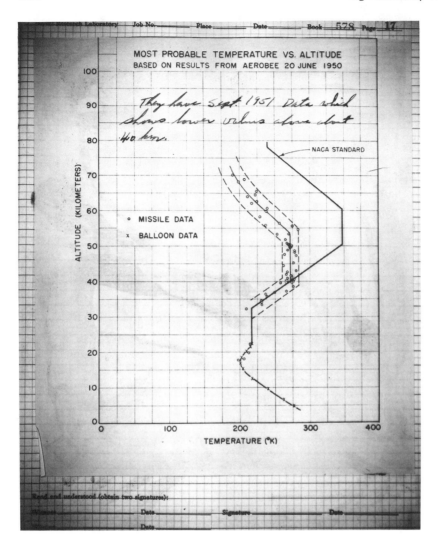

Throughout 1950 and 1951 there was continual comparison with the NACA standard. Herman LaGow at NRL kept track of progress in his registered laboratory notebook no. 578, p. 17. HONRL.

nate the final reduction of its combined data. The meeting at the University of Rhode Island was a coming to terms on many issues: chief was the UARRP's recognition that the divergence between the rocket data and Whipple's meteor data was caused by the seasonal and diurnal variability in temperature and pressure revealed by the meteors. Their combined results continued to differ with the NACA (figure, p. 290), but at least now the rocket data possessed a satisfying degree of internal coherence.[86]

Van Allen pushed for a consensus with a pep talk that celebrated "the wealth of experimental information . . . and the impressive degree of concordance of the results from the diverse, independent methods." "Yet," he added, "no such impression exists in the general literature." He urged each group to publish its results and also called again for a collective UARRP revision of the NACA standard: "The data now available far exceeds those on which the NACA tentative standard was based. In fact, marked revision of the NACA data seems to be required."[87]

In October, while the UARRP met again in Chicago to sift through the data once more and to decide where to publish, the NACA Special Subcommittee met again to ponder the situation. Of course, there was significant overlap in the two bodies. On 22—23 October, Van Allen, Ference, Wexler, Newell,

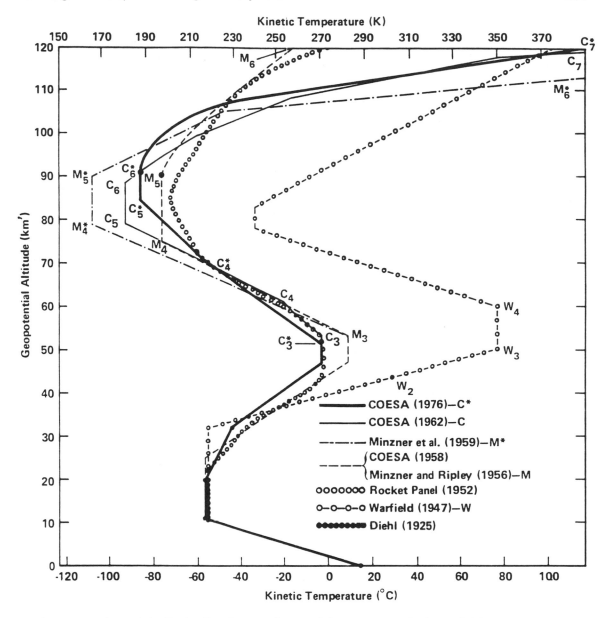

Modern comparison of the Rocket Panel atmosphere and the NACA standard [Warfield (1947)], with later temperature profiles, particularly those produced by the combined Weather Bureau—Air Force Committee on Extension to the Standard Atmosphere (COESA). From Minzner (1976), p. 57, fig. 32. NASA SP-398.

and Whipple attended the NACA subcommittee meeting, and then on the 24th Van Allen and Ference flew to Chicago to finalize the UARRP statement. It was a hectic whirlwind of meetings, and there were some tense moments. Reacting to Whipple's announcement that a "rocket atmosphere" was now at hand and that it differed significantly from the NACA curves, the NACA subcommittee "strongly recommended that the paper presenting the summary be so written and titled that it would be obvious that the information is a summary of data and not a proposed new standard for use in rocket work."[88]

Although the subcommittee endorsed both Whipple's continued meteor studies and rocket firings from different latitudes, it would not relinquish authority for the standard atmosphere.[89] So in Chicago, the UARRP took steps to seize it. Using a weighting system based on the relative number of flights from which data were returned and the internal consistency of each group's efforts, it assigned NRL and SCEL data the greatest weight and Michigan data intermediate weight, with Whipple's meteor data lowest weight.[90]

Whipple was not at the Chicago meeting, but he asked Van Allen to send him the original data from NRL and SCEL "in order to combine the data more effectively."[91] Van Allen and Ference agreed that Whipple should perform the final analysis, for greatest political effect: "There was some feeling on the part of NACA," Van Allen told his colleagues, "which is not shared by this Panel, that the rocket data are inadequate in coverage to justify revision of the NACA curve."[92] Whipple was the panel's key player in staking its claim.

## Whipple Changes his Mind

Why was Whipple sympathetic to the UARRP's low assessment of his own work? Whipple had earlier argued that the seasonal effects in his meteor data proved the veracity and sensitivity of his technique. Wexler agreed with this and searched for seasonal correlations, in general supporting Whipple's contention that they were real and simply lay beyond the capabilities of the rocket techniques.[93]

But Whipple's original skepticism in June 1948 of the Michigan and NRL data had softened by 1949, and by 1950, a deeper interest demanded that he change his mind altogether.[94] Whipple was an ardent student of cometary statistics and lifetimes, and the source of cometary material—classic astronomical problems that were then seeing renewed life.[95] Knowing that cometary statistics did not satisfy predictions for the number of comets that should be around, Whipple realized that comets must lose mass quicker than anyone previously considered possible.

Knowing something as well about the composition of comets from recent spectroscopic studies, he concluded in 1950 that cometary nuclei must therefore be extremely fragile, low-density solid composites held together by frozen gases: "a true conglomerate."[96] This was the genesis of his famous dirty snowball model, as well as the turning point for his upper atmospheric reconnaissance.

Whipple knew since 1947 that it was possible that the seasonal effects he had deduced were in reality caused by two classes of meteor swarms, each with different kinematics and structural characteristics. New analyses provided by Luigi Jacchia and Zdenek Kopal showed that the meteor trails his staff were photographing and analyzing were from objects wholly different from the dense nickel-iron or even stony meteorite samples found on earth. In September 1948, Whipple argued that his "photographic" meteors were almost entirely due to cometary debris: "As a working hypothesis the writer prefers to associate all or nearly all of the photographic and brighter visual meteors generically with comets rather than with meteorites and asteroids."[97] And further analysis of meteor trail heights in 1949 confirmed that "those meteoric bodies may have been unusually fragile or porous." Finally, by June 1950, astronomical observations of hydroxyl airglow emission convinced at least two prominent atmospheric physicists, David R. Bates and Marcel Nicolet, that the rocket data deserved greater weight than either Whipple's data or the NACA standard.[98]

By 1951, this set of circumstances prompted Whipple to reduce his assumed meteoritic densities. Whipple said nothing about it at the time in print; in fact, in the spring of 1951, as a member of the newly created Standing Committee on Problems of the Upper Atmosphere of the meteorology section within the American Geophysical Union (AGU), yet another group hastily formed in late 1950 to review and interpret the rapidly growing field, Whipple continued to support his view of seasonal variations. But this was the last time he would do so.[99]

Between the spring meeting of the AGU committee and the end of the year, Whipple changed his mind completely. The first Super-Schmidt was put into operation in June at Las Cruces and was showing that the older data were highly fragmentary. Also, Whipple started to find many consistencies in the rich results from the first T-Day in December 1950. Ironically, he was driven to these observations by being asked to review upper atmospheric research in the United States at an assembly of the International Union of Geodesy and Geophysics in Brussels, held in August 1951. In preparation, he tried out his new thoughts before a *fifth* body called the Permanent Committee on the Upper Atmosphere (beyond the UARRP, the RDB Panel on the Upper Atmosphere, the AGU Standing Committee, and the NACA Special Subcommittee), formed in early 1951 by the American Meteorological Society.[100]

Whipple reviewed all means of examining the upper atmosphere and weighted each in three classes, the lowest of which was worthy only of "after-dinner conversation."[101] The wealth of rocket data now dominated his discussion, and he hedged on the reality of the seasonal effect, which he now thought might be caused by "some peculiarity of the interaction of meteors with the atmosphere."[102] Above all, Whipple wished to highlight the immense effort that went into the rocket observations:

> I must apologize to the many investigators, administrators, technicians and members of the Armed Forces who have contributed so much to the upper atmospheric research with rockets. . . . I have been unable to present with any degree of verisimilitude the enormous work of planning, designing, constructing and firing the intricate equipment used in the rocket experiments, as well as the tedious observations and reduction required in many cases. This work represents the most extensive example of group research on problems of purely scientific interest with which I have come in close contact.[103]

Whipple's sympathies continued to grow as new analyses of the rocket data came pouring in. Through 1951, Spencer and Dow's measurements of pressure and temperature from continued Aerobee alphatron flights yielded data which by now included bow-shock observations that they concluded exhibited a satisfying degree of self-consistency.[104] Spencer and Dow also found, along with Whipple, that their results agreed with those from NRL and SCEL. Whipple was convinced.

From this point on, Whipple used rocket data from White Sands as a means of determining the densities of meteors observed from Harvard and White Sands. The resulting data for atmospheric temperatures and densities, Whipple contended, "now provides a completely independent check on the theory and numerical constants involved in the meteor theory."[105] Using the rocket data for the atmosphere and new experimental ballistic data on drag coefficients, Whipple found meteoritic densities to be from 0.3 to 1.0 grams per cubic centimeter, compared to the old value of 3.5 derived from terrestrial samples. "We are dealing here," Whipple concluded to a December 1951 astronomical audience, "almost entirely with cometary debris, which appears to be entirely different from meteorites physically, but not chemically." This was a conclusion, Whipple observed, that "would follow naturally from the writer's icy conglomerate model of comets."[106] In effect, Whipple realized that his dirty snowball comet model was supported by what the rocket observations said about the earth's upper atmosphere.

It took several years for Whipple to formally announce his reversal; this came in 1956, when Jacchia and Whipple reviewed the first decade of their work. Now, they said, atmospheric densities "from New Mexico meteors agree remarkably well with the density profile derived from V-2 rocket flights." The nagging seasonal effect disappeared with their early assumptions about meteor densities: "It would look as though the fault may lie with the meteors themselves, rather than with the theory. A widely variable density of the material from meteor to meteor appears as a probable explanation."[107]

## 'The Rocket Panel' Report

As a member of at least five interrelated panels on the upper atmosphere and with an elite international reputation as a spokesman for research on the high atmosphere, Whipple was indeed the best person to orchestrate the UARRP's report. He generated a set of temperature-altitude tables and presented these to the UARRP in April 1952 as a "best fit" of the available data.[108] By this time, the panel only differed over the question of mixing in the upper atmosphere.[109] But they all knew that solidarity was more important than detailed presentations of individual views; above all, they agreed to call for more observations, possibly a second T-Day, O'Day hoped, or observations at other latitudes, which might emerge from IGY plans.[110] Panel members liked Whipple's treatment and pushed for publication in the *Physical Review*.

Ignoring the NACA subcommittee's plea, the Rocket Panel's report, "Pressures, Densities, and Temperatures in the Upper Atmosphere," was hardly presented in tentative terms.[111] The report did not claim to be setting a new standard atmosphere, but its tone implied otherwise, as "a representation of the combined results to January, 1952." Data were presented, and deviations from the NACA standard noted, with the clear message that the rocket data were a foretaste of better data to come. Continued rocket flights were required, "if we are to master scientifically this domain such a few miles distant."[112]

With the Rocket Panel report in press, the UARRP minutes announced, "This publication represents an important milestone in upper atmospheric physics."[113] Indeed, its publication marked the first time a collective statement had been issued on conditions in the high atmosphere, as determined from rockets. The panel report also predicted that all subsequent efforts to create standardized models of the atmosphere would be based on data from rockets. Within a year, this prediction came true: Wexler combined forces with the AFCRC to improve the extension to the Standard Atmosphere, combining the Air Force's continuing stream of rocket data with theoretical modeling (figure, p. 291).

The Rocket Panel atmosphere was never adopted as a standard, even though it soon proved to be a necessary adjustment to Warfield's 1947 model. Thus while the NACA Special Subcommittee was disbanded after the NACA agreed to accept the International Civil Aviation Organization (ICAO) Standard Atmosphere, which "essentially eliminated the further use of the [1947] Warfield upper-atmosphere tables,"[114] the continuing efforts of members of the UARRP contributed significantly to later U.S. standard atmosphere development. A precedent had been set by the Rocket Panel report. As both a political and technical milestone, the panel report became a bench mark in the minds of its authors after which rocket data could no longer be ignored as an authority for the high atmosphere.[115]

## Notes to Chapter 15

1. Jacchia and Whipple (1956), p. 989.
2. On radio and radar meteor studies, see Whipple (1951); and McKinley (1961). And on ionospheric studies by radio, see Gillmor (1981, 1986, and 1989). Doel (1990a), chap. 1 reviews various traditional interests in studying meteors. See also Hoffleit (1988).
3. A. M. Skellett, response to AIP "Questionnaire for History of Modern Astronomy," 1979. SHMA/AIP. See also Skellett (1931), p. 1668; (1935); and (1938), p. 472.
4. Gilbert (1976), pp. 187–88; Doel (1990a); McKinley (1961), pp. 15–16; Hey (1973), pp. 19–23; phone conversation, Fred Whipple with the author, June 1980.
5. Whipple (1943), pp. 247; 258–59. See also Chapter 6.
6. Since positional accuracy depended on the spatial precision and temporal resolution of the photographic cameras, the cameras had to photograph both the meteor trails and the stars in as short a time as possible. Faster cameras, widely separated, would therefore yield better data. Whipple (1943) reported on earlier attempts at triangulation in the Harvard area. See also Whipple (1948a); and Kidwell (1991). On pioneering attempts at meteor height triangulation at Yale, see Hoffleit (1988). On Schmidt and his designs

and later modifications, see King (1979), pp. 356ff. On James G. Baker's modifications for Whipple's meteor cameras, see King, pp. 365–68. See also Mayall (1946); and James G. Baker OHI, pp. 120–30rd. SHMA/AIP.

7. Thomas (1947), p. 517.

8. Van Allen to E. J. Workman, 27 May 1946. Cabinet 8, JAVA/APL.

9. On Zwicky's program, see Fritz Zwicky et al., "Research on Rockets and Other Vehicles as Carriers of Scientific Information," 23 December 1948. Aerojet Report no. 346. JPL/L.

10. As both a theorist and experimentalist, Richard N. Thomas became part of Whipple's meteor project after the war, but his plans did not fully develop until 1950 when Thomas secured support from the Naval Ordnance Laboratory at Inyokern, California, to set up a small firing facility in Utah. Thomas had a long association with the NOL, remained a consultant to them, and continued through the early 1950s as an important contributor to the project. See Thomas (1947); and R. N. Thomas file in the Whipple Papers, specifically Thomas to Charles Federer, 8 December 1950. R. N. Thomas file, FLW/HUA.

11. The artificial meteor project combined the talents of APL, Harvard, Caltech, and the New Mexico School of Mines, where the grenades were actually prepared. Megerian to Bain, "V-2 Report no. 3," 2 May 1946, p. 10. On Whipple's advisory role, see "V-2 Report no. 6," 10 September 1946, p. 14. V2/NASM. See Whipple (1952a), pp. 22–23.

12. Whipple to D. T. Griggs, 2 August 1946, pp. 1–2. File 0812 05 00173XF, JPL/L.

13. [Harvard University News Office], 14 November 1946, p. 2. Science Service files, NASM.

14. [War Department Special Staff, Public Relations] to Watson Davis, 6 December 1946. Science Service files, NASM.

15. See F. Zwicky to Otto Struve, 3 July 1946. OS/AIP. Zwicky brought an ultrafast 20 centimeter F/1 Schmidt camera from Palomar for the flight and also conducted a series of ground-based firings of the rocket grenades to test the photographic network. He found that the luminous trails made by the vaporizing shards could be easily photographed from cameras up to 4 kilometers distant. Zwicky (1947a), p. 33.

16. Dorrit Hoffleit, "Two days and Thirteen Hours," 22 December 1946, pp. 1–2. Dorrit Hoffleit private papers, p. 8.

17. Ibid., p. 9.

18. Ibid., p. 10.

19. They were scheduled for firing 70, 80, and 90 seconds into the flight, when the missile was expected to be between 60 and 100 kilometers altitude. The V-2 exhaust was photographed from Tucson, some 450 kilometers away, and Zwicky also demonstrated that his F/1 Schmidt easily picked up the glowing carbon jet vanes after burnout at 40 kilometers. Zwicky (1947a), p. 33; (1947b), pp. 69–70.

20. Whipple to Watson Davis, 18 December 1946, from El Paso. Science Service File, NASM. Leon Davidson to Whipple, 19 November 1952; and Whipple to Davidson, 25 November 1952. Box 7, 1952 A-Q, DEF folder, FLW/HUA.

21. Zwicky (1986), pp. 330. There were later successful flights, starting in October 1957. See Zwicky (1986); and "American Astronomers Report—Artificial Meteor Observations," *Sky & Telescope* 18 (June 1959), p. 440; and "Artificial Meteors," *Sky & Telescope* 17 (January 1958), p. 111.

22. Whipple to Richard Pratt, 22 November 1947. HUG 4876.808, Correspondence special folders 1937–1942 (sic), "Prior Jan 1 1948 Official Communications," FLW/HUA.

23. Whipple saw this situation reversing in several years, however. Whipple to Terman, 23 May 1947. Box 5, Perkin Elmer folder 1940–1950 T-Z/ONR, FLW/HUA.

24. See Whipple to Gerard Kuiper, 19 June 1946. Box 3, 1940–1950, K-New Mexico Meteor project folder, FLW/HUA.

25. See, for instance, Whipple to George Gardiner (Physics Department, New Mexico School of Agriculture), 8 November 1947. FLW/HUA.

26. Whipple to Tombaugh, 18 April 1950. FLW/HUA.

27. Fred Whipple to Fritz Zwicky, 7 November 1946. Box 4, YZ folder, 1946–1950, FLW/HUA.

28. Doel (1990a), chap. 2, has examined this issue in the context of discipline formation in solar system astronomy.

29. Hok, Spencer, and Dow (1953), p. 235.

30. Best, Gale, and Havens (1946), pp. 47–48; Havens and LaGow (1946), pp. 44–46.

31. Spencer and Dow (1954), p. 82.

32. Tiny tubulations at the top of these tubes were broken when the payload reached operational heights. The heated filament generated electrons in the tube that were modulated by the grid and attracted toward the plate. The resulting grid and plate currents, suitably amplified, telemetered, and analyzed, revealed the pressure of the gas in the tube. Spencer and Dow (1954), p. 85.

33. The original Michigan November 1946 flight gave readings far below all the others, and thus were discounted later. Spencer and Dow (1954), p. 89, fig. 8; produced originally as University of Michigan report, Schultz, Spencer, and Reifman (1948), p. 10, fig 1.1.

34. Spencer recalls that they were interested more in building equipment than in analyzing data or publishing results. See Spencer OHI, p. 41rd.

35. Spencer, ibid. Spencer evidently used a revised version of Whipple's 1943 study in his own analysis. See Spencer to Dow, 20 October 1947, with appended Whipple manuscript. WGD/BHLUM.

36. Whipple to Spencer, 19 May 1948. Spencer OHI file, SAOHP/NASM.

37. Whipple to Dow, 9 June 1948. Spencer OHI file, SAOHP/NASM. Whipple also thought the *Astrophysical Journal* would be appropriate.

38. Ibid.

39. Schultz, Spencer, and Reifman (1948), p. 9. No journal publication would in fact appear presenting the pressure curves until the early 1950s, even though Dow and Reifman discussed how the same instrumentation could be used as a dynamic probe of the ionosphere in Reifman and Dow (1949a, 1949b). Hok, Spencer, and Dow (1953) discuss the results of their first season of flights, only in terms of the ionization densities encountered. Discussion of the ambient pressure results were made in 1954 at the Oxford Conference, using recent data, compared to their first season. Spencer and Dow (1954).

40. Havens OHI; La Gow (1954), p. 76.

41. Havens OHI; La Gow (1954), ibid.

42. Havens OHI, pp. 4–5.

43. Ibid.

44. LaGow (1954), p. 77.

45. Typically, a 200 microgram speck of either polonium or radium served as a source of alpha particles, which ionized the ambient gas in the gauge. The resulting ions were then collected on a separate electrode, to which had been applied appropriate polarizing potentials for the pressure range the gauge was intended to examine. G. Ellis Gray to W. G. Dow, "Vacuum Gauge Calibrations," 10 September 1948. Spencer OHI file, NASM.

46. N. W. Spencer and H. F. Schulte, Jr., "A Rocket-Borne Alphatron Atmospheric Pressure Measurement System," abstract, attached to Spencer to E. A. Blasi, 30 December 1950. Spencer OHI file, NASM; Spencer and Dow (1954), pp. 82–83.

47. The "Langmuir probe" is described in detail in Hok, Spencer, and Dow (1953); and in Hok and Dow (1954). It is compared to other probe designs in Spencer and Dow (1954). See also R. C. Edwards (of Cook) to Franklin Institute, 25 June 1948; Spencer to Dow, "Impressions of Cook Research Laboratory work on mass spectrometer subcontract," 10 May 1950. Spencer OHI file, NASM.

48. On DOVAP, see Hoffleit (1949), and Chapter 7.

49. Schultz, Spencer, and Reifman (1948), p. 13. See also M. H. Nichols, "Measurement of Ionization Density by Radio Methods," 29 January 1946. Box 3, folder 1, M668 "Notes and Reports," WGD/BHLUM.

50. See Bartman (1954); and Jones (1975), pp. 215–26.

51. See Marcel J. E. Golay to Harry Wexler, 19 July 1946; Wexler to Marcel Golay, 24 July 1946. HW/LC. Golay indicated that SCEL was indeed going to initiate studies of how sound measurements with rockets could be translated into atmospheric density profiles. He soon added smoke puff ejection devices for wind studies.

52. Stroud (1975), pp. 239–40, gives a general review of their later work.

53. A detailed description of the technique is in Weisner (1954). See also Newell (1953), pp. 128–29; and Stroud (1975), pp. 241–43.

54. See Warfield (1947); and press release draft, attached to R. G. Robinson to H. Wexler, 6 February 1947. HW/LC. On the history of NACA and its responsibilities, see Roland (1985a), and on the history of the Standard Atmosphere see Minzner (1976).

55. This was first suggested in L. E. Root to G. W. Lewis (rec'd) 1 March 1945. "Rocket and Satellite Panel 1945" folder, HN/NACA/NHO.

56. James M. Benson, 3 December 1945 memorandum for the file. "Rocket and Satellite Panel 1945" folder, HN/NACA/NHO.

57. The origins of the special subcommittee are taken partly from "Minutes of Meeting of the Special Subcommittee on the Upper Atmosphere," 2 May 1947, p. 2. HW/LC.

58. F. Reichelderfer to Oliver Wulf, 18 September 1946. HW/LC. The Committee on Aerodynamics contained the Subcommittee on Meteorological Problems. On NACA's committee structure, see Roland (1985a) vol. 2, especially pp. 461, 487.

59. The special subcommittee was at first called the NACA Panel on the Upper Atmosphere. See "Minutes of Meeting, Panel on the Upper Atmosphere, Committee on Aerodynamics," 4 March 1946, p. 5. NACA folder, HW/LC.

60. Ibid., pp. 2–4. Other members of the committee distrusted the German results, however.

61. L. C. Stevens to NACA, 3 April 1946. "Rocket and Satellite Panel, 1945-," HN/NACA/NHO.

62. They also provided access to classified materials, and provided the infrastructure needed for extended operations. B. Ames to T. P. Wright, 22 April 1946. NACA files, "Rocket and Satellite Panel, 1945-." See also "Minutes of Meeting, Special Subcommittee on the Upper Atmosphere," 24 June 1946, p. 3. "Rocket and Satellite Panel, 1945-," HN/NACA/NHO; and "Minutes of Meeting of the Special Subcommittee on the Upper Atmosphere," 2 May 1947, p. 2. HW/LC.

63. Newell "Record of Consultative Services," 27 June 1946, pp. 2–3. Box 98, folder 2, NRL/NARA. Unfortunately, Newell's notes do not explicitly describe Gutenberg's criticisms.

64. See "Minutes of Meeting of the Special Subcommittee on the Upper Atmosphere," 2 May 1947, rev. pp. 5–6. HW/LC.

65. Ibid., p. 2. The report was by William S. Aiken, "Standard Nomenclature for Air Speeds with Tables and Charts for Use in the Calculation of Airspeed," NACA TN-1120, September 1946. HW/LC. Also in September, Calvin Warfield's revised "Tentative Tables for the Properties of the Upper Atmosphere" were distributed for comments and criticisms. Robinson to members

66. Robinson, 24 September, ibid., p. 3. See also R. G. Robinson (asst. director of aeronautical research) to Wexler, 6 February 1947, with attached copy of the press release for the "tentative standard properties of the upper atmosphere recently published by the NACA." HW/LC.

67. Wexler, "Trip Report," 4 October 1946. HW/LC.

68. Wexler to Reichelderfer, "Trip Report—Boston, Massachusetts, September 5, 1946," 4 October 1946, pp. 2–3. HW/LC. See also Wexler notes on "V-2 Meeting AAF Field Station Cambridge 5 Sept. 1946," 5 September 1946. Box 34, UARRP folder no. 1, HW/LC.

69. "Minutes of the Meeting of the Subcommittee on Meteorological Problems, Committee on Operating Problems," 30 April 1947, p. 10. Box 32, NACA folder no. 3, HW/LC.

70. Ibid., p. 6

71. "Minutes of Meeting, Special Subcommittee on the Upper Atmosphere, Committee on Aerodynamics," 2 May 1947, p. 5. HW/LC.

72. They also suggested that in short order, the NACA should publish a statement on the limits of knowledge of conditions in the high atmosphere beyond 120 kilometers, and the resulting uncertainties of the aerodynamics of high-speed flight at these altitudes. Ibid., pp. 10–11.

73. See T. Smull to R. J. Havens, 18 August 1948. NACA files, "Rocket and Satellite Panel, 1945-," HN/NACA/NHO.

74. "Minutes of Meeting, of the NACA Special Subcommittee on the Upper Atmosphere," 17 September 1948, fig. 9. NACA files, "Rocket and Satellite Panel, 1945-," HN/NACA/NHO.

75. "Minutes of Meeting, Special Subcommittee on the Upper Atmosphere," 5 October 1950, p. 3. NACA files, "Rocket and Satellite Panel, 1945-," HN/NACA/NHO. The NACA subcommittee actually had two problems on its hands in 1950: (1) extending its atmospheric tables, and (2) reconciling differences between the NACA standard and the new European standard to 20 kilometers altitude. Concerning the latter, after a series

of meetings with ICAO, the NACA compromised in late 1950, giving authority to the International Commission for Aerial Navigation, which then resulted in the adoption by NACA of the ICAO standard in November 1952. The extension of the tables beyond that point was taken up by Wexler and the Air Force in 1953, partly as a result of the combined efforts of the UARRP. For a cogent review of problems of setting the international standard, see Minzner (1976).

76. "Panel Report no. 23," 14 February 1950, pp. 25–26. V2/NASM.

77. "Panel Report no. 24," 20 April 1950, pp. 14–15. V2/NASM.

78. The formative meeting took place in Van Allen's home in Silver Spring on 5 April 1950, where, stimulated by Chapman's desire to coordinate observations, they discussed an international cooperative effort reminiscent of the Polar Years in 1888 and 1932. See Newell (1980), pp. 50–51; James Van Allen OHI, pp. 170–72; and Chapman (1959).

79. "Panel Report no. 25," 13–14 June 1950. V2/NASM.

80. Ibid., p. 3.

81. "Minutes of Meeting, Special Subcommittee on the Upper Atmosphere," 5 October 1950, p. 3. NACA files, "Rocket and Satellite Panel, 1945-," HN/NACA/NHO. "Panel Report no. 26," 7–8 September 1950; "Panel Report no. 27," 31 January 1951; and minutes from "Committee on Coordination of Atmospheric Temperature and Composition Experiments," 2 August 1950. V2/NASM.

82. Whipple report, in "Panel Report no. 27," 31 January 1951, p. 11. V2/NASM.

83. "Panel Report no. 28," 25 April 1951, p. 14. V2/NASM.

84. Ibid.

85. Ibid., pp. 14–15. The NACA temperatures derived by Warfield were later shown to be "excessively high," whereas the UARRP values fell far closer to the international standards set in the 1960s and 1970s. See Minzner (1976).

86. "Panel Report no. 29," 14–15 August 1951, pp. 2–14; 21–24. V2/NASM. Those making presentations were: R. J. Havens (for NRL), N. W. Spencer (for the University of Michigan and for the AMC), R. A. Minzner (of AFCRL, for the ARDC), F. L. Bartman (for the University of Michigan and for the

SCEL), M. Ference (for SCEL), F. L. Whipple (for Harvard), W. W. Berning (for BRL).

87. "Panel Report no. 29," 14/15 August 1951, p. 14. V2/NASM.

88. NACA also "strongly recommended . . . that information on the extremes of temperature in addition to the average be included in the paper." "Minutes of Meeting, Special Subcommittee on the Upper Atmosphere," 23 October 1951, p. 6. NACA files, "Rocket and Satellite Panel, 1945-," HN/NACA/NHO.

89. W. W. Kellogg led the discussion that ended in a formal resolution by the special subcommittee to endorse the UARRP's continued efforts to improve its data, especially for different latitudes such as at a new arctic site at Fort Churchill, Manitoba, on the shore of Hudson Bay. Ibid., pp. 6–7.

90. "Panel Report no. 30," 24 October 1951. V2/NASM.

91. "Panel Report no. 31," 8 January 1952, p. 11. V2/NASM.

92. Ibid., p. 12. On the mildly contentious atmosphere that existed between the two bodies at this time, see Havens OHI, p. 17; and Van Allen OHI no. 4.

93. See Wexler (1950); and Doel (1990a), chap. 2, pt. 2.

94. Even though significant deviations remained, Whipple publicly argued, "To many of us, this deviation really measures agreement; only a few years ago the uncertainty was at least two orders of magnitude." Whipple (1949), p. 92.

95. See Doel (1990a), chap. 4.

96. Whipple (1950), p. 376.

97. Whipple (1948b), p. 53. Whipple, Jacchia, and Kopal were concerned in 1947 that winter showers were of low-velocity swarms, whereas summer swarms were comparatively higher in mean velocity. See Yerkes Conference: "Seasonal Variations in the Density of the Upper Atmosphere," in Whipple, Jacchia, and Kopal (1949). Whipple's comments on meteors and cometary debris appeared in [Whipple] (1948c), pp. 167–68.

98. Whipple (1950). p. 376. Whipple also knew in 1950 from his associate R. E. McCroskey's laboratory experiments that most cometary debris were of low density and easily crushed. This he felt was comforting news to designers of spacecraft, but meant

as well that his atmospheric density and temperature studies were seriously in error. Fred Whipple, "Some Results from the Harvard Photographic Meteor Program" [Working Paper, 1950] in Whipple (1972), vol. 1, pp. 222–23. Bates and Nicolet (1950), pp. 302–03.

99. Formed in December 1950, their first meeting report was dated 20 April 1951. Membership included some familiar names, such as W. W. Kellogg of RAND, and a member of the NACA Subcommittee, A. G. McNish of the NBS, Michael Ference from the Signal Corps, Joseph Kaplan of UCLA, and Franklin Roach from the Navy, as well as Whipple, Van Allen, Oliver Wulf, and Lyman Spitzer, who had provided RAND with a seminal study of the atmosphere above 300 kilometers in 1946. Reporting on their first meeting, Kellogg et al. (1951), p. 757, noted, "In recent years some question concerning the validity of the density determinations made by the photographic meteor techniques has arisen because of a discrepancy between density determinations by this method and those made by rockets." See also G. G. Holzman to Fred Whipple, 27 December 1950. Box 6, 1951 A-N folder, FLW/HUA.

100. The committee was chaired by Michael Ference and included Wexler, Whipple, Bernard Haurwitz, and other familiar players. See Charles Brooks to Whipple, 10 February 1951. Box 6, folder 1951 A-N, FLW/HUA.

101. Whipple (1952a), p. 14.

102. Ibid., p. 18. Whipple gave high ratings to Ference's grenade results, Dow's Mach angle probes, and NRL's Havens Cycle Gauge observations, along with APL's ozone profile observations made with photoelectric spectrometers. He had less confidence in the early mass spectrometer efforts and in composition results by direct sampling. Generally, he liked the correlations he was finding: Temple University's AMC contract study of acoustic noise on rocket skins indicated a high micrometeorite flux, which corresponded to the higher fluxes indicated by his first Super-Schmidt trials; ion density and refraction studies by NRL in a V-2 flight agreed strikingly well with radio observations that persistent ionization from meteors occurs in 5 kilometer layers in the ionosphere.

103. Ibid., pp. 24–25.

104. Spencer and Dow (1954), pp. 85–87.

105. Whipple (1952b), p. 28.

106. Ibid., p. 29. Doel has followed the complex changes in Whipple's thinking about cometary debris during this period, as well as his related reversals concerning the efficacy of his meteor data. See Doel (1990a), chap. 4, pt. 2.

107. Jacchia and Whipple (1956), p. 989. Doel (1990a), chap. 2, pp. 69–70, discusses Whipple's realization that meteors came in two density ranges, based on the better observational data, and R. N. Thomas' ballistic calculations and tests.

108. "Panel Report no. 32," 30 April 1952. V2/NASM.

109. Between 1950 and 1952, Leslie Jones' Michigan group detected diffusive separation through atmospheric sampling, whereas Newell's NRL mass spectrometer group could not at first confirm the detection. In 1951, even though Havens criticized reports by F. A. Paneth, who was working with Leslie Jones, the panel concluded, "In spite of Dr. Havens' objections it seems that the Michigan results do show that separation must be occurring in the atmosphere. . . ." "Panel Report no. 29," 14–15 August 1951, pp. 22–24; and Chackett, Paneth, Reasbeck, and Wiborg (1951), p. 358. The matter arose again, in "Panel Report no. 32," 30 April 1952, p. 2. V2/NASM. See Jones (1975), pp. 206–11.

110. The UARRP was now planning for a symposium in Oxford in the summer of 1953, where it would discuss plans for IGY flights from a wide range of latitudes. See "Panel Report no. 33," 7 October 1952.

111. [The Rocket Panel] (1952), pp. 1027–32.

112. Ibid., p. 1027.

113. "Panel Report no. 33," 7 October 1952. V2/NASM.

114. The NACA Special Subcommittee did not suffer at the hands of the UARRP but through negotiations the NACA conducted with the International Civil Aviation Organization. R. A. Minzner (1976), p. 58, provides a useful overview and also identifies the independence of the Rocket Panel report. Kellogg et al. (1953), p. 116, notes that the NACA subcommittee's dissolution was made without reconciling its tables with those obtained through rocketry. How and why

the NACA made this decision requires further study, as does how Wexler and the AFCRC combined forces to create the Committee on Extension to the Standard Atmosphere (COESA), which, through a series of working groups, eventually developed a series of improved tables under Air Force auspices. In June 1955, explaining why the AFCRC's Geophysics Research Directorate decided to develop the new standard, Minzner told the UARRP, "In spite of its increased reliability over the Warfield atmosphere, the Panel Atmosphere has not been generally used by aeronautical and military groups . . . [partly because] . . . this atmosphere failed to take into account the results of several ground based experiments which indicated a higher temperature in the 50 km region." [R. A. Minzner] "Extension to the U. S. Standard Atmosphere," p. 20, in "Panel Report no. 41," 2 June 1955. V2/NASM. Whipple, heading one of the working groups on the extension, asked Newell to collaborate with an AFCRC contract group in RAND Corporation's Missiles Division, Hilde Kallmann-Bijl and W. B. White, to provide a combined "speculative atmosphere" incorporating the latest dissociation theory with the rocket data. See Kallmann, White, and Newell (1956); and Kallmann (1958), p. 130. On the demise of the NACA Special Subcommittee, see B. Haurwitz to Wexler, 12 December 1951; Hunsaker to Wexler, 14 December 1951; Wexler to B. Haurwitz, 14 December 1951; and Hugh Dryden to Wexler, 4 January 1952. HW/LC.

115. These were mainly members connected with the Signal Corps or Air Force. Leslie Jones claimed in 1974 that the "Panel Atmosphere . . . evolved into what is now the U.S. Standard Atmosphere (1962)" and its supplements. Jones (1975), p. 204. This contention is supported by Stroud (1975), p. 250, as well as by inferences in Whipple OHI, p. 10, and in William Dow to the author, 30 July 1990, p. 62. D. R. Bates and others did accept the new data in the mid-1950s, at least to 80 kilometers. Bates (1957), p. 66.

# 16

# On the Periphery of Ionospheric Research

*Primarily what we are trying to do is to stimulate attention to a new type of instrumentation.*

—William Dow to D. R. Bates (1950)[1]

Radio probing has always been the dominant method for learning about the ionosphere and how it could be used for long-range radio communications; in the late 1920s, the pulse-echo method developed by Merle Tuve and Gregory Breit at the Department of Terrestrial Magnetism (DTM) became favored worldwide. Edward Appleton and J. A. Ratcliffe in England soon used the pulse-echo method to show that the ionosphere consisted of a succession of layers of ion concentration. The layers were given additional reality when, in 1931, Sydney Chapman, then at Imperial College, London, created a theoretical model for the electron density distribution that persuasively matched the observations by pulse-echo techniques.[2]

The radio technique, however, could detect only the relative maxima of the vertical ion distribution; concentration levels between the maxima were undetectable. Chapman and many of his British colleagues, along with Frederick E. Terman and Lloyd V. Berkner in America, all argued that the ion or electron concentration in the ionosphere never dropped to zero, but they had no observations to support this.[3] Nevertheless, the concept of discrete layers caught the imagination of radio technologists, largely because their radio tracings looked like disconnected layers,[4] (figure, p. 305) and because the model proved to be of great practical use.

Ionospheric physics at the end of World War II thus possessed a robust theoretical structure and a dominant observational technique. Even though Chapman's model was an idealization, its portrayal of discrete layers of ionization was eminently useful, especially to those interested in using the medium for long-range radio communications.[5] The model monopolized thinking among radio specialists and therefore set the landscape in which all the rocket groups worked at first. The contribution that rocket observations were ultimately to make did not come easily: they would be the first to reveal what existed between the maxima. But when they announced their findings in the mid-1950s, they garnered little attention.

## Who was Interested

After the war, radio propagation research was given highest priority by every military agency involved with scientific rocketry and was constantly promoted by major committees reporting to the Joint Research and Development Board (JRDB). In April 1947, the JRDB Panel on the Upper Atmosphere called for intensive research into the effect of ionospheric conditions and associated magnetic disturbances on long-range radio and wire communications and the radio detection, guidance, and control of missiles and aircraft.[6] The National Advisory Committee for Aeronautics (NACA) Special Subcommittee on the Upper Atmosphere, and the National Bureau of Standards Central Radio Propa-

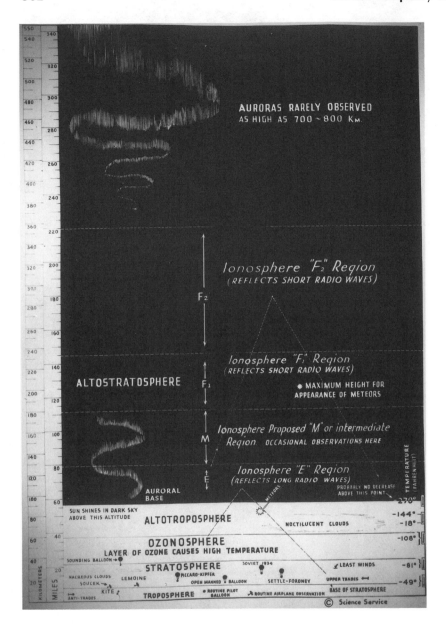

AURORAS RARELY OBSERVED
AS HIGH AS 700~800 Km.

*Ionosphere "F₂" Region*
(REFLECTS SHORT RADIO WAVES)

F₂

*Ionosphere "F₁" Region*
(REFLECTS SHORT RADIO WAVES)

ALTOSTRATOSPHERE      F₁
                                    ✳ MAXIMUM HEIGHT FOR
                                       APPEARANCE OF METEORS

*Ionosphere Proposed "M" or intermediate*
*Region.* OCCASIONAL OBSERVATIONS HERE

M

*Ionosphere "E" Region*
(REFLECTS LONG RADIO WAVES)

AURORAL          E          PROBABLY NO DECREASE
BASE                            ABOVE THIS POINT.                    270°

SUN SHINES IN DARK SKY
ABOVE THIS ALTITUDE        ALTOTROPOSPHERE                          -144°
                                    NOCTILUCENT CLOUDS                    -18°

OZONOSPHERE                                                          -108°
LAYER OF OZONE CAUSES HIGH TEMPERATURE
SOUNDING BALLOON

STRATOSPHERE        SOVIET 1934          ↙ LEAST WINDS               -81°
NACREOUS CLOUDS    LEMOINE    PICCARD-KIPFER
SOUCEK                      OPEN MANNED BALLOON    SETTLE-FORDNEY    UPPER TRADES        -49°
KITE                                                               BASE OF STRATOSPHERE
ANTI-TRADES        TROPOSPHERE    ROUTINE PILOT    ROUTINE AIRPLANE OBSERVATION
                                    BALLOON
                                                        © Science Service

Profile of the ionosphere in the early 1930s, as depicted by Science Service. The layers (depicted as regions here), were soon redefined as D, $E_1$, $E_2$, $F_1$, and $F_2$. The terms "layer" and "region" were loosely defined in the 1930s and 1940s. After World War II, there were attempts to clarify the terms, but the situation remained confused. See Mitra (1948), p. 210. Both promoted the concept of discreteness. Science Service collection. NASM.

gation Laboratory's Committee on Electronics and its Basic Research Panel identified many technical problems that could only be solved with the aid of rocket observations, including knowledge of antenna loading from ions in the upper atmosphere and interference from rocket exhaust in the ionosphere. They asserted that an improved understanding of the ionosphere was essential to the modern military.[7]

With such universal endorsement, it is not surprising to find that most of the military agency members of the V-2 Panel planned or attempted some form of radio propagation experimentation with rockets.[8] The members flew electronic detectors to determine ion densities and electron temperatures as a function of height, performed Doppler radar experiments, and examined the effect of the rocket exhaust on radio transmissions. The ionospheric groups that formed around the rocket, like their colleagues in other areas of scientific rocketry, first focused their attention on experimental designs and procedures. They worked within institutions devoted to practical ends and accordingly applied their technical experience and expertise to those ends. Also in consonance with the other rocketry groups, none of those who chose to study the ionosphere with rockets were a part of the prevailing community that used radio to probe the ionosphere; nor were they among those who employed meteor, auroral, and solar radio techniques, or laboratory investigations of the behavior of atmospheric gases under extreme conditions. They were even more distant from those who built theoretical models of the upper atmosphere based on these physical data.[9]

The marginality of the rocketry groups to the field of ionospheric research was at first not an issue. They had the support of their military patrons and the attention and advice of practical interests that dominated American institutions such as the Central Radio Propagation Laboratory (CRPL), or the NRL and DTM.[10] But when the rocket groups began to try to make sense of their data, after overcoming years of frustrating technical failure, they found less clear direction. As pioneer UCLA geophysicist Joseph Kaplan remarked in 1948: there was then a "need for intensifying theoretical and experimental upper atmosphere research in this country." Speaking as a member of the NACA Special Subcommittee on the Upper Atmosphere, Kaplan stressed "that research teams in Europe, and particularly in England, are extremely active on practically all phases of upper atmosphere research. . . . In this country, however, many aspects are neglected or given a minimum amount of attention. . . . no concerted coordinated effort is present."[11]

As Kaplan worked diligently to establish his new Institute for Geophysics at UCLA in the 1940s, he encouraged his younger students to follow the results of the rocket groups and to make a "concerted coordinated effort" to evaluate them in light of the best available theory. His students knew from his

counsel that, although the field had its trends and techniques, it still required a wide array of geophysical and physical input. As William W. Kellogg, a Kaplan graduate student and Project RAND employee, observed in late 1947, "It is this very diversity which presents a challenge to anyone interested in adding to our knowledge of this region, and perhaps helps to explain why to date so little is known about it."[12]

Taking the RAND perspective, Kellogg believed that the upper atmosphere "would be a theater of operations in any future war," which added "a note of urgency to the normal leisurely course of scientific research."[13] However this sense of urgency may have led the rocketry groups to overcome the technical hurdles of their enterprise, it did not prepare them for their ultimate test: demonstrating that they had established a new way of studying the ionosphere. We begin by examining these technical hurdles, and conclude by identifying subsequent conceptual and professional hurdles that all the rocketry groups faced as their data were scrutinized at Oxford in the summer of 1953.

## Studying the Ionosphere with Rocket-Borne Radio

### The Naval Research Laboratory

In the first several years of the V-2 era, NRL devoted by far the greatest effort to ionospheric radio propagation. Krause put T. Robert Burnight at the head of a large team that included J. Carl Seddon, Charles Y. Johnson, and about 10 others to fly sets of radio transmitters and receivers on as many flights as they could handle. They would measure the varying index of refraction of the medium with height to derive ion and electron densities and electron collisional frequencies. The techniques developed at NRL were similar to those suggested by Myron Nichols and taken over by Marcus O'Day's staff at the Air Force Cambridge Research Center (AFCRC).[14]

After extended consultation with ionospheric experts from the NBS and the DTM,

Carl Seddon, a physicist and radio engineer who had extensive practical experience building and testing radio transmitters and had worked on radio propagation problems during the war at NRL, decided that his team under Burnight would develop active means of radio probing. Phase differences between two or more harmonically related frequencies from rocket-borne transmitters would provide direct indications of the ion density of the medium between the rocket and the ground receiving stations.[15] At first, Seddon chose the higher frequency in a range in which velocity was unaffected by the ionosphere, and the lower frequency set at the lowest value for which the medium was still transparent—the critical frequency, a constantly changing quantity that Newbern Smith of the National Bureau of Standards' wartime Interservice Radio Propagation Laboratory was happy to provide for the V-2 flights.[16] DTM's Lloyd Berkner, who had been a central figure in radio propagation research since the 1930s and during the war had been head of the Electronics Materiels Branch of the Engineering Division of the Bureau of Aeronautics, advised Seddon to set the lower frequency just above the critical frequency so as to be able to test for the presence or absence of reflections from upper layers of the ionosphere (an indication of sporadic E-layer activity). This would also relax tolerances on the predicted value of the critical frequency, even though Berkner felt that the NBS could provide predictions one month in advance.[17]

Seddon's work for Burnight and Krause was part of a larger wartime effort at NRL to examine sources of radio interference from guided missiles, specifically from their exhaust gases, "to determine" as Seddon recorded in his registered laboratory notebook in February 1945, "suitable methods of controlling [guided missiles] smoothly, and to receive telemetry information back from the rocket."[18] As Seddon's team geared up for their work with the V-2s after the war, Krause continued wartime projects to study the effects of exhaust gases from large rocket motors under controlled conditions on the ground.[19]

The NRL team began their work in stages; as they attacked various technical problems, they also taught themselves ionospheric physics, aided at one point by a bibliography of standard references provided by Berkner.[20] More than any of the other types of experiments conducted with the V-2s in the first years, the ionospheric radio propagation effort required a high degree of coordination and cooperation between the participating agencies. The NBS field station team at White Sands provided much of the needed radio height-finding equipment and expertise, as well as a variable frequency unit for critical frequency determinations.[21] Both General Electric and APL provided space for the NRL experiment on their flights, and the Ballistic Research Laboratory (BRL) at Aberdeen was responsible for straightening out all radio interference problems and performing preliminary data analysis.[22]

The most daunting technical problem NRL faced was antenna design: what configuration would work best on a rocket, at the frequencies required. They first tried a number of simple antennae, including a suggestion by Myron Nichols to stretch wires from the midbody of the V-2 to two of the four V-2 fins, simply to see what would happen. After four flights of this design and a trailing wire system attached to the fins that deployed as the rocket rose from the pad (figure, p. 307), the NRL group realized that no simple solutions were at hand. A fifth flight on 30 July 1946, now with three trailing wires, produced only weak, erratic signals, "indicating the sudden development of trouble."[23]

The trailing wire antennae experienced enormous drag problems in the lower atmosphere, as did the delta-type antennae. Any object of extended dimension outside the skin of the rocket was subject to hostile treatment. The NRL team concluded in early October that "the antenna problem remains one of the most serious, and is receiving considerable attention."[24] Throughout the rest of 1946, the NRL field team gained practical diagnostic experience and collected some information on antennae radiating patterns, but no useful data on the ionosphere were gathered in the first year.

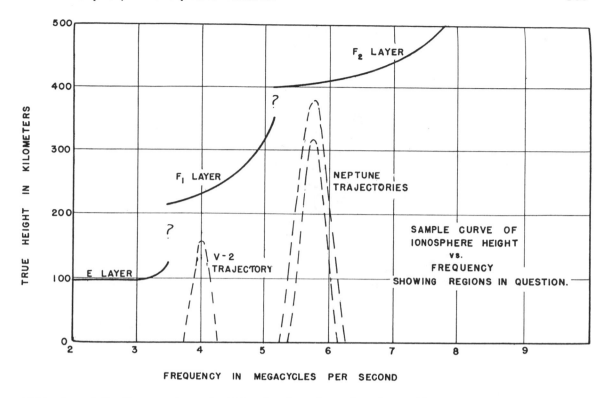

**NRL's view of the Chapman layers in 1947, showing schematic radio profiles for the E and F layers, and the limits of in situ measurement by V-2 and Neptune vehicles. Appendix C in [NRL]. The Navy Upper Atmosphere Research Program With Rocket Vehicles. Pt. 1. (10 April 1947).**

Seddon advised Krause and Burnight at the end of 1946 that drag antennae might work eventually, but they were impossible to tune or calibrate properly since they deployed only when the rocket was in flight. Searching for a better ionospheric antenna design, NRL and White Sands personnel performed pattern and impedance tests in the spring of 1947 on a V-2 airframe stuck nose down in the desert sands[25] (figure, p. 309).

After devoting a year to this effort and to transmitter and receiver refinements, NRL finally created a system that provided good signals during a 400 second flight on 7 March 1947, from which it made its first estimates of ion densities in the E layer. In addition to more powerful and stable rocket-borne transmitters and new ground-based superheterodyne receivers using harmonically related local oscillators, the 7 March flight used antennae that were stretched loosely between the fins of the rocket at its base. This antenna geometry seemed to be a compromise between radiating pattern and ruggedness.[26] The flight record was far from flawless. At 128 seconds, severe interference from a local communications signal degraded the data. Also, signal strength was erratic, but this time enough worked so that the team could conclude that they were measuring conditions in the ionosphere.

By the end of its fourth firing season in 1948, although there were still many technical problems, Seddon's NRL team was encouraged that it had found a reliable combination of transmitters and antennae that could return useful data. The most serious problems, such as antenna impedance stabilization, came under a degree of control in the first two years; a significant advance came to pass when they succeeded in recording phase beat frequencies independent of signal amplitude.[27]

But the way was far from smooth. Personnel departures stalled the work at times, even

Sydney Chapman attends the 25th meeting of the V-2 Panel, 13—14 June 1950, in Boulder, Colorado. First row: Capt. James C. Sadler, Sydney Chapman, Homer Newell, George Megerian, James Van Allen, Walter Orr Roberts. Second row: James Jackson(?), C. R. Briggs(?), Michael Ference, Marcus O'Day, Charles Green, Nelson Spencer, Raymond Minzner. Third row: Herb Karsch, W. B. Pietenpol, Thorp Walker, Ralph Havens, Fred Bartman, F. C. Walz. Nelson Spencer collection. NASM.

though Homer Newell pushed to maintain the pace set by Krause.[28] More frustrating, however, were conflicting priorities between the different research areas. In late 1947, the NRL group, along with APL and AFCRC, started to create standardized mounting systems for a wider array of instrumentation. Standardization was welcomed by the ionospheric groups, since they required systematic and precise antenna geometries.[29] But when the Army relaxed its constraints on the skin geometry of the V-2 warheads at the same time, all the groups began to modify their payloads, adding pressurized chambers, wholly aluminum skins, and protruding instruments. This new freedom changed the electrical characteristics of the rocket and meant that the ionospheric groups had to go through costly and tedious impedance calibration procedures each time.[30]

Problems such as these, along with flight failures in 1948 and 1949, reduced the number of successes the NRL ionospheric group could

point to when the time came to discuss ion densities in the high atmosphere. After an intensive flight schedule in the 1940s, flights of ionospheric experiments dwindled in the early 1950s, as the Viking ceased to gain favor as a guided missile. By this time, to Seddon's dismay, Homer Newell was giving less attention to ionospheric research. Krause had given ionospheric research "top priority" in 1946–1947, according to Seddon, but starting in August 1952, Newell pushed it back behind night sky-light, composition, and air density measurements. Seddon did not appreciate Newell's rationale that the "reversal was due to our present 'improved knowledge'," and told his boss, "In view of the military importance of the ionosphere, it seems to me somewhat ridiculous to put it last on the list."[31] He felt that the areas Newell emphasized certainly carried less military importance. In any case, Seddon could not count on unlimited support, and indeed, ionospheric research was curtailed on later Viking flights.[32]

V-2 launch showing trailing wire antennae. U.S. Navy photograph, Ernst Krause collection. NASM SI 83–13889.

## The Ballistic Research Laboratory

The aim of the NRL team was to assess the properties of the ionosphere for long-range communications and missile guidance, whereas the Ballistic Research Laboratory group, led by Louis A. Delsasso, wanted to improve its principal technique of trajectory analysis, DOVAP (Doppler Velocity and Position, see Chapter 7). Warren W. Berning, with training in physics and meteorology, had industry experience in aerodynamics before coming to BRL in 1947 as a meteorologist, where he became involved in testing and re-

fining DOVAP accuracy. He knew that phase errors in the high-frequency DOVAP method were caused by varying conditions in the ionosphere, and, as a result, "the rather meagre knowledge and large time variation of ionospheric structure have prevented the inclusion of the effects of this medium on phase wavelength in the DOVAP data reductions." Berning and others at BRL therefore used DOVAP to estimate charge density as a by-product of their need to assess DOVAP errors.[33]

The basic plan of the BRL team was to assess the influence of the ionosphere by testing actual trajectories against "vacuum trajectories" based on an ideal two-body solution for the rocket in the ionosphere after burnout. They would compare theoretical Doppler velocities at each point of the trajectory with the observed DOVAP velocities in the ionosphere. Divergences would yield the index of refraction, and, indirectly, charge density.

BRL's first flight was spectacular because it was part of the agenda of Bumper-WAC no. 5, flown from White Sands on 24 February 1949. DOVAP signals were received throughout the peak of the trajectory, to 390 kilometers, and thus were the first of their kind made deep into the F layer. Some, like Homer Newell, argued in the early 1950s that the BRL technique was overly simplified.[34] But by 1953, Berning felt that his analysis was sound for short ground-range trajectories where the altitude of the missile was less than 480 kilometers. He also examined earlier DOVAP data from V-2 flights during the late 1946 solar minimum, to show by comparison that under such conditions, the DOVAP technique was not sensitive enough to reveal low-level charge densities. On the other hand, during sunspot and solar activity maxima, such as the period 1949–50, he found that charge densities were high enough to measurably alter DOVAP trajectory calculations. Thus Berning had demonstrated his primary point: operational DOVAP applications for guided missile trajectory analysis required corrections for ionospheric effects.[35]

Although Berning had accomplished his primary operational goal, directed by the needs of his institution, he was also able to say something about the behavior of the E layer itself. Berning compared his data to NBS measures of the critical frequency of the atmosphere at White Sands on the dates of the firings he had analyzed and found considerable agreement for charge densities, especially for the peaks in the E and F layers.[36]

DOVAP transceivers were flown piggyback many times and Berning exploited their results. But because Bumper was a classified program, BRL's L. A. Delsasso, present at most meetings of the UARRP, never reported formally on Berning's progress. At least no mention of the BRL charge density work appeared in the UARRP minutes in the early 1950s.

## Scaling Up The Laboratory: Electronic Probes at Michigan

Instead of studying the properties of the ionosphere by observing how it affected radio transmissions, groups at the University of Michigan and NRL placed arrays of thermionic probes on rockets to sense the ionized medium directly. At Michigan, they had always explored the properties of ionized gases "from experiments with electrical discharges through gas-filled tubes. It [seemed] natural," therefore, "to try to carry over as much as possible of this knowledge and experience to the study of the ionized layers of the upper atmosphere."[37]

William Dow's group developed what they called a Langmuir probe, or ion probe, to measure both positive ion and electron charge densities. Irving Langmuir of General Electric, the man who succeeded in making the incandescent light bulb practical, continued through the 1920s and 30s to study the interaction of hot metals in gases and found a way to measure the temperatures and concentrations of electrons and ions in a gas by a voltage scanning probe. William Dow was familiar with Langmuir's techniques and used them as diagnostic elements in his own vacuum-tube technology research. Now he

planned to scale up his technique to study the ionosphere.

Dow's idea was to make the entire rocket a probe. Traveling through the ionized medium, the rocket would collect more electrons than ions, since the former have greater random velocities. This would result in a net negative charge on the rocket skin that would in turn inhibit subsequent negative ion flow to a point where a positive ion sheath existed around the rocket to a depth of several centimeters. This ion sheath could be sampled and evaluated by varying the potential of a small portion of the exposed tip of the rocket cone, Dow's Langmuir probe. In Dow's design, the controlled variations in cone tip current could be compared to the resulting potential difference between the main rocket body and the cone, which would in turn reveal the density and temperature of the free electrons in the ambient gas.[38]

Dow and his group were never fully satisfied with their own probe design, which was dictated by the geometry of the V-2 warhead. It was difficult to calibrate and had a number of characteristics that were hard to control. The collecting area was a small truncated cone, insulated from the rest of the nosepiece. It was set above a larger cone that was connected to the rest of the warhead and was in electrical contact with the missile (figure, p. 280). The nose tip commonly carried ionization gauges with their dynamic probes, initially for the determination of ion density, and then, when reliable density information proved elusive, for electron temperatures.[39]

By the early 1950s, after flying three times, Dow's group could point to only one V-2 flight, on 8 December 1947, on which their Langmuir probe provided "reducible data."[40] But even when the probe worked properly, supporting systems on the rocket or the rocket itself would fail, or its erratic motions would somehow interfere with the electrical measurements they were trying to make. There were simply too many variables, such as contamination and mechanical disruption, in what was typically a rough flight.[41] This did not frustrate the Michigan group; it was just what Dow relished, a good fight.

Inverted V-2 airframe created to test ionospheric antennae designs in the spring of 1947. NRL was testing specifically for antenna impedance characteristics. The configuration being tested here was adopted by NRL in later flights. Burnight, Fry, and Seddon (1947), p. 100. U.S. Navy photograph.

### Establishing a Capability

As the V-2s dwindled and ran out completely in 1952, Dow's group, now under Spencer (figure, p. 310) as its project engineer, continued an active atmospheric temperature and pressure program using Aerobees.[42] Influenced heavily by Dow, their efforts continued to be experimental rather than programmatically observational. As he had pointed out frequently throughout the V-2 era, Dow viewed this activity as experimentation and engineering development. Just as Berning at BRL wished to establish that the ionosphere had to be factored into any ballistics trajectory calculations, Dow regarded his work as

leading to the establishment of a capability in instrumentation, which he felt was as important as learning about the ionosphere itself.[43] Dow thus introduced Belfast atmospheric physicist David R. Bates to what they were up to at Michigan:

*Primarily what we are trying to do is to stimulate attention to a new type of instrumentation. The ionosphere exhibits many of the properties of the plasma of an ionized gas, with the exception of a very large scale factor and marked changes as to the origin of the sustaining energy. We should like to stimulate study of the ionosphere by*

Nelson Spencer inspecting the remains of the Blossom IV (V-2 47) rocket body on 15 June 1949, a day after its flight to 133 kilometers. Nelson Spencer collection. NASM SI 89–1128.

*various methods that have been extremely successful in the study of gas discharges.*[44]

Dow and his colleagues represented by the UARRP in the early 1950s found that what they were doing attracted more academic interest in England than about anywhere else. Bates and Oxford geophysicist Sydney Chapman, both leading contributors to ionospheric research, helped foster closer cooperation between the rocket groups and British centers for upper atmospheric research. Dow and his rocketry colleagues wanted to show that their new techniques and instruments yielded observations of temperature, pressure, density, and composition that were essential for advancing ionospheric physics. In particular, Dow's development of the means for determining electron temperatures could be used to check theoretical reaction and re-

combination rates of the dominant elements supposed to exist in the region. But his staff needed help to perform these analyses.

In 1950, Chapman was already a frequent visitor to UARRP meetings as he toured U.S. research sites from his temporary base at Caltech, and Dow hoped that Bates as well could spend a few days with his staff at Michigan, especially with Gunnar Hok, to further their understanding of the "analytical" aspects of their research. Dow felt that they had gone about as far as they could with theory but needed help with "some other aspects we are somewhat critical of at the moment."[45] What they needed, above all, was contact with theoretically strong specialists like Bates, to show them how to make sense of their observations. They were finding it difficult to reconcile their data with an idealized theory.

## The 1953 Oxford Conference

Unlike the enthusiastic but cursory treatment of rocket observations of the solar spectrum at the 1947 Yerkes Symposium, where only two V-2 Panel members attended and gave short papers, and in contrast to paper sessions on upper atmospheric research sponsored by the American Geophysical Union, the American Physical Society, or by the International Union of Geology and Geophysics, the 1953 Oxford Conference was a truly integrated effort. The UARRP was a major participant and its observations were closely evaluated and critiqued. This was, in many respects, the first attempt by an intellectual body to address the results of the American rocket experimenters.

In April 1952, at Sydney Chapman's urging, Harrie Massey, of the Gassiot Committee of the Royal Society of London, invited the UARRP to co-organize an international conference on upper atmosphere physics at Oxford that summer. Van Allen and the panel accepted, but asked for more time to prepare, so Massey set August 1953 as the new target.[46] The Gassiot Committee, formed in the late 19th Century to oversee the Kew Observatory, was asked by the British Air Council during the Second World War to examine equilibrium processes in the atmosphere, mainly for photochemical and meteorological interests, but also to continue research into what made the ionosphere work. They were naturally interested in the results of all types of *in situ* observations of atmospheric composition, as well as in the nature of the ultraviolet solar spectrum.[47]

Massey and Chapman's plan for Oxford was to ask each V-2 Panel member to prepare a draft manuscript in advance, for distribution to all intended participants. The British, Whipple gently advised in April 1953, "will be interested more in the scientific aspects of instrumentation rather than techniques and actual operational details."[48] The Americans had differing views of the purpose of the Oxford meetings. Newell, Van Allen, and others wanted to discuss the way rocket experimentation would be integrated into the IGY and wanted Oxford to become a forum.[49] Ralph Havens, Nelson Spencer, and J. R. Lien from AMC, members of an ad hoc UARRP committee that tried unsuccessfully to stimulate interest in a symposium on technical problems surrounding ionospheric studies with rockets in 1952, hoped that the Oxford Conference might meet their needs.[50] But William Dow wanted to use Oxford both as a means to promote his younger colleagues and as an excuse to visit his old wartime industrial contacts in British vacuum tube research; Dow in fact became preoccupied with the latter, wanting to catch up on the latest military and civilian activities.[51]

In any case, the Oxford Conference was seen on both sides of the Atlantic as a great opportunity to share techniques and results; it also stimulated many UARRP members to get caught up with long overdue data analysis. S. Fred Singer, who accepted a scientific liaison position with the ONR office in London after Van Allen's APL group folded, provided logistical support for the Americans. He also presented analyses of his old work at APL.[52]

The conference was held from 24 to 26 August in the University Examination Schools Lecture Room opposite Queen's College. Sessions began in the midafternoon and lasted into the night with breaks for tea and meals.[53] UARRP members found a critical but congenial forum through which they could better appreciate each other's results, as they were discussed and critiqued by their British audience. Although Dow may have been distracted by other interests, he certainly warmed to the proceedings. His scribbled notes on the printed agenda reminded him to tell Gunnar Hok, "Newell's *electron* density data is beautiful. . . . His negative ion density results are not worth the breath it takes to talk about them."[54] Dow's strongly partisan views were shared by other conferees, each testing the rocket data in terms of their favorite theories.[55]

The most detailed critiques were leveled at the ionospheric data and results presented by NRL and Michigan. R. L. F. Boyd, from University College, London, was concerned about the gross differences manifest between the Langmuir probe's use in the laboratory and on a "gassy rocket during its transient encounter with the ionosphere."[56] Boyd was particularly interested in experimental studies of collision processes in gases and complained that those using Langmuir probes on rockets, such as Dow's group at Michigan, had not taken into account the many possible sources of error in the activity of collecting positive ions in an electronegative plasma.[57]

D. R. Bates of Queen's University, Belfast, criticized the rocket data, even though he admitted that they were becoming easier to reconcile with theory than older ground-based indirect studies.[58] He focused on several of the greatest concerns voiced at the conference, such as J. C. Seddon's observations of the distribution of electron densities, which were confined to a single 29 September 1949 flight that returned data "reliable only within a factor of 2."[59] Seddon had argued that one fact stood out, well above the uncertainties in their data: they had detected a great concentration of electrons at the base of the E layer. Bates, however, could not reconcile Seddon's results with the accepted theory of dissociative recombination for the formation and destruction of ions that he and Massey had developed. He admitted that "the concentration inferred by Seddon is not completely excluded by [Bates and Massey's] work but it necessitates certain assumptions which are so extreme as to be scarcely credible."[60] He argued that an extremely high ionization rate was required to support Seddon's conclusions, and that no mechanism for it was then known.[61]

Adding to Bates' critique was the point that the NRL data went against the dominant Chapman discrete layer model, which required relatively low concentrations just where Seddon had found high values. Although nothing happened explicitly at Oxford to change anyone's views of the nature of the atmosphere and the stratification of the ionosphere, Seddon's data were hard to refute. But much of what was presented there also supported the Chapman model; all the rocket experimenters worked within its framework,

finding dips and peaks in the ion distribution that, although not consistent, kept to the general picture.[62]

But Oxford did act as an important watershed to bring two distinct communities together. At the time, Massey, Boyd, and Bates were beginning to plan for sounding rocket activities in Britain, and Chapman was anxious to establish a strong international agenda for ionospheric research during the IGY.[63] They benefited greatly from close contact with the comparatively affluent American rocket experimenters who had spared little expense to solve the myriad technical problems that now might not have to be repeated from scratch in Britain.

The members of the UARRP also found what they needed most: a critically receptive audience that could take their work seriously, evaluate it fully, and recommend refinements or corrections to methodology or analysis. This need may have been greatest at Michigan. Based on what happened at Oxford and the opportunity there to at last encounter and compare the thoughts of a wide range of ionospheric researchers, Dow recommended to both Gunnar Hok and Nelson Spencer that they take pains to reexamine their data and techniques carefully and to devour the recent literature on the ionosphere.[64] Accordingly, Spencer built stronger ties with the Physics Department on the university campus, and searched for ways to integrate his program into the university curriculum "as an educational facility of the EE Department."[65] Spencer's motives may have been as much political as they were practical. But whatever the motive, he now knew that, in addition to building instruments useful for ionospheric physics, they had to understand and make a concerted attempt to contribute to knowledge of that medium as well.

A new visibility also emerged from the Conference. The *New York Times* reported that "eighteen high-ranking rocket scientists from the United States" had been invited to Oxford to present their results.[66] At the 36th UARRP meeting, held partly to evaluate the conference, Van Allen concluded that as a forum, Oxford provided the most comprehensive summary of work thus far. And Chapman, attending the UARRP meeting en route to a new position at the University of Alaska, added that the conference had stimulated interest in the United Kingdom for initiating sounding rocket research in atmospheric physics.[67] He knew, along with Homer Newell, that ionospheric theorists like Bates were still hesitant to accept the rocket data, but panel members all felt satisfied that they had been given their day in court and that as a result more attention would be paid to incorporating rocketry into geophysical research. At least, according to Newell and Van Allen, many of the foreign theorists seemed to warm to the "UARRP atmosphere." They were gratified that "the Oxford Conference had aided a great deal in bringing such matters to general attention."[68]

## A New Picture Emerges

Although the rocket groups worked in the territory defined by the model created by Chapman, it was their data that would soon bring about what they thought was a significant change, both in their own thinking and in a new model created by a student of Joseph Kaplan at UCLA.

In the wake of the Oxford Conference, NRL staff pooled their resources to take a fresh look at the structure of the ionosphere. Seddon, with John E. Jackson, Ralph Havens, Herbert Friedman, and E. O. Hulburt, attacked the problem on two fronts. Seddon and Jackson reevaluated their data, while the others determined the specific solar sources causing ionization in the E and F regions, using their own data for the altitude variation of solar ultraviolet and X-ray radiation to reconcile their measurements of pressure, temperature, and atmospheric composition.

By April 1954, Jackson, analyzing radio propagation data from a recent Viking flight, concluded that a large dip in ion density that Seddon had found at 145 kilometers on a previous flight, thought to fit a region of low ionization between maxima, was spurious. By June, Seddon modified his ion profiles to agree

with Jackson and now expressed doubt in the reality of the $F_1$ "layer," which he put in quotation marks because the general appearance of the profile (without the dip) now indicated the presence of a "thick continuous region" where the layer itself was "merely a low-gradient shelf."[69] This doubt strengthened to a conviction by September when Seddon and Jackson completed a detailed reevaluation of electron densities to 200 kilometers. Chapman's "layers" were gone, replaced by "regions" that now reflected Seddon's and Jackson's view that "the ionosphere above 100 km is a heavily ionized continuum": "The idea of separate layers must be replaced by the more appropriate concept of adjacent $E_1$, $E_2$, $F_1$, and $F_2$ regions blending gradually into one another."[70]

What David Bates found difficult to reconcile with theory in August 1953 found a greater voice in September 1954. Stimulated by the precedent set by the Oxford Conference, the British Physical Society sponsored a conference on the physics of the ionosphere at Cambridge, in which the NRL group participated.[71] Havens and Friedman represented NRL and presented their new quantitative arguments identifying the specific portions of the solar spectrum that were responsible for ionization in the various overlapping regions. With this, they derived an electron distribution profile that agreed with Seddon and Jackson's new analysis.[72]

NRL's combined attack on the ionosphere paralleled almost exactly the sole effort of Hilde Kallmann-Bijl, working both as a graduate student at UCLA and as a RAND staff member. She in fact had the idea first and was aided by Kaplan and her RAND contacts, but she could not obtain NRL's data as quickly as NRL members could to elaborate the new picture. Although the majority of her work appeared in internal RAND documents, she also attended APS and AGU meetings in 1952 and 1953, where she presented some early results. Her first published statement of her new model appeared in April 1953, where, using Tousey's data for the variation of solar intensity with altitude, along with rocket data that allowed her to determine how composition and molecular weight varied with height, she found a continuous ionospheric electron distribution. This paper gained immediate praise from Fred Whipple, who wrote her stating that it was "*THE* paper on the ionosphere," but it garnered little public attention.[73]

Nevertheless, upon completing her Ph.D. thesis in December 1954, just before Seddon's and Jackson's papers appeared in the *Journal of Geophysical Research* (JGR) and before the publication of the proceedings of the Cavendish conference, Kallmann had come to the same conclusions and stated them in no uncertain terms. Using improved recombination and dissociation rates for atmospheric constituents derived by Nicolet, Bates, and others, together with the latest published rocket data, Kallmann created a theoretical profile for electron concentration. She then showed that her theoretical profile matched the revised density profiles Seddon and Jackson published in December along with Warren Berning's results from Bumper-WAC. Her model supported a "continuous increase of electron density with altitude" and led her to argue, "It is not necessary to imagine the ionosphere to be formed by . . . layers of electron densities with zero electron densities between them."[74] In a postscript, written after she had read Seddon's revised paper in the December JGR, Kallmann found more areas in which her theory predicted NRL's experimental results. She concluded: "Thus the agreement between the theoretical and experimental results is rather good indeed."[75]

NRL was quick to claim discovery of the continuous distribution and maintained the claim over the next several years.[76] Kaplan did the same for Kallmann's work, getting an early endorsement from Sydney Chapman that Kallmann's work was more complete. While in Iowa visiting Van Allen, Chapman read her thesis and quietly approved it in late January 1955. He told Kaplan that it was the first statement he had read showing his own model to be an oversimplification. But he argued as well that "the idealized theories still have a place" and have defined a phase in ionospheric research "which will not soon be outgrown."[77] Chapman reminded Kaplan that:

*The picture of the ionosphere as a superposition of Chapman layers is one that has never been thought of by any well-instructed person as more than a crude approximation. But it is good to have a more realistic version of the ionosphere in the E and F (tho' not $F_2$) regions.*[78]

Chapman, along with other "well-instructed" students of the ionosphere and radio propagation, like Frederick E. Terman and Lloyd V. Berkner, certainly knew the limitations of the idealized model. The rocket observations did not change their minds, they merely confirmed their suspicions. They also knew that stratification was a hard concept to shake among radio propagation specialists who were not in the mainstream of ionospheric physics. Some, like those at NRL who conducted the observations, were very excited by Kallmann's work.[79] But at best these efforts were regarded as confirmations, not as discoveries, and thus were quickly forgotten.[80] The real payoff came when those in control of ionospheric physics acknowledged that, after 1955 at least, rocket and satellite studies of the ionosphere were necessary components of future research programs.[81] Although the rocket scientists may have remained marginal through the mid-1950s, their technology took hold.

## The Marginality of the Effort

The Oxford conference helped to increase the visibility not only of the techniques of scientific rocketry, but of the results. At the time of the conference, ionospheric groups using rockets were highly marginal to both the domestic and international ionospheric physics communities. Among the some 1,500 scientific and technical papers produced by more than 100 groups identified in Laurence A. Manning's *A Survey of the Literature of the Ionosphere* in 1955, less than a dozen referred to observations from instruments on rockets.[82] Few radio propagation specialists, moreover, cited the work of the rocket groups during this time.[83]

The initial marginality of rocket-based ionospheric research can be appreciated using almost the same arguments we have identified for those who pursued solar spectroscopy or cosmic-ray research. The NRL and Michigan ionospheric groups, both well-versed in practical radio engineering and vacuum-tube technology, best fit in the community devoted to the practical improvement of radio propagation techniques. But unlike the dominant community using ground-based radio to probe the ionosphere, the rocketry groups were faced with establishing a new technique, one that required highly focused interests.

The sheer technical complexity faced by the rocketry groups demanded that they devote all their energy to making the experiments work. After a two-year reconnaissance of the territory, in 1949 RAND meteorologist William W. Kellogg found little of value beyond potential. He knew of the NRL program, but remained unaware of what was being attempted by the others.[84] Kellogg was a conscientious and energetic student of the ionosphere; he sensed in 1949 that the rocket researchers he interviewed were devoted more to solving technical problems than to providing data. Homer Newell much later pointed to the early obstacles faced by Seddon and Spencer as especially challenging. The techniques they adapted harbored sources of error that took many years to overcome. In the end, Newell argued, it was worth it, but the process did not encourage participation by traditional ionospheric researchers.[85]

The Oxford conference therefore marked the occasion on which established physicists like Chapman, Boyd, Bates, and Massey confronted the efforts of the rocket groups. Suitably stimulated by this attention from British physicists, the rocket groups that had assimilated the picture of the ionosphere created by the British community were now faced with constructive criticism by its leading lights. The discordance the earliest rocket data posed was first rationalized as error in the observations rather than in the model. Only after completely reevaluating their data and technique, stimulated in part by competition from a Ka-plan student, did the rocketry groups attempt to alter what they believed was the dominant model of the ionosphere.

# Notes to Chapter 16

1. Dow to D. R. Bates c/o William H. Pickering, 21 March 1950. M824 files, "technical," WGD/BHLUM.
2. Chapman examined how solar radiation would be absorbed by an isothermal atmosphere composed of atomic and diatomic oxygen and ozone. Breit and Tuve's work is discussed in the context of NRL's early interests in ionospheric research by Allison (1981), pp. 56ff.; in the context of the research agenda of the DTM by Dennis (1990), pp. 147–51; and in the context of the history of ionospheric physics by Gillmor (1981), pp. 103–05.
3. See Terman (1937), p. 600; and Berkner (1939), p. 443.
4. By the late 1930s, through the efforts of Lloyd V. Berkner, DTM was providing a practical systematized multifrequency radio sounder to observatories world-wide. Needell, Berkner draft biography, chap. 2.
5. Gillmor (1981), pp. 102; 106–08, has discussed the seductiveness of the visual radio trace, supporting the concept of discrete layers. He cites a letter from Appleton to Ratcliffe in 1932 cautioning against the literal acceptance of layers, suggesting the use of the more general term "region." But apparently in the 1940s, even Appleton recalled thinking of the layers as "discrete strata," according to Silberstein (1959), p. 382. On the range of thinking in the 1930s, see Berkner (1939), pp. 450–52 for the view at DTM, which regarded layering only as an idealized model. On the other hand, Harvard's Harry Rowe Mimno (1937) treated stratified layering as an experimental fact. See also Nicolet (1968), pp. 35–37; Bates (1968), pp. 31–34.
6. The JRDB Panel concluded, "The ionosphere has a profound effect upon radio propagation and terrestrial magnetism; it may also be considered as a gigantic laboratory on the fringes of the atmosphere, wherein solar and cosmic radiation may be studied without hindrance from atmospheric absorption." From "Preliminary Statement of Objectives in Upper Atmospheric Research," 10 April 1947, p. 5. Annex "A" to the Agenda for the third

meeting of the Geophysical Committee. UAT 19/2 (Panel on the Upper Atmosphere), JRDB/NARA. This statement was endorsed by the JRDB Committee on Electronics, its Basic Research Panel, and the Committees on Geophysics and Guided Missiles.

7. See CRPL statement on UAR needs (by Newbern Smith, NBS), Annex G, 12 March 1947 in "Minutes, Second Meeting of the Panel on the Upper Atmosphere," 20 March 1947. UAT 3/2, Box 239, folder 3, JRDB/NARA. On the NACA committee, see Chapter 15, and "NACA Special Subcommittee on the Upper Atmosphere," meeting of 17 September 1948. NACA files, HN/NACA/NHO. Harry Wells, an ionospheric physicist at the Department of Terrestrial Magnetism, standing in for Lloyd Berkner at the NACA meeting of this date, argued strongly that rocket observations might shed light on sporadic phenomena in the E layer. Even though the CRPL was supportive, it apparently rejected an early request to fund ionospheric research with rockets. See William Dow to the author, 13 August 1990, p. 44.

8. See Ripley, (1955), pp. 86–87.

9. Gillmor (1981), pp. 101–02, has identified three overlapping communities in ionospheric research: radio engineers who probed the ionosphere in order to find means of improving medium- and high-frequency transmission techniques; atmospheric physicists using nonelectronic means to better understand the structure and dynamics of the upper atmosphere; and physicists who explored the use of radio as a tool to see how well propagation experiments fit their mathematical and physical models of the upper atmosphere.

10. Interested in all aspects of correlated magnetic activity, the DTM best represented how ionospheric research was pursued in the United States. Starting in 1935, DTM called together representatives from government, military, university, and industrial laboratories, believing that the study of geomagnetic influences on radio propagation was a worthwhile practical effort. At the fourth conference in 1938, it was suggested that "large commercial organizations would not be spending time and money in studying the effects if it had not been found that information derivable from magnetic data is of definite engineering value in radio communication." "Report on Fourth Conference on Ionospheric Research, Washington DC, April 30, 1938," p. 1. "Ionosphere Conference Folder," (folder 51) in Cabinet 86882, DTM attic. See also Lloyd V. Berkner to Charles R. Burrows, 9 January 1948. UAT 43/1, Box 240, JRDB/NARA.

11. Remarks of Joseph Kaplan, cited in "Minutes of Meeting of the NACA Special Subcommittee on the Upper Atmosphere, Committee on Aerodynamics," 17 September 1948, p. 9. NACA files, HN/NACA/NHO.

12. William W. Kellogg, "Preliminary Proposal for Research on the Structure of the Stratosphere and Ionosphere" (n.d.), attached to Joseph Kaplan to Wexler, 28 October 1947. HW/LC.

13. Kellogg added that all available means of sensing the ionosphere, including rocketry, had to be correlated and evaluated; nothing could be ignored. Ibid., p. 2. In his evaluation, Kellogg viewed the upper atmosphere as a problem for operations research analysis, RAND's raison d'être. On operations research at RAND, see Smith (1966), p. 65. The origins of this connection require further attention, especially the role of the RAND Corporation in planetary atmospheres research.

14. Lien et al. (1954) identify the types of experiments devised in the AFCRC and the Air Materiel Command.

15. Seddon recorded his design in his registered laboratory notebook no. 304, p. 70B. HONRL. According to Charles Johnson, Seddon developed the basic technique of pulse beats for the NRL team. See C. Y. Johnson OHI, no. 1, p. 34rd. See also Burnight and Seddon (1947), and J. Carl Seddon "Record of Consultative Services," 12 March 1946. Box 98, folder 1, NRL/NARA.

16. "Record of Consultative Services," ibid. The critical frequency, also known as the penetrating frequency, of a medium is that frequency below which radio waves are reflected by an ionospheric layer, and above which radio waves are able to penetrate vertically.

17. Although many ionospheric radio propagation specialists used the term "reflection" they operated under the model that understood the process as one of refraction. In choosing these frequencies in this manner, Berkner knew that the unaffected high frequency would experience an index of refraction of unity, whereas the lower one would approach an index of

refraction of zero. As the rocket traveled through the ionized layer, the velocity of propagation of the lower frequency would be greatly affected, creating a constantly changing phase beat signal between the two, as seen from the ground. Analysis of this phase beat signal would then provide the average index of refraction for the lower frequency as a function of altitude. Berkner had been sought out by Seddon for advice. J. Carl Seddon "Record of Consultative Services," 27 March 1946 (13 March meeting). Box 98, folder 1, NRL/NARA. On Berkner's early interest in the use of satellites for communications, see Hall (1977), p. 254, n. 9.

18. J. C. Seddon registered laboratory notebook no. 304, p. 67, circa February 1945. Project A140R-S. HONRL. Seddon, with Shipman, Davison, and Anderson, was trying to determine how a hot gas-air jet affected reception at various frequencies.

19. In March 1946, he, Milton Rosen, and C. H. Smith visited the Naval Ordnance Test Station in Inyokern to examine facilities for such experiments, and, as a result, NRL asked the Bureau of Ordnance to establish an NRL test unit at NOTS for this purpose. Seddon's group also conducted in-flight tests with radio transmitters to compare with the NOTS data. This work was classified. [D. W. Atchley] to Chief of Naval Operations, 10 February 1945 [receipt date], p. 8. Box 44, "Jet Missiles Volume 2," RG 298/NARA; Capt F. W. MacDonald to Chief of the Bureau of Ordnance, 15 April 1946. NRL/NARA. On classification, see Code 3420 to All Section Members, 2 September 1948. Box 100, folder 16, NRL/NARA.

20. J. Carl Seddon "Record of Consultative Services," 27 March 1946 (13 March meeting). Box 98, folder 1, NRL/NARA.

21. The NBS gear was secured after the NRL team petitioned the Joint Wave Propagation Committee. Capt. F. W. MacDonald to Lt. Col. P. J. Greven (Chair, Joint Wave Propagation Committee, Joint Communications Board), 24 April 1946; E. U. Condon to NRL, 10 May 1946. Box 98, NRL/NARA.

22. The latter two problems lingered for quite some time, and at several points became contentious issues when the heavy demand for radio telemetry during a V-2 flight caused serious interference problems. See Chapters 7 and 8.

23. These scenarios are taken from Clark et al. (1946), pp. 57–58.

24. Ibid., p. 60. The trailing wires also tended to be vaporized at launch. See also C. Y. Johnson OHI no. 1, p. 34rd. Warren Berning recalled that NRL's antenna problems were, to an outsider, "almost an impossibility" to overcome, given the antenna patterns they tried to develop. Berning to the author, 25 June 1990.

25. The use of models is noted by Krause (1946), p. 146; and in Burnight, Fry, and Seddon (1947), p. 100. Seddon's registered laboratory notebook no. 304, pp. 72–73, records the process through which the trailing wire antennae were shown to be poor, and his 15 April 1947 entry identifies the inverted V-2 as a mock-up built in the late spring of 1947: "It is hoped that a dummy V-2 will be erected nose-down at WSPG so that 'free-space' impedances can be measured." Ibid., p. 74. HONRL.

26. Signal frequencies of 25.644 megacycles and 4.272 megacycles were received by two ground stations and produced a phase beat frequency that indicated the rate of change of the phase difference between the two transmitted frequencies as a function of height in the atmosphere. Burnight, Clark, and Seddon (1947), p. 85. NRL had the assistance of specialists at BuShips, White Sands, the NBS, and Aberdeen. On the detailed design of the electronics and antennae, see Clark, Miller, and Spaid (1947), pp. 100–02.

27. The phase beat frequency data were recorded by motion picture cameras directed onto the faces of oscilloscope screens in the receiving stations. The NRL White Sands field team still found that the impedance of the antenna system changed constantly; its value in the assembly building was different from what was measured from it when the V-2 was being carried to the launch site on its German Meillerwagon, or when the V-2 was standing upright on the launch pad. But they also found that these effects were repeatable, which gave them confidence in the design. Burnight, Fry, and Seddon (1947), p. 97.

28. After Krause's departure, the subsection consisted of Burnight, Seddon, Strain, Bourdeau, and Spaid. This was less than a third of their original manpower, considering that Strain and Burnight had other responsibilities. J. C. Seddon registered notebook no. 304, pp. 74, 92,

entries for 19 September 1947 and 28 August 1952. HONRL.

29. Both Krause and then Newell wanted the ionospheric work to look like routine drill, wishing to establish the fact that vehicles like the V-2 and its trimmer replacement, the Viking, were appropriate for research activities useful for the fleet; above all, the most practical application lay in the realm of radio propagation for missile guidance and control, as was emphasized in Chapter 10. The sense of standardization and normalization derives from inspecting how the NRL technical reports changed in structure and delivery from late 1946 through 1950. By 1948, only new instrumentation was discussed, and they concentrated on improving the V-2 as a research vehicle. See, for instance, Newell and Siry (1948), p. 29ff.

30. In March 1950, Seddon reported that the problem of calibration "has been an expensive one." It was justified if the procedure had to be done once, but "with rocket diameters changing and radio interference causing frequency shifts, the contracts would be neverending." J. C. Seddon, "Consultative Services Record," 10 March 1950. Box 101, folder 27, NRL/NARA.

31. J. C. Seddon registered notebook no. 304, p. 92, entry dated 28 August 1952. HONRL.

32. See J. C. Seddon registered notebook no. 8344, starting on page 5. HONRL.

33. Berning (1951, p. 262; 1954).

34. In what was inherently a complicated process fraught with both instrument error and many unknowns in the ionizing medium, Berning had apparently not taken into account what effect the rocket's departure from a vertical trajectory had on the results, in light of the simplifications that he had made. See Newell (1953), p. 220. Berning (1951).

35. Berning (1954), p. 262.

36. Ibid., p. 273.

37. Hok, Spencer, and Dow (1953), p. 235.

38. Among the many practical problems involved with measuring ion or electron densities in this fashion from a rocket is that of determining the reference potential. The rocket surface itself constantly encounters the ionized realm, and therefore the entire system adjusts its reference potential as it flies. Dow's group attempted to solve this problem by building a bi-polar circuit between the probe and the missile, using the body of the missile as a ref-

erence. For a full description, see Hok, Spencer, Reifman, and Dow (1951); Hok, Spencer, and Dow (1953); and Hok and Dow (1954). Their technique is also discussed in Dow to the author, 23 September 1990. APL, Michigan, and NRL developed mass spectrometers, gas samplers, and a rank of thermal ionization density gauges to gather corroborating evidence. Dow's planned program for 1948 included at least two and possibly three AMC flights of pressure and variable voltage probe instrumentation, shock-wave angle detectors, and a small mass spectrometer. See [N. W. Spencer], "Program of Work for 1948," 27 February 1948. UM UAR folder, WGD/BHLUM.

39. Hok and Dow (1954); Hok, Spencer, and Dow (1953); Dow to the author, 13 August 1990, p. 45.

40. Hok and Dow (1954), p. 243. The first discussion of the probe is in Reifman and Dow (1949a, 1949b). During the 1946 and early 1947 flights, F. V. Schultz of the Michigan research staff worried about influences from sporadic ionospheric activity. He asked Newbern Smith of the National Bureau of Standards for help, but Smith could not provide unequivocal evidence that could explain the confusing data they had obtained. F. V. Schultz (Res Eng UM) to N. Smith, NBS, 16 October 1947. WGD/BHLUM.

41. Their 20 February 1947 AMC Blossom flight, for instance, was most frustrating. O'Day had also arranged for mortar detonations for sound and smoke experiment tests, and other environment-altering experiments, including the ejection of the warhead. The missile spun rapidly, and the ionosphere beacon and much of the telemetry failed. Although Dow reported in March that their Langmuir probe and ionization gauges worked, they were unable to analyze the data. "V-2 Report no. 9," 31 March 1947, p. 8. V2/NASM; and Dow to Spencer, 5 June 1947, clipped to Spencer to Dow n.d. WGD/BHLUM.

42. Nelson W. Spencer to R. G. Folsom, W. G. Dow, and M. H. Nichols (Project 2096 Advisory Committee), 4 December 1953. Box 4, WGD/BHLUM. The AFCRC contract project under Spencer was coordinated by the University's Research Institute and the Electrical Engineering Department. It was for research on winds, temperatures, and related "Properties of the Atmosphere and Ionization

in the High Atmosphere" at a rate of $75,000 per year 1952/1953.

43. See Dow to Cmdr., AFCRC (Attn W. Pfister), 14 October 1955. Project 2096 folder, WGD/BHLUM.

44. Dow to D. R. Bates, c/o William H. Pickering, 21 March 1950. M824 files, "technical," WGD/BHLUM.

45. Ibid.

46. "Panel Report no. 32," 30 April 1952, p. 13. V2/NASM. See also Massey and Robins (1986), pp. 6–7.

47. Gassiot, a wealthy wine merchant, was the original patron of the committee. See Massey and Robins (1986), pp. 4–6.

48. All panel member papers had to be sent through military censors. See "Panel Report no. 35," 29 April 1953, p. 3. V2/NASM.

49. Ibid.

50. "Panel Report no. 33," 7 October 1952, p. 14. V2/NASM.

51. Dow reported to a contact at General Electric, "During the week beginning August 24, I will be at Oxford, where the Upper Atmosphere Research Panel of our Armed Services is meeting with a similar British group. The rest of my time in England will be spent chiefly in visiting people who are interested in microwave tubes and tube reliability." Dow to H. C. Steiner, 18 May 1953. He told another GE officer that these visits were to inform his GE contract activities and that he would have military clearance as an official representative of the Signal Corps. Dow to Guy Suits, 18 May 1953. See also Spencer to Dow, 17 December 1952; Dow to H. Landsberg, 3 April 1953; and A. L. Samuel, RDB Panel On Electron Tubes, to Dow, 7 April 1953. WGD/BHLUM.

52. S. Fred Singer OHI no. 1, and Papers. Box 1108, no. 2, ONR London file, SFS/NASM.

53. "Conference on Rocket Exploration of the Upper Atmosphere," Oxford agenda folder dated 24 August 1953. WGD/BHLUM.

54. Ibid.

55. For example, E. Vassy of the University of Paris commented briefly on the ozone results of Tousey's group, presented by F. S. Johnson. Although he agreed in general with their methods and reduction techniques, he suggested that they might improve their analysis by using more recent absorption coefficient estimates and contemporary terminology. Both Sydney Chapman and Julius Bartels of Göttingen commented on the geomagnetic observations of Singer, Maple, and Bowen, observing that their rocket flights took place during unusual geomagnetic conditions. Sydney Chapman talked at length about additional ways that rockets might explore currents in the earth's magnetic field, building on Singer's Oxford review of their detection of a concentrated equatorial current. See relevant chapters by Vassy, Chapman, and Bartels in Boyd and Seaton (1954). The work of Singer, Bowen, and Maple on the equatorial electrojet was discussed in Chapter 14.

56. Boyd (1954), p. 336.

57. Ibid., p. 338. Boyd argued that the potential of the rocket would vary not only with the ambient ionization encountered, but with increased frictional heating and electrification. Even though Dow and his group had designed their bi-polar probe to account for such problems, Boyd still worried that the varying potential of the rocket created a situation that severely limited the effectiveness of the Langmuir probe. One of Boyd's suggestions for further development of ion probes was that "closely integrated experimental and theoretical research into the operation of probes in plasmas in thermal equilibrium in the laboratory would be most valuable." On Boyd's interests, see Massey and Robins (1986), p. 17.

58. Bates (1954), p. 355.

59. Seddon (1954a), p. 222. Although Seddon was listed as the author of this report, he was not listed as one of the attendees at Oxford. Havens, Hulburt, F. S. Johnson, LaGow, and Newell attended. Presumably one of them read the report. Entries in Seddon's registered laboratory notebook during this time (no. 8344, pp. 2A through 3) reveal that he was working on signal generators and amplifiers for radio propagation experiments. HONRL.

60. Bates, with Harrie Massey in 1947, developed a theory of dissociative recombination in the upper atmosphere and was then continuing in a widening range of ionospheric studies that built on that theory. See Bates and Massey (1951); Bates (1954), p. 347; and Massey and Robins (1986), pp. 3–4.

61. In his critique, Bates did not consider other hints of a negative ion excess, specifically how the photon counter observations of the solar ultraviolet and X-ray flux by NRL's Byram, Chubb, and Friedman, reported as well at the

Conference (see Chapter 13), altered the picture. Friedman felt that his observations could account easily for all E-layer ionization and that enough was left over in the X-ray spectrum to support "a proportionally high ratio of negative ions to electrons." Byram, Chubb, and Friedman (1954a), p. 275.

62. When they found agreement between regions of high ion concentration and the location of the accepted Chapman layers, those like Berning, Lien, and Seddon were happy to point out how their efforts yielded evidence of ionospheric structure, in addition to the primarily practical aims of the research. See, for instance, Lien, et al. (1954); Berning (1954); and Seddon (1954a).

63. See Massey and Robins (1986), chaps. 1, 2, and 3. Oxford-style conferences were repeated the following year in Cambridge, and in 1956 at an AFCRC-sponsored conference on chemical aeronomy, where the Air Force brought together aeronomists, rocket scientists, atmospheric physicists, and aerothermochemistry specialists to address problems of human space travel. See Zelikoff (1957), and [Physical Society] (1955).

64. Dow to Spencer and Hok, 20 July 1954. Box 4, Project 2096 folder, WGD/BHLUM.

65. Nelson W. Spencer to Professors R. G. Folsom, W. G. Dow, and M. H. Nichols, 4 December 1953. Box 4, Project 2096 folder, WGD/BHLUM.

66. Highlighting an interview with Van Allen, the *Times* reported that "twenty tons" of instruments flown on the German V-2s yielded the scientific results that were being discussed and that the activity would continue with American Vikings, Aerobees, and Van Allen's Rockoons. "German V-2 Rockets Used Up in U.S. Tests," *New York Times*, 25 August 1953. Fragment in NASM Technical Files.

67. "Panel Report no. 36," 7 October 1953. V2/ NASM. This had been, of course, both Chapman's and Van Allen's ulterior motive all along. See Chapman (1954). On the influence of the conference, see also Massey and Robins (1986), pp. 6–8, 18.

68. Ibid., "Panel Report," p. 3. Megerian recorded in the UARRP minutes that "Professor Chapman indicated that there was general acceptance of the UARRP atmosphere up to 100 km, but there was still some reservation in accepting the higher altitude (and admittedly less reliable) data."

69. Seddon (1954b), p. 463.

70. Seddon, Pickar, and Jackson (1954), p. 523.

71. S. A. Bowhill (1974), p. 2236, has as well argued that the Oxford Conference "was the ancestor of the series of annual meetings of the Inter-Union Committee on Space Research (COSPAR)."

72. Havens, Friedman, and Hulburt (1955), p. 237; Seddon, Pickar, and Jackson (1954), p. 523.

73. Fred Whipple to H. K. Kallmann, 28 April 1953. Box 3065, folder 20, HKK/NASM. Kallmann (1953); Tousey, Watanabe, and Purcell (1951).

74. Kallmann (1955), pp. 54, 62.

75. Ibid., p. 64.

76. See Leak (1954), pp. 20–23; Townsend, Friedman, and Tousey (1958), p. 11.

77. Sydney Chapman to J. Kaplan, 24 January 1955. Box 3065, folder 6, HKK/NASM.

78. Ibid.

79. In spite of Terman (1937), p. 600; and Berkner (1939), p. 443, in 1957, Louis Gold, a staff member of MIT's Lincoln Laboratory, believed that he had discovered the continuous nature of the ionosphere, finding no indication of prior detection even after "an intensive search of the literature." Upon hearing indirectly of Kallmann's earlier work, he wrote her stating that all recent papers he had read in the *Proceedings of the Institute of Radio Engineers*, and papers presented at a recent URSI meeting, left him with the impression that the "belief in stratification was unshakable." Louis Gold to H. K. Kallmann, 4 October 1957. Box 3065, folder 20, HKK/ NASM.

80. Gillmor (1981), pp. 110–12, has presented the two views prevalent in the 1970s that, on the one hand, gave credit to the rocket observations for changing the picture, and on the other, claimed that the new picture was largely in place by 1955 and that rocket and satellite observations were merely confirmations. He argues that both views were consistent with the backgrounds of those holding them, which is confirmed here: Beyond mention by Kaplan at various conferences, Kallmann's papers announcing her new model were not cited through 1964, although her general reviews of the properties of the atmosphere received about 60 citations in that time, and Friedman did give her credit in 1960. See Friedman (1960), p. 193; and Barth and

Kaplan (1957), pp. 6–8. Seddon's paper with Jackson in December 1954, although it garnered two dozen citations through 1961, peaked at six citations in 1959 and was cited mainly by NRL colleagues in the first two years. This was hardly a significant reaction by a rapidly growing community that was publishing some 200 papers per year in 1955, and 500 in 1960. See Gillmor and Terman (1984), p. 90.

81. J. A. Ratcliffe (1974), p. 2095, identified 1955 as the end of the period when "nearly all methods were ground-based." By 1957, one influential reviewer argued that rocket observations of oxygen composition "have solved and verified" problems that were important in determining electron densities, and that the direct techniques, like those of Seddon and Jackson, placed ionospheric physics with rockets at the "threshold of the solution of many problems concerning the lower ionosphere which ionospheric physics have sought for many years." Waynick (1957), p. 749.

82. Warren Berning's contributions to knowing the charge density in the vicinity of a V-2 were noted, as were Seddon's contributions to knowledge of propagation characteristics. Members of Dow's group were cited, too, but primarily for their establishment of the dynamic probe as a potential means of measuring ion densities. Manning's review (1955), an update of a 1947 Stanford University survey of activities relevant to ionospheric radio propagation, was reportedly confined to research bearing only on radio observations. The light treatment of rocket work thus demonstrates its marginality. Although the AFCRC facility was identified as one of the institutions engaged in ionospheric research, no citations were found that dealt directly with contributions from rockets.

83. Based on citation studies using the *Science Citation Index, 1945–1954* (1988), Volumes 1, 4, and 6. As a member of a radio propagation group at AFCRC, J. R. Lien's work was cited prior to 1955 in the radio engineering literature, but Seddon's and Berning's studies were not. A 1957 review by A. H. Waynick, of the Penn State Ionospheric Research Laboratory, included rocket data, primarily pressure, temperature, density, and composition profiles. He felt that the Seddon and Jackson electron density data were highly promising, but only at the threshold. Waynick (1957).

84. According to T. R. Burnight, "Record of Consultative Services," 8 July 1949. Box 100, folder 23/24, NRL/NARA.

85. Newell (1980), chap. 6, pp. 78–79. Donald Menzel and William H. Pickering are good examples. Pickering was well placed at the Jet Propulsion Laboratory and Caltech as a member of the UARRP and maintained strong interests in many aspects of upper atmosphere and ionospheric research problems. But he confined himself to ground-based radio reconnaissance of the ionosphere in the early 1950s. Bergman, MacMillan, and Pickering (1954). On his return to Harvard from active duty, Menzel maintained an interest in ionospheric physics and radio propagation, but never considered exploiting his considerable connections to pursue rocketry to that end. See, for instance, Lt. Cmdr. D. H. Menzel to Capt. H. T. Engstrom, n.d. (circa October 1945). DHM/HUA.

*The rocket sonding program generates new techniques, instruments, and skilled personnel. . . . These often have important practical applications.*

—Position Description, Branch Head, Rocket-Sonde Branch
Atmosphere and Astrophysics Division, NRL, October 1955.[1]

Part 2 of this book has shown that the various groups devoted to upper atmospheric research with rockets created a new way to do science. They did so by overcoming technical and conceptual obstacles, migrating to those areas best suited for study from rockets, gaining attention for their techniques, and establishing the authority of their hard-won data. Thus, by the onset of the International Geophysical Year (IGY), scientific rocketry had taken its place alongside the traditional sites of upper atmospheric observational research in the geophysical sciences: the laboratory, observatory, and field station. The pioneer rocket groups demonstrated that this new mode of scientific research had to be embraced. To remain competitive, institutions interested in upper atmospheric research had to employ the technology associated with rocketry, which had been made possible and was ultimately defined by patterns of military patronage.[2]

Just how did members of each group get the word out about what they were capable of doing? Two factors in particular played a critical role. First, UARRP members became agents for IGY planning, and second, the identity of each member gradually changed as he chose new audiences in the 1940s and 1950s.

## Planning for the IGY

The idea of an IGY was the result of many parallel interests converging from international relations, national security, and the geophysical sciences. Each of these perspectives has been examined elsewhere, but as yet little effort has been made to integrate them in history.[3] Lloyd Berkner seems to have had the idea first, while listening to a conversation during a dinner party in Sydney Chapman's honor given by James Van Allen in April 1950. Chapman was looking for ways to coordinate geophysical research worldwide, and Berkner suggested a formal approach, as was done for the International Polar Years of 1888 and 1932.[4]

By the end of 1951, after the International Council of Scientific Unions endorsed the plan, Chapman and Berkner, along with the Belgian atmospheric physicist Marcel Nicolet, became president, vice-president, and secretary general, respectively, of the Special Committee for the International Geophysical Year (CSAGI).[5] In late 1952, the National Academy appointed a U.S. National Committee for the IGY, headed by UCLA's Joseph Kaplan and managed by Hugh Odishaw, which lobbied for support and set policy for the distribution of that support.[6]

Following the August 1953 Oxford Conference, both Chapman and Kaplan encouraged the UARRP to start organizing for the IGY. Whipple advised that "because of [the] necessity of obtaining funds and the necessity of a long period of preparation," they had better make some definite moves now.[7] As a result, Van Allen attended formative IGY meetings in Washington, representing the UARRP, which hoped that it would soon be designated an official coordinating body for upper atmospheric research with rockets. This did not happen, even though many UARRP members became active on the plethora of committees and panels that existed by 1955, where they lobbied both for an active rocketry program and a role in defining a satellite program.[8] The panel created to handle UARRP interests, the IGY Technical Panel on Rocketry, included Fred Whipple as chairman, Newell, as executive vice-chairman, and Van Allen.[9]

Although it did not become an official planning agent for the IGY, the UARRP still performed the same function, as its meetings became planning sessions for making scientific rocketry a part of the IGY. In January 1954, the UARRP was the only body able to provide rational cost estimates for a sounding rocket program. By that summer, at CSAGI meetings in Rome, Newell learned that the Working Group on Rockets agreed that IGY flights would "adhere to proven experiments in the various fields of interest" rather than try to develop new types of experiments. He found as well, since the Russians were reluctant to discuss what they had been up to, that the European members of the Working Group looked to the UARRP as a guiding agent.[10] Thus when Congress approved the fiscal 1955 budget, including IGY appropriations, the U. S. National Committee (USNC) turned to the UARRP to provide a revised budget for rocketry, accepted its estimate, and then funneled the money through the panel with the State University of Iowa acting as fiscal agent for Rockoons and NRL for Aerobees, "by common agreement within the UARRP and the USNC."[11]

These factors support the UARRP's contention in early 1955 that, even with official IGY centers residing elsewhere, the panel was still "a forum for the discussion of general scientific and technical aspects of all U. S. rocket research in the upper atmosphere, as an organizer of scientific conferences, as a correlator of scientific results, and as an advisory agency on related matters."[12] Thus, as a body, the UARRP's self-image in 1955 remained that of facilitator, coordinator, and promoter, for the scientific application of rocketry and whatever portion of satellite science it could capture.

Members of the UARRP certainly maintained special scientific interests as they planned programs for the IGY, but as before, their interests were circumscribed by the use of sounding rockets, and now, of satellites. Newell, for instance, as a member of the Special Committee for the IGY (SCIGY), and especially as head of NRL's Rocket-Sonde Branch, remained strongly devoted to supporting sounding rocket activities, while the UARRP, led by Van Allen, began to promote satellite science.[13]

Particularly revealing of UARRP member priorities as they planned for the IGY was the fact that, to honor the tenth anniversary meeting of the panel in January 1956, they decided to organize a symposium on the *Scientific Uses of Earth Satellites*, rather than a retrospective on the science they had accomplished in their first decade.[14] There, in consonance with its history, the UARRP reviewed more of the technical issues surrounding the doing of science on a satellite, rather than the science itself. The original plan for the tenth anniversary meeting, held in Michigan, was that it would be a highly visible public forum. But the national officers of the IGY felt otherwise, fearing that "a public symposium of quasiofficial nature was politically premature." The panel's entrepreneurial spirit was not appreciated by the National Committee, which above all desired control of all political issues, and the eventual U.S. satellite program was the most politically sensitive of all.[15] Nevertheless, the panel's actions gave its members the centrality they desired as they assumed a wide array of IGY responsibilities. The effect was to further heighten the importance of rocket research and broaden its application both nationally and internationally.

## Audience

As the preceding chapters amply demonstrate, the UARRP and its member agencies had good reason to align themselves with both the national and international geophysical communities. This was the culture that most warmly welcomed their existence and efforts. The process of alignment was not an easy task, however, mainly because in the early years the primary objective was to make an instrument work on a rocket; Van Allen's stated "strong back, weak mind" syndrome pervaded all areas of scientific rocketry in the first five years of effort (figure, p. 336). Only in the early 1950s did members of the UARRP attempt to establish themselves, and their chosen techniques, within the communities of science.

We can gauge the audiences the rocket scientists had in mind when announcing the results of their work by looking at the publications that reported their technical methods and their scientific results. During the first 10 years of scientific rocketry in the United States, upward of 65 people published more than once as authors and coauthors of laboratory reports or publications in the open literature. Compilations of where they chose to publish appear in tables 17.1 and 17.2, and contractor and laboratory reports are identified in table 17.3. The information is based on bibliographic surveys, contractor files, and the publication lists of prominent players. The tabulated data reveal a number of striking trends.[16]

Nearly all the early articles were concentrated in the *Physical Review*, as well as in agency and contractor reports and in technical journals. This preference for the *Physical Review* might be interpreted in a number of ways. For physicists, it remained the most prestigious journal of record and the one reaching the widest audience of traditional peers. It was the first choice for Krause, Golian, Perlow, and Van Allen. The next publication category in rank in the first five to six years were agency technical reports and a few technical journals, such as *Electronics* and the *Journal of the Optical Society of America*, which were active and enthusiastic technical audiences for instrumentation, again reflecting the realms of technical expertise inhabited by these tool-building groups. These technical papers are among the most highly

cited of all publications at the time. A 1951 paper in the *Journal of the Optical Society of America* by F. S. Johnson, Tousey, and K. Watanabe, on a means to sensitize photoelectric detectors for the ultraviolet, garnered more than 100 citations through 1964.[17]

Together with laboratory and contractor reports, many consisting of chapters by laboratory members, technical reports far outweighed the attention given to scientific results. If we were to add interim laboratory and contractor reports, the number of technical reports would be increased by a factor of two. Without doubt, scientific research with rockets was a highly specialized technical activity requiring subgroups to solve practical problems in electronics, communications, mechanical and optical fabrication, data processing, and logistics. This was, in itself, not much of a change from the way group efforts in the observational or experimental sciences were pursued in the past, but the scale of the enterprise and the very unusual technical obstacles that had to be overcome focused most of the attention within the rocket groups on problems in mechanical structures, optics, and electronics. No wonder then that the most common publication, in the aggregate, was the technical report, or descriptive review in a related technical journal like *Electronics* and *Journal of the Optical Society of America*. Insofar as they were interested in reaching a wider audience, they felt that the *Physical Review* would suffice.

After 1949—1950, attention to geophysical journals increases until, in the mid-1950s, the majority of scientific papers begin to appear in geophysics-related publications. This gradual redirection of attention toward geophysics was certainly stimulated by the enthusiastic reception given members of the UARRP at Oxford, largely by geophysicists, and the increased attention of a receptive geophysics community stimulated and defined by the IGY. Citations to all of Van Allen's publications show, in particular, an early clustering in physics journals, which then becomes dwarfed by citations in geophysics-related journals starting in 1959.[18] The "Rocket Panel Atmosphere," published in the

TABLE 17.1  Publication Trends, Scientific Journals, 1946–57

| Journal | 1946 | 1947 | 1948 | 1949 | 1950 | 1951 | 1952 | 1953 | 1954 | 1955 | 1956 | 1957 |
|---|---|---|---|---|---|---|---|---|---|---|---|---|
| **Geophysics** | | | | | | | | | | | | |
| Am. Geophys. Union Trans. | | | | | 1 | | 3 | | | | 1 | 3 |
| Am. Met. Soc. Bull. | | | | 1 | 1 | | 1 | 1 | | | | 1 |
| Ann. Geophys. | | | | | 1 | | 1 | 2 | | 3 | 1 | 2 |
| Ann. Meteorology | | | | 2 | 1 | | | | | | | |
| J. Atmos. Terres. Phys. | | | | | 1 | | | | | | 4 | 2 |
| JGR | | | | | 3 | 2 | 3 | 1 | 4 | 4 | 15 | 4 |
| J. Meteorology | | | | | | 1 | | | 1 | | 4 | 1 |
| Nature | | | | 1 | | 1 | 1 | 3 | 1 | 1 | 1 | 3 |
| Subtotal (geophysics) | 0 | 0 | 0 | 4 | 8 | 4 | 9 | 7 | 6 | 8 | 26 | 16 |
| **Physics** | | | | | | | | | | | | |
| Am. J. Phys. | | | 1 | | | | | | | | | |
| Am. Phys. Soc. Bull. | | | | | | | | | 2 | | 2 | |
| Phys. Rev. | 4 | 4 | 4 | 7 | 9 | 10 | 2 | 6 | | 9 | 1 | |
| Astron. J. | | 1 | | | | | | | | 1 | 1 | 1 |
| Astrophys. J. | | | | 1 | | | | 1 | 2 | | 2 | |
| Subtotal (physics) | 4 | 5 | 5 | 8 | 9 | 10 | 2 | 7 | 4 | 10 | 6 | 1 |
| Percentage (physics) | 100 | 100 | 100 | 66 | 53 | 71 | 21 | 50 | 40 | 55 | 19 | 6 |
| **Total** | 4 | 5 | 5 | 12 | 17 | 14 | 11 | 14 | 10 | 18 | 32 | 17 |

*Note:* Publication trends demonstrating an initial predominance in physics journals, with significant reversals coming only in the mid-1950s. *Prior to 1949 the *Journal of Geophysical Research* was published as *Terrestrial Magnetism and Atmospheric Electricity*; for the period 1940–1948 no research from rocket observations appeared there.

TABLE 17.2  Publication Trends, Technical Journals, 1946–57

| Journal | 1946 | 1947 | 1948 | 1949 | 1950 | 1951 | 1952 | 1953 | 1954 | 1955 | 1956 | 1957 |
|---|---|---|---|---|---|---|---|---|---|---|---|---|
| *Aero. Eng. Rev.* | | 1 | | | | 1 | 1 | | | | 1 | |
| *Am. Rock. Soc. J.* | | | | | | | | | | | | |
| *App. Mech.* | | | 1 | | | | | 2 | | | | |
| *Astronautics* | | | | | | | | | | | | 4 |
| *Coast. Artill. J.* | | 3 | | | | | | | | | | |
| *Elect. Eng.* | | | | | | | | | | | 1 | 2 |
| *Electronics* | | 2 | 1 | 1 | 2 | | 1 | 1 | 4 | | | 1 |
| *Engineering* | | | | | | | 1 | 3 | 1 | | 2 | 1 |
| *Instruments* | | | | 1 | | | | | | | | |
| *Inst. Radio Engin. Proc.* | | | | | | | | | 1 | | 1 | 1 |
| *Jet Prop.* | | | | | | | | | | | 4 | 12 |
| *J. Op. Soc. Am.* | | 3 | 4 | 1 | | 3 | 3 | 5 | 4 | 2 | 3 | 2 |
| *Mech. Eng.* | | 1 | | 2 | 2 | 1 | | | | 2 | | |
| *Nav. Av. Nws.* | 1 | | | | | | | | | | | |
| *Ordnance* | 1 | | | | | | | | | | | |
| *Photo Engin.* | | | | | 2 | | | | | | | |
| *Rev. Sci. Instr.* | 1 | | 1 | | 1 | | 2 | | 1 | 1 | 3 | 1 |
| *Soc. Motion Pix. & TV Eng.* | | | | | | | | | 1 | | 1 | |
| **Total** | **3** | **10** | **7** | **5** | **7** | **5** | **8** | **11** | **12** | **3** | **16** | **24** |

*Note:* Publication trends in technical journals demonstrating major attention to electronics and optics. The 111 technical publications constitute less than the total of 158 publications in scientific journals during the same period, but technical publications dominate when supplemented by laboratory and contractor reports identified in Table 17.3

TABLE 17.3   Publication Trends, Formal Technical Reports (unclassified), 1946–57

| Institution | 1946 | 1947 | 1948 | 1949 | 1950 | 1951 | 1952 | 1953 | 1954 | 1955 | 1956 | 1957 |
|---|---|---|---|---|---|---|---|---|---|---|---|---|
| AFCRC* | | | | | | 2 | | 3 | 1 | | | 2 |
| APL | | 2 | 2 | | | 1 | | | | | | |
| BU* | | | | | [ | 3] | | | | | | |
| CNO | | | | 1 | | | | | | | | |
| Colorado* | | | | 1 | 2 | 1 | 1 | 3 | 3 | 2 | 2 | 2 |
| Denver* | | | | [ | | 1] | | | | | | |
| Michigan* | 1 | | 3 | | 3 | 1 | | 2 | 3 | | 1 | |
| NRL | 4 | 5 | 4 | 5 | 5 | 3 | 6 | 3 | 4 | 3 | | 7 |
| Rhode Island* | | | | | [ | 1] | | | | | | |
| Temple* | | | [ | | | 1] | | | | | | |
| **Total** | 5 | 8 | 9 | 7 | 11 | 14 | 7 | 11 | 11 | 5 | 3 | 14 |

*Note:* Some 106 separate publications and reports have been found, not including far more numerous "interim reports" provided by most of the Air Force contractors. The NRL, APL, and Colorado reports included detailed descriptions of instruments and techniques. Brackets indicate the reporting periods of the smaller Air Force contractors, working under the Blossom umbrella. *Air Force contractors.

*Physical Review* in December 1952, was cited equally in physics and geophysics journals in the first two years, but in the next decade it gathered 69 citations in major American and British geophysical journals, and only 14 in physics journals.[19] The redirection was also due to the definite failure of rocket-borne devices to make contributions to cosmic-ray physics or astrophysics, especially nuclear and high-energy physics, at a time when the accelerator laboratory was replacing the field observatory in observational particle physics, and when the resolution of the early spectra did not meet astronomers' requirements.

The redirection was, of course, mutual, as the citation trends noted above reveal, and many in the geophysics community began to pay more attention to the use of rockets. Starting in 1950, as outlined in table 17.4, various professional organizations took steps to incorporate the new activity into their own activities, either to coordinate or just to keep track of the progress of upper air research with rockets. The succession of committees and panels that formed in the United States were, without question, based in geophysical, rather than physical or astrophysical organizations. This trend was amplified by planning for the IGY. The JRDB, the NACA, national geophysical associations, and international co-ordinating bodies for the IGY all looked to those who had developed rocket techniques to provide the basis for upper air observations. No major scientific bodies composed of physicists or astronomers had made this gesture by the mid-1950s. Although here and there a rocketeer would win a society prize—Tousey being nominated for one by the American Astronomical Society in late 1946 for the first ultraviolet spectrum of the sun—nowhere was the reaction as positive as from the geophysical sciences.

Nevertheless, in the 1950s few of the rocket scientists migrated wholly into geophysics. The publication patterns of the leaders—Tousey, Van Allen, and Friedman—reveal first that Tousey (table 17.5) directed his remarks mainly to the optical sciences community, less to generalist physicists, and a little to astrophysicists. He felt ambivalent toward the astronomical and geophysical communities through the 1950s; in Tousey's multi-au-

---

TABLE 17.4   Technical and Scientific Committees and Panels

| | |
|---|---|
| Early 1946 | **V-2 Panel** (then Upper Atmosphere Rocket Research Panel then Rocket and Satellite Research Panel) 1946–1960. |
| Late 1946 | **JRDB Panel on the Upper Atmosphere**, within the Committee on Geophysics of the JRDB. |
| 1948 | **Subcommittee on the Upper Atmosphere** (NACA) to review data and determine best standard model for the upper atmosphere. |
| Late 1950 | **Standing Committee on Problems of the Upper Atmosphere** of the meteorology section within the American Geophysical Union. "To review and interpret the rapidly growing field." |
| Early 1951 | **Permanent Committee on the Upper Atmosphere** of the American Meteorological Society (Michael Ference, chair, and including Wexler, Bernard Haurwitz, Charles Brooks, Fred Whipple). |
| 1953 | Recognition by the **Gassiot Committee**. |
| 1954 | **Planning Panels for the IGY**<br>Tech Panel on Earth Satellite<br>Tech Panel on Rocketry<br>Tech Panel on Cosmic Rays<br>Tech Panel on Aurora and Airglow<br>Working Group on Internal Instrumentation |

*Note*: Series of panels and committees on which members of the UARRP served, showing a predominance of geophysicists. Even though the Joint Research and Development Board placed its Panel on the Upper Atmosphere logically within its Geophysics Committee, the panel had strong ties to the Guided Missiles Committee.

TABLE 17.5  Tousey's Publication History, with Collaborators, 1946–57

| Journal | 1946 | 1947 | 1948 | 1949 | 1950 | 1951 | 1952 | 1953 | 1954 | 1955 | 1956 | 1957 |
|---|---|---|---|---|---|---|---|---|---|---|---|---|
| *Phys. Rev.* | 1 | 1 | 1 | 1 | | 1 | | | 1 | | | |
| *Op. Soc. Am. J.* | | 2 | 3 | 1 | | 3 | 2 | 4 | 1 | 1 | 1 | 2 |
| *Trans. AGU* | | | | | 1 | | | | | | | |
| *Ap. J.* | | | | 1 | | | | 1 | 1 | | | |
| *JGR* | | | | | | 1 | 1 | | | | 1 | |
| *Astron. J.* | | 1 | | | | | | | | | | |

*Note:* Richard Tousey's publication history indicates highly focused attention on the optical sciences community, and less on generalist physicists. There is also a general indifference to astronomical and geophysical communities.

thored papers, only those headed by Eric Durand appeared in the *Astrophysical Journal*, since astronomy appealed more to Durand.[20] Van Allen (table 17.6) certainly did not make a clear transition to geophysics, maintaining consistent attention toward the physics community, as we noted in Chapter 14. But as an entrepreneur, he also looked for a broader audience than did Tousey. Many of Van Allen's publications promoted both rocket and satellite research, especially his *Scientific Uses of Earth Satellites*. The entrepreneurial spirit is strongest in Herbert Friedman's publication history (table 17.7; figure, p. 334); although his initial results were directed toward the physics and optics communities, he cast his net as wide as possible during the IGY.

As group leaders, Tousey and Friedman represent widely differing styles, both in their adopted instrumental technique and in their professional outlook. When asked about people who left his staff early in the program, Tousey felt that the extreme ultraviolet "wasn't part of [their] career." Tousey lived and worked in the world of ultraviolet spectroscopy: he was out to obtain the best spectrum of the sun possible, identify the lines, measure the ozone distribution in the upper atmosphere, and obtain a measure of the intensity distribution of solar energy in the rocket ultraviolet. He attacked these problems with the tools and interests of the experimental physicist.[21]

Friedman worked in the world of solid-state electronics, and unlike Tousey, sought a larger role in the International Geophysical Year; his interest in high-energy solar phenomena was defined by the Navy's interest in ionospheric research, but it led him into the IGY and into coordinating solar-terrestrial studies. As his publication history attests, he also found the IGY a springboard to a far wider audience. Tousey, on the other hand, insulated himself from large-scale coordinated studies, preferring the laboratory and launch pad to the meeting room.

The lack of attention to the astronomical community by both Tousey and Friedman was also a reflection of institutional interests at NRL. None of Tousey's or Friedman's peo-

ple were trained in astronomy, whereas a few, like F. S. Johnson, had some experience in physics and meteorology and developed strong interests in ozone absorption and atmospheric ultraviolet corrections to the solar constant. Kenichi Watanabe, for example, came to NRL with prior training and experience under John Strong, studying the infrared spectrum of ozone, and continued the same line of research under Tousey in the ultraviolet.[22] Thus although the sun was the main focus of attention, it was not at first the primary interest of either of the early NRL groups.[23]

Throughout the 1950s, both Tousey and Friedman continued to study the high-energy spectrum of the sun, finding it both a technical challenge and one that could yield information about the nature of the ionosphere. Each employed different techniques, for different energy regimes, and each sought out the communities that were either most receptive to their work or to which they felt some affinity. As we noted, Tousey identified himself with optical technology and vacuum ultraviolet spectroscopy, whereas the papers that appeared in geophysical and astrophysical publications were co-authored by his staff. Although Friedman recalls difficulty getting his first publication into the *Physical Review*, he chose to concentrate his attention there until the IGY.[24] Neither he nor his group paid any serious attention to an astronomical audience until 1956, although they did direct their attention earlier to audiences concerned with geophysical and solar-terrestrial relations.[25]

Van Allen's publication history reveals a complex interplay of interests and activities. As a program director and then university professor, Van Allen maintained a strong identity with the physics community, contributing to cosmic-ray studies while making forays into nuclear physics and fusion research at Brookhaven and Princeton. Until Sputnik, he remained loyal to cosmic-ray research, but this loyalty ultimately led him to geophysics: by the mid-1950s, his primary role in planning for the IGY and his continuing dedication to both rocket and satellite re-

TABLE 17.6 Van Allen's Publication History, with Collaborators, 1946–57

| Journal | 1946 | 1947 | 1948 | 1949 | 1950 | 1951 | 1952 | 1953 | 1954 | 1955 | 1956 | 1957 |
|---|---|---|---|---|---|---|---|---|---|---|---|---|
| *Phys. Rev.* | 1 | 1 | 3 | 1 | 3 | 1 | | | | 2 | | |
| *Il. Nov. Cim.* | | | | | | | | 1 | | | | |
| *Science* | | | 1 | | | | | | | | | |
| *Nature* | | | | | | | 1 | | | | | |
| *J. Aero. Sci.* | | | 1 | | | | | | | | | |
| *Am. Phys. Soc. Bull.* | | | | 1 | | | | | | | | |
| *Echo Lake Conf.* | | | | 2 | | | | | | | | |
| *Aero. Digest* | | | | | 1 | | | | | | | |
| *Int. Assoc. Terrestr. Mag. & Atmos. Elec.* | | | | | 1 | | | | | | | |
| *Terr. Optics* | | | | | | 1 | | | | | | |
| *JGR* | | | | | | | | | | | 1 | |
| *S&T* | | | 1 | | | | | | | | | |
| *Terr. Atm.* | | | | | | 1 | | | | | | |
| *Phys. & Med. UA* | | | | | | | 2 | | | | | |
| *Boyd & Seaton* | | | | | | | | | 1 | | | |
| *Sci. Uses of E Sats.* | | | | | | | | | | | ed. | |
| *NAS Proceedings* | | | | | | | | | | | | 1 |

*Note:* James Van Allen's publications are directed largely to a physics community, but also reflect as effort to reach a wider audience. Many of these publications were entrepreneurial, calling attention to the application of rocketry to the solution of scientific problems.

TABLE 17.7  Friedman's Publication History, with Collaborators, 1946–57

| Journal | 1946 | 1947 | 1948 | 1949 | 1950 | 1951 | 1952 | 1953 | 1954 | 1955 | 1956 | 1957 |
|---|---|---|---|---|---|---|---|---|---|---|---|---|
| Phys. Soc. (U.K.) | | | | | | | | | | 2 | | |
| Jet Prop. | | | | | | | | | | | | 1 |
| Elec. Eng. | | | | | | | | | | | | 1 |
| Yale Sci. | | | | | | | | | | | | 1 |
| Astronautics | | | | | | | | | | | | 1 |
| Phys. Rev. | | | | | | 1 | | | | | | 2 |
| Op. Soc. Am. J. | | | | | | | 1 | 2 | 1 | 2 | 2 | |
| Ann. Geophys. | | | | | | | | | 2 | 1 | | |
| Rev. Sci. Instr. | | | | | | | | | | 1 | | |
| JGR | | | | | | | | | | | 1 | 1 |
| Ap. J. | | | | | | | | | | | 1 | 1 |
| Science | | | | | | | | | | | 1 | |
| ICSU Rpt. | | | | | | | | | 1 | | | 1 |
| IAGA Bull. | | | | | | | | | | | | 1 |
| Mil. Elect. | | | | | | | | | | | | 1 |
| A.J. | | | | | | | | | | | | 1 |
| Nature | | | | | | | | | | | | 2 |
| AFCRC Chem. Aeronomy | | | | | | | | | | | | 2 |

Note: Herbert Friedman was engaged in classified research prior to 1950, and then for some time in instrument and program development. Primary focus at first is the *Physical Review*, and then an explosive search for a larger audience begins in 1956–57.

Friedman's entrepreneurial spirit and the personalities of his group members and Army staff at White Sands, as depicted by Gilbert Moore in the mid-1950s. E. T. Byram recently interpreted his cartoon character as someone who was never satisfied. Electron Optics Branch folder, Friedman VHI, Smithsonian Archives.

search, ultimately brought a change in his perspective. By the end of the decade, he no longer thought of the earth's geophysical environment as a gigantic mass spectrometer with which to examine the cosmic ray, but rather he looked to the cosmic ray as a probe of the earth's magnetosphere.

In contrast to Van Allen, Tousey, and Friedman, one of the clearest patterns of migration was established by Charles Y. Johnson, an NRL staff member who acted as instrumentation group leader in Newell's branch (table 17.8). In the late 1940s, Johnson was a member of the cosmic-ray group at NRL, trying out a large array of detectors without any noticeable scientific success, although his group was the only one to build cloud chambers and make them work on a V-2. The cosmic-ray group disbanded in 1950–51, and Johnson moved into a section newly formed to develop ways to study the ion and neutral composition of the high atmosphere. Several years elapsed as they built radio-frequency mass spectrometers to make the ion and neutral composition observations, which began to produce useful data only in the pre-IGY period. Johnson's publication history reflects this migration, with an early cluster in the *Physical Review*, a hiatus during which the new group was developing its instruments and a reemergence wholly within geophysics.

People like Charles Johnson and APL's Lorence Fraser were the ones with the most adaptable talents and training. They were also

TABLE 17.8  C. Johnson's Publication History, with Collaborators, 1946–57

| Journal | 1946 | 1947 | 1948 | 1949 | 1950 | 1951 | 1952 | 1953 | 1954 | 1955 | 1956 | 1957 |
|---|---|---|---|---|---|---|---|---|---|---|---|---|
| *Phys. Rev.* | | | 1 | 1 | 1 | | | | | | 1 | |
| *JGR* | | | | | | | | | | 2 | 1 | |
| *Ann. Geophys.* | | | | | | | | | | 1 | | |
| *Trans. AGU* | | | | | | | | | | | 1 | 1 |

*Note*: Charles Y. Johnson's publication history identifies his migration pattern from physics to geophysics after the NRL cosmic-ray group disbands, and Johnson is assigned to a new group developing mass spectrometers to determine the ion composition of the upper atmosphere. The hiatus in publications reflects the time required for instrument development.

Van Allen's "strong back" syndrome is well-illustrated in this scene of Naval Research Laboratory personnel improvising to find the center of gravity of a warhead suspended in the White Sands assembly building, circa Spring 1946. After the warhead was weighed on the scale beneath it, it was hoisted and tilted by Ralph Havens, kneeling on another scale. Thor Bergstralh (obscured by the scale, left center background) is preparing to read the scale to determine the tension that Havens is exerting on the rope. Serge Golian (at the left on the stand) is keeping tension on the rope to keep the pulley directly above Havens. F. S. Johnson is holding a ruler to let Krause (at the right, background) read the tilt of the nosecone from the vertical, defined by the cables holding the warhead. Johnson kindly provided the explanation. U.S. Navy photograph, Ernst Krause collection. NASM SI 83–13934.

the ones without strong preexisting scientific research interests. Fraser never thought of himself as a cosmic-ray physicist. He gained that title at APL in an ad hoc social process. He looked on himself as an engineer, doing "engineering in the true sense of the word, translating the ideas of physicists and experimenters into reality."[26]

People like Fraser, Johnson, J. C. Seddon, and others among the member groups of the UARRP tended to work for institutions that wanted to establish the upper atmosphere as an operational theater for war. Institutional identities, therefore, were not with any one

audience or discipline for these people, but with any discipline that could possibly inform their work in the context of the rocket.

Alignment with geophysics was also an alignment with military needs. The Joint Research and Development Board placed upper atmospheric research within its Geophysics Committee (table 17.4). Both the Navy and the Air Force were always strong supporters of geophysical research, as they both required information on the upper atmosphere. When the Office of Naval Research formed as the Office of Research and Invention in May 1945, geophysics was part of ORI's military program branch, and not part of its scientific branch, which contained all the other sciences.[27] Thus military support in abundance was available for geophysical research at the end of World War II to examine problems in cosmic-ray physics, UV solar spectroscopy, and ozone research, air composition, and constitution studies, and all forms of ionospheric research that might shed light on radio propagation. What the military needed was the expertise to perform that research. It also wished to maintain that expertise within military laboratories, as well as extend it into civilian laboratories as far as possible. With both patrons and audience thus pushing for geophysics, it is no surprise that the rocket experimenters moved in that direction.

The publication trends reviewed here and the early centrality in planning for the IGY of all the major groups associated with scientific rocketry mark as well the emergence of a new technical culture that defined its science in terms of the rocket.[28] Most of the group leaders we have identified, such as Van Allen, Friedman, and Dow, never migrated far from the shop floor even as they became advocates and entrepreneurs for additional launch vehicles and for establishing and maintaining the considerable infrastructure required to perform research with rockets. Although they felt pressure to disseminate their publications widely, they, along with Tousey, Rense, Seddon, Spencer, F. S. and C. Y. Johnson, and their many colleagues, remained part of the hands-on technical culture devoted first and foremost to instrument and vehicle development and to the certification of a new way to conduct scientific research.[29]

Thus it was instrument development that defined the first decade of research with rockets, and it was the character and applicability of these new instruments that defined their ultimate audiences. Even though specific scientific problems had been defined before the rocket era, they were redefined not only by the new horizons created by the institutions interested in rockets, but also by the success of the instrument builders in identifying problems that were best handled by the tools they were able to build.

Inevitably, however, as each of the rocket groups grew larger, and as their newer or more junior members became more sequestered from the forces that made their work possible in the first place, many of them began to specialize in, and wholly identify with, the scientific problem areas that were being addressed.[30] Even so, they and their elders all remained under an umbrella of effort defined by patrons who were concerned first and foremost with national security needs.

# Notes to Chapter 17

1. Box 38, biographical files, HN/NHO.
2. Military interest and support extended beyond the vehicles to the extensive logistical infrastructure required to launch them from land and sea, to receive their telemetry signals, and to attempt physical recovery. Two examples are the use of naval vessels and the establishment of a firing facility in the auroral zone at Fort Churchill. A description of the latter, prior to the establishment of a rocket-firing facility, is in [Army Ordnance]. *An Introduction to Churchill, for Churchill, and Surrounding Area* (7099th ASU, U.S. Army: July 1950). OCO/NARA.
3. Among many other sources, the IGY has been reviewed by its scientific leaders, such as Chapman (1959), and Wilson (1961). The majority of historical attention has been confined to the political decision to launch an artificial satellite and the choice of launch vehicle. See, for instance, McDougall (1985), chap. 5. Hall (1963, 1977) has discussed early satellite proposals that set the stage for the political decisions made before the IGY. See also Green and Lomask (1970).
4. Chapman (1959); Green and Lomask (1970), pp. 19–23; Van Allen OHI, pp. 170–72.
5. Wilson (1961), p. 32; Sullivan (1961), pp. 26–27.
6. Odishaw convinced the White House to boost the IGY budget to $13 million, with $2 million for fiscal 1955. See "Panel Report no. 40," 3 February 1955, p. 14. V2/NASM; and McDougall (1985), p. 118.
7. Commentary noted in "Panel Report no. 36," 3 October 1953, p. 16. V2/NASM.
8. Although Kaplan was happy to draw on UARRP members to staff many of the IGY's

panels, the National Committee's policy was not to "adopt any existing organization to serve the function of these quasi-official panels." See "Panel Report no. 40," pt. 1, 3 February 1955, p. 13. V2/NASM.

9. Ibid. Other members included N. C. Gerson, a prominent member of the AFCRC rocketry group, Bernard Haurwitz from New York University, Kaplan, and S. Fred Singer. In February 1955, the UARRP suggested to Kaplan that Dow and Stroud be added to represent Michigan and the Signal Corps, Berning to represent BRL, and P. H. Wychoff to strengthen AFCRC representation.

10. Newell Papers. Box 28, green notebooks; and Box 36, IGY CSAGI/Rome folders, HN/NHO. See also "Panel Report no. 40," 3 February 1955, p. 14. V2/NASM. The Gassiot Committee received approval from the Royal Society Council to form a sub-committee to advise on a rocket program on 3 May 1955. It was "the nearest analogue to the American Rocket Research Panel . . . " Massey and Robins (1986), p. 20.

11. The amount appropriated was $1.7 million, close to the total initial allotment for fiscal 1955 IGY planning. See "Panel Report no. 40," p. 1, 3 February 1955, p. 14. V2/NASM; McDougall (1985), p. 118.

12. "Panel Report no. 40," ibid.

13. Newell was among a small group of advocates for a satellite program during the IGY, led by Lloyd Berkner at the summer 1954 CSAGI meetings in Rome, but it appears that he was less enthusiastic than others in the group, such as S. Fred Singer and Athelstan Spilhaus. See Green and Lomask (1970), pp. 19; McDougall (1985), pp. 118–19. S. Fred Singer notes that the Navy did not want to become subject to international oversight. Singer OHI no. 3, pp. 22–27rd., 29rd.

14. The symposium was published as Van Allen (1956).

15. "Panel Report no. 42," 27 October 1955, p. 20. V2/NASM. Louis Delsasso from Aberdeen argued that the symposium nevertheless should take place, but that it should be a closed meeting by invitation only. Just why the USNC objected to a public forum has not been investigated. Eisenhower announced the U.S. satellite program in July 1955, and the Stewart Committee chose Vanguard by August. McDougall (1985), pp. 121–22; Green and Lomask (1970), pp. 19–23.

16. Bibliographic sources include Benton (1959); Townsend, Friedman, and Tousey (1958); contractor reports in the AFGL Library; and publications lists for Tousey, Friedman, and Van Allen. See also [Tousey] (1961), pp. 382–83.

17. Johnson, Watanabe, and Tousey (1951); citations identified in [Institute for Scientific Information] (1984); [Institute for Scientific Information] (1988).

18. These are citation counts to 1964, based on his publications only to 1959. During this period there was some overlap: his 1950 and 1951 papers were cited 15 times each in physics journals, and only once each in geophysics journals through 1964; his 1957 papers in following years garnered 9 citations in geophysics journals and only 2 in physics journals. Starting in 1959, however, geophysics always dominated. His 1959 papers were cited 30 times *within that year* in geophysics journals, and 81 times through 1964, whereas physics journals cited the same papers only 28 times through 1964. Counts taken from [Institute for Scientific Information] (1984); [Institute for Scientific Information] (1988).

19. In the first two years, the *British Journal for Atmospheric and Terrestrial Physics* paid the most attention with 6 citations, followed by the *Journal of Geophysical Research* and the *Physical Review*. Counts taken from [Institute for Scientific Information] (1984); [Institute for Scientific Information] (1988).

20. Tousey OHI, p. 9rd.

21. Tousey OHI, p. 110rd.

22. F. S. Johnson OHI, pp. 47–48; Tousey OHI, p. 86.

23. Hevly (1987), p. 212, recounts a telling exchange between Burnight and astronomer Aden B. Meinel in 1950. Meinel had asked Burnight if his X-ray observations could distinguish sources on the sun: did they arise in the corona or photosphere? He was taken aback when Burnight replied that the angular resolution of his detector was 90 degrees. Hevly relates this interchange as an example of the limitations of Burnight's photographic technique, but it is also an example of institutional priorities and experimental style.

24. On the *Physical Review*'s rebuke of Friedman's first paper on X-ray solar radiation, see Friedman OHI, p. 62. Group publication data derived from appendixes to Townsend, Friedman, and Tousey (1958).

25. Townsend, Friedman, and Tousey (1958) list one paper from Friedman's group in the *Astrophysical Journal* in 1956, and two abstracts of talks in the *Astronomical Journal* in 1957.

26. Lorence Fraser OHI, p. 44.

27. Sapolsky (1990), fig. 3–1; NRL correspondence files, circa May 1945. NRL/NARA. The alignment process in oceanography is examined in Mukerji (1989).

28. Howard E. McCurdy (1989) has employed the term 'technical culture' to describe the first decade of NASA employees who built Apollo, and his observations are relevant here.

29. Preferences for instrument building were voiced consistently by those interviewed. See especially DeWitt Purcell OHI, p. 27; and Lorence Fraser OHI, pp. 5–13, 44–46, as well as comments in Charles Johnson and Fred Wilshusen interviews, and the more recent videohistory studies of members of Herbert Friedman's and Charles Johnson's groups at NRL.

30. The formation of specialized subgroups within each of the larger organizations that we have examined, especially at NRL and AFCRL, led to specialty identities as well as an integration of engineering and scientific types, much like those Peter Galison has found for large bubble chamber groups in the 1950s. See Galison (1985), p. 356.

# 18
# Conclusions

*We shape our tools, and then our tools shape us.*

—attributed to Marshall McLuhan.[1]

*. . . the interest developed . . . because of the opportunity.*

—F. S. Johnson.[2]

## The V-2 as Symbol

The space sciences in the United States would not have emerged in the way they did without the existence of the V-2. The V-2 created a surge of activity that extended the intensity and urgency of wartime research into peacetime. Those who made the commitment to fly on the first (and possibly the only) space vehicles that might ever become available in their professional lifetimes—Dow, Friedman, Krause, Tousey, and Van Allen—had to develop instruments and logistical systems with a vengeance reminiscent of the wartime effort.

The V-2 thus became a symbol, a challenge, engendered in the postwar world of preparing for the next war. Even though the military services did not at first wholeheartedly support guided missile development, factions within each of the services vigorously campaigned for the new technology. The V-2 demonstrated what was possible; new tactical missiles had to be developed to match or surpass them, and means to defend against them also had to be found. Both goals encouraged scientific and engineering groups to use the missiles at hand both to determine what the upper atmosphere was all about and to act as new specialty groups that might provide advice useful for the continued development of missiles. A viable missile system required knowing how to make delicate devices work during flight, as well as knowing the medium in which the vehicle operated. The military services had little problem with relevance because they had redefined upper atmosphere research goals as national security goals.

## Tool Building

Those who remained in scientific rocketry, either as entrepreneurs or on the shop floor, devoted the substance of their professional energies in the first decade to tool building and program development. They thus maintained activities that were useful to their institutional patrons. Just as Holger Toftoy did not want to build any one missile, but improve the breed in general, NRL's E. O. Hulburt strove to master a broad range of optical and electro-optical technology, wanting to establish and maintain a technical culture capable of wresting greater control over the principles of physical optics, for the benefit of the Navy. Hulburt's vision was more than met in the efforts of Tousey's and Friedman's groups, as they explored ways to make optical and electronics devices work on missiles.

Those who engaged in scientific rocketry, however, did not seek out new instrument or detector technologies, but applied the techniques they had already mastered or were already committed to through training or ex-

341

perience. Tousey, Hopfield, and Rense, all trained in the laboratory study of the vacuum ultraviolet, adapted its most common tools—the Rowland grating and the specially sensitized photographic plate—to the spectroscopic study of the sun. Herbert Friedman applied his knowledge of solid-state physics and experience developing Geiger counters to the study of the absorption edges of metals, and then at NRL refined and extended this technology not only to many wartime applications, but to the study of the ultraviolet and X-ray spectrum of the sun.

In atmospheric physics, William Dow applied his established expertise in vacuum-tube technology to rocket instrumentation and brought back a number of his former students to assist, who were likewise familiar with the technology. J. C. Seddon at NRL applied his wartime experience in radar countermeasures directly to his multiple-frequency ionospheric probes. William Stroud, who trained at Chicago under Marcel Schein, applied the cosmic-ray counter telescope techniques he had learned there to the design of some of Princeton's V-2 payloads.

Style, or technical practice, also migrated into the rocket work. The highly simplified and spare instrument styles favored at first by James Van Allen and later by Herbert Friedman stemmed from their parallel experiences building proximity fuzes during the war at Johns Hopkins and at APL, whereas NRL's attempts at building pointing controls created large systems that mirrored their technical origins. Indeed, next to their shared experience and training in radio-related electronics, the pioneer rocket experimenters examined here are best typified by their ability to adapt wartime experiences and contacts to the pursuit of upper-atmospheric studies with rockets.

Although William Dow continued to focus on vacuum-tube research and development, just as Tousey remained in optics, both drew others into rocketry who did develop strong interests in the upper atmosphere and the sun. But most remained at institutions favoring a technical focus on vehicle development. Nelson Spencer stayed at Michigan until NASA drew him to the Goddard Space Flight Center. Many who had worked for Krause, Tousey, and Friedman, from Homer Newell on down, leapt to the new agency in 1958 not only to continue space research with rockets and satellites, but to obtain greater flexibility and upward mobility, whereas others, like Francis Johnson, William Baum, or Gilbert Perlow, who chose academic futures, ultimately distanced themselves from rocketry.

Younger staff from every one of the groups identified here who migrated to NASA in the wake of Sputnik became part of the technical and administrative infrastructure that made space science a possibility in the first years of the Glennan administration. In a sense, they were the harvest of a government policy, formed at the dawn of the Cold War, that promoted a continuing relationship with science to improve the breed of missile it would want to use in the next war. Thus when it called for it, the nation found a ready and eager pool of scientific workers available in 1957, at a time when political necessity and 0public policy dictated that the United States form a civilian space program.

## The Establishment of a Technical Culture

The community of workers devoted to scientific rocketry established means both for formal and informal communications ranging from the V-2 Panel to the Oxford Conference. In all of these efforts, those who focused on scientific rocketry placed themselves in the dual role of contributors to science and inventors of a new way to conduct science. The Oxford Conference was in large part stimulated by the growing spirit of the International Geophysical Year—and indeed led the many participating American groups into highly productive sounding rocket research programs during that period. But the UARRP looked forward to the IGY as preparation for the next step into space; as its commemorative tenth anniversary meeting in 1956 demonstrated, this was a time to discuss how to build scientific satellites, rather than review contributions to science. Clearly, the die had been cast, for the community of workers who had grown up around the V-2 as a vehicle for research now thought mainly of a future circumscribed by the rocket-launched satellite. These workers concentrated on solving technical problems; the science they tackled was defined as much by the technology developed as it was by the influence of institutional missions. Thus, Tousey's devotion to photographic techniques led him into spectral line identification, whereas Friedman's broad-band solid-state detectors prompted him to search for the source of the ionosphere in the deep ultraviolet and X-ray regions of the solar spectrum.

Competition was a natural by-product of the time and effort each group invested in the technical effort. In solar spectroscopy, the race was defined as a competition to be the first to obtain an ultraviolet solar spectrum, and then to be the first to reach Lyman Alpha. Van Allen made clear that although APL was not the first to obtain a solar spectrum, its results were of better quality. But competition also bred liaisons through shared experience. Van Allen likes to recall the V-2 era as giving rise to a fraternity of workers, rather than creating a discipline. Indeed, as a tool-building technical culture focused on the use of rockets as vehicles for research, the community of workers shared many experiences and frustrations. The community competed within itself to overcome the daunting technical problems it faced, quite thoroughly enjoying what it was doing. They all clearly were excited by the experimentation and the technical challenge. Van Allen correctly perceived that this distinction promoted their "strong back— weak mind" reputation, and made them second-class citizens in the class-conscious world of science.

Shared experiences also promoted substantial communication and migration between the groups. More often than not, technical information gained through both success and failure was communicated quickly and openly through technical reports and meetings of the UARRP. Members met at each other's laboratories, communicated their problems constantly through the V-2 Panel reports, and shared each other's technical documentation and reports to the limit allowed by the censors.

## Those on the Sidelines

Not everyone in a position to take advantage of the existence of rockets as scientific vehicles was equally willing to have their scientific careers so defined, largely because the new mode of doing science was too peculiar and required extreme dedication and technical focus. Some, such as W. H. Pickering, found that their institutions were moving in other directions and followed that course, in his case concentrating on building the missiles themselves. Pickering knew that although the pace of research with rockets was enormously demanding, real scientific payback was going to be thin in the first years. Goldberg, Greenstein, Spitzer, Wexler, Whipple, Zwicky, and members of the Princeton physics group entered the arena with great enthusiasm. But none persevered to the point of achieving any degree of success, and all save Whipple left by 1950. As with Pickering, neither their institutions nor their personal scientific agendas required devotion to missile-related basic research.

The spectrum of intent among astronomers who initially wanted to get involved ranged from opportunism to sincere personal excitement. J. Allen Hynek thought the whole thing a "screwy program" but one "we should do something about" as he convinced Jesse Greenstein to get on board, whereas Goldberg was interested in solar research, and was willing to shave his head and live in a cell in order to make it happen. Goldberg as well as Spitzer and Menzel, also saw the rocket as a ticket to create a new form of military patronage for astronomy, as did John Wheeler for physics. But few remained easy with the style of research required; the frustration experienced by Greenstein was more than matched by Goldberg's distaste for flying expensive instruments that would inevitably be destroyed upon use, a situation that military-based entrepreneur scientists like Ernst Krause found exhilarating.

Few scientists shied away from the military aspects of the work, or from building new instruments. Indeed, Wheeler was deeply involved in nuclear research, extending his wartime activities, and Spitzer later joined Wheeler in developing Princeton's applied plasma physics facility. Neither left rocketry because they felt uncomfortable with developing large-scale government-sponsored high technology. They, as well as Harry Wexler of the Weather Bureau, left because the rocketry, although promising, held out little immediate hope of revolutionary research. Each of them had other programs to attend to, such as reestablishing and maintaining observatories, and found these problems more compelling than the rewards they could expect from rocket flight.

## Institutional Imperatives

Although individuals drawn into scientific rocketry may not have been previously involved in the problems they addressed, to a certain extent the institutions they served were. The case is strongest for NRL, where E. O. Hulburt, as mentor to Tousey and Friedman, had pursued ionospheric research before the war. Hulburt's personal interests and vision were also in consonance with the mission of his institution, which was to study radio propagation, wherein he was prompted to find the physical mechanisms in the atmosphere that affected its ability to transmit and absorb radio energy. Before the war, he searched for the source of ionization along mainly theoretical lines, even though extensions of his research interests can be seen in the observational efforts in the NRL groups engaged in rocketry. Whereas the style of his research may well have been reminiscent of the basic scientific researcher, the research itself was directly related to the mission of the institution he served.

Similarly, at APL, we find the strong tradition of Carnegie's Department of Terrestrial Magnetism (DTM) in instrument building and geophysical research redefined in the warhead of a rocket. Harry Vestine, Howard Tatel, and others who remained at or returned to DTM after the war, were among Van Allen's most valued mentors and colleagues. But Van Allen also valued the help of those who knew from their wartime experience how to make proximity fuzes work after they were fired from a gun at accelerations many thousands of times greater than that due to gravity. Bringing the two together was at first the idea of Merle Tuve, when he and especially Lawrence Hafstad saw that instrumenting V-2 rockets was a viable means of participating in the development of guided missile control systems, such as Van Allen's altimeters. But when the connections with DTM weakened after Tuve left APL, and its parental influence waned as the Cold War deepened and long-range goals became short-term necessities in the Korean War, upper atmospheric research ceased and its staff dispersed. Van Allen's choice of returning to Iowa, when he clearly had alternatives such as Princeton or Brookhaven to consider, is a reflection of both the institutional landscape and the intellectual direction he preferred. Both led to a purer form of basic research.

Priorities and institutional traditions among the many Air Force contractors Marcus O'Day recruited on university campuses were less well defined than within the Navy laboratories. The most active Air Force contractors, at Michigan and Colorado, were tool-oriented and so built devices that could perform observations of interest to the Air Force. These were contracts for well-defined military purposes, not grants for pure nondirected research reflecting prior interests on these university campuses. The AFCRC's interest in an omnibus program of ionospheric research that might shed light on radio propagation and missile guidance and control thus extended to university campuses nationwide and established new institutional imperatives for research where none existed before.

## The Military Context

Beyond establishing a scientific agenda for rockets, the most striking similarity between Erich Regener's experience in Germany and that of the groups in the United States (and no doubt those in the Soviet Union) was that rockets were made available only through a single source that both promoted their development and made their scientific use possible. The priorities of the postwar world, based on the lessons of wartime, placed national security preparedness uppermost, and science was transformed in the process as it became a part of this process, in which guided missile development was but one activity.

Both the American and German plans were conducted wholly within a military context, and both were led by civilian scientists. Whether instigated by von Braun at Peenemünde or by Toftoy at the Pentagon, the science that was supported was defined in terms of the needs of a rocket development program.

## Migration, Marginality, and Authority

Military context, institutional imperatives, tool-building technical culture—all helped to insulate the groups that formed to conduct research with rockets, at least until the IGY. Insularity was primarily a result of the daunting technical obstacles that both defined the early challenge of space research and the nature of the players who met the challenge, players who were reluctant at first to identify themselves as anything other than experimental physicists. But insularity was also a condition imposed by the outside scientific world.

Just as Tousey had initially rejected the intrusion of astronomers, annoyed with their arrogance and fearing they would exploit his hard-won rocket data, astronomers argued that they were the only ones capable of exploiting the data fully and properly. Yet when the data first started to flow, they did not meet astronomers' needs. No astronomer could make his mark identifying lines and blends, whereas for Tousey, they were hard-won confirmation that his technical choices were a first sound step. Accordingly, even though APL's Harold Clearman was invited to Yerkes to learn astrophysical theory, he was not accepted there as a thesis student in astrophysics, but rather became a graduate student in laboratory spectroscopy at Princeton.

Tousey, J. J. Hopfield, and their younger associates like F. S. Johnson, found a warmer reception for what their spectra could say about the vertical distribution of atmospheric ozone and the solar constant. Naturally attracted to warmer climates, each rocket group adjusted itself accordingly: Van Allen ultimately decided that more was to be gained in geophysics than in nuclear physics; and Herbert Friedman found his first successes in ionospheric physics.

Of all the disciplines served in the process, none was as quickly affected as atmospheric physics. Here, the Michigan, NRL, and Signal Corps groups united in a cooperative scheme to make simultaneous observations of pressure, temperature, and density that would overcome both the dissonant results they obtained as they flew their instruments piecemeal in the 1940s and the discordances with established ground-based techniques used by Fred Whipple, and codified in the NACA tables. The T-Day effort promoted by the UARRP was not only an attempt to overcome these discordances, but also gave the rocket groups a collective identity as they published their combined results. The Rocket Panel atmosphere, although published in the *Physical Review*, became a milestone that was most often cited in geophysical journals as establishing the importance and authority of in situ measurements.

Unlike atmospheric physics, ionospheric physics had long been dominated by a ground-based probing technique that was both the vehicle for study, and the object of interest. Radio propagation observations created a picture of the ionosphere that the rocket groups at first supported, and later found difficult to change, especially by unfamiliar techniques at the limit of their sensitivity. Thus ionospheric research with rockets was not only impeded by severe technical obstacles and by the marginality of the contributors, but also by the seductive practicality of the traditional model and technique.

The authority of in situ data in the atmospheric sciences was just emerging as the IGY was being planned. Indeed, the instigators of the IGY movement were also among those most sympathetic to the continued and expanded application of rocketry. They saw the advantages of continuing a relationship with government and military agencies who could support an infrastructure capable of performing global atmospheric and geophysical research. Future investigations will have to show how the national security interests that established upper atmospheric research with rockets in the V-2 era were carried over and redefined in the context of civilian space research during the IGY, as well as in the NASA era.

## Notes to Chapter 18

1. Quoted in a review of Michael Schrage, *Shared Minds: The New Technologies of Collaboration*. Random House, 1991. The flavor of the passage attributed here is shared in McLuhan (1962), "Prologue"; and McLuhan (1964), pp. 55–56.
2. F. S. Johnson OHI, pp. 27–28. SAOHP/NASM.

# 19
# Sources

## Archival Sources

Archival sources cited herein are identified by the following shorthand notations:

AAS/AIP: Papers of the American Astronomical Society, American Institute of Physics Center for History of Physics.

ATW/LC: Alan T. Waterman Papers, Library of Congress Manuscripts Division.

CDS/LO: C. D. Shane Papers, Mary Lea Shane Archives of the Lick Observatory (courtesy of Dorothy Schaumberg).

CF/AFGL: Contractor's files, Air Force Geophysics Laboratory Library.

DHM/HUA: Donald H. Menzel Papers, Harvard University Archives, HUG 4567; HUG 4567.11.

DMA: Deutsches Museum Archives.

DTM: Department of Terrestrial Magnetism files, Carnegie Institution of Washington. See also: MAT/DTM: Tuve files, DTM.

EHK/NASM: Ernst H. Krause papers, selections at National Air and Space Museum.

ER/MPG: Erich Regener Papers, Archiv zur Geschichte der Max-Planck-Gesellschaft, Munich (courtesy Marion Kazemi).

FLW/HUA: Fred L. Whipple Papers, Harvard University Archives HUG 4876.808; HUG 4876.810.

FLW/SIA: The Papers of Fred L. Whipple, Smithsonian Institution Archives.

FSJ/NASM: Francis Johnson file, Space Astronomy Oral History Project, National Air and Space Museum.

GEH/CIT: George Ellery Hale Papers microfilm edition, California Institute of Technology Archives.

HKK/NASM:        Hilde Korf Kallmann-Bijl Collection (Acc 1989–0042), National Air and Space Museum Archives.

HNR/P:           Henry Norris Russell Papers, Princeton University Library.

HONRL:           History Office, Naval Research Laboratory.

HT/ASRC:         Holger Toftoy Papers, Alabama Space and Rocket Center, Huntsville, Alabama.

HT/DTM:          Howard Tatel File, Department of Terrestrial Magnetism files, Carnegie Institution of Washington.

HVN/CIT:         H. Victor Neher Papers, California Institute of Technology Archives.

HW/LC:           Harry Wexler Papers, Library of Congress Manuscripts Division.

HN/NHO:          Homer Newell Papers, NASA History Office. Housed at the Washington National Records Center, Suitland, Maryland. Newell's NACA-era files are included, and some are noted as HN/NACA/NHO.

ISB/CIT:         Ira S. Bowen Papers, Caltech Archives.

JB/NASM:         Julius Braun files, National Air and Space Museum, Department of Space History.

JG/CIT:          Jesse Greenstein Papers, California Institute of Technology Archives.

JPL/L:           JPL historical files, maintained by the JPL Library. These files, culled from JPL records, are being reevaluated by JPL archival staff.

JRDB/NARA:       Records of the Joint Research and Development Board, NARA Record Group 330, entry 341; stack area 13W4 Row 11, Compartment 7, Shelf C, e-341: Boxes 198, 227, 239, and 240.

JVA/APL:         Research, High Altitude (RHA) files, of the APL High-Altitude Research Group, specifically those of James A. Van Allen, courtesy Lorence Fraser. Copies in APL File SAOHP.

JVA/UI:          James A. Van Allen Papers, University of Iowa, copies provided by Van Allen.

JZ/MIT:          Jerrold Zacharias Papers (MC 31, Box 1), MIT Institute Archives. Copies courtesy Paul Forman.

K/DMA:           Ernst Klee Collection, Deutsches Museum Archives.

| | |
|---|---|
| KP/UA: | Kuiper Papers, University of Arizona Special Collections. |
| LG/HUA: | Leo Goldberg Papers, Harvard University Archives, copies provided by Goldberg to the National Air and Space Museum, Department of Space History, SAOHP files. |
| LSP/P: | Lyman Spitzer Jr. Papers, Princeton University Library. |
| MAT/DTM: | Merle A. Tuve records, Department of Terrestrial Magnetism files, Carnegie Institution of Washington. |
| MAT/LC: | Merle A. Tuve Papers, Library of Congress Manuscripts Division. |
| MOD/AFGL: | Marcus O'Day Papers, History Files, Air Force Geophysics Laboratory Library. |
| NACA/NHO: | National Advisory Committee for Aeronautics files, NASA History Office. |
| NBS/NARA: | Records of the National Bureau of Standards, Record Group 167. |
| NRL/NARA: | Rocket-Sonde Research Section, Naval Research Laboratory Records, National Archives and Records Administration. Accession number 11704 S78–1(119). Other designations include R320– 35/46 RF42–1/84, and F42–1/84(1322) 1320– 84/46rl, and other variations. All located at the Washington National Records Center, Suitland, Maryland. |
| OCO/NARA: | Records of the Office of the Chief of Ordnance, Department of the Army. Service Forces (ASF), RG 156. Records housed at the Washington National Records Center, Suitland, Maryland. |
| OS/AIP: | Otto Struve Papers, Yerkes Observatory, University of Chicago. Microfilm edition prepared by the American Institute of Physics, New York City. |
| PH/NHO: | Project Hermes File, NASA History Office. |
| PP/DMA: | Peenemünde Papers, Deutsches Museum Archives. |
| PP/P: | Princeton Physics Department Records, Princeton University Archives. |
| RAM/CIT: | Robert A. Millikan Papers microfilm edition, California Institute of Technology Archives. |
| RG/CIWA: | Robert Goddard file, Carnegie Institution of Washington Archives. |

RRM/BHL:         Robert R. McMath Papers, Bentley Historical Library, University of Michigan.

SAOHP:           Space Astronomy Oral History Project, Department of Space History, National Air and Space Museum. Working files from the project.

SFS/NASM:        S. Fred Singer Papers, National Air and Space Museum Archives.

SIVP/SIA:        Smithsonian Institution Videohistory Program, Smithsonian Institution Archives.

VB/LC:           Vannevar Bush Papers, Library of Congress Manuscripts Division.

V2/NASM:         Reports of the V-2 Panel, from 1946 though 1962, portions collected from the NASA History Office, Homer Newell Papers; from the James Van Allen Papers, University of Iowa; from the Harry Wexler Papers, Library of Congress Manuscripts Division; and from the personal collections of Nelson Spencer and William Stroud. Complete copy dating from January 1946 through 1962 housed in the SAOHP files, National Air and Space Museum.

WAR/SAOHP:       William A. Rense Files, Space Astronomy Oral History Project.

WGD/BHLUM:       The Papers of William G. Dow, College of Engineering, University of Michigan. Bentley Historical Library.

WFGS/APS:        W. F. G. Swann Papers, American Philosophical Society Library.

WOR/UC:          Walter Orr Roberts Papers, University of Colorado Archives.

WSA/H:           Walter Sydney Adams papers, Huntington Library, Pasadena, California.

## Oral Histories

Oral histories and videohistories cited herein are identified by shorthand notations OHI and VHI. All of these interviews have been transcribed, although not all were in final form at the time they were examined. Page numbers marked "rd" in the endnotes indicate a fully edited rough draft. Interview series paginated consecutively are identified by page number only. Interviews not otherwise credited were conducted by the author, and were conducted under the auspices of the Space Astronomy Oral History project, NASM.

DSH/NASM:        Department of Space History, National Air and Space Museum.

GSFC/NASA:       Goddard Space Flight Center, National Aeronautics and Space Administration.

NASM:      National Air and Space Museum Archival facility.

SAOHP/NASM:      Space Astronomy Oral History Project, National Air and Space Museum, Department of Space History.

SHMA/AIP:      Sources for the History of Modern Astrophysics, American Institute of Physics Center for History of Physics.

SIVP/SIA:      Smithsonian Institution Videohistory Program, Smithsonian Institution Archives.

Aarons, Jules, 12 December 1983, SAOHP/NASM.

Baker, James G., 9 June 1980, SHMA/AIP.

Ball Brothers Research Corporation (Fred C. Dolder, Ruben H. Gablehouse, R. A. Gaiser, R. C. Mercure), 26 July 1983, SAOHP/ NASM.

Baum, William A., 12 January 1982, SAOHP/NASM.

Bergstralh, Thor A., 1 August 1983, SAOHP/NASM.

Chandrasekhar, Subrahmanyan (by Spencer R. Weart), 17 May 1977, AIP/SHMA.

Dolder, Fred C., 13 January 1982, SAOHP/NASM.

Fraser, Lorence, 9 March 1983, SAOHP/NASM.

Friedman, Herbert, 2 September 1983, SAOHP/NASM.

Friedman, Herbert, Talbot A. Chubb, E. T. Byram, and Robert Kreplin, "Early X-Ray Astronomy: Session One," VHI no. 1, 12 December 1986, SIVP/SIA.

Goldberg, Leo (by Spencer R. Weart) no. 1, 16 May 1978, SHMA/ AIP.

Goldberg, Leo, no 2, 22 February 1983, SAOHP/NASM.

Hamill, James, 23 August 1967, NASM Library.

Havens, Ralph, 6 October 1983, SAOHP/NASM.

Hoffleit, Dorrit, 9 August 1979, SHMA/AIP.

Hoffleit, Dorrit, 26 April 1980, SAOHP/NASM.

Hulburt, E. O. (by David K. Allison), 22 August and 8 September 1977, HONRL.

Johnson, Charles Y., no. 1, 3 March 1982, SAOHP/NASM.

Johnson, Charles Y., no. 2 (by DeVorkin and Margaret Shea), 21 April 1982, SAOHP/NASM.

Johnson, Francis S., 23 June 1982, SAOHP/NASM.

Jursa, Adolph, 13 December 1983, SAOHP/NASM.

Krause, Ernst H., no. 1, 10 August 1982, SAOHP/NASM.

Krause, Ernst H., no. 2, 1 July 1983, SAOHP/NASM.

Krause, Ernst H., no. 3, 29 April 1986, SAOHP/NASM.

Newell, Homer E. (by Richard Hirsh), 17 July 1980, SHMA/AIP.

Newkirk, Gordon, 1 June 1983, SAOHP/NASM.

Ney, Edward P., 29 February 1984, SAOHP/NASM.

Pickering, William H., no. 1 (by Allan Needell and Joseph Tatarewicz), 14 December 1982, SAOHP/NASM.

Pickering, William H., no. 2, 4 August 1983, SAOHP/NASM.

Porter, Richard W., 16 April 1984, SAOHP/NASM.

Purcell, DeWitt, 28 October 1982, SAOHP/NASM.

Reisig, Gerhard, 27 June 1985, SAOHP/NASM.

Rense, William A., 27 July 1983, SAOHP/NASM.

Roberts, Walter Orr, 26 July 1983, SAOHP/NASM.

Rosen, Milton, 25 March 1983, SAOHP/NASM.

Schopper, Erwin, 24 November 1988, SAOHP/NASM.

Schwarzschild, Martin (by Spencer R. Weart and David DeVorkin), 16 December 1977, AIP/SHMA.

Silberstein, Richard, 28 July 1983, SAOHP/NASM.

Simpson, John, 28 July 1983, SAOHP/NASM.

Simpson, John A., no. 1 (by Allan Needell), 28 August 1986, DSH/NASM.

Simpson, John A., no. 2 (by Allan Needell), 21 March 1988, DSH/NASM.

Singer, S. Fred, no. 1, 11 February 1991, DSH/NASM.

Singer, S. Fred, no. 2, 4 March 1991, DSH/NASM.

Singer, S. Fred, no. 3 (by DeVorkin and Allan Needell), 23 April 1991, DSH/NASM.

Sitterly, Charlotte Moore, 15 June 1978, SHMA/AIP.

Spencer, Nelson W., 22 December 1986, SAOHP/NASM.

Spitzer, Lyman Jr., 17 June 1982, SAOHP/NASM.

Strong John, 20 April 1984, SAOHP/NASM.

Stroud, William G. (by James Capshew), 6 December 1988; 5 January 1989. Combined abstract, 11 January 1989, GSFC/NASA.

Tombaugh, Clyde, 17 February 1982, SAOHP/NASM.

Tousey, Richard, no. 1, 17 November 1981, SAOHP/NASM.

Tousey, Richard, no. 2, 8 January 1982, SAOHP/NASM.

Tousey, Richard, no. 3, 6 April 1982, SAOHP/NASM.

Van Allen, James A., no. 1, 18 February 1981, SAOHP/NASM.

Van Allen, James A., no. 2, 12 June 1981, SAOHP/NASM.

Van Allen, James A., no. 3, 18 June 1981, SAOHP/NASM.

Van Allen, James A., no. 4, 22 June 1981, SAOHP/NASM.

Van Allen, James A., no. 5, 15 July 1981, SAOHP/NASM.

Van Allen, James A., no. 6, 16 July 1981, SAOHP/NASM.

Van Allen, James A., no. 7, 28 July 1981, SAOHP/NASM.

Van Allen, James A., no. 8, 6 August 1981, SAOHP/NASM.

Whipple, Fred (by Pamela Henson), 24–25 June 1976, Smithsonian Institution Archives.

Whipple, Fred, 1977, SHMA/AIP.

## Published Sources

A few of the following sources—such as APL and NRL laboratory reports—are not readily available. These, along with the far more numerous Air Force contractor reports, can be found in the libraries at NRL, APL, and the Air Force Geophysics Laboratory, Hanscom Field, Massachusetts.

Akasofu, Syun-Ichi, Benson Fogle, and Bernhard Haurwitz. 1968. *Sydney Chapman, Eighty*. NCAR, private printing.

Allison, David Kite. 1981. *New Eye for the Navy: The Origin of Radar at the Naval Research Laboratory.* Washington, D.C.: Naval Research Laboratory, Report 8466, 29 September.

Alpher, Ralph A. 1950. "Theoretical Geomagnetic Effects in Cosmic Radiation." *Journal of Geophysical Research* 55 no. 4 (December):437–71.

[APL]. 1946. "Papers Presented at: I. Princeton Telemetry Symposium, [and] II. APL Telemetering Conference." *Bumblebee Report* no. 42 (December).

[APL]. 1948. *So Columbus Was Right!* Silver Spring: Applied Physics Laboratory, n.d.

[APL] 1983. *The First Forty Years.* Baltimore: Applied Physics Laboratory.

[Army Ordnance]. 1945. *Peenemünde East, through the Eyes of 500 Detained at Garmish.* Known also as *The Story of Peenemünde, of What Might Have Been.* Typescript, JPL Library file no. 519.652–1, 4055–6A.

Augenstein, Bruno W. 1982. *Evolution of the U.S. Military Space Program 1945–1960: Some Key Events in Study Planning and Program Development.* Santa Monica: RAND.

Auger, Pierre. 1945. *What Are Cosmic Rays?* Trans. M. M. Shapiro. Chicago.

[Auger, Pierre]. 1949. *Repertory of Cosmic Rays Laboratories and Physicists.* UNESCO: International Union of Pure and Applied Physics, R.C. 49–1.

Bartels, Julius, and Alfred Ehmert. 1961. "Max-Planck-Institut für Aeronomie in Lindau." *Jahrbuch der Max-Planck-Gesellschaft zur Förderung der Wissenschaften e.v.,* pt. 2:16–30.

Barth, C. A. and J. A. Kaplan. 1957. "Chemistry of an Oxygen-Nitrogen Atmosphere." In Zelikoff, 1957:3–13.

Bartman, F. L. 1954. "Falling Sphere Method for Upper Air Density and Temperature." In Boyd and Seaton 1954:98–107.

Bates, Charles C., and John F. Fuller. 1986. *America's Weather Warriors 1814–1985.* Austin: University of Texas Press.

Bates, David R. 1954. "Consideration of the Results Obtained by Rockets." in Boyd and Seaton 1954:347–56.

Bates, David R. 1957. "Exploration of the Upper Atmosphere." chap. 4, in D. R. Bates, ed., *Space Research and Exploration.* London: Eyre and Spottiswoode.

Bates, David R. 1968 "Ionospheric Physics and Aeronomy." In Akasofu 1968:31–34.

Bates, D. R., and Harrie S. W. Massey. 1946. "The Basic Reactions in the Upper Atmosphere I." *Proceedings of the Royal Society* 187 (November):261–96.

Bates, David R., and Harrie S. W. Massey. 1951. "The Negative-Ion Concentration in the Lower Atmosphere." *Journal of Atmospheric and Terrestrial Physics* 2:1–13.

Bates, David R., and Marcel Nicolet. 1950. "The Photochemistry of Atmospheric Water Vapor." *Journal of Geophysical Research* 55 no. 3:301–27.

Baum, William A., F. S. Johnson, J. J. Oberly, C. C. Rockwood, C. V. Strain, and R. Tousey. 1946. "Solar Ultraviolet Spectrum to 88 Kilometers." *Physical Review* 70:781–82.

Baxter, James Phinney. 1946. *Scientists Against Time*. Little, Brown.

Beard, Edmund. 1976. *Developing the ICBM: A Study in Bureaucratic Politics*. Columbia.

Behring, W. E., J. M. Jackson, S. C. Miller Jr., and William A. Rense. 1954. "Monochromatic Camera for Photography in the Far Ultraviolet." *Journal of the Optical Society of America* 44 no. 3:229–31.

Benton, Mildred. 1959. *The Use of High-Altitude Rockets for Scientific Investigations: An Annotated Bibliography*. Washington D.C.: Naval Research Laboratory.

Bergman, C. W., R. S. MacMillan, and W. H. Pickering. 1954. "A New Technique for Investigating the Ionosphere at Low and Very Low Radio Frequencies." In Boyd and Seaton 1954:247–55.

Berkner, L. V. 1939. "Radio Exploration of the Earth's Outer Atmosphere." in Fleming 1939, chap. 9.

Berning, Warren W. 1951. "Charge Densities in the Ionosphere from Radio Doppler Data." *Journal of Meteorology* 8 (June):175–81.

Berning, Warren W. 1954. "The Determination on Charge Density in the Ionosphere by Radio Doppler Techniques." In Boyd and Seaton 1954:261–73.

Best, N. R., D. I. Gale, and R. J. Havens. 1946. "Pressure Measurements in the V-2." in Garstens, Newell, and Siry 1946, chap. 3, pt. E:47–48.

Beyerchen, Alan D. 1977. *Scientists Under Hitler*. Yale.

[The Bird Dogs]. 1981. "The Evolution of the Office of Naval Research." *Physics Today* (August). Reprinted in Spencer R. Weart and Melba Phillips, eds., *History of Physics*. New York: American Institute of Physics, 1985.

Bishop, Amasa A. 1958. *Project Sherwood: The U.S. Program in Controlled Fusion*. Addison-Wesley.

Bowen, I. S., R. A. Millikan, and H. V. Neher. 1938. "New Light on the Nature and Origin of the Incoming Cosmic Rays." *Physical Review* 53:855–61.

Bower, Tom. 1987. *The Paperclip Conspiracy*. London: Michael Joseph.

Bowhill, S. A. 1974. "Investigations of Ionosphere by Space Techniques." *Journal of Atmospheric and Terrestrial Physics* 36:2235–43.

Boyce, J. C. 1941. "Spectroscopy in the Vacuum Ultraviolet." *Reviews of Modern Physics* 13 no. 1:1–57.

Boyce, J. C., ed. 1947. *New Weapons for Air Warfare*. OSRD: Science in World War II Series. Little, Brown.

Boyd, R. L. F. 1954. "Some Problems in the Use of Ion and Electron Probes." In Boyd and Seaton 1954:336–38.

Boyd, R. L. F., and M. J. Seaton, with H. S. W. Massey. 1954. *Rocket Exploration of the Upper Atmosphere*. London: Pergamon.

Braddick, H. J. J. 1939. *Cosmic Rays and Mesotrons*. Cambridge University Press.

Breit, Gregory, and Merle Tuve. 1926. "A Test of the Existence of the Conducting Layer." *Physical Review* 28:554–75. This paper appeared after two notes (in *Terrestrial Magnetism* 1925 30:15–16; and *Nature* 1925 116:357) by Tuve and Breit announcing the technique itself.

Bromberg, Joan Lisa. 1982. *Fusion: Science, Politics, and the Invention of a New Energy Source*. MIT Press.

Brown, Eunice H., James A. Robertson, John W. Kroehnke, Charles R. Poisall, and Lt. E. L. Cross. 1959. *White Sands History: Range Beginnings and Early History*. Public Affairs Office, White Sands Missile Range.

Brown, Laurie, and Lillian Hoddeson. 1983. "Birth of Elementary Particle Physics: 1930–1950." in Brown and Hoddeson 1983:8–9.

Brown, Laurie, and Lillian Hoddeson, eds., 1983. *Birth of Particle Physics*. Cambridge University Press.

Burnight, T. R. 1949. "Soft X-Radiation in the Upper Atmosphere." *Physical Review* 76 (July):165. Abstract.

Burnight, T. R. 1952. "Ultraviolet Radiation and X-Rays of Solar Origin." In White and Benson 1952, chap. 13:226–38.

Burnight, T. R., and J. C. Seddon. 1947. "The Ionosphere." In Newell and Siry 1947a, chap. 2, pt. C:17–23.

Burnight, T. R., J. F. Clark Jr., and J. C. Seddon. 1947. "The March 7 Ionosphere Experiment." In Newell and Siry 1947b, chap. 6, pt. A:85–88.

Burnight, T. R., W. F. Fry, and J. C. Seddon. 1947. "Rocket Antennas for Ionosphere Research." In Newell and Siry 1947b, chap. 6, pt. C:97–100.

Burrows, William E. 1986. *Deep Black: Space Espionage and National Security*. Random House.

Bush, Vannevar. 1970. *Pieces of the Action*. Morrow.

Bushnell, David. 1962. *The Sacramento Peak Observatory 1947–1962*. Office of Aerospace History OAR-5.

Byram, E. T., T. A. Chubb, and H. Friedman. 1954a, "Solar X-rays and E-Layer Ionization." in Boyd and Seaton 1954:274–75.

Byram, E. T., T. A. Chubb, and H. Friedman. 1954b, "The Study of Extreme Ultraviolet Radiation From the Sun With Rocket-Borne Photon Counters." In Boyd and Seaton 1954:276–78.

Carmichael, Hugh. 1985. "Edinburgh, Cambridge and Baffin Bay." In Sekido and Elliot 1985:99–113.

Chackett, K. F., F. A. Paneth, P. Reasbeck, and B. S. Wiborg. 1951. "Variations in the Chemical Composition of Stratosphere Air." *Nature* 168 (1 September):358.

Chapman, Sydney. 1931. "The Absorption and Dissociative or Ionizing Effect of Monochromatic Radiation in an Atmosphere on a Rotating Earth." *Proceedings of the Physical Society London* 43:26–45; 433–501.

Chapman, Sydney. 1953. "The International Geophysical Year 1957–58." *Nature* 172 (22 August):327–29.

Chapman, Sydney. 1954. "Rockets and the Magnetic Exploration of the Ionosphere." In Boyd and Seaton 1954:292–305.

Chapman, Sydney. 1959. *IGY, Year of Discovery*. Ann Arbor, Mich.

Chapman, Sydney, and J. Bartels. 1940. *Geomagnetism*. Oxford.

Christman, Albert B. 1971. *Sailors, Scientists, and Rockets: Origins of the Navy Rocket Program and of the Naval Ordnance Test Station, Inyokern*. vol. 1. Naval History Division: History of the Naval Weapons Center, China Lake, California.

Chubb, Talbot A., and Herbert Friedman. 1955. "Photon Counters for the Far Ultraviolet." *Review of Scientific Instruments* 26:493–98.

Clark, Harry L. 1949. "A Sun-Follower for the V-2 Rockets." *Upper Atmosphere Research Report*. no. ix. NRL Report 3522 (11 August).

Clark, Harry L. 1950. "Sun Follower for V-2 Rockets." *Electronics* 23 (October):71–73.

Clark, Harry L. 1953. "A Lightweight Azimuth-Correcting Sun Follower." *Upper Atmosphere Research Report* no. xx. NRL Report 4267 (17 December).

Clark, J. F. Jr., T. M. Moore, and J. C. Seddon. 1946. "The Ionosphere Experiment." In Newell and Siry 1946, chap. 4, pt. E:91–99.

Clark, J. F. Jr., C. W. Miller, and G. H. Spaid. 1947. "The Ionosphere Receiving and Recording System." In Newell and Siry 1947b, chap. 6, pt. D:100–10.

Clark, J. F. Jr., C. Y. Johnson, T. M. Moore, and J. C. Seddon. 1946. "Development of Equipment and Techniques for the Ionosphere Experiment." In Garstens, Newell, and Siry 1946, chap. 3, pt. G:57–62.

Clarke, Arthur C. 1945. "V-2 for Ionospheric Research?" Letters to the Editor, *Wireless World* 51 no. 2 (February):58.

Clearman, Harold. 1949. "The Ultraviolet Solar Spectrum." In Kuiper 1949:125–34.

Clearman, Harold. 1953. "The Solar Spectrum from 2285 to 3000Å." *Astrophysical Journal* 117:29–40.

Cochrane, Rexmond C. 1966. *Measures for Progress: A History of the National Bureau of Standards*. Washington, D.C.: Department of Commerce.

Compton, A. H. 1933. "A Geographical Study of Cosmic Rays." *Physical Review* 43:387–403.

Compton, A. H. 1936. "Cosmic Rays as Electrical Particles." *Physical Review* 50:1130.

Condon, E. U. 1952. "Some Thoughts on Science in the Federal Government." *Physics Today* (April). Reprinted in Weart and Phillips 1985:130–37.

Corliss, William R. 1971. *NASA Sounding Rockets, 1958–1968: A Historical Summary*. Washington, D.C.: NASA SP-4401.

Corn, Joseph J. 1983. *The Winged Gospel: America's Romance with Aviation, 1900–1950*. New York: Oxford.

Davies, Merton E., and William R. Harris. 1988. *RAND's Role in the Evolution of Balloon and Satellite Observation Systems and Related U.S. Space Technology*. Santa Monica: RAND.

Davis, Vincent. 1967. "The Politics of Innovation: Patterns in Navy Cases." In *The Social Science Foundation and Graduate School of International Studies Monograph Series in World Affairs* 4 no. 3. University of Denver.

Dellinger, J. H. 1937. "Sudden Disturbances of the Ionosphere." *Proceedings of the Institute of Radio Engineers* 25 no. 2:1253–1290.

Delsasso, L. A., L. G. de Bey, and D. Reuyl. 1947. *Full-Scale Free-Flight Ballistic Measurements of Guided Missiles.* Ballistic Research Laboratory, n.d. circa 1947.

De Maria, Michelangelo, and Arturo Russo. 1989. "Cosmic Ray Romancing: The Discovery of the Latitude Effect and the Compton-Millikan Controversy." *Historical Studies in the Physical and Biological Sciences* 19:211–66.

De Maria, Michelangelo, Mario Grilli, and Fabio Sabastiani, eds. 1989. *The Restructuring of Physical Sciences in Europe and the United States 1945–1960.* London: World Scientific.

Dennis, Michael Aaron. 1986. "Making Space: Sounding the Territory of the Upper Atmosphere Research Archipelago, 1944–1946." Paper presented at the Workshop on the Military and Post-War Academic Science, 17–18 April.

Dennis, Michael Aaron. 1990. *A Change of State: The Political Cultures of Technical Practice at the MIT Instrumentation Laboratory and the Johns Hopkins University Applied Physics Laboratory, 1930–1945.* Ph.D. diss., The Johns Hopkins University. University Microfilms, 1991.

Dennis, Michael Aaron. 1991. "Reconstructing Technical Practice: The Johns Hopkins University Applied Physics Laboratory and the Massachusetts Institute of Technology Instrumentation Laboratory after World War II." In David van Keuren and Nathan Reingold, eds., *Science and the Federal Patron.* Forthcoming.

DeVorkin, David H. 1975. "A. A. Michelson and the Problem of Stellar Diameters." *Journal for the History of Astronomy* 6:1–18.

DeVorkin, David H. 1980. "An Astronomer Responds to War: Otto Struve and the Yerkes Observatory During World War II." *Minerva* 18 no. 4 (Winter):595–623.

DeVorkin, David H. 1984. "The Harvard Summer School in Astronomy." *Physics Today* 37 no. 7 (July):48–55.

DeVorkin, David H. 1985. "Electronics in Astronomy: Early Applications of the Photoelectric Cell and Photomultiplier for Studies of Point Source Celestial Phenomena." *Proceedings of the IEEE* 73 (July):1205–20.

DeVorkin, David H. 1986. "The Dawn of Balloon Astronomy." *Sky & Telescope* 72 (December):579–81.

DeVorkin, David H. 1987a. "Organizing for Space Research: The V-2 Panel." *Historical Studies in the Physical and Biological Sciences* 18, pt. 1:1–24.

DeVorkin, David H. 1987b. "John Strong's First Aluminized Mirror." *Rittenhouse* 2:1–10.

DeVorkin, David H. 1989a. *Race to the Stratosphere: Manned Scientific Ballooning in America.* Springer.

DeVorkin, David H. 1989b. "Along for the Ride: The Response of American Astronomers to the Possibility of Space Research, 1945–1950." In De Maria, Grilli, and Sabastiani 1989:55–74.

DeVorkin, David H. 1990 "Defending a Dream: Charles Greeley Abbot's Years at the Smithsonian." *Journal for the History of Astronomy* 21:121–36.

DeVorkin, David H. 1991. "Back to the Future: American Astronomers' Response to the Prospect of Federal Patronage, 1947–1955: The Origin of the ONR Program in Astronomy." In David van Keuren and Nathan Reingold, eds., *Science and the Federal Patron*. Forthcoming.

Dewey, Anne Perkins. 1962. *Robert Goddard, Space Pioneer*. Little, Brown.

Doel, Ron. 1990a. *Unpacking a Myth: Interdisciplinary Research and the Growth of Solar System Astronomy, 1920–1958*. Ph.D. diss., Princeton. University Microfilms.

Doel, Ron. 1990b. "Redefining a Mission: The Smithsonian Astrophysical Observatory on the Move." *Journal for the History of Astronomy* 21:137–53.

Dornberger, Walter R. 1958. *V-2*. Viking.

Dornberger, Walter R. 1963. "The German V-2." *Technology and Culture* 4 no. 4:393–408.

Dow, W. G., A. Reifman, and F. V. Schultz. 1947. *Experimental Pressure Data Obtained from V-2 Rocket Flights of 21 November 1946 and 20 February 1947 with a Preliminary Analysis of these Data*. Department of Engineering Research, University of Michigan, 3 December.

Dupré, J. Stefan, and Sanford A. Lakoff. 1962. *Science and the Nation*. Prentice-Hall.

Dupree, A. Hunter. 1957. *Science in the Federal Government: A History of Policies and Activities*. Harvard, 1957; 2nd edn., Johns Hopkins, 1986.

Dupree, A. Hunter. 1960. "Influence of the Past: An Interpretation of Recent Development in the Context of 200 Years of History." *Annals of the American Academy of Political and Social Science* 327:19–26.

Dupree, A. Hunter. 1965. "Paths to the Sixties." In D. L. Arm, ed., *Science in the Sixties*:1–9. University of New Mexico Press.

Dupree, A. Hunter. 1972. "The Great Instauration of 1940: The Organization of Scientific Research for War." In Gerald Holton, ed., *The Twentieth Century Sciences: Studies in the Biography of Ideas*:443–67. Norton.

Durand, Eric. 1949. "Rocket Sonde Research at the Naval Research Laboratory." In Kuiper 1949:134–48.

Durand, Eric, and R. Tousey. 1947. "Objectives, Accomplishments and Proposed Immediate and Long Range Plans in the Basic Research Program A. Astrophysics and Atmospheric Composition." In Newell and Siry 1947a:7–13.

Durand, Eric, J. J. Oberly, and Richard Tousey. 1947. "Solar Absorption Lines between 2950 and 2200 Angstroms." *Physical Review* 71:827.

Durand, Eric, J. J. Oberly, and Richard Tousey. 1949. "Analysis of the First Rocket Ultraviolet Solar Spectra." *Astrophysical Journal* 109:1–16.

Durant, F. C. III. 1974. "Robert H. Goddard and the Smithsonian Institution." in F. C. Durant and George S. James, eds., *First Steps Toward Space*. chap. 5:59.

Dyson, Freeman. 1979. *Disturbing the Universe*. Harper and Row.

Edge, David O., and Michael Mulkay. 1976. *Astronomy Transformed: The Emergence of Radio Astronomy in Britain*. Wiley.

Ehricke, Krafft A. 1950. "The Peenemünde Rocket Center." *Rocketscience* 4:17–22; 31–34; 57–63. Journal of the Cleveland Rocket Society.

Ehmert, A. 1953. "The Measurement of Cosmic-Ray Time Variations with Ion Chambers and Counter Telescopes." In Kuiper 1953:711–15.

England, J. Merton. 1982. *A Patron for Pure Science*. Washington, D.C.: National Science Foundation.

Evans, J. W. 1953. "The Coronograph." In Kuiper 1953:635–44.

Fleming, J. A., ed. 1939. *Physics of the Earth VIII: Terrestrial Magnetism and Electricity*. McGraw-Hill.

Forbush, S. E., T. B. Stinchcomb, and M. Schein. 1950. *Physical Review* 79:501.

Forman, Paul. 1987. "Behind Quantum Electronics: National Security as Basis for Physical Research in the United States, 1940–1960." *Historical Studies in the Physical and Biological Sciences* 18 pt. 1:149–229.

Fraser L. W. 1948. "Aerobee High Altitude Sounding Rocket: Design, Construction and Use." *APL Bumblebee Report* no. 95. December.

Fraser L. W. 1951. "High Altitude Research at the Applied Physics Laboratory." *APL Bumblebee Report* no. 153. May.

Fraser L. W., and E. H. Siegler. 1948. "High Altitude Research Using the V-2 Rocket, March 1946–April 1947." *APL Bumblebee Report* no. 81. July.

Fraser, L. W., R. P. Peterson, H. E. Tatel, and J. A. Van Allen. 1947. "Methods in Cosmic-Ray Measurement in Rockets." Abstract. *Physical Review* 72:173.

Fraunberger, F. 1975. "Regener, Erich Rudolph Alexander." *Dictionary of Scientific Biography* 11:347–48. Scribners.

Freier, Phyllis, E. J. Lofgren, E. P. Ney, F. R. Oppenheimer, H. L. Bradt, and B. Peters. 1948a. "Evidence for Heavy Nuclei in the Primary Cosmic Radiation." *Physical Review* 74:213–17.

Freier, Phyllis, E. J. Lofgren, E. P. Ney, and F. R. Oppenheimer. 1948b. "The Heavy Component of Primary Cosmic Rays." *Physical Review* 74:1818–27.

Friedman, Herbert. 1960. "The Sun's Ionizing Radiations." In Ratcliffe 1960, chap. 4.

Friedman, Herbert, S. W. Lichtman, and E. T. Byram. 1951. "Photon Counter Measurements of Solar X-Rays and Extreme Ultraviolet Light." *Physical Review* 83:1025–30.

Friedman, Robert. 1989. *Appropriating the Weather.* Cornell.

Galison, Peter. 1985. "Bubble Chambers and the Experimental Workplace." In Achinson and Hannaway 1985, chap. 10:309–73.

Galison, Peter. 1987. *How Experiments End.* University of Chicago.

Garber, Sheldon, and Felicia Holton. 1960. "Voyage of the 'Valley Forge'." *University of Chicago Reports* 10 (January-February 1960):nos. 4, 5.

Ghosh, S. N., and S. R. Das Gupta. 1978. "M. N. Saha's Work on the Effect of Solar UV Radiations on Upper Atmosphere and Subsequent Developments." *Science and Culture* 44:437–452.

Garstens, M. A., H. E. Newell Jr., and J. W. Siry, eds. 1946. *Upper Atmosphere Research Report* no. I. Naval Research Laboratory Report R-2955, October.

Gibson, R. E. 1976. "Reflections on the Origin and Early History of the Applied Physics Laboratory." *APL Technical Digest* 15 no. 2 (April-June):2–30.

Gieryn, Thomas F., and Richard F. Hirsh. 1983. "Marginality and Innovation in Science." *Social Studies of Science* 13:87–106.

Gieryn, Thomas F., and Richard F. Hirsh. 1984. "Marginalia: Reply to Simonton and Handberg." *Social Studies of Science* 14:624.

Gilbert, G. Nigel. 1976. "The Development of Science and Scientific Knowledge: The Case of Radar Meteor Research." In Lemaine, Macleod, Mulkay, and Weingart 1976:187–204.

Gillmor, C. Stewart. 1981. "Threshold to Space: Early Studies of the Ionosphere." In Hanle and Chamberlain 1981:101–14.

Gillmor, C. Stewart. 1986. "Federal Funding and Knowledge Growth in Ionospheric Physics, 1945–81." *Social Studies of Science* 16:105–33.

Gillmor, C. Stewart. 1989. "Geospace and Its Uses: The Restructuring of Ionospheric Physics Following World War II." In De Maria, Grilli, and Sabastiani 1989:75–84.

Gillmor, C. Stewart, and C. J. Terman. 1984. "Communication Modes of Geophysics: The Case of Ionospheric Physics." In Gillmor, ed., *History of Geophysics* 1:89–97. American Geophysical Union.

Gimbel, John. 1986. "U.S. Policy and German Scientists: The Early Cold War." *Political Science Quarterly* 101 no. 3:433–51.

Gimbel, John. 1990. "German Scientists, United States Denazification Policy, and the '*Paperclip* Conspiracy,'"*International History Review* 12 (August):441–65.

Gingrich, Curvin. 1944. "Review of Willy Ley, *Rockets, the Future of Travel Beyond the Stratosphere.*" *Popular Astronomy* 52 (August):360.

Ginzburg, V. L., and S. I. Syrovatskii. 1964. Trans. H. S. H. Massey, and ed. D. ter Haar. *The Origin of Cosmic Rays.* Macmillan.

Gerwin, Robert. 1986. "75 Jahre Max-Planck-Gesellschaft." *Naturwissenschaftliche Rundschau* 39:1–10; 49–62; 97–109.

Goddard, Robert H. 1919. "A Method of Reaching Extreme Altitudes." *Smithsonian Miscellaneous Collections* 71 (2 November). Reprinted in [Goddard]. 1919:340–41.

[Goddard, Robert H.]. 1970. *The Papers of Robert H. Goddard*. vol. 1. Esther C. Goddard and G. Edward Pendray, eds. McGraw-Hill.

Goldberg, Leo. 1981. "Solar Physics." In Hanle and Chamberlain 1981:15–29.

Goldberg, Stanley. 1989. "Between Old and New: Goudsmit at Brookhaven." In De Maria, Grilli, and Sabastiani 1989:117–37.

Goldstine, Herman H. 1973. *The Computer from Pascal to von Neumann*. Princeton University Press.

Golian, S. E., and E. H. Krause. 1947. "Further Cosmic-Ray Experiments Above the Atmosphere." *Physical Review* 71:918–19.

Golian, S. E., E. H. Krause, and G. J. Perlow. 1946. "Cosmic Radiation Above 40 Miles." *Physical Review* 70:223–24.

Golian, S. E., C. Y. Johnson, E. H. Krause, and M. L. Kuder. 1948. "The Cloud Chamber." In Newell and Siry 1948, chap. 4, pt. D:58–64.

Golian, S. E., C. Y. Johnson, E. H. Krause, M. L. Kuder, G. J. Perlow, and C. A. Schroeder. 1949. "V-2 Cloud Chamber Observation of a Multiply Charged Primary Cosmic Ray." *Physical Review* 75:524–25.

Goudsmit, Samuel A. 1947. *Alsos*. Schuman.

Green, Charles F. 1954. "Utilization of the V-2 A-4 Rocket in Upper Atmosphere Research." In Boyd and Seaton 1954:28–45.

Green, Constance McLaughlin, and Milton Lomask. 1970. *Vanguard: A History*. Washington, D.C.: NASA.

Greenough, M. L., J. J. Oberly, and C. C. Rockwood. 1947. "The Photoelectric Spectrometer." In Newell and Siry 1947b, chap. 5, pt. C:78–84.

Greenstein, J. L. 1984. "Optical and radio astronomers in the early years." In W. T. Sullivan III ed., *The Early Years of Radio Astronomy*:43–82. Cambridge University Press.

Hagen, John P. 1963. "The Viking and the Vanguard." *Technology and Culture* 4 no. 4:435–51.

Hall, R. Cargill. 1963. "Early U.S. Satellite Proposals." *Technology and Culture* 4 no. 4:410–34.

Hall, R. Cargill. 1977. "Earth Satellites, A First Look by the United States Navy." In Hall 1986:253–77.

Hall, R. Cargill, ed. 1986. *History of Rocketry and Astronautics*. AAS History Series 7 pt. 2. Originally printed with same pagination as: *Essays on the History of Rocketry and Astronautics: Proceedings of the Third through the Sixth History Symposia of the International Academy of Astronautics*. vol. 2. NASA Conference Publication 2014: 1977.

Hallion, Richard P. 1977. *Legacy of Flight: The Guggenheim Contribution to American Aviation*. University of Washington.

Handberg, Roger. 1984. "Response to Gieryn and Hirsh." *Social Studies of Science* 14:622–24.

Hanle, Paul A., and Von Del Chamberlain, eds. 1981. *Space Science Comes of Age*. Smithsonian.

Harwit, Martin. 1981. *Cosmic Discovery: The Search, Scope and Heritage of Astronomy*. Basic Books.

Haurwitz, Bernhard. 1936. "The Physical State of the Upper Atmosphere." *Journal of the Royal Astronomical Society of Canada* 30:315–330; 349–366; 396–415; 31 (1937):19–42; 76–92.

Havens, R. J., and H. E. LaGow. 1946. "Temperature Measurements in the V-2." In Garstens, Newell, and Siry 1946, chap. 3, pt. D:44–46.

Havens, R. J., R. Koll, and H. E. LaGow. 1952. "The Pressure, Density, and Temperature of the Earth's Atmosphere to 160 Kilometers." *Journal of Geophysical Research* 57:59–72.

Havens, R. J., Herbert Friedman, and E. O. Hulburt. 1955. "The Ionospheric $F_2$ Region." In [Physical Society]:237–44.

Heeren, V. L., C. H. Hoeppner, J. R. Kauke, S. W. Lichtman, and P. R. Shifflet. 1947. "Telemetering from V-2 Rockets." *Electronics* 20 (April):pt 1, 100–05; pt. 2, 124–27. Based on NRL Report R-3013, 1 October 1946.

Henry, R. C., Peter Beer, and D. H. DeVorkin, eds. 1986. *H. A. Rowland and Astronomical Spectroscopy*. Vistas in Astronomy 29 no. 2:115–236.

Hevly, Bruce William. 1987. *Basic Research Within a Military Context: The Naval Research Laboratory and the Foundations of Extreme Ultraviolet and X-Ray Astronomy*. Ph.D. diss. The Johns Hopkins University.

Hey, J. S. 1973. *The Evolution of Radio Astronomy*. Neale Watson.

Hinsley, F. H. 1988. *British Intelligence in the Second World War*. vol. 3, pt. 2. London.

Hinteregger, H. E., and Kenichi Watanabe. 1953. "Photoelectric Cells for the Vacuum Ultraviolet." *Journal of the Optical Society of America* 43 no. 7:604–08.

Hinteregger, H. E., K. R. Damon, L. Heroux, and L. A. Hall. 1960. "Telemetering Monochromater Measurements of Solar 304Å Radiation and its Attenuation in the Upper Atmosphere." In H. K. Kallmann-Bijl, ed., *Space Research* 1:615–27.

Hintsches, Eugen. 1988. "Space Exploration Begins with Regener's Tonne." ("Mit der Regener-Tonne beginnt die Weltraumforchung.") *Forschungsberichte und meldungen aus der MPG*. Max Planck Gesellschaft, PRI SP 4/88 (25), 24 November.

Hirsh, Richard. 1983. *Glimpsing an Invisible Universe: The Emergence of X-Ray Astronomy*. Cambridge University Press.

Hoffleit, Dorrit. 1949. "DOVAP–-A Method for Surveying High-Altitude Trajectories." *Scientific Monthly* 68 no. 3 (March):172–78.

Hoffleit, Dorrit. 1949b. "Rockets in Exploration and Science." *Popular Astronomy* 57 no. 1:1–8.

Hoffleit, Dorrit. 1988. "Yale Contributions to Meteoric Astronomy." *Vistas in Astronomy* 32:117–43.

Hok, Gunnar, and W. G. Dow. 1954. "Exploration of the Ionosphere by Means of a Langmuir-Probe Technique." In Boyd and Seaton 1954:240–46.

Hok, Gunnar, N. W. Spencer, and W. G. Dow. 1953. "Dynamic Probe Measurements in the Ionosphere." *Journal of Geophysical Research* 58 no. 2:235–42.

Hok, Gunnar, N. W. Spencer, A. Reifman, and W. G. Dow. 1951. "Dynamic Probe Measurements in the Ionosphere." In *Upper Air Research Program Report* no. 3. University of Michigan: Engineering Research Institute, February 1951.

Holliday, C. T. 1950a. "Preliminary Report on High Altitude Photography." *Photographic Engineering* 1:16–26.

Holliday, C. T. 1950b. "Seeing the Earth from 80 Miles Up." *National Geographic Magazine* (October):511–28.

Hölsken, Heinz Dieter. 1984. *Die V-Waffen: Entstehung–Propaganda–Kriegseinsatz.* Stuttgart: Deutsche Verlagsanstalt.

Hopfield, J. J. 1922. "Spectra of Hydrogen, Nitrogen and Oxygen in the Extreme Ultraviolet." *Physical Review* 20:573–88.

Hopfield, J. J. 1946. "Ultraviolet Absorption Spectrum of Air in the Region 600–2000Å." *Astrophysical Journal* 104:208–10.

Hopfield, J. J., and H. Clearman. 1948. "The Ultraviolet Spectrum of the Sun from V-2 Rockets." *Physical Review* 73:877–84.

Howland, B., G. J. Perlow, and J. D. Shipman Jr. 1948. "The Counter Telescope Experiment of May 15, 1947." In Newell and Siry 1948, chap. 4, pt. A:45–49.

Hoyle, Fred, and D. R. Bates. 1948. "The Production of the E-Layer." *Terrestrial Magnetism and Atmospheric Electricity* 53:51–62.

Hubert, L. F., S. Fritz, and H. Wexler. 1960. "Pictures of the Earth from high altitudes and their meteorological significance." In Kallmann-Bijl 1960:3–7.

Hufbauer, Karl. 1991. *Exploring the Sun: Solar Science Since Galileo.* Johns Hopkins.

Hufbauer, Karl. 1992. "Breakthrough on the Periphery: Bengt Edlén and the Identification of the Coronal Lines." In Svante Lindqvist, ed., *Center on the Periphery.* Forthcoming.

Hulburt, E. O. 1938. "Photoelectric Ionization in the Ionosphere." *Physical Review* 53:344–51.

Hulburt, E. O. 1939. "The E Region of the Ionosphere." *Physical Review* 55:639–47.

Hunt, Linda. 1985. "U. S. Coverup of Nazi Scientists." *Bulletin of the Atomic Scientists* 41 (April):16–24.

Hunter, A. 1942–43. "The Spectrum of the Sun in the Far Ultraviolet." *Reports on Progress in Physics* 9:5–9.

Huzel, Dieter K. 1962. *Peenemünde to Canaveral.* Prentice-Hall.

Irving, David. 1965. *The Mare's Nest.* Little, Brown.

[Institute for Scientific Information]. 1984. *Science Citation Index 1955–1964.* Philadelphia.

[Institute for Scientific Information]. 1988. *Science Citation Index 1945–1954.* Philadelphia.

Ivanov-Kholodnyi, G. S., and G. M. Nikol'skii. 1969. *The Sun and the Ionosphere: Short-Wave Solar Radiation and its Effect on the Ionosphere.* Washington, D.C.: Department of Commerce NTIS, 1972, NASA TT F-654.

Jacchia, Luigi G., and Fred L. Whipple. 1956. "The Harvard Photographic Meteor Programme." *Vistas in Astronomy* 2:982–994.

Jackson, John E. 1954. "Measurements in the E-Layer with the Navy Viking Rocket." *Journal of Geophysical Research* 59 no. 3:377–90.

Jacobs, David Michael. 1975. *The UFO Controversy in America*. Indiana University Press.

Jánossy, L. 1948. *Cosmic Rays*. Clarendon Press.

Johnson, C. Y., L. R. Davis, and J. W. Siry. 1954. "Cosmic Ray Measurements in Rockets." In Boyd and Seaton 1954, sec. 5, 306–12.

Johnson, F. S. 1954. "The Solar Constant." *Journal of Meteorology* 11:431.

Johnson, F. S., DeWitt Purcell, and Richard Tousey. 1951. "Measurements of the Vertical Distribution of Atmospheric Ozone from Rockets." *Journal of Geophysical Research* 56 no. 4:583–94.

Johnson, F. S., Kenichi Watanabe, and Richard Tousey. 1951. "Fluorescent Sensitized Photomultipliers for Heterochromatic Photometry in the Ultraviolet." *Journal of the Optical Society of America* 41:702–08.

Johnson, F. S., DeWitt Purcell, Richard Tousey, and N. Wilson. 1953. "A New Photograph of the Mg II Doublet at 2800 in the Sun." *Astrophysical Journal* 117:238–39.

Johnson, F. S., DeWitt Purcell, Richard Tousey, and K. Watanabe. 1952. "Direct Measurements of the Vertical Distribution of Atmospheric Ozone to 70 Kilometers Altitude." *Journal of Geophysical Research* 57:157–176.

Johnson, T. H. 1939. "Composition of Cosmic Rays: Evidence that Protons are the Primary Particles of the Hard Component." *Reviews of Modern Physics* 11:208.

Johnson, T. H., and S. A. Korff. 1939. "Geiger Counter Measurements in the Upper Atmosphere Bearing upon the Nature of the Radiation from Solar Flares and Radio Fade-Outs." *Terrestrial Magnetism* 44:23–27.

Jones, Bessie Zaban. 1965. *Lighthouse of the Skies*. Smithsonian.

Jones, Bessie Zaban, and Lyle Gifford Boyd. 1971. *The Harvard College Observatory*. Harvard.

Jones, Leslie M. 1954. "The Measurement of Diffusive Separation in the Upper Atmosphere." In Boyd and Seaton 1954:143–56.

Jones, Leslie M. 1975. "Structure of Neutral Upper Atmosphere Air Samples and Falling Sphere." In Ordway 1989:201–29.

Jones, R. V. 1978. *Most Secret War*. London: Hamish Hamilton.

Kallmann, H. K. 1958. "Properties of the Atmosphere Between 90 and 300 KM." *Vistas in Astronautics* 1:130–34.

Kallmann-Bijl, H. K. 1953. "The Continuous Layer Formation in the Atmosphere." *Physical Review* 90:153–54.

Kallmann-Bijl, H. K. 1955. *A Study of the Structure of the Ionosphere*. Ph.D. Diss., UCLA.

Kallmann-Bijl, H. K, ed. 1960. *Space Research: Proceedings of the First International Space Science Symposium*. Interscience.

Kallmann-Bijl, H. K., W. B. White, and H. E. Newell Jr. 1956. "Physical Properties of the Atmosphere from 90 to 300 Kilometers." *Journal of Geophysical Research* 61 no. 3:513–24.

Kargon, Robert H. 1981. "Birth Cries of the Elements: Theory and Experiment Along Millikan's Route to Cosmic Rays." In Harry Woolf, ed., *The Analytic Spirit*:309–25. Cornell.

Kargon, Robert H. 1982. *The Rise of Robert Millikan: Portrait of a Life in American Science*. Cornell.

Kellogg, W. W. et al. 1951. "Report of the Standing Committee on Problems of the Upper Atmosphere." *Transactions of the American Geophysical Union* 32 no. 5 (October):755–59.

Kellogg, W. W. et al. 1953. "Report of the Standing Committee on Problems of the Upper Atmosphere, 1951–1952." *Transactions of the American Geophysical Union* 34 no. 1 (February):115–21.

Kenat, Ralph, and D. H. DeVorkin. 1990. "Quantum Physics and the Stars III: Towards a Rational Theory of Stellar Spectra." *Journal for the History of Astronomy* 21:157–86.

Kennedy, Gregory P. 1983. *Vengeance Weapon 2: The V-2 Guided Missile*. Smithsonian.

Kevles, Daniel. 1975. "Scientists, the Military, and the Control of Post War Defense Research: The Case of the Research Board for National Security." *Technology and Culture* 16:20–47.

Kevles, Daniel. 1978. *The Physicists: The History of a Scientific Community in Modern America*. Alfred A. Knopf.

Kidwell, Peggy Aldrich. 1990. "Harvard Astronomers in the Second World War." *Journal for the History of Astronomy* 21:105–06.

Kidwell, Peggy Aldrich. 1991. "Harvard Astronomers in World War II–-Disruption and Opportunity." In Margaret W. Rossiter and Clark A. Elliott, eds., *Science at Harvard University: Historical Perspectives*. Forthcoming.

Kiepenheuer, K. O. 1948. "Solar-Terrestrische Erscheinungen." In P. ten Bruggencate, ed. *Naturforschung und Medizin in Deutschland 1939–1946*. Review of German Science 20 "Astronomie, Astrophysik und Kosmogonie," chap. 9. Wiesbaden: Field Information Agency, Technical.

King, Henry C. 1979. *The History of the Telescope*. Dover. Reprint of 1955 edition.

Klee, Ernst, and Otto Merk. 1965. *The Birth of the Missile: The Secrets of Peenemünde*. Dutton. Translation of *Damals in Peenemünde*. Oldenburg, 1963.

Knight, David. 1986. *The Age of Science*. Blackwell.

Komons, Nick A. 1966. *Science and the Air Force: A History of the Air Force Office of Scientific Research*. Arlington, VA: Historical Division, Office of Information, Office of Aerospace Research.

Kopal, Zdenek. 1986. *Of Stars and Men*. Adam Hilger.

Koppes, Clayton. 1982. *JPL and the American Space Program*. Yale.

Koppes, Clayton. 1989. "National Security Perceptions and the Jet Propulsion Laboratory." In De Maria, Grilli, and Sabastiani 1989:138–49.

Koppl, Frederick. 1947. "Recent Progress in the Measurement of Atmospheric Pressure." *Review of Scientific Instruments* 18 no. 11 (November):850–51.

Krause, Ernst H. 1946. "Future Research." In Newell and Siry 1946, chap. 6:145–47.

Krause, Ernst H. 1949. "High-Altitude Research with V-2 Rockets." Proceedings of the American Philosophical Society 91 no. 5 December 1947. Reprinted in *The Smithsonian Report for 1948*: 189–208.

Krause, E. H., and G. J. Perlow. 1947. "Objectives, Accomplishments and Proposed Immediate and Long Range Plans in the Basic Research Program." In Newell and Siry 1947a, chap. 2:14–15.

Kuiper, Gerard P. 1946. "German Astronomy During the War." *Popular Astronomy* 54 no. 6 (June):263–87.

Kuiper, Gerard P. ed. 1949; 1952. *The Atmospheres of the Earth and Planets*. University of Chicago. Rev. 1952.

Kuiper, Gerard P. ed. 1953. *The Sun*. University of Chicago.

LaGow, Herman E. 1954. "Physical Properties of the Atmosphere up to the $F_1$-Layer." In Boyd and Seaton 1954:73–81.

Lang, Daniel. 1948. *Early Tales of the Atomic Age*. Doubleday.

Lasby, Clarence G. 1975. *Project Paperclip*. Athenaeum.

Lattu, Kristan R. ed. 1989. *History of Rocketry and Astronautics*. San Diego: Univelt. American Astronautical Society History Series, vol. 8. Symposium date, 1974.

Leak, William M. 1954. *Rocket Research at NRL*. NRL Report 4441 (12 November).

Leitch, Alexander. 1978. *A Princeton Companion*. Princeton.

Lemaine, Gerard, Roy Macleod, Michael Mulkay, and Peter Weingart, eds. 1976. *Perspectives on the Emergence of Scientific Disciplines*. The Hague.

LePrince-Ringuet, Louis. 1950. *Cosmic Rays*. Trans. Fay Ajzenberg. Prentice-Hall.

Leslie, Stuart. 1987. "Playing the Education Game to Win: The Military and Interdisciplinary Research at Stanford." *Historical Studies in the Physical and Biological Sciences* 18 pt. 1:55–88.

Ley, Willy. 1944. *Rockets: the Future of Travel Beyond the Stratosphere*. New York.

Liebowitz, Ruth. 1985. *Chronology: From the Cambridge Field Station to the Air Force Geophysics Laboratory 1945–1985*. Special report no. 252 AFGL-TR-85–0201. Air Force Geophysics Laboratory, Hanscom Air Force Base, Bedford, Mass.

Lien, J. R., R. J. Marcou, J. C. Ulwick, J. Aarons, and D. R. McMorrow. 1954. "Ionosphere Research with Rocket-borne Instruments." In Boyd and Seaton 1954:223–39.

Liller, William. 1961. *Space Astrophysics*. McGraw-Hill.

Lin, Leslie, and Marlene Bunch. 1984. "The University of Michigan in the American Space Program." *Chronicle*. Quarterly Magazine of the Historical Society of Michigan 20 no. 2 (Summer):11–15; no. 3 (Fall):10–12.

Lindemann, F. A. and G. M. B. Dobson. 1923. "A Theory of Meteors and the Density and Temperature of the Outer Atmosphere to Which it Leads." *Proceedings of the Royal Society of London Series A* 102 (March):411–37.

Linsley, John. 1963. "Evidence for a Primary Cosmic-Ray Particle with Energy $10^{20}$ eV." *Physical Review Letters* 10 no. 4 (15 February), A521 1–3–A521 3–3.

Longmate, Norman. 1985. *Hitler's Rockets: The Story of the V-2s.* London: Hutchinson.

Lovell, Bernard. 1973. *The Origins and International Economics of Space Exploration.* Edinburgh University Press.

[McCrosky, Richard E.]. 1959. "American Astronomers Report—Artificial Meteor Observations." *Sky & Telescope* 18 (June):440.

McCurdy, Howard E. 1989. "The Decay of NASA's Technical Culture." *Space Policy* 5:301–10.

McDougall, Walter A. 1985. *The Heavens and the Earth: A Political History of the Space Age.* Basic Books.

McGovern, James. 1964. *Crossbow and Overcast.* Paperback Books.

Mack, Pamela E. 1983. *The Politics of Technological Change: A History of Landsat.* Ph.D. diss., University of Pennsylvania. University Microfilms Edition. Revision published as *Viewing the Earth: The Social Construction of the Landsat Satellite System.* MIT Press, 1990.

MacKenzie, Donald. 1990. *Inventing Accuracy: An Historical Sociology of Nuclear Guidance.* MIT Press.

McKinley, D. W. R. 1961. *Meteor Science and Engineering.* McGraw Hill.

McLuhan, Marshall. 1962. *The Gutenberg Galaxy.* 1966 ed., Toronto.

McLuhan, Marshall. 1964. *Understanding Media: The Extensions of Man.* 2nd ed., Mentor.

Macrakis, Kristie Irene. 1989. *Scientific Research in National Socialist Germany: The Survival of the Kaiser Wilhelm Gesellschaft.* Ph.D. diss., Harvard University. University Microfilms, 1990

Malina, Frank J. 1971. "America's First Long-Range-Missile and Space Exploration Program: The ORDCIT Project of the Jet Propulsion Laboratory, 1943–1946: A Memoir." In Hall 1986:339–83.

Mann, Martin. 1946. "Rocket Camera to Shoot Sun." *Popular Science* (October):78–80.

Manning, Laurence A. 1955. *A Survey of the Literature of the Ionosphere.* Stanford: Radio Propagation Laboratory, 31 July.

Marshak, Robert E. 1983. "Particle Physics in Rapid Transition: 1947–1952." In Brown and Hoddeson 1983:376–401.

Martyn, D. F., G. H. Munro, A. J. Higgs, and S. E. Williams. 1937. "Ionospheric Disturbances, Fadeouts and Bright Hydrogen Solar Eruptions." *Nature* 140 (9 October):603–05.

Massey, Harrie, and M. O. Robins. 1986. *History of British Space Science.* Cambridge University Press.

Mayall, N. U. 1946. "Bernhard Schmidt and his Coma-Free Reflector." *Publications of the Astronomical Society of the Pacific* 58:282–90.

Medaris, Maj. Gen. J. B., and Arthur Gordon. 1960. *Countdown for Decision*. Putnam.

Meeter, George F. 1967. *The Holloman Story*. University of New Mexico Press.

Menzel, Donald H. 1931. "A Study of the Solar Chromosphere." *Lick Observatory Publications* 17.

Menzel, Donald H. 1939a. "Problems of the Solar Atmosphere." *Proceedings of the American Philosophical Society* 81 no. 2 (June):107–24.

Menzel, Donald H. 1939b. "Some Unresolved Problems in Astrophysics." *Astronomical Society of the Pacific Leaflet* no. 125 (July).

Menzel, Donald H. 1950. *Our Sun*. Blakiston.

Michel, Jean. 1973. *Dora*. Paris; New York: Holt, Rinehart, and Winston, 1980.

Miller, Robert Ryal. 1968. *For Science and National Glory: The Spanish Scientific Expedition to America, 1862–1866*. Oklahoma.

Mimno, Harry Rowe. 1937. "The Physics of the Ionosphere." *Reviews of Modern Physics* 9 (January):1–43.

Minzner, R. A. 1976. "The 1976 Standard Atmosphere and its Relationship to Earlier Standard Atmospheres." Appendix B to [NASA Task Group II to COSEA]. *The 1976 Standard Atmosphere Above 86-Km Altitude*. NASA SP-398.

Mitra, S. K. 1948. *The Upper Atmosphere*. Calcutta.

Moore, C. B. 1952. "Plastic Balloons: A Platform for Experiments in the Upper Atmosphere." In White and Benson 1952:395–404.

Moore, C. E. 1945. "A Multiplet Table of Astrophysical Interest." *Contributions of the Princeton University Observatory* no. 20.

Moulton, Forest Ray. 1935. *Consider the Heavens*. Doubleday, Doran.

Mukerji, Chandra. 1989. *A Fragile Power: Scientists and the State*. Princeton University Press.

[National Bureau of Standards]. 1945. *The Radio Proximity Fuzes for Bombs, Rockets and Morters*. Washington, D.C.: Ordnance Development Division, NBS.

Naugle, John E. 1987. "An Irreverent History of Space Science." In R. Ramaty, T. L. Cline, and J. F. Orme. *Essays in Space Science*:413–423. NASA Conference Publication 2464.

Needell, Allan A. 1987. "Preparing for the Space Age: University-based Research, 1946–1957." *Historical Studies in the Physical and Biological Sciences* 18 pt. 1:89–109.

Nelkin, Dorothy. 1972. *The University and Military Research: Moral Politics at MIT*. Cornell.

Nelkin, Dorothy. 1990. "Scientists in a Golden Cage." Review of Mukerji 1990. *New York Times Book Review* (8 April):18.

Neher, H. Victor. 1985. "Some of the Problems and Difficulties Encountered in the Early Years of Cosmic-Ray Research." In Sekido and Elliot 1985:91.

Neufeld, Jacob. 1990. *Ballistic Missiles in the United States Air Force 1945–1960*. Washington: Office of Air Force History.

Neufeld, Michael J. 1989. "Peenemünde-Ost: The State, The Military, and Technological Change in the Third Reich." Paper presented at the International Congress of the History of Science, Hamburg, 2 August.

Neufeld, Michael J. 1990. "Weimar Culture and Futuristic Technology: The Rocketry and Spaceflight Fad in Germany, 1923–1933." *Technology and Culture* 31:725–52.

Neufeld, Michael J. 1991. "The Guided Missile and the Third Reich: Peenemünde and the Forging of a Technological Revolution." In Monika Renneberg and Mark Walker, eds., *Science, Technology and National Socialism*. Cambridge, forthcoming.

Newell, Homer E. 1953. *High Altitude Rocket Research*. Academic Press.

Newell, Homer E. ed. 1959a. *Sounding Rockets*. McGraw Hill.

Newell Homer E. 1959b. "Viking." In Newell 1959a, chap. 13.

Newell, Homer E. 1980. *Beyond the Atmosphere: Early Years of Space Science*. Washington, D.C.: NASA, SP-4211.

Newell, Homer E. and J. W. Siry, eds. 1946. *Upper Atmosphere Research Report* no. 2. NRL Report no. 3030, December.

Newell, Homer E. and J. W. Siry, eds. 1947a. *Upper Atmosphere Research Report* no. 3. NRL Report no. 3120, April.

Newell, Homer E. and J. W. Siry, eds. 1947b. *Upper Atmosphere Research Report* no. 4. NRL Report no. 3171, October.

Newell, Homer E. and J. W. Siry, eds. 1948. *Upper Atmosphere Research Report* no. 5. NRL Report no. 3358, June.

Nicolet, Marcel. 1968. "Ionospheric Physics and Aeronomy." In Akasofu 1968:35–38.

Noble, David E. 1984. *Forces of Production: A Social History of Industrial Automation*. Knopf. Reprinted, Oxford, 1986.

Oberth, Hermann. 1972. *Wege zur Raumschiffahrt*. Munich, 1929. Translated as *Ways to Spaceflight*. NASA TT F-622.

O'Day, Marcus. 1954. "Upper Air Research by Use of Rockets in the U.S. Air Force." In Boyd and Seaton 1954:1–10.

Ordway, Frederick I., ed. 1959. *Advances in Space Science I*. Academic Press.

Ordway, Frederick I., ed. 1960. *Advances in Space Science II*. Academic Press.

Ordway, Frederick I. ed. 1989. *History of Rocketry and Astronautics*. San Diego: Univelt; American Astronautical Society History Series, vol. 9. Symposium date, 1977.

Ordway, Frederick I., and Mitchell R. Sharpe. 1979. *The Rocket Team*. Crowell.

Paetzold, H. K., G. Pfotzer, and E. Schopper. 1974. "Erich Regener als Wegarbiter der Extraterrestrischen Physik." In H. Brett, et al. eds., *Zur Geschichte der Geophysik*:167–188. Springer-Verlag.

Pendray, G. Edward. 1945. *The Coming Age of Rocket Power*. Harper.

Perlow, Gilbert J. 1949. "Cosmic Ray Measurements in Rockets." *Scientific Monthly* (December):382–85.

Perlow, Gilbert J., L. R. Davis, C. W. Kissinger, and J. D. Shipman Jr. 1952. "Rocket Determination of the Ionization Spectrum of Charged Cosmic Rays at [geomagnetic latitude] 41° N." *Physical Review* 88:321–25.

Perry, Robert L. 1963. "The Atlas, Thor, and Titan." *Technology and Culture* 4:466–77.

Pfotzer, Georg. 1936. *Zeitschrift fur Physik* 102:23.

Pfotzer, Georg. 1972. "History of the Use of Balloons in Scientific Experiments." *Space Science Reviews* 13:199–242.

Pfotzer, Georg. 1985. "On Erich Regener's Cosmic Ray Work in Stuttgart and Related Subjects." In Sekido and Elliot 1985:75–89.

[Physical Society]. 1955. *Report of the Physical Society Conference on the Physics of the Ionosphere held at the Cavendish Laboratory, Cambridge, September 1954*. London: The Physical Society.

[Pickering, William H.]. 1946. *Abstracts of Papers Presented at the Guided Missiles and Upper Atmosphere Symposium*. GALCIT-Jet Propulsion Laboratory Publication no. 3.

Pickering, William H. 1947. *Study of the Upper Atmosphere by Means of Rockets*. Jet Propulsion Laboratory Publication no. 8.

Pickering, William H., and James H. Wilson. 1972. "Countdown to Space Exploration: A Memoir of the Jet Propulsion Laboratory, 1944–1958." In Hall 1986:385–421.

Pietenpol, W. B., W. A. Rense, F. C. Walz, D. S. Stacey, and J. M. Jackson. 1953. "Lyman Alpha-Line Photographed in Sun's Spectrum." *Physical Review* 90 no. 1:156.

Pohl, R. W. 1938. "Zusammenfassender Bericht uber Elektronenleitung und Photochemische Vorgange in Alkalihalogenidkristallen." *Physikalishe Zeitschrift* 39:36–54.

Possony, Stefan T. 1949. *Strategic Air Power: The Pattern of Dynamic Security*. Washington, D.C.: Infantry Journal Press.

Preston, William H. 1940. "The Origin of Radio Fade-Outs and the Absorption Coefficient of Gases for Light of Wavelength 1215.7Å." *Physical Review* 57:887–94.

Price, Don K. 1954. *Government and Science*. New York University Press.

Ratcliffe, J. A., ed. 1960. *Physics of the Upper Atmosphere*. Academic Press.

Ratcliffe, J. A. 1970. *Sun, Earth and Radio*. McGraw-Hill.

Ratcliffe, J. A. 1974. "Experimental Methods of Ionospheric Investigation 1925–1955." *Journal of Atmospheric and Terrestrial Physics* 36:2095–103.

Rearden, Steven L. 1984. *The Formative Years, 1947–1950*. vol. 1 of Alfred Goldberg, ed., *History of the Office of the Secretary of Defense*. Historical Office, Office of the Secretary of Defense, pts. 1 and 2.

Regener, Erich. 1939. "Aufbau und Zussamensetzung der Stratosphäre." *Schriften der Deutschen Akademie der Luftfahrtforschung* no. 5:7–48. Translated as "The Structure and Composition of the Stratosphere." ATSC 509 F-TS609-RZ, January 1946.

Regener, Erich. 1949. "Ozonschicht und Atmosphärische Turbulenz." *Berichte des Deutschen Wetterdienstes in der US-Zone* no. 11, Bad Kissingen. Reprinted from *Forschungsund Erfahrungsberichte des Reichswetterdienstes* no. 9, 1941.

Regener, Erich, and Victor Regener. 1934a. "Aufnahmen des ultravioletten Sonnenspektrums in der Stratosphäre und vertikale Ozonverteilung." *Physikalische Zeitschrift* 35:788–793.

Regener, Erich, and Victor Regener. 1934b. "Ultra-Violet Solar Spectrum and Ozone in the Stratosphere." *Nature* 134 (8 September):380.

Regener, Erich, H. K. Paetzold, and A. Ehmert. 1954. "Further Investigations on the Ozone Layer." In Boyd and Seaton 1954:202–07.

Reifman, A., and W. G. Dow. 1949a. "Theory and Application of the Variable Voltage Probe for Exploration of the Ionosphere." *Physical Review* 75:1311–12.

Reifman, A., and W. G. Dow. 1949b. "Dynamic Probe Measurements in the Ionosphere." *Physical Review* 76:987–88.

Reingold, Nathan. 1987. "Vannevar Bush's New Deal for Research: or The Triumph of the Old Order." *Historical Studies in the Physical Sciences* 17 pt. 2:299–344.

Rense, William A. 1953. "Intensity of Lyman-Alpha Line in the Solar Spectrum." *Physical Review* 91:299–302.

Rense, William A., and T. Violett. 1959. "Method of Increasing Speed of Grazing-Incidence Spectrograph." *Journal of the Optical Society of America* 49 (February 1959):139–141.

Ripley, William S. 1955. "U.S. Rocket Research of the Upper Atmosphere, A Tabulation and Bibliography." AFCRC-TN-55–205.

Roberts, Walter Van Braam. 1990. *Autobiography of Walter Van Braam Roberts.* Private printing.

[The Rocket Panel]. 1952. "Pressures, Densities, and Temperatures in the Upper Atmosphere." *Physical Review* 88:1027–32.

Roland, Alex. 1985a. *Model Research: A History of the National Advisory Committee for Aeronautics 1915–1958.* Washington, D.C.: NASA SP-4103.

Roland, Alex. 1985b. "Science and War." *Osiris* 1 2d ser., 247–72.

Rosen, Milton. 1955. *The Viking Rocket Story.* Harper.

Rosen, Milton W., and J. Carl Seddon. 1951. "Conversion of Viking Into a Guided Missile." *Rocket Research Report* no. 6. NRL Report no. 3829.

Rosen, Milton W., M. L. Kuder, and E. N. Pettitt. 1945. *Development of Remote Control for the JB-2 Flying Bomb.* NRL Report R-2616.

Rosseland, Svein. 1936. *Theoretical Astrophysics.* Oxford.

Rossi, Bruno. 1948. "Interpretation of Cosmic-Ray Phenomena." *Review of Modern Physics* 20: 537–83.

Rossi, Bruno. 1953. "Where Do Cosmic Rays Come From?" *Scientific American* 189 (September):64–70.

Rossi, Bruno. 1964. *Cosmic Rays.* McGraw Hill.

Rossi, Bruno. 1981. "Early Days in Cosmic Rays." *Physics Today* 34 (October):34–41.

Rossi, Bruno. 1985. "Arcetri, 1928–1932." In Sekido and Elliot 1985:53–73.

Russell, Henry Norris. 1934. "Could a Manned Rocket Reach Mars?" *Scientific American* (December):294–95.

Saha, M. N. 1921. "On a Physical Theory of Stellar Spectra." *Proc. Royal Society London, Ser. A* 99. See also *Collected Works*:58–59.

Saha, M. N. 1937a. "A Stratosphere Solar Observatory." *Harvard College Observatory Bulletin* no. 905:1–7. Reprinted in *Collected Works*:240–3.

Saha, M. N. 1937b. "On the Action of Ultra-Violet Sunlight Upon the Upper Atmosphere." *Proc. Royal Society London, Ser. A* 160:155. Reprinted in *Collected Works*:254.

[Saha, M. N.]. 1982. *Collected Works of Meghnad Saha*. vol. 1 (Santimay Chatterjee, ed.). Calcutta: Orient Longman.

Sandström, A. E. 1965. *Cosmic Ray Physics*. Wiley.

Sapolsky, Harvey M. 1972. *The Polaris System Development: Bureaucratic and Programmatic Success in Government*. Harvard.

Sapolsky, Harvey M. 1979. "Academic Science and the Military." In Nathan Reingold, ed., *The Sciences in the American Context: New Perspectives*:379–99. Smithsonian.

Sapolsky, Harvey M. 1990. *Science and the Navy: The History of the Office of Naval Research*. Princeton.

Satterly, John. 1923. "The Upper Atmosphere." *Journal of the Royal Astronomical Society of Canada* 17:291–304.

Saward, Dudley. 1984. *Bernard Lovell: A Biography*. London: Robert Hale.

Schade, Henry A. 1946. "German Wartime Technical Developments." *Society of Naval Architects and Marine Engineers Transactions* 54:83–111.

Schein, Marcel, William P. Jesse, and E. D. Wollan. 1940. "Intensity and Rate of Production of Mesotrons in the Stratosphere." *Physical Review* 57:847–54.

Schein, Marcel, William P. Jesse, and E. D. Wollan. 1941. "The Nature of the Primary Cosmic Radiation and the Origin of the Mesotron." *Physical Review* 59:615.

Schultz, F. V., N. W. Spencer, and A. Reifman. 1948. *Upper Air Research Program Report* no. 2. Engineering Research Institute, University of Michigan, 1 July.

Seddon, J. Carl. 1953. "Propagation Measurements in the Ionosphere with the Aid of Rockets." *Journal of Geophysical Research* 58 no. 3:323–35.

Seddon, J. Carl. 1954a. "Propagation Measurements in the Ionosphere with the Aid of Rockets." In Boyd and Seaton 1954:214–22.

Seddon, J. Carl. 1954b. "Electron Densities in the Ionosphere." *Journal of Geophysical Research* 59 no. 4:463–66.

Seddon, J. C., A. D. Pickar, and J. E. Jackson. 1954. "Continuous Electron Density Measurements up to 200 KM." *Journal of Geophysical Research* 59 no. 4:513–524.

Seidel, R. W. 1983. "Accelerating Science: The Postwar Transformation of the Lawrence Radiation Laboratory." *Historical Studies in the Physical Sciences* 13:375–400.

Sekido, Yataro, and Harry Elliot, eds. 1985. *Early History of Cosmic Ray Studies*. Reidel.

Shapiro, Maurice M., and Rein Silberberg. 1970. "Heavy Cosmic Ray Nuclei." *Annual Review of Nuclear Science*:323–92.

Shapley, Harlow. 1969. *Through Rugged Ways to the Stars*. Scribners.

Shenstone, Allan G. 1973. "Rudolf Ladenburg." *Dictionary of Scientific Biography* 7:552–56. Scribners.

Sherry, Michael S. 1977. *Preparing for the Next War*. Yale.

Sherry, Michael S. 1987. *The Rise of American Air Power*. Yale.

Shternfeld, Ari. 1959. *Soviet Space Science*. Basic Books.

Sigethy, Robert. 1980. *The Air Force Organization for Basic Research 1945–1970: A Study in Change*. Ph.D. diss., American University. University Microfilms.

Silberstein, R. 1959. "The Origin of the Current Nomenclature for the Ionospheric Layers." *Journal of Atmospheric and Terrestrial Physics* 13:382.

Simonton, Dean Keith. 1984. "Is the Marginality Effect all that Marginal?" *Social Studies of Science* 14:621–22.

Simpson, Christopher. 1988. *Blowback*. Weidenfeld and Nicolson.

Simpson, John A. 1985. "Cosmic Ray Astrophysics at Chicago 1947–1960." In Sekido and Elliot 1985:385–409.

Simpson, John A. 1986. *To Explore and Discover*. Ryerson Lecture, University of Chicago.

Singer, S. Fred. 1954. "Rocket Exploration of Magnetic Fields and Electric Currents in the Upper Atmosphere." In Boyd and Seaton 1954:256–60.

Singer, S. Fred. 1958. "The Primary Cosmic Radiation and its Time Variations." *Progress in Elementary Particle and Cosmic Ray Physics* 4:205–335.

Singer, S. Fred, E. Maple, and W. A. Bowen. 1951. "Evidence for Ionospheric Currents from Rocket Experiments Near the Geomagnetic Equator." *Journal of Geophysical Research* 56:265–81.

Siry, Joseph W. 1950. "The Early History of Rocket Research." *Scientific Monthly* 71:326–332; 408–421.

Skellett, A. M. 1931. "The Effect of Meteors on Radio Transmission through the Kennelly-Heaviside Layer." *Physical Review* 37:1668.

Skellett, A. M. 1935. "The Ionizing Effect of Meteors." *Proceedings of the Institute of Radio Engineers* 23:132–49.

Skellett, A. M. 1938. "Meteoric Ionization in the E-region of the Ionosphere." *Nature* 141:472.

Smith, Charles P. Jr., 1954. "Summary of Upper Atmosphere Rocket Research Firings." *Upper Atmosphere Research Report* no. 21. Washington, D.C.: NRL Report 4276, February. Updated in February 1958 and January 1959 by Eleanor C. Pressly, and reissued as NRL Report 4276. Cited herein as Smith and Pressly 1959.

Smith, Bruce L. R. 1966. *The RAND Corporation: Case Study of a Non-profit Advisory Organization*. Harvard.

Smith, Robert W. 1989. *The Space Telescope: A Study of NASA, Science, Technology, and Politics*. Cambridge University Press.

Snyder, Wilbert F., and Charles L. Bragaw. 1986. *Achievement in Radio: Seventy Years of Radio Science, Technology, Standards, and Measurement at the National Bureau of Standards*. National Bureau of Standards Special Publication no. 555.

Spencer, Nelson W., and W. G. Dow. 1954. "Density-gauge Methods for Measuring Upper-air Temperature, Pressure, and Winds." In Boyd and Seaton 1954:82–97.

Spitzer, Lyman Jr.. 1946. *Physics of Sound in the Sea*. Columbia University Press.

Staab, Heinz A. 1986. *Kontinuität und Wandel einer Wissenschaftsorganisation: 75 Jahre Kaiser-Wilhelm-/Max-Planck-Gesellschaft*. Munich.

Stacey, David. 1953. "Instrumentation Manual for Biaxial Pointing Control." *University of Colorado Upper Air Laboratory Special Report* no. 6 (30 April).

Stares, Paul B. 1985. *The Militarization of Space: U.S. Policy, 1945–1984*. Cornell.

Steelman, John R. 1947. *Administration for Research*. vol. 3 of [The President's Scientific Research Board], *Science and Public Policy* (4 October).

Sterne, Nancy B. 1981. *From ENIAC to UNIVAC*. Bedford, Mass: Digital Press.

Stetson, H. T. 1934. *Earth, Radio, and the Stars*. McGraw Hill.

Stewart, Irvin. 1948. *Organizing Scientific Research for War: The Administrative History of the OSRD*. Little, Brown.

Stewart, John Q., and Harold A. Zahl. 1947. "Radar Observations of the Draconids." *Sky and Telescope* no. 65:3–5.

Stroud, William G. 1975. "Early Scientific History of the Rocket Grenade Experiment." In Ordway 1989:237–52.

Sullivan, Walter. 1961. *Assault on the Unknown: The IGY*. New York.

[Tabanera, T. M., et al.]. 1964. *Advances in Space Research*. Pergamon.

Tatarewicz, Joseph N. 1990. *Space Astronomy and Planetary Astronomy*. Indiana University Press.

Tatel, Howard E., and James A. Van Allen. 1948. "Cosmic-Ray Bursts in the Upper Atmosphere." *Physical Review* 73:87–88.

Terman, Frederick E. *Radio Engineering*. 2nd. edn. McGraw-Hill, 1937.

Thomas, R. N. 1947. "Meteorics and Ballistics." *Popular Astronomy* 55 (December):517–29.

Thomas, Shirley. 1960. *Men of Space*. vol. 1. Chilton.

Thorp, Willard, Minor Myers Jr., and Jeremiah Stanton Finch. 1978. *The Princeton Graduate School*. Princeton.

Tousey, Richard. 1936. "Optical Constants of Fluorite in the Extreme Ultraviolet." *Physical Review* 50:1057.

Tousey, Richard. 1939. "On Calculating the Optical Constants from Reflection Coefficients." *Journal of the Optical Society of America* 29:235–39.

Tousey, Richard. 1953a. "Rocket Spectroscopy." *Journal of the Optical Society of America* 43 no. 4:245–49.

Tousey, Richard. 1953b. "Solar Work at High Altitude from Rockets." In Kuiper 1953:658–676.

[Tousey, Richard]. 1961. "Richard Tousey: Frederic Ives Medalist for 1960." *Journal of the Optical Society of America* 51 no. 4:379–83.

Tousey, Richard. 1963. "The Extreme Ultraviolet Spectrum of the Sun." *Space Science Reviews* 2:3–69.

Tousey, Richard. 1964. "The Spectrum of the Sun in the Extreme Ultraviolet." *Quarterly Journal of the Royal Astronomical Society* 5:123–44.

Tousey, Richard. 1967. "Some Results of Twenty Years of Extreme Ultraviolet Solar Research." *Astrophysical Journal* 149 no. 2:239–52.

Tousey, Richard. 1986. "Solar Spectroscopy from Rowland to SOT." In Henry, Beer, and DeVorkin 1986:175–99.

Tousey, Richard, Kenichi Watanabe, and Dewitt Purcell. 1951. "Measurements of Solar Extreme Ultraviolet and X-rays from Rockets by Means of a $CaSO_4$:Mn Phosphor." *Physical Review* 83:792–97.

Townsend, J. W. Jr., H. Friedman, and R. Tousey. 1958. "History of the Upper-Air Rocket-Research Program at the Naval Research Laboratory 1946–1957." *Upper Atmosphere Research Report* no. 32. NRL Report 5087.

Truax, Capt. Robert C. 1959. "The Use of Sounding Rockets for Military Research." In Newell 1959, chap. 3.

Tuve, Merle. 1974. "Early Days of Pulse Radio at the Carnegie Institution." *Journal of Atmospheric and Terrestrial Physics* 36:2079–84.

Tuve, Merle A., and Gregory Breit. 1925. "Note on a Radio Method of Estimating the Height of the Conducting Layer." *Terrestrial Magnetism and Atmospheric Electricity* 30 nos. 15, 16.

[U.S.S.R. Academy of Sciences]. 1958. *The Russian Literature of Satellites* pt. 2. New York: International Physical Index, Inc.

Vaeth, J. Gordon. 1951. *200 Miles Up*. Ronald.

Van Allen, James A. 1948. "Exploratory Cosmic Ray Observations at High Altitudes by Means of Rockets." *Sky and Telescope* 7 no. 7 (May):171–75.

Van Allen, James A. 1952. "The Nature and Intensity of the Cosmic Radiation." In White and Benson 1952:239–66.

Van Allen, James A. ed. 1956. *Scientific Uses of Earth Satellites*. London: Chapman and Hall.

Van Allen, James A. 1957. "Direct Detection of Auroral Radiation with Rocket Equipment." *Proceedings of the National Academy of Sciences* 43:57–62.

Van Allen, James A. 1983. *Origins of Magnetospheric Physics*. Smithsonian.

Van Allen, James A. 1990. "What is a Space Scientist? An Autobiographical Example." *Annual Reviews of Earth and Planetary Sciences* 18:1–26.

Van Allen, James A., and H. E. Tatel. 1948. "The Cosmic-Ray Counting Rate of a Single Geiger Counter from Ground Level to 161 Kilometers Altitude." *Physical Review* 73:245–51.

Van Allen, James A., and Albert V. Gangnes. 1950. "The Cosmic Ray Intensity Above the Atmosphere at the Geomagnetic Equator." *Physical Review* 78:50–52.

Van Allen, James A., and J. J. Hopfield. 1952. "Preliminary Report on Atmospheric Ozone Measurements from Rockets." In "L'Etude Optique de l'Atmosphere terrestre." *Liege Symposium*:181. Louvain.

Van Allen, James A., and Melvin B. Gottlieb. 1954. "The Inexpensive Attainment of High Altitudes with Balloon-Launched Rockets." In Boyd and Seaton 1954:53–64.

Van Allen, James, A., L. W. Fraser, and J. F. R. Floyd. 1948. "The Aerobee Sounding Rocket––A New Vehicle for Research in the Upper Atmosphere." *Science* 108 (31 December):746–47.

Van Allen, James A., John W. Townsend Jr., and Eleanor C. Pressly. 1959. "The Aerobee Rocket." In Homer Newell 1959, chap. 4.

Vassy, E. 1954. "Remarks on the Paper by F. S. Johnson, J. D. Purcell, and R. Tousey." In Boyd and Seaton 1954:200–01.

Vegard, Lars. 1938. "Vorgänge und Zustand in der Nordlichtregion." *Geofysiske Publikasjoner* 12 no. 5:18.

Vegard, Lars. 1939. "The Auroral Spectrum and the Upper Atmosphere." In *Relations entre les Phénomnes Solaires et Terrestres*:136–49. Conseil International des Unions Scientifiques, *Cinquime Rapport de la Commission . . .* Firenze.

Vegard, Lars. 1957. "Solar and Terrestrial Phenomena Resulting from Photoelectric Effect of Solar X-Rays." In Zelikoff, 1957:22–31.

Vernov, A. E. Chudakov. 1960. "Study of Cosmic Rays by Rockets and Sputniks in the U.S.S.R." *Transactions of the International Astronomical Union* 10:710–12. Cambridge University Press.

Villard, O. G. 1976. "Ionospheric Sounder and its Place in the History of Radio Science." *Radio Science* 11:847–60.

von Braun, Wernher, and Frederick I. Ordway. 1975. *History of Rocketry and Space Travel*. 3rd ed. Crowell.

Walker, Mark. 1989. *German National Socialism and the Quest for Nuclear Power, 1939–1949*. Cambridge University Press.

Warfield, Calvin N. 1947. "Tentative Tables for the Properties of the Upper Atmosphere." Washington, D.C.: NACA TN-1200.

Waynick, A. H. 1957. "The Present State of Knowledge Concerning the Lower Ionosphere." *Proceedings of the Institute for Radio Engineers* 45 (June):741–49.

Weart, Spencer R., and Melba Phillips, eds. 1985. *History of Physics*. New York: American Institute of Physics.

Weart, Spencer R. 1988. *Nuclear Fear*. Harvard.

Weiner, Charles. 1969. "A New Site for the Seminar: The Refugees and American Physics in the Thirties." In Donald Fleming and

Bernard Bailyn, *The Intellectual Migration*:190–234. Harvard, Belknap.

Weisner, Allan G. 1954. "The Determination of Temperatures and Winds Above Thirty Kilometers." In Boyd and Seaton 1954:133–42.

Weisz, Paul. 1941. "The Geiger-Müller Tube: An Electronic Instrument." *Electronics* (December):18–21.

Wexler, Harry. 1950. "Annual and Diurnal Temperature Variations in the Upper Atmosphere." *Tellus* 2:262–74.

Wexler, Harry. 1964. "Interpretation of Observations Received from Meteorological Rockets and Satellites." In [Tabanera, et al.] 1964:403–409.

Whipple, Fred L. 1938. "Photographic Meteor Studies, I." *Proceedings of the American Philosophical Society* 79 (November):499–548.

Whipple, Fred L. 1943. "Meteors and the Earth's Upper Atmosphere." *Reviews of Modern Physics* 15 (October):246–64.

Whipple, Fred L. 1948a. "Meteors and the Upper Atmosphere." *Centennial Symposia*:349–52. Harvard College Observatory.

Whipple, Fred L. 1948b. "The Orbits of Meteors Photographed at Two Stations." Abstract. *Astronomical Journal* 54:53.

[Whipple, Fred L.]. 1948c. "Meteor Astronomy." *The Collected Contributions of Fred L. Whipple*. vol. 1:167–68. Smithsonian Astrophysical Observatory, 1972. Excerpted from *Observatory* 68:226–32.

Whipple, Fred L. 1949. "The Harvard Photographic Meteor Program." *Sky & Telescope* 8 no. 4:90–93.

Whipple, Fred L. 1950. "A Comet Model. I. The Acceleration of Comet Encke." *Astrophysical Journal* 111:375–94.

Whipple, Fred L. 1951a. "Meteors as Probes of the Upper Atmosphere." In *Compendium of Meteorology*:356–65. American Meteorological Society.

Whipple, Fred L. 1952a. "Results of Rocket and Meteor Research." *Bulletin of the American Meteorological Society* 33 (January):13–25.

Whipple, Fred L. 1952b. "On meteor masses and densities." Abstract, read late December 1951. *Astronomical Journal* 57:28–29.

Whipple, Fred L. 1954. "Photographic Meteor Orbits and their Distribution in Space." *Astronomical Journal* 59:207–17.

[Whipple, Fred L.]. 1972. *The Collected Contributions of Fred L. Whipple*. Smithsonian Astrophysical Observatory.

Whipple, Fred L., Luigi Jacchia, and Zdenek Kopal. 1949. "Seasonal Variations in the Density of the Upper Atmosphere." In Kuiper 1949:149–58.

White, Clayton S., and Otis O. Benson Jr. 1952. *Physics and Medicine of the Upper Atmosphere*. University of New Mexico Press.

White, L. D. 1952. *Final Report, Project Hermes V-2 Missile Program*. General Electric Report no. R52A0510.

Wilson, J. Tuzo. 1961. *IGY: The Year of the New Moons*. New York.

Winckler J. R., and D. J. Hoffman. 1966. *Cosmic Rays.* "Resource Letter CR-1 on Cosmic Rays." American Association of Physics Teachers, Reprint.

Winter, Frank H. 1983. *Prelude to the Space Age: The Rocket Societies, 1924–1940.* Smithsonian.

Wise, George. 1985. "Science and Technology." *Osiris* 1 2d ser., 229–46.

Wright, Helen. 1966. *Explorer of the Universe.* Dutton.

Zahl, Harold A. 1952. "Physics in the Signal Corps." *Physics Today* 5 (February):16–19.

Zahl, Harold A. 1968. *Electrons Away, or Tales of a Government Scientist.* Vantage Press.

Zelikoff, M., ed. 1957. *The Threshold of Space: The Proceedings of the Conference on Chemical Aeronomy.* Pergamon.

Ziegler, Charles A. 1986. "Ballooning and the Birth of Cosmic Ray Physics." *NASM Research Report*:69–92. Smithsonian.

Ziegler, Charles A. 1988. "Waiting for Joe-1: Decisions Leading to the Detection of Russia's First Atomic Bomb Test." *Social Studies of Science* 18 no. 2:197–229.

Ziegler, Charles A. 1989. "Technology and the Process of Scientific Discovery: The Case of Cosmic Rays." *Technology and Culture* 30:939–63.

Zwicky, Fritz. 1946. "On the Possibility of Earth-Launched Meteors." *Publications of the Astronomical Society of the Pacific* 58:260–1.

Zwicky, Fritz. 1947a. "The First Night-Firing of a V-2 Rocket in the United States." *Publications of the Astronomical Society of the Pacific* 59:32–3.

Zwicky, Fritz. 1947b. "Research with Rockets." *Publications of the Astronomical Society of the Pacific* 59:64–8.

Zwicky, Fritz. 1986. "A Stone's Throw Into the Universe: A Memoir." In Hall 1986:325–37.

# Index